T0215774

Communications in Computer and Information Science 874

Commenced Publication in 2007
Founding and Former Series Editors:
Phoebe Chen, Alfredo Cuzzocrea, Xiaoyong Du, Orhun Kara, Ting Liu,
Dominik Ślęzak, and Xiaokang Yang

More information about this series at http://www.springer.com/series/7899

Kangshun Li · Wei Li
Zhangxing Chen · Yong Liu (Eds.)

Computational Intelligence and Intelligent Systems

9th International Symposium, ISICA 2017
Guangzhou, China, November 18–19, 2017
Revised Selected Papers, Part II

 Springer

Editors
Kangshun Li
College of Mathematics and Informatics
South China Agricultural University
Guangzhou
China

Wei Li
Jiangxi University of Science
   and Technology
Ganzhou, Jiangxi
China

Zhangxing Chen
Chemical and Petroleum Engineering
University of Calgary
Calgary, AB
Canada

Yong Liu
School of Computer Science
   and Engineering
The University of Aizu
Aizu-Wakamatsu, Fukushima
Japan

ISSN 1865-0929   ISSN 1865-0937 (electronic)
Communications in Computer and Information Science
ISBN 978-981-13-1650-0   ISBN 978-981-13-1651-7   (eBook)
https://doi.org/10.1007/978-981-13-1651-7

Library of Congress Control Number: 2018948807

This Springer imprint is published by the registered company Springer Nature Singapore Pte Ltd.
The registered company address is: 152 Beach Road, #21-01/04 Gateway East, Singapore 189721, Singapore

Preface

Volumes CCIS 873 and CCIS 874 comprise proceedings of the 9th International Symposium on Intelligence Computation and Applications (ISICA 2017) held in Guangzhou, China, during November 18–19, 2017. ISICA 2017 successfully attracted over 180 submissions. After rigorous reviews and plagiarism checking, 51 high-quality papers are included in CCIS 873, while another 50 papers are collected in CCIS 874. ISICA conferences are one of the first series of international conferences on computational intelligence that combines elements of learning, adaptation, evolution, and fuzzy logic to create programs as alternative solutions to artificial intelligence.

ISICA 2017 featured the most up-to-date research in analysis and theory of evolutionary computation, neural network architectures and learning, neuro-dynamics and neuro-engineering, fuzzy logic and control, collective intelligence and hybrid systems, deep learning, knowledge discovery, learning, and reasoning. ISICA 2017 provided a venue to foster technical exchanges, renew everlasting friendships, and establish new connections. Prof. Yuanxiang Li, one of the pioneers in parallel and evolution computing at Wuhan University, wrote a beautiful poem in Chinese for the ISICA 2017 event. It is our pleasure to translate his poem with the title of "Computational Intelligence Debate on the Pearl River":

> Wear a smile on a bright face;
> Under the night light on the Pearl River;
> You are like star and moon shining on the Tower Small Slim Waist;
> Ride waves on the cruise ship;
> Leave bridges behind in a boundless moment.
>
> You are from far away;
> A journey of thousand miles;
> Meet in the Guangzhou City;
> Brighten up the field of intelligent evolution;
> Explore the endless road to intelligence.

Prof. Li's poem points out one of ISICA's missions of pursuing the truth that a complex system inherits the simple mechanism of evolution, while simple models may lead to the evolution of complex morphologies. Following the success of the past eight ISICA events, ISICA 2017 continued to explore the new problems emerging in the fields of computational intelligence.

On behalf of the Organizing Committee, we would like to thank warmly the sponsors, South China Agricultural University, who helped in one way or another to achieve our goals for the conference. We wish to express our appreciation to Springer for publishing the proceedings of ISICA 2017. We also wish to acknowledge the dedication and commitment of both the staff at the Springer Beijing Office and the CCIS editorial staff. We would like to thank the authors for submitting their work, as well as the Program Committee members and reviewers for their enthusiasm, time, and expertise. The invaluable help of active members from the Organizing Committee, including Wei Li, Hui Wang, Lei Yang, Yan Chen, Lixia Zhang, Weiguang Chen, Zhuozhi Liang, Junlin Jin, Ying Feng, and Yunru Lu, in setting up and maintaining the online submission systems by EasyChair, assigning the papers to the reviewers, and preparing the camera-ready version of the proceedings is highly appreciated. We would like to thank them personally for helping to make ISICA 2017 a success.

March 2018   Kangshun Li
                                                                      Yong Liu
                                                                       Wei Li
                                                              Zhangxing Chen

Organization

Honorary Chairs

Hisao Ishibuchi	Osaka Prefecture University, Japan
Qingfu Zhang	City University of Hong Kong, SAR China
Yang Xiang	Deakin University, Australia

General Chairs

Kangshun Li	South China Agricultural University, China
Zhangxing Chen	University of Calgary, Canada
Yong Liu	University of Aizu, Japan

Program Chairs

Aniello Castiglione	University of Salerno, Italy
Jing Liu	Xidian University, China
Han Huang	South China University of Technology, China
Hailin Liu	Guangdong University of Technology, China

Local Arrangements Chairs

Wei Li	South China Agricultural University, China
Yan Chen	South China Agricultural University, China

Publicity Chairs

Lei Yang	South China Agricultural University, China
Lixia Zhang	South China Agricultural University, China

Program Committee

Aimin Zhou	East China Normal University, China
Allan Rocha	University of Calgary, Canada
Dazhi Jiang	Shantou University, China
Dongbo Zhang	Guangdong University of Science and Technology, China
Ehsan Aliabadian	University of Calgary, Canada
Ehsan Amirian	University of Calgary, Canada
Feng Wang	Wuhan University, China
Guangming Lin	Southern University of Science and Technology, China
Guoliang He	Wuhan University, China

Hailin Liu	Guangdong University of Technology, China
Hu Peng	Jiujiang University, China
Hui Wang	Nanchang Institute of Technology, China
Iyogun Christopher	University of Calgary, Canada
Jiahai Wang	Sun Yet-Sen University, China
Jing Wang	Jiangxi University of Finance and Economics, China
Jun He	Aberystwyth University, UK
Jun Zou	The Chinese University of Hong Kong, SAR China
Kangshun Li	South China Agricultural University, China
Ke Tang	Southern University of Science and Technology, China
Kejun Zhang	Zhejiang University, China
Lingling Wang	Wuhan University, China
Lixin Ding	Wuhan University, China
Lu Xiong	South China Agricultural University, China
Maoguo Gong	Xidian University, China
Mohammad Zeidani	University of Calgary, Canada
Rafael Almeida	University of Calgary, Canada
Sanyou Zeng	China University of Geosciences, China
Shenwen Wang	Shijiazhuang University of Economics, China
Wayne Li	University of Calgary, Canada
Wei Li	South China Agricultural University, China
Wensheng Zhang	Chinese Academy of Sciences, China
Xiangjing Lai	University of Angers, France
Xin Du	Fujian Normal University, China
Xinyu Zhou	Jiangxi Normal University, China
Xuesong Yan	China University of Geosciences, China
Xuewen Xia	East China Jiaotong University, China
Ying Huang	Gannan Normal University, China
Yong Liu	The University of Aizu, Japan
Zahra Sahaf	University of Calgary, Canada
Zhangxing Chen	University of Calgary, Canada
Zhun Fan	Shantou University, China

Contents – Part II

Complex Systems Modeling – Multimedia Simulation

Intelligent Information Systems – Information Retrieval

Intelligent Information Systems – E-commerce Platforms

Artificial Intelligence and Robotics – Query Optimization

Artificial Intelligence and Robotics – Intelligent Engineering

Virtualization – Motion-Based Tracking

Virtualization – Image Recognition

Contents – Part I

Evolutionary Multi-objective and Dynamic Optimization
– Optimal Control and Design

Evolutionary Multi-objective and Dynamic Optimization
– Hybrid Methods

Data Mining – Association Rule Learning

Data Mining – Data Management Platforms

Cloud Computing and Multiagent Systems – Service Models

Cloud Computing and Multiagent Systems – Cloud Engineering

Everywhere Connectivity – IoT Solutions

Everywhere Connectivity – Wireless Sensor Networks

Swarm Intelligence – Cooperative Search

Differential Opposition-Based Particle Swarm

Lanlan Kang[1(✉)], Wenyong Dong[2], Shanni Li[3], and Jianxin Li[4]

[1] Jiangxi University of Science and Technology, Ganzhou 341000, China
victoryk11@163.com
[2] Wuhan University, Wuhan 430072, China
[3] Southern Capital Management Co., Ltd., Shenzhen 518000, China
[4] Dongguan Polytechnic, Dongguan 523808, Guangdong, China

Abstract. Particle Swarm Optimization (PSO) is slow but steady learner although it exhibits strong competence in solving complicated problems. However, during the course of searching process, the particles gradually gather into the vicinity of the best particle found so far. Furthermore, some evidences show that the unreasonable setting of its inertial term in the kinetic equations may lead to slow convergence of PSO. Thus, a differential opposition-based particle swarm optimization with adaptive elite mutation (DOPSO) is presented to overcome these drawbacks in this paper. There are two strategies are introduced into DOPSO to balance the contradiction between exploration and exploitation during its searching process: (1) Firstly a new particle's position update rule in which differential term replaces the inertia term is designed to accelerate its convergence; (2) Secondly an adaptive elite mutation strategy (AEM) is included to avoid trapping into local optimum. Experimental results show that the proposed method has a significant improvement in performance compared with some state-of-art PSOs.

Keywords: Particle swarm optimization
Generalized opposition-based learning
Adaptive elite mutation · Differential term

1 Introduction

Particle swarm optimization (PSO), one of the popular Evolutionary algorithm (EA) [1], firstly proposed by Kennedy and Eberhart in 1995. Inspired by the simulation of bird flocks and fish schooling foraging behavior, particles in PSO could search for global optimum through cooperation between particles [2]. At present, PSO has been successfully applied to a number of applications, such as resource allocation [3], prognostic model design for metabolic syndrome [4], DNA sequence compression [5] and so on. However, it can be found that the more complex problems, the slower PSO perform and the worse search in efficiency. As a result, the weakness can cause performance degradation of PSO, such as overmuch computation overhead, premature convergence.

There are many strategies to improve the performance of basic PSO, such as parameters adaptation tricks, opposition-based searching methods and hybridization of PSO with other intelligent algorithm.

© Springer Nature Singapore Pte Ltd. 2018
K. Li et al. (Eds.): ISICA 2017, CCIS 874, pp. 3–15, 2018.
https://doi.org/10.1007/978-981-13-1651-7_1

Because the performance of basic PSO highly relies on its parameters, such as inertial term, social learning coefficient, some strategies have been proposed to adaptively adjust these parameters for different problem [8]. Zhang et al. [9] employ the evolutionary state estimation to identify the evolutionary states of the swarm as exploration, exploitation, convergence and jumping out, and propose and adaptive control mechanism to change inertial weight and acceleration coefficients. Hu et al. [10] further put forward a parameter control mechanism to adaptively adjust parameters at iteration by subgradient method to close to the global best position and thus improve the robustness of PSO-MAM.

In order to enhance local search ability, Opposition-based learning (OBL) was firstly introduced into PSO [11]. The empirical researches prove that OBL is effective strategy to enhance local search ability and increase convergence rate. Generalized OBL (GOBL) was developed on the basis of OBL in literature [12].

The hybridization of PSO with other intelligent algorithm has been proved to be a promising technique that combines the desirable properties of these methods to improve individual weakness. For example, A hybrid operation combining DE and PSO is devised, where each individual is sequentially carried out DE and PSO operators in order to make each subpopulation quickly track the moving peaks [13]. Multi-frequency vibrational PSO employ periodic mutation application strategy and diversity variety based on an artificial neural network to avoid premature convergence [14].

For purpose of performance, a differential opposition-based particle swarm optimization with adaptive elite mutation, termed as DOPSO, is proposed in this paper. The algorithm is composed of a novelty velocity update formula (VD), GOBL and AEM strategy. VD formula differs from traditional PSO, contains a new momentum component instead of inertial motion component in basic PSO. The new velocity update formula can efficiently accelerate the convergence rate and enhance exploitation ability of algorithm by more widely learning from other individuals. With the introduction of generalized opposition-based Learning (GOBL) strategy in this paper, an adaptive elite mutation strategy (AEM) is applied to avoid particles, especially elite particles, trapping into local optimum [15]. Experiment results show that DOPSO has significantly improved convergence speed compared with the other OBL-based PSOs.

The rest of this paper is organized as follows. Section 2 briefly introduces the basic PSO algorithm and GOBL strategy. VD formula and AEM strategy are presented in Sect. 3 including detailed analysis of thought originated and search behaviors. Ultimately, Sect. 3 gives detailed procedure of DOPSO. Section 4 experimentally compares the DOPSO with various opposition-based learning PSO algorithms on 13 benchmark problems. Furthermore, the optimal parameter settings are suggested. Finally, some conclusions and discussions on the future work are given in Sect. 5.

2 Related Work

2.1 The Basic PSO

PSO is modeled on an abstract framework of "collective intelligence" in social animals. Each particle of a swarm represents a candidate solution, has two basic properties, i.e.

velocity and position [16]. It is supposed that velocity $v_{i,j}$ and position $x_{i,j}$ of the jth dimension of the ith particle are updated according to the following equations:

$$v_{i,j}(t+1) = w \cdot v_{i,j}(t) + c_1 rand_1(pbest_{i,j} - x_{i,j}(t)) + c_2 rand_2(gbest_j - x_{i,j}(t)) \quad (1)$$

$$x_{i,j}(t+1) = x_{i,j}(t) + v_{i,j}(t+1) \quad (2)$$

Where, $i = 1, 2, \cdots, N$, $j = 1, 2, \cdots, D$, N is the size of swarm, *pbest* is particle's historical best position, *gbest* is the swarm's global best position, c_1 and c_2 are two acceleration coefficients which control the influence of the cognitive and social components and guide each particle toward the *pbest* and the *gbest*, respectively. w is inertial weight. In general, $c_1, c_2 \in [0, 2]$. $rand_1$ and $rand_2$ are two uniformly distributed random variables in the interval $[0, 1]$.

2.2 A Generalized Opposition-Based Learning (GOBL)

A generalized opposition-based learning, called GOBL [17]. GOBL applied to PSO is defined as follows:

Let $X_i = (x_{i,1}, x_{i,2}, \cdots, x_{i,D})$ is the ith particle in a D-dimension space. The opposite particle $\breve{X}_i = (\breve{x}_{i,1}, \breve{x}_{i,2}, \cdots, \breve{x}_{i,D})$ is defined by:

$$\breve{x}_{i,j} = k \cdot (da_j + db_j) - x_{i,j} \quad (3)$$

Where, $j = 1, 2, \cdots, D$, k is a random number drawn from the uniform distribution in the interval $[0, 1]$. da_j and db_j are the minimum and maximum values of the jth dimension of the ith particle, respectively:

$$da_j = \min(x_{i,j}), \quad db_j = \max(x_{i,j}) \quad (4)$$

3 DOPSO Algorithm

This paper proposes DOPSO algorithm, which incorporates a new velocity formula and two strategies to enhance the global search ability as well as accelerate the convergence rate of DOPSO. First, GOBL strategy is introduced to generate initial population and select superior particles in each generation by given probability. Second, we adopt a new velocity formula that contains no inertial component to generate the next particle positions. Third, AEM strategy is designed to augment search space and avoid trapping into local optimum.

In the following sections, the differential velocity formula (VD) will firstly be elaborated, including its original intention, design principle and formula expression. Then, AEM strategy will be introduced in detail, including its design motivations and purposes. Finally, the detailed design steps of DOPSO algorithm are given.

3.1 A New Update Equation of Velocity

There is broad recognition that evolution is a slow learner, but it is steady increase in computing power and is able to find the optimum in the amount of practical problems if computation budget is sufficient. Therefore, how to increase convergence speed under the condition of ensuring the accuracy of solutions is to be worth researching. Next, there is a thorough analysis about PSO for its high search efficiency than most of EAs.

As mentioned, each particle in swarm has a velocity and position. The flight direction of each particle is modified in search space depend on its velocity formula constantly updated in each iteration (see Eq. (1)). The velocity formula is composed of three components, i.e. inertial component, cognitive component and social component.

Although the inertial term in Eq. (1) can bring in diversity of swarm, it can also lead to slow convergence. Can we find some new strategies to take the role of inertial term while enhance the performance of PSO?

To answer the above questions, the velocity formula of PSO is depicted as shown in Fig. 1(a) for Eq. (1). We will divide all the terms in Eq. (1) into two parts: (1) the first part (shown as red-dotted box) stands for the self-related information part which is composed of the inertial component and the cognition component; and (2) the second part stands for the social cognition part which is utilized only by *gbest* in social component (shown in blue-dotted box). The illustration shows particles may not obtained sufficient information from current environment or global information may not be fully used to lead particles' flight in the next time.

To address above question, inspired by differential evolution, the velocity formula is modified using the difference between two individuals randomly sampled in swarm in Eq. (5), called VD.

$$vd_{i,j}(t+1) = s \cdot (x_{r1,j}(t) - x_{r2,j}(t)) + c_1 rand_1(pbest_{i,j} - x_{i,j}(t)) + c_2 rand_2(gbest_j - x_{i,j}(t)) \quad (5)$$

Where, $i = 1, 2, \cdots, N$, $j = 1, 2, \cdots, D$, s is a differential factor, used to control the scope of the search. r_1 and r_2 are two different integers selected from a uniform distribution in the interval [1, N], $r_1 \neq r_2 \neq i$. Other parameters setting, such as c_1, c_2, are the same as Eq. (1).

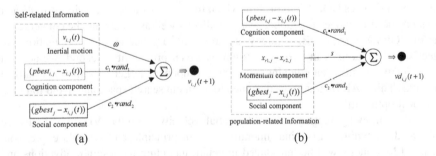

Fig. 1. Decomposition diagram of update equation of velocity (Color figure online)

Obviously, the first part is displaced by a differential mutation operator inspired by differential evolution scheme which further strengthen the search abilities by learn from a pair of particles randomly sampled in swarm, and thus it is called momentum component and not inertial motion. The structure of new formula shows that particles have transformed the thinking mode and depends on not only self-experience but also others experience when updating fly direction in the next time. The momentum component mostly focuses on population information instead of self-related information (as shown in red-dotted box of Fig. 1(b)). Improved formula (VD) could augment the search range of PSO and produce some additional exploration ability of the search space by adaptively changing flying direction with the behavior of the swarm, and then efficiently improve the speed of convergence and calculation accuracy.

3.2   Adaptive Elite Mutation Selection Strategy

In the process of evolution, capacity of searching space of each particle is not the same. In order to full motivate particles' activity, "elite strategy" is adopted to speed up the convergence through more evolution is carried out by elite particles whose fitness values in swarm is better. In this paper, global optimum (*gbest*) is regarded as elite particle. Through adaptive mutation operation for *gbest* in every generation, a new particle is generated for further exploration in search space. If the fitness of it is better than original *gbest*, the new particle will substitute for *gbest* and then enters into the next generation. Above thought of elite mutation is called adaptive elite mutation strategy, shortly AEM [15].

In AEM, generated factor *xm* for mutation in the ith dimension is defined by:

$$xm(i) = \exp\left(-\lambda \cdot t/t_{\max}\right) \cdot (1 - r(i)/r_{\max}) \tag{6}$$

Where, λ is a constant, t is iteration times, $t_{\max}$ is the maximum of t, $r(i)$ is the distance from particles to *gbest* in ith dimension in a generation. $r_{\max}$ is the maximum of r vector. $r(i)$ is defined as follows:

$$r(i) = |gbest(i) - avg\_pbest(i)| \tag{7}$$

Where, $i = 1, 2, \cdots, D$, $avg\_pbest(i)$, a mean value of the *pbest* of N particles, is formulated as follows:

$$avg\_pbest(i) = \left(\sum_{j=1}^{N} pbest[j][i]\right) \Big/ N \tag{8}$$

Where, $pbest[j][i]$ is the value of pbest in the ith dimension of the jth population. N is the size of population.

xm as a generated factor inserts into mutation function (9):

$$F(xm) = sign \cdot \left(\frac{1}{\pi}\arctan(xm) + C\right) \tag{9}$$

Where, symbol variable $sign \in \{-1, 1\}$, C is an undetermined constant which will get different value as Eq. (9) according to clustering situation of population.

$$C = \begin{cases} 1.5, & st\_d < 10^{-2} \\ 1.0, & 10^{-2} \le st\_d < 10^{-1} \\ 0.5, & other \end{cases} \tag{10}$$

In Eq. (10), st_d which means standard deviation of fitness is defined as Eq. (11) in this paper and used to measure the clustering degree of individuals. The value of st_d is divided into three sections: $(0, 10^{-2}]$, $(10^{-2}, 10^{-1}]$ and $(10^{-1}, +\infty)$ (the distribution of fitness value within segmentation points is studied in analysis of strategy).

$$st\_d = \sum_{i=1}^{N} \left| \frac{f_i - f_{gbest}}{f_{gbest}} \right| \tag{11}$$

Where, f_i is the fitness of the ith individual, f_{gbest} is the fitness of current global best position. Equation (11) shows st_d is less when all particles are closer to the gbest, and vice versa.

According to the above mentioned analysis, an adaptive elite mutation strategy (AEM) is defined as follows:

$$gbest^* = gbest + F(xm) \tag{12}$$

In each generation, recorded the position of the ith dimension after mutation operation using AEM, the global best mutation position is be generated, that is $gbest^*$.

3.3   DOPSO Algorithm

A new opposition-based PSO inspired by differential thought, called DOPSO, is composed of differential velocity update formula (VD) and adaptive elite mutation (AEM) applied to opposition-based learning PSO. The main steps of DOPSO algorithm are shown as Table 1. Where jr is the probability of using GOBL operator.

4   Experiments

In this section, two parts of experiment are carried out based on a set of benchmark problems. The one part is performance comparison, the performance of DOPSO is compared with a set of OBL-based PSOs. The second part is parameter sensitivity study, some suggestions about parameter settings are given via sensitivity analysis of key parameters.

4.1   Benchmark Problems

All experiments are conducted by MATLAB2012 on 13 benchmark problems with dimension $D = 30$ and the size of population $N = 40$. According to their properties,

Table 1. The main steps of DOPSO algorithm.

---

1 randomly initializes N particles in the population P;

2 generate the opposite solutions of N particles to construct opposite population OP according to Eq.(5);

3 calculate the fitness of P and OP;

4 select N best fittest solutions from $P \cup OP$ as an initial population P;

5 update the fitness of P;

6 while the stopping criterion is not meet do

7   If rand(0,1)<jr then

8   update the dynamic interval boundaries $[da_j, db_j]$ according to Eq.(5);

9   generate the new opposite population OP of N particles according to Eq.(4);

10   calculate the fitness of OP ;

11   select N best fittest solutions from $P \cup OP$ as a new current population P;

12   update *pbest* vector and *gbest* if needed;

13   else

14   for i=1 to N do

15   calculate the velocity of the ith particle according to Eq.(6) ;

16   update the position of the ith particle according to Eq.(2);

17   update the fitness the ith particle;

18   update *pbest,* if needed;

19   end for;

20   update *gbest* if needed;

21   end if

22   for j=1 to D do

23   generate $gbest^*$ via mutation operation Eq.(13);

24   end for

25   if the fitness of $gbest^*$ is better than gbest

         $gbest = gbest^*$

26   end if

27 end while

---

the problems are divided into two categories, which are unimodal problems f_1-f_7 and multimodal problems f_7-f_{13}. Compared with unimodal, multimodal problem is difficult since the number of local optima grows exponentially with the increase of dimensionality. All test functions are to be minimized in the following experiments. A brief description of these benchmark problems is listed in Table 2.

Noted, in rotated functions $f_{11}-f_{13}$ and f_7, M is an $D \times D$ orthogonal matrix, and $x = (x_1, x_2, \cdots, x_D)$ is a D-dimensional row vector.

4.2 Parameter Settings

The selection of the parameters which is very important to bionic algorithm can greatly influence the performance of PSO. In comparison between algorithms section, the parameter settings in this paper are consistent with the original algorithms as far as possible so as to better compare the performance of DOPSO with other similar

Table 2. The 13 benchmark functions in the experiments, where D is the dimension, f_{min} is the global minimum value of the test function.

Test function		D	Search space	f_{min}	Function
Unimodal	$f_1(x) = \sum_{i=1}^{D} x_i^2$	30	$[-100, 100]^D$	0	Sphere
	$f_2(x) = \sum_{i=1}^{D} [(x_i + 0.5)]^2$	30	$[-100, 100]^D$	0	Step
	$f_3(x) = \sum_{i=1}^{D-1} [100 \cdot (x_{i+1} - x_i^2)^2 + (1 - x_i)^2]$	30	$[-30, 30]^D$	0	Rosenbrock
	$f_4(x) = \sum_{i=1}^{D} (\sum_{j=1}^{i} x_j)^2$	30	$[-100, 100]^D$	0	Quadric
	$f_5(x) = \sum_{i=1}^{D} |x_i| \prod_{i=1}^{D} |x_i|$	30	$[-10, 10]^D$	0	Schwefel2.22
	$f_6(x) = \sum_{i=1}^{D} (10^6)^{\frac{i-1}{D-1}} x_i^2$	30	$[-100, 100]^D$	0	Elliptic
	$f_7(x) = f_6(z), \quad z = x * M$	30	$[-5.12, 5.12]^D$	0	Rotated elliptic
Multimodal	$f_8(x) = \sum_{i=1}^{D} [x_i^2 - 10\cos(2\pi x_i) + 10]$	30	$[-5.12, 5.12]^D$	0	Rastrigin
	$f_9(x) = -20 \cdot \exp\left(-0.2 \cdot \sqrt{\left(\sum_{i=1}^{D} x_i^2 / D\right)}\right)$ $- \exp\left(\left(\sum_{i=1}^{D} \cos(2\pi x_i)\right) / D\right) + 20 + e$	30	$[-32, 32]^D$	0	Ackley
	$f_{10}(x) = \frac{1}{4000} \sum_{i=1}^{D} x_i^2 - \prod_{i=1}^{D} \cos\left(\frac{x_i}{\sqrt{i}}\right) + 1$	30	$[-600, 600]^D$	0	Griewank
	$f_{11}(x) = f_8(z), \quad z = x * M$	30	$[-32, 32]^D$	0	Rotated Rastrigin
	$f_{12}(x) = f_9(z), \quad z = x * M$	30	$[-600, 600]^D$	0	Rotated Ackley
	$f_{13}(x) = f_{10}(z), \quad z = x * M$	30	$[-32, 32]^D$	0	Rotated Griewank

Table 3. The specific parameter settings of DOPSO.

c_1	c_2	s	jr	λ	N	D
1.49618	1.49618	0.2	0.3	10	40	30

algorithms. The velocity v is limited to the half range of the search space on each dimension. Experiments are always performed in running 30 times. The default parameter settings of DOPSO are listed as Table 3.

4.3   Performance Comparison Between OBL-Based PSO

The performance of DOPSO is compared with OBL-based PSOs including EOPSO [18], GOPSO [12], OVCPSO [19] and OPSO [11] in this paper. Every algorithm runs 10000 generations every time, all the experiments were conducted 30 times and then recorded the mean value of the global best optimum of each algorithm as Table 4. Where Symbol "+" denotes the performance of DOPSO is better than other algorithm, symbol "−" denotes worse and "∼" denotes equal to other algorithm. The best results in six algorithms aim at the thirteen benchmark problems are shown in bold.

It can be concluded from Table 4 that DOPSO has almost achieved good results in all benchmark problems. Experimental results show that DOPSO outperforms all test

Table 4. Mean value of the global optimum in 30 runs among six OBL-based PSO aiming at the 13 benchmark functions.

Funs.		OPSO		GOPSO		OVCPSO		EOPSO		DOPSO
Multimodal	f_1	4.59E−36	+	**0.00E+00**	~	5.00E−01	+	**0.00E+00**	~	0.00E+00
	f_2	2.09E−35	+	3.02E−321	+	6.67E−2	+	1.97E−323	+	0.00E+00
	f_3	7.18E+00	−	2.82E+01	+	8.70E+01	+	**1.57E−24**	−	5.30E+00
	f_4	4.13E+04	+	1.32E+04	+	2.94E+01	+	5.01E−03	+	0.00E+00
	f_5	−6.51E−12	+	−6.98E−162	+	−2.49E+00	+	3.01E−162	+	0.00E+00
	f_6	1.43E−32	+	**4.21E−316**	+	1.39E−04	+	9.88E−324	+	0.00E+00
	f_7	9.82E+06	−	**1.96E+06**	+	3.06E+00	+	5.08E+02	+	0.00E+00
	f_8	1.51E+01	+	**0.00E+00**	~	7.57E−01	+	2.09E+00	+	0.00E+00
	f_9	1.85E+00	+	**0.00E+00**	~	9.99E−01	+	1.86E−01	+	0.00E+00
	f_{10}	3.83E−01	+	**0.00E+00**	~	2.05E−02	+	1.26E−02	+	0.00E+00
	f_{11}	1.51E+01	+	**0.00E+00**	~	4.14E+01	+	4.11E+00	+	0.00E+00
	f_{12}	2.98E+00	+	**0.00E+00**	~	1.97E+00	+	3.41E−01	+	0.00E+00
	f_{13}	2.33E−02	+	**0.00E+00**	~	6.14E−01	+	1.22E−02	+	0.00E+00
Total	+	12		6		13		10		−
	~	0		7		0		1		−
	−	1		0		0		1		−

problems than OVCPSO and obviously superior to OPSO and PSO besides f_3. Conversely, optimal value of f_3 is obtained by EOPSO. It is worth noting that GOPSO, like DOPSO, obtained the optimal value in seven test functions and near-optimal solution in two test functions (i.e. f_2 and f_6). However, these two algorithms are all trapped into local optima at Rosenbrock function (f_3).

Further experiment was conducted to compare the comprehensive performance between DOPSO and GOPSO. Average number of fitness calls were shown in Fig. 2 after 30 times running DOPSO and GOPSO on the condition that precision is 10^{-16} and maximum iterations is 10000, a striking result in Fig. 2 is that average number of function calls of DOPSO less significantly than GOPSO in almost test functions, which proves that DOPSO can quickly converge to optimal value.

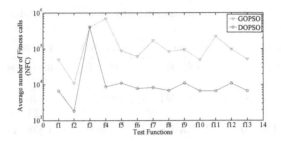

Fig. 2. Comparisons of average number of fitness calls (NFC) in 30 times running between DOPSO and GOPSO. The horizontal axis is 13 benchmark functions and the vertical axis is NFC.

4.4   Parameter Sensitivity Study

In order to investigate what factors may affect the performance of the algorithm, the two parameters including λ, s, set different value in DOPSO, are carried out experiment separately on the thirteen test functions. Besides, it has been verified in literature [11] that Opposition-based strategy will get best performance when $jr = 0.3$. This paper has followed this conclusion.

As mentioned above, DOPSO is composed of AEM, improved update equation of velocity (VD) and GOBL strategy, the value of λ in AEM strategy may be a critical factor about whether the algorithm can be smoother and faster convergence to the global optimal value. Due to limited space, Fig. 3 demonstrates the trend of the algorithm convergence to the global optimum only aiming at four test functions through λ is set different values in {10, 20, 30, 40, 50, 60}. From Fig. 3, it can be seen that the performance of DOPSO is not affected too much by using different values of λ.

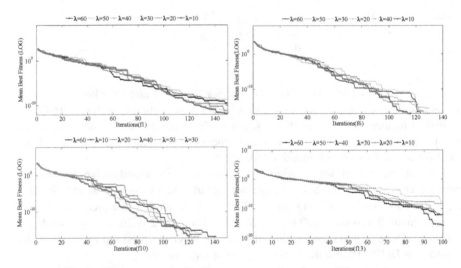

Fig. 3. The process of global convergence when λ gets different value

The results also explain that the choice of λ is not crucial to the performance of algorithm. However, we still suggest that the value of 10 for λ should be used in DOPSO since it achieved relatively better and steady result in almost all test functions than other values for λ.

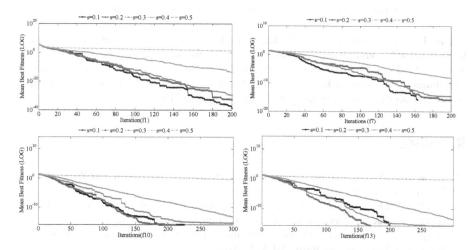

Fig. 4. The process of global convergence when s gets different value

In addition, differential factor s in the improved update equation of velocity is another critical factor to performance of algorithm. The experiments are carried out by setting different values in interval [0.1, 0.5]. In theory, the larger the value of s, the more powerful ability of exploration in search space, however, the experimental results indicate the performance of DOPSO is significantly degraded if the value of s is set to too large. The reason may be that, large scaling momentum component prevent population from convergence to global optima. Through specific analysis of experiment, the range of parameter s is suggested from 0.1 to 0.2 that be validated in Fig. 4.

5 Conclusion and Future Work

In this paper, we presented a differential opposition-based particle swarm optimization with adaptive elite mutation algorithms, termed DOPSO. Several new techniques have been incorporated into PSO to substantially accelerate the convergence speed as well as improve the precision. The first strategy is the new learning mechanism to modify velocity formula of basic PSO, which thoroughly get rid of inertial component from canonical velocity formula, termed VD. The second strategy is GOBL in the light of the fact that the probability of opposite individuals superior to original individuals accounts for 50%. The last strategy is adaptive elite mutation, which can adaptively adjust the mutation scale according to clustering degree of individuals. Compared with opposition-based PSO and its varieties on 13 benchmark functions, the experimental results show DOPSO algorithm has stronger competitive ability in both calculation accuracy and computation cost.

In the future, there are several issues need pursue. Firstly, we should study the theoretical foundation for a new learning mechanism appears in DOPSO to find why they can do well and which should do better than the others. Furthermore, we should generalize this learning mechanism to other evolutionary computing family.

Acknowledgment. The authors would like to thank the editors and the anonymous reviewers for their valuable comment and suggestions. This work was supported by the National Natural Science Foundation of China (No. 61170305), Science and Technology Research Project in Department of Education of Jiangxi Province (NO. GJJ161568), and Social Science and Technology Development Project of Dongguan City (No. 2017507156388).

References

1. Eiben, A.E., Smith, J.: From evolutionary computation to the evolution of things. Nature **521**, 476–482 (2015)
2. Bonyadi, M.R., Michalewicz, Z.: Particle swarm optimization for single objective continuous space problems: a review. Evolut. Comput. **25**(1), 1–54 (2017)
3. Gong, Y.J., et al.: An efficient resource allocation scheme using particle swarm optimization. IEEE Trans Evol Comput **16**(6), 801–816 (2012)
4. Dehuri, S., Roy, R., Cho, S.-B.: An adaptive binary PSO to learn bayesian classifier for prognostic modeling of metabolic syndrome. In: Genetic and Evolutionary Computation Conference (GECCO), 12–16 July 2011, pp. 495–501 (2011)
5. Zhu, Z., Zhou, J., Ji, Z., Shi, Y.-H.: DNA sequence compression using adaptive particle swarm optimization-based memetic algorithm. IEEE Trans Evol Comput **15**(5), 643–658 (2011)
6. Ismail, A., Engelbrecht, A.P.: Self-adaptive particle swarm optimization. In: Bui, L.T., Ong, Y.S., Hoai, N.X., Ishibuchi, H., Suganthan, P.N. (eds.) SEAL 2012. LNCS, vol. 7673, pp. 228–237. Springer, Heidelberg (2012). https://doi.org/10.1007/978-3-642-34859-4_23
7. Zhan, Z.H., Zhang, J., Li, Y., Chung, H.S.H.: Adaptive particle swarm optimization. IEEE Trans. Syst. Man Cybern.-Part B: Cybern. **39**, 1369–1381 (2009)
8. Shi, Y., Eberhart, R.C.: Parameter selection in particle swarm optimization. In: Porto, V.W., Saravanan, N., Waagen, D., Eiben, A.E. (eds.) EP 1998. LNCS, vol. 1447, pp. 591–600. Springer, Heidelberg (1998). https://doi.org/10.1007/BFb0040810
9. Zhang, W., Liu, Y., Clerc, M.: An adaptive PSO algorithm for reactive power optimization. In: Proceedings of 6th International Conference Advances in Power System Control, Operation and Management, pp. 302–307, November 2003
10. Hu, M., Wu, T., Weir, J.D.: An adaptive particle swarm optimization with multiple adaptive methods. IEEE Trans Evol Comput **17**, 705–720 (2013)
11. Wang, H., Li, H., Liu, Y., et al.: Opposition-based particle swarm algorithm with cauchy mutation. In: Proceedings of IEEE Congress on Evolutionary Computation, Tokyo, pp. 356–360 (2007)
12. Wang, H., Wu, Z., Rahnamayan, S., et al.: Enhancing particle swarm optimization using generalized opposition-based learning. Inf Sci **181**, 4699–4714 (2011)
13. Zuo, X., Xiao, L.: A DE and PSO based hybrid algorithm for dynamic optimization problems. Soft Comput **18**, 1405–1424 (2014)

14. Pehlivanoglu, Y.V.: A new particle swarm optimization method enhanced with a periodic mutation strategy and neural networks. IEEE Trans Evol Comput **17**(3), 436–452 (2013)
15. Dong, W.Y., Kang, L.L.: Opposition-based particle swarm optimization with adaptive elite mutation and nonlinear inertia weight. J. Commun. **37**(12), 1–10 (2016)
16. Shi, Y., Eberhart, R.C.: A modified particle swarm optimizer. In: Proceedings of IEEE World Congress Computational Intelligence, pp. 69–73 (1998)
17. Wang, H., Wu, Z., Liu, Y., Wang, J., Jiang, D., Chen, L.: Space transformation search: a new evolutionary technique. In: 2009 Proceedings of World Summit Genetic Evolution Computer, pp. 537–544 (2009)
18. Zhou, X.Y., Wu, Z.J., Wang, H., et al.: Elite opposition-based particle swarm optimization. Acta Electron. Sin. **41**(8), 1647–1652 (2013)
19. Shahzad, F., Baig, A.R., Masood, S., Kamran, M., Naveed, N.: Opposition-based particle swarm optimization with velocity clamping (OVCPSO). In: Yu, W., Sanchez, E.N. (eds.) Advances in Computational Intelligence. AINSC, vol. 116, pp. 339–348. Springer, Berlin (2009). https://doi.org/10.1007/978-3-642-03156-4_34

Research on Hierarchical Cooperative Algorithm Based on Genetic Algorithm and Particle Swarm Optimization

Linrun Qiu[✉]

Department of Computer Science, Guangdong University of Science and Technology,
Dongguan, China
79713159@qq.com

Abstract. In this paper, a hierarchical cooperative algorithm based on the genetic algorithm and the particle swarm optimization is proposed that utilizes the global searching ability of genetic algorithm and the fast convergence speed of particle swarm optimization. The proposed algorithm starts from Individual organizational structure of subgroups and takes full advantage of the merits of the particle swarm optimization algorithm and the genetic algorithm (HCGA-PSO). The algorithm uses a layered structure with two layers. The bottom layer is composed of a series of genetic algorithm by subgroups that contributes to the global searching ability of the algorithm. The upper layer is an elite group consisting of the best individuals of each subgroup and the particle swarm algorithm is used to perform precise local search. The experimental results demonstrate that the HCGA-PSO algorithm has better convergence and stronger continuous search capability, which makes it suitable for solving complex optimization problems.

Keywords: Genetic algorithm · Particle swarm optimization
Cooperative algorithm · Hybrid algorithm

1 Introduction

The genetic algorithm (GA) and the particle swarm optimization (PSO) are both evolutionary computation techniques based on population [1]. Both have their own characteristics and advantages, but also have some shortcomings and deficiencies. The genetic algorithm has a strong global search ability, but its local search ability is poor, which makes a simple genetic algorithm more time-consuming and reduce its late evolutionary search efficiency. Whereas, the particle swarm optimization can achieve a simple and fast convergence. However, the fast convergence also leads to fast population decline reducing the overall search ability prone to premature convergence. In the face of small-scale optimization problems, the GA and the PSO both have promising performance [2]. However, as practical optimization problems have become more complex, their defects are becoming increasingly prominent and the optimization efficiency (time) and the quality of the solution are "powerless". Therefore, improvement in their optimization performance is imminent. In view that the genetic algorithm and the particle swarm

© Springer Nature Singapore Pte Ltd. 2018
K. Li et al. (Eds.): ISICA 2017, CCIS 874, pp. 16–25, 2018.
https://doi.org/10.1007/978-981-13-1651-7_2

optimization have almost complementary advantages [3], researchers suggest the combination of the two so that they can learn from each other in order to develop a better performance algorithm. A common approach is to mix the two algorithms in the same position. There are two main mixing methods that can be summed up from some of the proposed hybrid algorithms: and parallel refer to the genetic process and the particle swarm optimization for all individuals in the evolution of each generation. Parallel is the individual that divides the population into two generations in the evolution of each generation: the evolution of the genetic algorithm and the particle swarm optimization. If the algorithm uses the same population size, the series mixing will be more than twice the amount of parallel mixing [4]. In both mixing methods, the resultant algorithm's search ability will be stronger than a single genetic algorithm or particle swarm optimization [5]. However, both of these mixing methods face the same problem, that is, the division of labor between the genetic algorithm and the particle swarm optimization is not clear when both are in the same position, which makes their respective advantages unaffected. Therefore, there is a lot of space to be excavated in the fusion of genetic algorithm and particle swarm optimization [6].

In this paper, a new genetic algorithm and particle swarm optimization cooperative algorithm is proposed. The proposed algorithm uses a hierarchical structure, the bottom consists of a series of subgroups using the genetic algorithm evolution that contributes to the global search ability of the algorithm; the upper layer is composed of the optimal individuals of each subgroup of elite groups using the PSO algorithm for accurate local search to speed up the convergence. This paper starts from the organizational structure of the population, and separates the global search from the local search, which can speed up the convergence speed and avoid the decrease of the diversity caused by the convergence. The global search ability and the convergence rate both are effectively improved.

2   Introduction of Related Algorithm

2.1   Genetic Algorithm

The genetic algorithm (GA) is a stochastic search method based on the natural evolution of Biology [7]. It was first proposed by Professor Holland in the United States in 1975. The algorithm adopts the evolution of "survival of the fittest" (including selection, crossover and mutation), so that the population evolves in order to meet the requirements of the optimal solution [8]. Currently, the genetic algorithm is being successfully applied to various areas, such as optimizing the design, neutral network training, pattern recognition, timing prediction, etc. The genetic algorithm can use the genetic phenomenon as a prototype to solve many problems. Using the genetic algorithm to solve the problem includes the following steps:

(1) The problems to be solved are analyzed and translated into all aspects of the genetic phenomenon.
(2) Encoding.
(3) The fitness value of each individual is calculated by using objective function.

(4) A certain number of individuals in the population are selected for free random hybridization, resulting in the next generation. The process of breeding the next generation will be genetic, mutation, and so on. This process will produce more adaptive value high individual until the request is reached.

(5) Once the individual that meets the requirements is found, the whole problem is solved and the individual that meets the requirements is the optimal solution for the problem.

2.2 Particle Swarm Algorithm

Particle swarm optimization (PSO) is a new intelligent optimization algorithm proposed by Kennedy and Eberhart in 1995 [9]. PSO is based on the study of foraging behavior of birds. Particle swarm optimization is simple and easy to implement, it has fast convergence speed and has been widely used in engineering optimization, image processing and other fields. In the PSO algorithm, the position of each particle represents a candidate solution of the search space, and the particle has the characteristics of position and velocity. The particle function coordinates correspond to the objective function as the fitness of the particle. The algorithm first initializes a group of random particles, and then iteratively determines the optimal solution. In each iteration, the particle updates itself by tracking two "extremes": one is the current optimal position found by the particle itself, the individual extremum pib (t); the other is the current optimal position found by the entire population, i.e., the global extremum pgb (t). The updated equation for the speed and the position of the standard PSO algorithm is as follows:

$$v_i(t+1) = \omega v_i(t) + c_1 r_1 \left(p_{ib}(t) - x_i(t) \right) + c_2 r_2 \left(p_{gb}(t) - x_i(t) \right) \tag{1}$$

$$x_i(t+1) = x_i(t) + v_i(t+1) \tag{2}$$

where v_i and x_i represent the velocity vector and the position vector of the i-th particle, respectively, w is the inertia weight, c_1 and c_2 are the acceleration constants, r_1 and r_2 are both in [0, 1] within the scope of the uniform distribution of random variables. Particle swarm algorithm is affected by factors such as particle velocity, particle current position, particle self-learning and social learning.

When the particle swarm algorithm is used to solve the problem, the problem is described first, and a certain number of initial particles are generated [10]. Then, the fitness value of these particles is calculated by the objective function. The particles with the best fitness are selected and recorded as the temporary solutions. The best available solution of the solution to the relevant information will be shared for other particles reference use. Thereafter, in the given iteration times, the search space is constantly searched in the definition domain. When the particle with better fitness is found, it is recorded as a new optimal feasible solution and replaces the old solution. The process is repeated until the resulting optimal feasible solution satisfies the preset condition or reaches the maximum number of iterations to find the final optimal solution.

3   Hierarchical Cooperative Algorithm of Genetic Algorithm and Particle Swarm Optimization

The organization of the subgroups in the HCGA-PSO algorithm is shown in Fig. 1. The bottom layer is composed of N independent subgroups. The genetic algorithm is used to evolve for each subgroup. It is the cornerstone of the entire algorithm. Exceptional individuals constitute the upper elite group using the PSO algorithm evolution, which is mainly responsible for the elite of the local search in order to speed up the algorithm convergence rate.

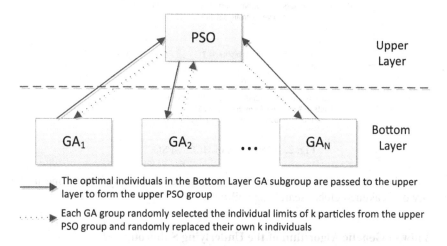

The optimal individuals in the Bottom Layer GA subgroup are passed to the upper layer to form the upper PSO group

Each GA group randomly selected the individual limits of k particles from the upper PSO group and randomly replaced their own k individuals

Fig. 1. Organization structure of individuals of HCGA-PSO

The algorithm first randomly initializes the N subgroups and records them as GA_i, $i = 1, 2, \ldots, N$. Each subgroup runs the respective genetic algorithm independently and the respective optimal individuals are taken out after some algebra. The PSO algorithm is used to evolve the elite group, and after some algebra, it is judged whether the stop criterion is satisfied. If the criterion is satisfied, the result is produced and the algorithm is stopped. Otherwise, each GA subgroup is randomly obtained from the upper elite group. The N GA subgroups are terminated at the end of the first round of the HCGA-PSO algorithm. The N GA subgroups start the genetic algorithm operation and continue to cycle until the stop criterion is satisfied [11]. The flow chart of the HCGA-PSO algorithm is shown in Fig. 2.

Compared with other hybrid algorithms of genetic algorithm and particle swarm optimization, the HCGA-PSO algorithm takes the genetic algorithm as the main body to ensure the global search ability of the algorithm. The multi-subgroup structure can better maintain the population diversity and benefit the global search. Moreover, it is also suitable for coarse-grained parallel computing. Particle swarm algorithm is used to search the elite individuals locally. The convergence speed of the HCGA-PSO algorithm is fast. The use of the hierarchical structure separates the individuals responsible for the global search from the individuals searching for the upper part. The hierarchical structure

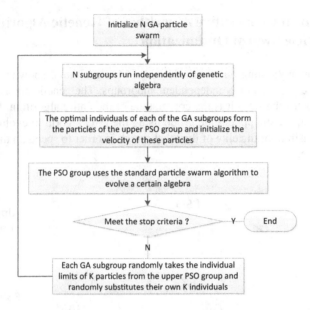

Fig. 2. Flow chart of HCGA-PSO

speeds up the optimization and avoids the convergence caused by the decline in the diversity of weakened global search capabilities.

3.1 Flow of Genetic Algorithm of the Underlying Subgroup

This algorithm is mainly aimed at continuous quantity optimization, and uses real number coding for realization. The number of individuals is n, the individual dimension is d, and the individual is $x_i(i = 0, 1, \ldots, n - 1)$. The middle generation is $x_i'(i = 0, 1, \ldots, n - 1)$, and a certain dimension of an individual is $x_j(j = 0, 1, \ldots, d - 1)$. The individual fitness value is $f_i(i = 0, 1, \ldots, n - 1)$, the history of the best individual is I_{best}. The fitness value and mutation probability of the best individual are F_{best} and p_m, respectively. Assuming the iterative algebra t = 0, the iteration T is stopped [12]. The concrete steps are as follows:

(1) If the iteration algebra t = T, the algorithm stops; otherwise proceed to step 2.
(2) The use of unselected league selection method. Select the middle of the generation of n individuals, the equation is as follows:

$$x_i' = \begin{cases} x_i, & f_i \leq f_k \\ x_k, & \text{Others} \end{cases} \tag{3}$$

where $i = 0, 1, \ldots, n - 1$, k are randomly generated integers belonging to [0, n − 1], and k ≠ i.

(3) Completely cross on the intermediate generation (the probability of crossing 100%) to generate a new generation of individuals, the equation is as follows:

$$\begin{cases} x_i = rand1 \cdot x_i' + (1 - rand2) \cdot x_{i+1}' \\ \quad i = 0, 1, \ldots, n-2 \\ x_i = rand1 \cdot x_i' + (1 - rand2) \cdot x_0' \\ \quad i = n-1 \end{cases} \tag{4}$$

where rand1, and rand2 are random variables that are uniformly distributed within the range [0, 1].

(4) Use non-uniform variation, according to the probability p_m for each individual component of the variation, the variance equation is as follows:

$$x_j = \begin{cases} x_j + \Delta\left(t, U_{max}^j - x_j\right), & rand \leq 0.5 \\ x_j - \Delta\left(t, x_j - U_{min}^j\right), & \text{Others} \end{cases} \tag{5}$$

where
$rand \sim Uniform(0, 1)$, $\Delta(t, y) = y \times \left(1 - r^{(1-t/T)^b}\right)$, $r \sim Uniform(0, 1)$, $x_j \in \left[U_{min}^j, U_{max}^j\right]$. The constant b is a parameter that determines the degree of nonconformity. It plays an important role in adjusting the local search area. The general range of b is. The smaller the value is, the larger the variation is, that is, the local search area that is based on the global search ability of genetic algorithm. Thus, the variation range of genetic algorithm is relatively large. Therefore, the constant b is set to 2.

(5) Calculate the fitness value of the individual.
(6) In the current generation, if the individual is superior to the historical optimal individual, both the historical optimal individual I_{best} and its fitness value F_{best} are updated; otherwise, any one of the current generation is randomly replaced by the historical optimal individual.
(7) $t = t + 1$, go to step 1.

3.2 Velocity Initialization of Elite Particle Swarm

In the standard PSO algorithm, there is a velocity interval $\left[-v_{max}^j; v_{max}^j\right]$ for each dimension component v_j ($j = 0, 1, \ldots, d - 1$) of the particle velocity vector vi, where v_{max}^j is based on the search space set experimental value. At the time of initialization, let $v_j = r \cdot v_{max}^j$, where r is a random variable that is uniformly distributed within the range of $[-1, 1]$. The algorithm uses the standard PSO algorithm to optimize the upper elite group initialization speed. Some changes in the vicinity of the elite can be an accurate local search that will improve the optimization speed. If the velocity interval $\left[-v_{max}^j; v_{max}^j\right]$ is still used to initialize the particles, the velocity of the particles is different and divergent and cannot achieve the purpose of focusing on the current solution. Therefore, when the particle velocity of the elite group is initialized, the direction of each dimension velocity of the clamp particle is positive or both, that is, the velocity interval is $\left[0, v_{max}^j\right]$ or $\left[-v_{max}^j, 0\right]$. When the velocity of all the particles will be concentrated in the space of

$1 = 2^d$, the degree of freedom of the particles will be greatly reduced, the motion will be more aggregated and it will be easier to cooperate with each other for accurate local search.

3.3   Convergence Analysis of HCGA-PSO Algorithm

Gunter Rudolph has proved that as long as the historical optimal solution is contained in each generation, the classical genetic algorithm included in each generation can converge to the global optimal solution before or after the selection operator, which is called the optimal reservation strategy. The genetic algorithm of the evolutionary subgroups in the proposed algorithm uses the optimal retention strategy and finds the historical optimal solution before the selection operator. If the solution is not in the current population, it randomly replaces any one of the current population. Therefore, the genetic algorithm used by the underlying subgroups has global convergence. In particle swarm optimization, it is assumed that $p_{ib}(t)$, and $p_{gb}(t)$ remain unchanged in the evolution, then when w, c1, and c2 satisfy:

$$\sqrt{2(1 + \omega - \varphi)^2 - 4\omega} < 2 \tag{6}$$

The $x_i(t)$ of the PSO algorithm converges to the weighted center of $p_{ib}(t)$, and $p_{gb}(t)$, i.e., $\varphi = \varphi_1 + \varphi_2, \varphi_1 = c_1 r_1, and \varphi_2 = c_2 r_2$.

$$x_i(t) \rightarrow \frac{\varphi_1 p_{ib}(t) + \varphi_2 p_{gb}(t)}{\varphi} \tag{7}$$

Considering the HCGA-PSO algorithm, the global optimal position is p_{gbset}. Since the genetic algorithm of the underlying subgroup has global convergence, all the particles in the upper elite group will be in the global maximum once all the subgroups converge to the global optimal solution. In the optimal position, the individual extremes of the particles are the same as the global extremes and remain invariant in the evolutionary process, all of which are p_{gbset}. At this point, Eq. (7) can be changed:

$$x_i(t) \rightarrow \frac{\varphi_1 p_{ib}(t) + \varphi_2 p_{gb}(t)}{\varphi} = \frac{(\varphi_1 + \varphi_2) p_{gbest}}{\varphi} = p_{gbest} \tag{8}$$

That is, the upper particle swarm optimization algorithm will converge to the global optimal. Therefore, when the underlying subgroups adopt the genetic algorithm with the optimal retention strategy evolution, the upper particle swarm optimization should choose w, c_1, and c_2. Thus, the HCGA-PSO algorithm has global convergence.

4 Experimental Results and Analysis

4.1 Tests of Typical Function

In order to verify the performance of the proposed algorithm, experiments are performed on typical continuous functions: single and double pole values, six hump functions, and 0/1 knapsack problem. Comparative analysis is executed on the HCGA-PSO algorithm with a single GA and PSO (GA-PSO) algorithm.

(1) The unicorn point of the six peaks hump function is: $F_1 = 10 + \sin(1/x) \div (0.1 + (x - 0.16)^2)$, $0 < x <$. The function has a global maximum: $f(0.1275) = 19.8949$.

(2) The two-point point of the six peaks hump function is: $F_2 = (4 - 21x_1^2 + (1/3)x_1^3)x_1^2 + x_1x_2 + (-4 + 4x_2^4)x_2^2$, $-3 \leq x_i \leq 3, i = 1, 2$. The function has six local minimum points, but there are two global minimum points:
$f(-0.0898, 0.7126) = -1.031628$,
$f(0.898, 0.216) = -1.031628$.

The parameters of the algorithm and the function in the test are set as follows: The population size is 10, the chromosome of the function F_1 is 9 bits and the number of iterations is 100. The chromosome of the function F_2 is 30 bits, the number of iterations is 300, and the average is the mean value of the optimal solution. The GA-PSO algorithm and the HCGA-PSO algorithm comparison results are shown in Table 1. With same number of iterations, the HCGA-PSO algorithm takes on average 6 generations to find the optimal solution, while the GA-PSO algorithm optimizes the average algebra in 40 or above generations. Moreover, the optimal value variance of the HCGA-PSO algorithm is lower than that of the GA-PSO algorithm. The results clearly show that the HCGA-PSO algorithm is better and more stable than the GA-PSO algorithm.

Table 1. Typical test results and comparison of functions

Function	Theoretical optimal solution	Algorithm	Average value	The optimal value	Average generation	Optimal variance
F_1	18.8949	GA-PSO	18.82122	18.89396	42.3	2.13700
		HCGA-PSO	18.88663	18.89484	6.1	0.00246
F_2	−1.051326	GA-PSO	−0.98984	−1.03748	43.1	0.00967
		HCGA-PSO	−1.03864	−1.04262	4.7	0.00132

4.2 Knapsack Problem Experiment

Knapsack problem is a typical combination of optimization problems [13]. In this experiment, the HCGA-PSO is used to solve the 0/1 knapsack problem and is compared with GA-PSO and GA. The population size is 10, and the chromosome coding is 20-bit. The convergences of the algorithms are shown in Figs. 3 and 4.

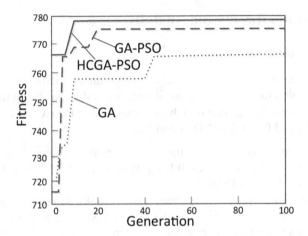

Fig. 3. Adaptability evolution curves of HCGA-PSO and GA-PSO and GA

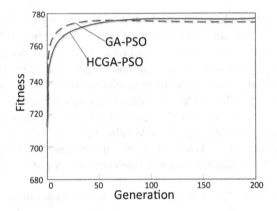

Fig. 4. HCGA-PSO and GA-PSO run 20 times the average best fitness evolution curve

In Fig. 3, all the algorithms are iterated for 100 generations. The HCGA-PSO, GA-PSO and GA find their optimal solutions near 10, 20, and 40 generations, respectively, and the optimal solution decreases. The convergence speed and the results of the HCGA-PSO and the GA-PSO are better than those of GA. The HCGA-PSO algorithm has the best convergence rate and the optimal solution before the 10th generation.

In Fig. 4 shows a schematic diagram of the solution of the mean fitness with the evolutionary algebraic change when the HCGA-PSO and GA-PSO are run independently 20 times and the iterative algebra is 200. When the evolutionary algebra is small, the optimal fitness of the GA-PSO is better than that of HCGA-PSO. With the increase of the evolutionary algebra, the HCGA-PSO achieves better optimal fitness than the GA-PSO from about 60 generations, that is, the HCGA-PSO can converge to the optimal solution. The optimal solution of the HCGA-PSO algorithm is better and more stable than that of the PSO-GA.

5 Conclusion

In this paper, a hierarchical cooperative algorithm based on genetic algorithm and particle swarm optimization (HCGA-PSO) is proposed. The proposed algorithm adopts the layer structure and introduces the global convergence of the genetic algorithm to evolve the bottom group, with global search ability. The optimal group composed of the optimal individuals of each subgroup is evolved by the particle swarm optimization algorithm with fast convergence speed. The particle swarm optimization algorithm can be used to search the elite individual accessories accurately. The algorithm separates the subgroups responsible for the global search from the elite groups that are responsible for the local search, which can improve the convergence speed of the algorithm and avoid the diversity of the convergence. The global convergence, the superior performance in global search ability, the searching speed and the stability of the proposed HCGA-PSO algorithm are demonstrated and verified experimentally using the test functions.

References

1. Jiang, Q., Wang, Y.: Research on optimizing dynamic pricing based on evolutionary computation techniques. Comput. Eng. Appl. **46**(24), 229–232 (2010)
2. Chang, J.X., Bai, T., Huang, Q., et al.: Optimization of water resources utilization by PSO-GA. Water Resour. Manag. **27**(10), 3525–3540 (2013)
3. Rao, D.T., Kumar, P.R., Rajeswari, K.R.: Range resolution of pulse compression using genetic algorithm and particle swarm optimization. Int. J. Appl. Eng. Res. **10**(16), 37255–37260 (2015)
4. Wan, W., Birch, J.B.: An improved hybrid genetic algorithm with a new local search procedure. J. Appl. Math. **3**, 4334–4347 (2013)
5. Jiang, X., Fan, Y., Wang, W., et al.: BP neural network camera calibration based on particle swarm optimization genetic algorithm. J. Front. Comput. Sci. Technol. **8**(10), 1254–1262 (2014)
6. Dai, S.P., Song, Y.D.: Parameter selection of support vector machines based on the fusion of genetic algorithm and the particle swarm optimization. Comput. Eng. Sci. **34**(10), 113–117 (2012)
7. Yang, D., Rao, K., Xu, B., et al.: PIR sensors deployment with the accessible priority in smart home using genetic algorithm. Int. J. Distrib. Sens. Netw. **11**, 1–10 (2015)
8. Feng, G., Liu, M., Guo, X., et al.: Genetic algorithm based optimal placement of PIR sensor arrays for human localization. Optim. Eng. **15**(3), 643–656 (2014)
9. Naruse, H., Olariu, C.: Research on glowworm swarm optimization with ethnic division. J. Netw. **9**(2), 305–314 (2014)
10. Chen, R.Z.: Improved self-adaptive glowworm swarm optimization algorithm. Appl. Mech. Mater. **19**(1), 798–801 (2014)
11. Li, N., He, P., Zhao, Q.: Face recognition classifier design based on the genetic algorithm and neural network. Adv. Mater. Res. **10**, 869–872 (2014)
12. Huang, L., Huang, G., Lebeau, R.P., et al.: Optimization of aifoil flow control using a genetic algorithm with diversity control. J. Aircr. **44**(4), 1337–1349 (2015)
13. Dean, B.C., Goemans, M.X., Vondrdk, J.: Approximating the stochastic knapsack problem: the benefit of adaptivity. Math. Oper. Res. **33**(4), 945–964 (2008)

An Adaptive Particle Swarm Optimization Using Hybrid Strategy

Peng Shao[1]($\boxtimes$), Zhijian Wu[2], Hu Peng[3], Yinglong Wang[1],
and Guangquan Li[1]

[1] School of Computer and Information Engineering,
Jiangxi Agricultural University, Nanchang 330045, Jiangxi, China
sp198310@163.com
[2] Computer School, Wuhan University, Wuhan 430072, Hubei, China
[3] School of Information Science and Technology, Jiujiang University,
Jiujiang 332005, Jiangxi, China

Abstract. As an intelligent algorithm inspired by the foraging behavior in nature, particle swarm optimization (PSO) is famous for its few parameters, easy to implement and higher convergence accuracy. However, PSO also has a weakness over the local search, also called the prematurity, which resulted in the convergence accuracy reduced and the convergence speed slowed. For this, extremal optimization (EO), an excellent local search algorithm, has been introduced to be improved (CEO) and enhance the local search of PSO. Meanwhile, for improving its global search further, an improved opposition-based learning based on refraction principle (UOBL) has been chosen to enhance the global search of PSO, which is a better global optimization algorithm. In order to balance both of PSO to improve its optimization performance further, an adaptive hybrid PSO based on UOBL and CEO (AHOPSO-CEO) is proposed in this article. The large number of experiment results and analysis reveals that AHOPSO-CEO achieves better performance with other algorithms on the convergence speed and convergence accuracy for optimization problems.

Keywords: Particle swarm optimization · Opposition-based learning
Extremal optimization

1 Introduction

Evolutionary algorithms (EA), which mainly mimics the social behavior of biology such as foraging in nature, increasingly have aroused researchers' interest in recent years due to their excellent optimization performance for solving many problems with complexity, high dimension and so on. PSO, one of the excellent evolutionary algorithms, originally put forward by Kennedy [1], is a global optimization algorithm that mimics bird foraging behavior. It is because of its some merits such as few parameters and easy to implement that PSO has been paid increasing attention in optimization fields, which is particularly suitable for continuous optimization problems and obtains more excellent optimization performance [2–7]. However, what is regrettable is that

© Springer Nature Singapore Pte Ltd. 2018
K. Li et al. (Eds.): ISICA 2017, CCIS 874, pp. 26–39, 2018.
https://doi.org/10.1007/978-981-13-1651-7_3

PSO has some disadvantages over optimization such as slower convergence speed in its latter period, easy to falling into the local optimal solution. In other word, it is weaker for its local search. In order to enhance its local search further, EO is introduced to improve its performance further.

EO, a more excellent heuristic local search algorithm, is proposed by Boettcher and Percus [8], which is inspired by recent progress in understanding far-from-equilibrium phenomena in accordance with the Bak-Sneppen (BS) model [9] which says that the emergence of self-organized criticality (SOC) [10] in ecosystems where almost all individuals have reached a certain threshold (called a fitness) and for those weakest individuals and their nearest neighbors are forced to be mutated. EO, luckily, makes full use of these weakest individuals to improve performance of algorithms and has been successfully used to solve complex problems such as image processing [11], production scheduling [12], multi-objective optimization problems [13], traveling salesman [14], and Dynamic combinatorial optimization problem [15] and so on.

It is because EO has very strong local search that other algorithms with weaker local search but stronger global search make full use of this advantage to hybridize with EO, which balances global search and local search of these algorithms to enhance their performance further. To the best of our knowledge, there has been some related intelligence algorithms hybridized with EO so far, including artificial bee colony (ABC), shuffled frog-leaping algorithm (FLA), glowworm swarm optimization (GSO) and PSO and so forth. For instance, for hybridization of ABC and EO, Azadehgan et al. [16] puts forward a new hybrid algorithm for optimization based on ABC and EO (EABC). Chen et al. [17] proposes a novel ABC algorithm by integrating of EO to solve numerical optimization problems (ABC-EO). For the hybridization of FLA and EO, Li et al. [18] proposes an improved shuffled frog-leaping algorithm by integrating with EO to solve continuous optimization (MSFLA-EO). For the hybridization of GSO and EO, Ghandehari et al. [19] proposes hybrid EO and glow-worm swarm optimization (HEGSO). For the hybridization of PSO and EO, Chen et al. [20] proposes a novel particle swarm optimizer hybridized with EO (PSO-EO). For PSO, in order to enhance its optimization performance, there are some works mentioned above for hybridizing EO to improve the local search, but PSO hybridized with EO has still weaker global search relatively. In order to overcome its disadvantage to enhance performance of EO, CEO improved EO is chosen in PSO. Meanwhile, for the sake of obtaining better optimization performance of PSO, the improved opposition-based learning (UOBL) [21], a better global search optimization strategy, is chosen to hybridize with CEO in PSO. The purpose of introducing CEO and UOBL in PSO is that CEO enhances its local search and UOBL improves its global search, which makes both balance in some condition. Hence an improved PSO introduced EO and UOBL, called AHOPSO-CEO, is proposed by this paper.

The rest of this paper is constructed as follows. In Sect. 2, related works, including PSO, PSO based on opposition-based learning, extremal optimization, are described. Section 3 designs the process of the proposed algorithm and the experimental results are listed in Sect. 4. The conclusion is described in Sect. 5.

2 Related Works

2.1 Overview of PSO

PSO [1] is a global optimizing algorithm which stems from the inspiration of bird foraging behavior in nature. Each particle of PSO represents a candidate solution and adjusts its searching strategies according to its own empirical value which is individual extremum called *pbest* and other particle empirical value which is global optimum called *gbest*. In a space with D dimension, n particles are generated randomly and the position of the ith particle is represented by $X_i = (x_{i1}, x_{i2}, \cdots, x_{iD})$, $i = 1, 2, \cdots, n$ and the velocity of the ith particle in each iteration is expressed by $V_i = (v_{i1}, v_{i2}, \cdots, v_{iD})$, $i = 1, 2, \cdots, n$. Each particle updates its position and velocity by tracking the *pbest* and *gbest*. Their models are formulated respectively as follows.

$$v_{id}(t+1) = \omega v_{id}(t) + c_1 r_1 (pbest_{id} - x_{id}(t)) + c_2 r_2 (gbest_d - x_{id}(t)) \tag{1}$$

$$x_{id}(t+1) = x_{id}(t) + v_{id}(t+1) \tag{2}$$

Where $v_{id}(t+1)$ is the velocity of the $(t+1)$ generation of the ith particle, $x_{id}(t+1)$ is the position of the $(t+1)$ generation of the ith particle, ω is the inertia weight; r_1 and r_2 are random numbers distributed uniformly with [0, 1].

2.2 PSO Based on Opposition-Based Learning

Opposition-Based Learning. Opposition-based learning (OBL) proposed by Tizhoosh [21] is a new scheme for machine intelligence, which has been applied to various optimization algorithms and proven to be effective. The basic idea of OBL is that by computing opposite solutions of its candidate solutions will enhance the probability of searching a candidate solution closer to the global optimum for a given problem. The two definitions of OBL are described as follows.

Definition 1 [21, 27]: *Let x be a real solution defined on a certain interval [a, b]. The opposite solution x_1 of x is defined as follows.*

$$x_1 = a + b - x \tag{3}$$

Analogously, the opposite solution is extended to the n dimension and it is defined as follows.

Definition 2 [21, 27]: *Let $P(x_1, x_2, \ldots, x_n)$ be a point in n dimension, and x_i be a real solution defined in interval $[a_i, b_i]$. The opposite solution x^* of x_i is defined as follows.*

$$x^* = a_i + b_i - x_i, i = 1, 2, 3, \cdots, n \tag{4}$$

The basic idea of OBL [21, 27] is described in detail: let $f(x)$ be a function of a certain problem and $g(\bullet)$ be a proper evaluation function. Hypothesis that x is a candidate solution generated randomly and x_1 is its opposite solution, and their function values $f(x)$ and $f(x_1)$ are calculated in each iteration cycle respectively. When $g(f(x)) \geq g(f(x_1))$, OBL continues performing with x, otherwise with x_1. Its sketch chart is shown as follows (Fig. 1).

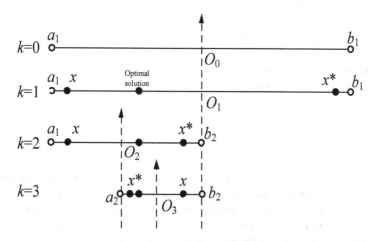

Fig. 1. The sketch chart of OBL

Improved PSO Based on OBL. Due to some merits of PSO such as few parameters, easy to implement and so forth, it has very wide application on many fields. However, PSO has relative weaker global search, especially very weaker local search, which results in slower convergence speed and lower convergence accuracy. In order to enhance its global search, OBL has been chosen to improve the performance of PSO by some related scholars. For example, Wang et al. introduces OBL in [22] and improves OBL to propose generalized OBL and applies it to PSO in [23]. Shao et al. proposes an improved OBL based on refraction principle which is applied to improve PSO in [24].

2.3   Extremal Optimization

EO, a robust and excellent heuristic local search optimization algorithm, is proposed by Boettchert [8] according to self-organized criticality (SOC) of complex systems and Bak-Sneppen biological evolution model (BS model) which is used to search high quality solutions for complex problems. It is because EO has strong local search that it is successfully applied to many engineering optimization problems and gets excellent optimization performance.

The idea of EO is that a fitness value with uniform distribution in the range [0, 1] is assigned to each species (called individuals in intelligent algorithms) randomly. The worst species with the worst fitness is forced to be mutated and a new random value is assigned to it. By this way makes individuals interact with each other and evolve together so that the algorithm approaches the optimal solution or find the optimal solution. The implementation process of EO is described as follows.

Algorithm 1: The EO algorithm[8]
Step 1: Initialize current solution S;
Set optimal solution $S_{best}=S$;
Step 2: For the current solution:
1) Computing fitness λ_i of x_i according to fitness function $f(X)$, $X=x_1,x_2,...,x_N$;
2) For all i, find j satisfying $\lambda_i \geq \lambda_j$;
3) Choose S' in a neighborhood $N(S)$ of S, and x_j must be changed;
4) Accept $S'=S$ unconditionally;
5) If $f(S)<f(S_{best})$, set $S_{best}=S$;
Step 3: Repeat Step 2 as long as desired;
Step 4: Return S_{best} and $f(S_{best})$.

From the above Algorithm 1, it can be seen that EO operates on a solution and has only a mutation operation for each iteration, which makes EO have few parameters and implement easily. It shows better performance over problems such as graph partitioning, but what's more, when applied to other types of problems, no any parameter is used to control the choose of optimal solution, which results in a deterministic search. That is to say, it is of easiness to trap into local optimal solution and finds global optimization solution more difficultly.

3   The Adaptive Particle Swarm Optimization Using Hybrid Strategy

3.1   The Proposed Algorithm Design

For an adaptive hybrid PSO algorithm designed, EO and UOBL are chosen to enhance its optimization performance. For EO, it is performed at each generation, which increases its time complexity and deteriorates the rate of convergence to some extent. In PSO-EO, the parameter INV is introduced to control the probability of EO performed by control artificially. Some researchers suggest the value of INV is between 50 and 100, which reduces its time complexity. There are some disadvantages for the INV by set artificially for different functions, which is of difficulty to find the appropriate INV value to make the time complexity reduce the lowest. In AHOPSO-CEO, therefore, the parameter rnd is introduced to control the performing probability of different strategies.

As mentioned above, the mutation operation of EO is performed against the worst individual who is chosen by given each dimension of the solution a random fitness and then ordered by fitness found. There is a shortage of ways to choose the worst individual. It is because the worst individual is chosen to be mutated each time that the way results in the other individual to have any chance of being chosen to be mutated, which reduces the probability to find optimal solution.

3.2 Mutation

For EO, the only operation is the mutation, of which the mutation process is random mutation strategy. For enhancing the performance of EO further to improve optimization performance of PSO, the chaotic mutation strategy, called logistic mutation [19], is introduced to improve its performance, called CEO, which stems from non-linear dynamical system and which is formulated as follows.

$$L = \mu x_t (1 - x_t) \tag{5}$$

Where the parameter μ stands for a constant, of which excellent value is 4. The parameter x is a random value between 0 and 1. The t is the number of iterations. The algorithm process of CEO is described as follows.

Algorithm 2: The CEO algorithm
Step 1: Set current particle i=1;
Step 2: For the ith particle with D dimension, its position vector X_i=(x_{i1}, x_{i2},$\cdots$, x_{iD}) is operated as follows.
(1) Each dimension of the X_i is mutated according to L in proper order and the other dimensions remain unchanged. Consequently, the new D positions X_{ik}(k=1,2,$\cdots$,D) are generated;
(2) Compute $F(X_{ik})$, which are the fitness values of these new positions, and compute λ_{ik} (λ_{ik}= $F(X_{ik})$ - $F(gbest)$) to find the individual corresponding to the worst dimension w according to λ_{ik} by order;
(3) If $F(X_{iw})$<$F(X_i)$, X_i= X_{iw} and $F(X_i)$=$F(X_{iw})$, and go to (4), or go to Step 3;
(4) Update P_{best} and $gbest$.
Step 3: If the value of i reaches the number of swarms, the results are printed, or i=i+1 and go to Step 2.

3.3 Integration with EO and UOBL

Because of some disadvantages of OBL, UOBL [24] shows excellent performance to improve PSO and it is introduced to enhance optimization performance of AHOPSO-CEO. For EO, improved EO algorithm based on Logistic mutation is used for enhancing the performance of AHOPSO-CEO. How to hybridize with the both is a problem to solve next.

The way to solve the integration with the both above is introduced a probability parameter rnd. The probability of performing the CEO is p and op as performing the improved OBL based on refraction principle. Order

$$p + op = 1 \tag{6}$$

As can be seen from the above formula, when the $rnd < op = 1 - p$, the improved OBL based refraction principle is performed and when the $1 - p \leq rnd \leq p$, PSO is performed. Hence AHOPSO-CEO algorithm is described as follows.

Algorithm 3: The AHOPSO-CEO algorithm
Step 1 Initiate related parameters : current iteration number e, the maximum number $MAXe$ and so on;
Step 2 while ($e{\leq}MAXe$)
Step 3 $L=\mu*x*(1-x)$;
Step 4 $rnd=rand()$;
Step 5 if ($rnd{<}op$)
Step 6   Perform the improved OBL based on refraction principle;
Step 7 elseif ($rnd{>}p$)
Step 8   Pserform the CEO algorithm;
Step 9   else
Step 10   Perform the PSO algorithm;
Step 11   end if
Step 12 Do not meet the maximum iteration numbers，$e=e+1$，go to Step 2;
Step 13 end while(when meets maximum iteration numbers)
Step 14 Print the optimization solution and the algorithm is over.

4   Numeric Experimental Results and Discussion

4.1   Parameters Setting and Benchmark Functions

For verifying optimization performance of the AHOPSO-CEO algorithm, next the 12 benchmark functions with 30 dimension (Sphere, Quadric, Rosenbrock, Rastrigin, Griewank, Ackley, Schaffer, Schwefel, Weighted Sphere, Quartic, Nocontinous Rastrigin, Weierstrass) are introduced by the experiment and they are marked as F_1–F_{12} by order respectively which are no peak or multi-peak functions. The optimal solution of all functions is 0 except for F_8. To compare the optimization performance of the AHOPSO-CEO algorithm, four algorithms (PSO [1], PSO-EO [20], GOPSO [23], refrPSO [24]) are used as compared algorithms. The PSO-EO is EO-based, but GOPSO and refrPSO are OBL-based. The optimal parameter setting of five algorithms is listed as follows.

Note that '-' represents the parameter is non-existent and the parameters referent corresponding documents are not listed by Table 1.

Table 1. Parameter setting of five algorithms

Parameters	PSO	PSO-EO	GOPSO	refrPSO	AHOPSO-CEO
Population size	10	10	10	10	10
Maximal iterations	100000	100000	100000	100000	100000
$c_1 = c_2$	1.496,	1.496	1.496	1.496	1.496
ω	0.79	0.79	0.79	0.79	0.79
V_i^{max}, v_i^{min}	$-2,2$	$-2,2$	$-2,2$	$-2,2$	$-2,2$
k	-	-	-	0.75	0.75
Ft	-	-	-	0.001	0.001
α	-	-	-	$\pi/8$	$\pi/8$
p	-	-	-	-	0.7
μ	-	-	-	-	4
INV	-	100	-	-	-

4.2 Experimental Results and Analysis

Performance Analysis. For verifying the optimization performance of AHOPSO-EO algorithm, the experiment results of five algorithms are listed by Table 2 with 12 functions and the performed times are 30.

From the Table 2, it can be seen that the AHOPSO-CEO has the same convergent accuracy compared with the other four algorithms for the functions such as F_1, F_2, F_4, F_5, F_6, F_7, F_{12}. For F_3, F_8, F_9, F_{11}, the convergent accuracy of AHOPSO-CEO is higher than the other four algorithms evidently, especially for F_3, which is a morbid non-convex function and the minimum is very difficult to be found, but the AHOPSO-CEO shows the very excellent optimization performance, which illustrates the stronger local search of AHOPSO-CEO. For F_{10}, compared with refrPSO, the convergent accuracy of AHOPSO-CEO is slightly lower but higher than the other three algorithms.

To sum up, in the convergent accuracy, the designed AHOPSO-CEO overwhelms the other four compared algorithms.

Average Iteration Times and Success Rate. Average iteration times (AIT) and success rates (SR) are introduced to verify the optimization performance of the proposed AHOPSO-CEO. The less AIT is and the higher the SR is, the better the optimization performance of the AHOPSO-CEO is. The experiment results of AIT and SR are showed in the Table 3. The referent accuracy of F_1–F_{12} is as follows: 0, 10^{-200}, 10^{-2}, 10^{-25}, 0, 10^{-25}, 10^{-25}, -12569.4, 10^{-200}, 10^{-25}, 10^{-25}, 10^{-200}.

Note that '-' represents that AIT is 100000 in the Table 3, which means that they do not meet the referent accuracy.

As is shown in the Table 3, for AIT, AHOPSO-CEO is higher than refrPSO when they test the functions of F_2, F_4, F_5, F_8, F_9, but lower than the other three algorithms. For these functions, due to EO algorithm is introduced, and a part of iteration times are consumed, which result in the increment of AIT. For SR, we can see from the Table 3 that five PSO algorithms don't meet referent accuracy for F_6, F_7, F_{10}, F_{11}. The proposed AHOPSO-CEO and refrPSO meets referent accuracy for F_1, F_2, F_4, F_5, F_9, F_{12}

Table 2. The results of five PSO algorithms

Functions	Algorithms	Optimal solutions	Ave Err	Std Dev	The worst values
F_1	PSO	5.85e−015	7.20e−006	3.38e−005	1.85e−004
	PSO-EO	3.18e−016	3.89e−002	1.95e−001	1.06e+000
	GOPSO	8.79e−162	1.75e−132	9.59e−132	5.25e−131
	refrPSO	0	0	0	0
	AHOPSO-CEO	0	0	0	0
F_2	PSO	4.63e−002	7.34e−002	4.01e−001	2.20e+000
	PSO-EO	1.17e−010	4.88e−003	2.66e−002	1.46e−001
	GOPSO	2.97e−024	4.12e−003	1.71e−002	9.20e−002
	refrPSO	0	0	0	0
	AHOPSO-CEO	0	0	0	0
F_3	PSO	7.02e+000	2.61e+001	1.22e+001	7.78e+001
	PSO-EO	6.66e−002	6.92e+001	5.44e+001	2.47e+002
	GOPSO	1.97e+001	2.07e+001	4.22e−001	2.18e+001
	refrPSO	0	8.46e−001	4.63e+000	2.68+001
	AHOPSO-CEO	0	0	0	0
F_4	PSO	3.08e+001	5.82e+001	1.25e+001	7.95e+001
	PSO-EO	3.88e+001	6.84e+001	1.41e+001	9.05e+001
	GOPSO	0	1.38e+001	5.14e+001	5.26e+002
	refrPSO	0	0	0	0
	AHOPSO-CEO	0	0	0	0
F_5	PSO	2.16e−011	5.98e−001	8.84e−001	4.02e+000
	PSO-EO	1.19e−007	6.9e−001	1.68e+000	8.33e+001
	GOPSO	0	0	0	0
	refrPSO	0	0	0	0
	AHOPSO-CEO	0	0	0	0
F_6	PSO	3.46e+000	7.68e+000	2.35e+000	1.21e+001
	PSO-EO	5.05e+000	7.71e+000	2.14e+000	1.35e+001
	GOPSO	9.58e−013	9.58e−013	1.45e−015	9.58e−013
	refrPSO	**4.44e−016**	**4.44e−016**	0	**4.44e−016**
	AHOPSO-CEO	**4.44e−016**	**4.44e−016**	0	**4.44e−016**
F_7	PSO	1.02e−001	1.49e−001	2.27e−002	2.03e−001
	PSO-EO	1.01e−001	1.49e−001	3.71e−002	2.50e−001
	GOPSO	2.45e−003	2.45e−003	2.78e−011	2.45e−003
	refrPSO	**2.45e−003**	**2.45e−003**	**3.23e−017**	**2.45e−003**
	AHOPSO-CEO	**2.45e−003**	**2.45e−003**	**1.12e−012**	**2.45e−003**
F_8	PSO	−7614.57	−6819.17	548.65	−5323.52
	PSO-EO	−9243.31	−8235.13	475.625	−7321.61
	GOPSO	−11621.5	−11018.3	385.244	−10227.9
	refrPSO	−12569.5	−9667.72	2702.34	−6055.3
	AHOPSO-CEO	**−12569.5**	**−11090.8**	**816.25**	**−9713**

(*continued*)

Table 2. (*continued*)

Functions	Algorithms	Optimal solutions	Ave Err	Std Dev	The worst values
F_9	PSO	1.21e−014	5.28e−005	2.31e−004	1.25e−003
	PSO-EO	6.65e−017	4.47e−002	2.61e−001	1.43e+000
	GOPSO	2.91e−160	1.14e−040	5.80e−040	3.17e−039
	refrPSO	**0**	3.67e−284	**0**	1.01e−282
	AHOPSO-CEO	**0**	**0**	**0**	**0**
F_{10}	PSO	1.64e−002	8.71e−002	9.18e−002	4.37e−001
	PSO-EO	7.78e−003	8.39e−002	5.53e−002	2.38e−001
	GOPSO	3.77e−003	2.26e−002	1.13e−002	5.36e−002
	refrPSO	**1.11e−015**	**5.58e−006**	**1.21e−005**	**6.33e−005**
	AHOPSO-CEO	5.44e−006	5.10e−004	3.11e−004	1.28e−003
F_{11}	PSO	6.60e+000	1.16e+001	3.49e+000	1.77e+001
	PSO-EO	6.37e+000	1.15e+001	2.55e+000	1.61e+001
	GOPSO	7.10e−014	3.98e−009	7.38e−009	2.92e−008
	refrPSO	2.13e−014	2.10e−009	2.44e−009	8.21e−009
	AHOPSO-CEO	**7.10e−015**	**4.26e−013**	**4.35e−013**	**2.12e−012**
F_{12}	PSO	3.6e+001	1.18e+002	5.8e+001	2.96e+002
	PSO-EO	4.00e+001	1.40e+002	6.20e+001	3.32e+002
	GOPSO	0	1.34e+002	7.84e+001	2.15e+002
	refrPSO	**0**	**0**	**0**	**0**
	AHOPSO-CEO	**0**	**0**	**0**	**0**

Table 3. The results of AIT and SR of algorithms

Algorithms	Metrics	F_1	F_2	F_3	F_4	F_5	F_6	F_7	F_8	F_9	F_{10}	F_{11}	F_{12}
PSO	AIT	-	-	-	-	-	-	-	-	-	-	-	-
	SR	0	0	0	0	0	0	0	0	0	0	0	0
PSO-EO	AIT	-	-	-	-	-	-	-	-	-	-	-	-
	SR	0	0	0	0	0	0	0	0	0	0	0	0
GOPSO	AIT	-	-	-	54462	17307	-	-	-	-	-	-	93231
	SR	0	0	0	0.86	1	0	0	0	0	0	0	0.16
refrPSO	AIT	3420	9699	16862	3739	393	-	-	66994	3157	-	-	3742
	SR	1	1	0.87	1	1	0	0	0.43	1	0	0	1
AHOPSO -CEO	AIT	1175	17957	5037	19969	3199	-	-	93338	46502	-	-	3235
	SR	1	1	1	1	1	0	0	0.46	1	0	0	1

completely. But for F_3, F_8, only AHOPSO-CEO meets the referent accuracy completely and for a part of functions refrPSO meets the referent accuracy, and the other algorithms do not.

The Analysis of Convergence. From the Fig. 2, it can be seen that the convergent chart of AHOPSO-CEO and other improved PSO.

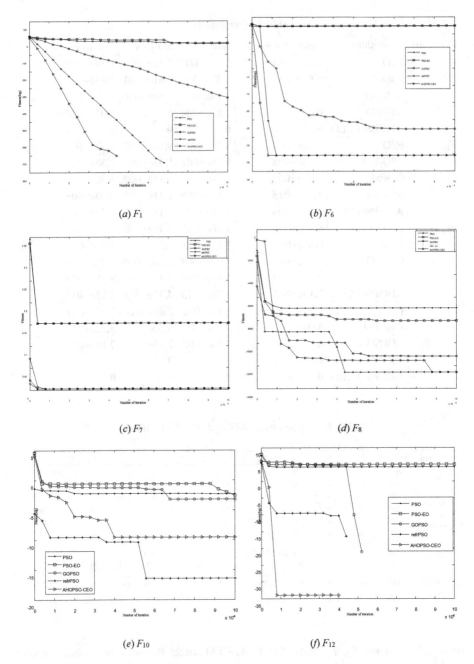

(a) F_1

(b) F_6

(c) F_7

(d) F_8

(e) F_{10}

(f) F_{12}

Fig. 2. The convergence chart of five algorithms

From the Fig. 2, it can be seen that the convergence speed of AHOPSO-CEO overwhelms the other algorithms and gets higher convergent accuracy for F_1, F_6, F_{12}. For F_7, F_8, the convergence speed of AHOPSO-CEO approaches or slightly slows the refrPSO algorithm but excesses the other algorithms, and meanwhile it can be seen that the convergent accuracy of AHOPSO-CEO is higher than the other algorithms. For F_{10}, the convergent speed of AHOPSO-CEO is not a patch on refrPSO but optimizes the other algorithms and the convergence accuracy is in accordance with the other algorithms.

For the convergence of the refrPSO, it has proved that the improved PSO algorithm (AHOPSO-CEO) is convergent in [24], and next the convergence proof of the AHOPSO-CEO is descried as follows.

Theorem 1. *If the PSO based on the opposition-based learning of refraction principle (refrPSO) is convergent, the AHOPSO-CEO is also convergent.*

Proof. It is known from the above design process description of AHOPSO-CEO that AHOPSO-CEO consists of three parts of the implementation process of refrPSO, the execution process of EO based on chaotic mutation and the implementation process of PSO.

For the first part, the refrPSO has proved to be convergent in the literature [24]. For PSO, it has proved to be convergent in the literatures [25, 26]. For the third part of CEO, from the previous algorithm description it can be seen that its essence is that through the mutation of each dimension of each particle position, it finds out the local optimum, and then the local optimum is added to the original population in iteration. This process is actually the process of searching for local optimal solution, which has no influence over the convergence of the AHOPSO-CEO algorithm, but only affects the convergence speed of it. □

In summary, it can be concluded that the AHOPSO-CEO algorithm is also convergent.

5 Conclusions

To overcome the imbalance between the local search and global search of PSO, this paper improves its optimization performance from two aspects of global search and local search by introducing UOBL which is a strategy with excellent global search and CEO which is an optimization algorithm with excellent local search. Based on CEO and UOBL, AHOPSO-CEO is proposed. A large number of experiment results and analysis show that AHOPSO-CEO has higher convergent accuracy and faster convergent speed compared with other algorithms.

Acknowledgments. This work was supported by the National Natural Science Foundation of China (No. 70971043 and 61763019), the Science and Technology plan project of Jiangxi Province (No. GJJ160409, GJJ161076).

References

1. Kennedy, J., Eberhart, R.C.: Particle swarm optimization. In: Proceedings of the IEEE International Joint Conference on Neural Networks, pp. 1942–1948. IEEE Press (1995)
2. Wang, L., Yang, B., Chen, Y.: Improving particle swarm optimization using multi-layer searching strategy. Inf. Sci. **274**(8), 70–94 (2014)
3. Tran, D.C., Wu, Z., Wang, H.: A new approach of diversity enhanced particle swarm optimization with neighborhood search and adaptive mutation. In: Loo, C.K., Yap, K.S., Wong, K.W., Teoh, A., Huang, K. (eds.) ICONIP 2014. LNCS, vol. 8835, pp. 143–150. Springer, Cham (2014). https://doi.org/10.1007/978-3-319-12640-1_18
4. Zuo, X.Q., Xiao, L.: A DE and PSO based hybrid algorithm for dynamic optimization problems. Soft Comput. **18**(7), 1405–1424 (2014)
5. Elsayed, S.M., Sarker, R.A., Mezura-Montes, E.: Self-adaptive mix of particle swarm methodologies for constrained optimization. Inf. Sci. **277**, 216–233 (2014)
6. Cheng, R., Jin, Y.C.: A social learning particle swarm optimization algorithm for scalable optimization. Inf. Sci. **291**, 43–60 (2015)
7. Schmitt, M., Wanka, R.: Particle swarm optimization almost surely finds local optima. Theor. Comput. Sci. **561**, 57–72 (2015)
8. Boettcher, S., Percus, A.G.: Optimization with extremal dynamics. Phys. Rev. Lett. **86**, 5211–5214 (2001)
9. Bak, P., Sneppen, K.: Punctuated equilibrium and criticality in a simple model of evolution. Phys. Rev. Lett. **71**(24), 4083–4086 (1993)
10. Bak, P., Tang, C., Wiesenfeld, K.: Self-organized criticality: an explanation of the 1/f noise. Phys. Rev. Lett. **59**(59), 381–384 (1987)
11. Boettcher, S., Percus, A.G.: Extremal optimization at the phase transition of the three-coloring problem. Phys. Rev. E Stat. Nonlinear Soft Matter Phys. **69**(6Pt2), 66703 (2004)
12. Chen, Y.W., Lu, Y.Z., Yang, G.K.: Hybrid evolutionary algorithm with marriage of genetic algorithm and extremal optimization for production scheduling. Int. J. Adv. Manuf. Technol. **36**(9), 959–968 (2008)
13. Chen, Y.W., Lu, Y.Z., Chen, P.: Optimization with extremal dynamics for the traveling salesman problem. Phys. A Stat. Mech. Appl. **385**(1), 115–123 (2007)
14. Chen, M.R., Lu, Y.Z., Yang, G.K.: Multi-objective extremal optimization with applications to engineering design. J. Zhejiang Univ. - Sci. A: Appl. Phys. Eng. **8**(12), 1905–1911 (2007)
15. Paczuski, M., Maslov, S., Bak, P.: Avalanche dynamics in evolution, growth, and depinning models. Phys. Rev. E Stat. Phys. Plasmas Fluids Relat. Interdisc. Top. **53**(1), 414–443 (1996)
16. Azadehgan, V., Jafarian, N., Jafarieh, F.: A new hybrid algorithm for optimization based on artificial bee colony and extremal optimization. In: IEEE Conference Anthology, pp. 1–6. IEEE (2014)
17. Chen, M.R., Zeng, G.Q., Zeng, W., et al.: A novel artificial bee colony algorithm with integration of extremal optimization for numerical optimization problems. In: Evolutionary Computation, pp. 242–249. IEEE (2014)
18. Li, X., Luo, J., Chen, M.R., et al.: An improved shuffled frog-leaping algorithm with extremal optimisation for continuous optimisation. Inf. Sci. Int. J. **192**(6), 143–151 (2012)
19. Ghandehari, N., Miranian, E., Maddahi, M.: Hybrid extremal optimization and glowworm swarm optimization. In: Das, V. (ed.) Proceedings of the Third International Conference on Trends in Information, Telecommunication and Computing. LNEE, vol. 150, pp. 83–89. Springer, New York (2013). https://doi.org/10.1007/978-1-4614-3363-7_10

20. Chen, M.R., Li, X., Zhang, X., et al.: A novel particle swarm optimizer hybridized with extremal optimization. Appl. Soft Comput. **10**(2), 367–373 (2010)
21. Tizhoosh, H.R.: Opposition-based learning: a new scheme for machine intelligence. In: Proceedings of International Conference on Intelligent Agent, Web Technologies and Internet Commerce, pp. 695–701. IEEE Press, Vienna (2005)
22. Wang, H., Li, H., Liu, Y., et al.: Opposition-based particle swarm algorithm with cauchy mutation. In: IEEE Congress on Evolutionary Computation, pp. 4750–4756. IEEE Press, Singapore (2007)
23. Wang, H., Zhijian, W., Rahnamayan, S., et al.: Enhancing particle swarm optimization using generalized opposition-based learning. Inf. Sci. **181**(20), 4699–4714 (2011)
24. Shao, P., Wu, Z., Zhou, X., et al.: Improved particle swarm optimization algorithm based on opposition learning of refraction. Acta Electronica Sin. **43**(11), 2137–2144 (2015)
25. Zeng, J.C., Cui, Z.H.: A guaranteed global convergence particle swarm optimizer. J. Comput. Res. Dev. **3066**(8), 762–767 (2004)
26. Lu, R.F., Wang, X.Y.: Convergence analysis of particle swarm optimization algorithm. Sci. Technol. Eng. **4**(14), 25–32 (2008)
27. Shao, P., Wu, Z., Zhou, X., et al.: FIR digital filter design using improved particle swarm optimization based on refraction principle. Soft Comput. **21**(10), 2631–2642 (2017)

ITÖ Algorithm with Cooperative Coevolution for Large Scale Global Optimization

Yufeng Wang[1,2](✉), Wenyong Dong[1], and Xueshi Dong[1]

[1] School of Computer Science, Wuhan University, Wuhan 430072, Hubei, China
wangyufeng@whu.edu.cn
[2] School of Software, Nanyang Institute of Technology,
Nanyang 473000, Henan, China

Abstract. Problem decomposition and subcomponent optimization play a key role in cooperative coevolution (CC) for large scale global optimization. In this paper, we firstly introduce a new variable interactions identification (VII) method to recognize the indirect decision variables. Then, we proposed a new reallocate computational resources method, aims to give more computational resources to the more important subcomponents. Hence, a novel ITÖ algorithm with cooperative coevolution (CCITÖ) strategy based on above two strategies is proposed. In order to understand the characteristics of CCITÖ, we have carried out extensive computational studies on the CEC'2010 benchmark function. Experimental results show that our algorithm achieves competitive results compared with other four state-of-the-art algorithms in the large scale global optimization problems.

Keywords: Large scale global optimization · ITÖ algorithm
Cooperative coevolution · Variable interactions identification
Reallocate computational resources

1 Introduction

Large scale global optimization (LSGO) problems have attracted an enormous amount of researchers' attention in the past two decades [1,2]. These problems become difficult to be solved when a large number of variables involved in the complicated interactions. There are two key factors make LSGOs extremely difficult to handle. The first one is the well-known curse of dimensionality, that is, the search space grows exponentially with the size of dimensions increases [3]. Another one is that a large number of dimensions leads to the emergence of complex behaviors and characteristics in its landscape, such as multimodal landscape and singularity [3].

In order to address LSGO, many approaches have been proposed to improve the scalability of evolutionary optimizers. Cooperative Coevolution (CC) is one

© Springer Nature Singapore Pte Ltd. 2018
K. Li et al. (Eds.): ISICA 2017, CCIS 874, pp. 40–51, 2018.
https://doi.org/10.1007/978-981-13-1651-7_4

of the most effective strategies among these approaches, and it attracts more and more researchers to study this strategy. CC is a divide-and-conquer strategy that divides the high-dimensional problem into several subcomponents with low dimensions, and then cooperatively solves them by using different meta-heuristic algorithms, such as: cooperative particle swarm optimizer (CPSO) [4], cooperative coevolving PSO (CCPSO2) [5], covariance matrix adaptation evolution strategy with CC (CC-CMA-ES) [6], CC orthogonal artificial bee colony (CCOABC) [7], and fast evolutionary programming with CC (FEPCC) [8], Differential Evolution (DECC-G) [9], etc.

Problem decomposition plays a significant role in CC strategy. These have two decompose strategies: static grouping and dynamic grouping [10]. Static grouping divides the high dimension problems into some fixed value of the subcomponent size, but this method loses its effectiveness for handing partially-separable and fully-nonseparable problems. Dynamic grouping can dynamically change the grouping structure. Such as: random grouping [9], multilevel CC (MLCC) [11] and differential grouping (DG) [12]. DG improved the performance of CC on nonseparable problems with direct interactive variables, but it ignored indirect interactive variables.

In this paper, in order to overcome the above mentioned shortcomings of CC, a new variable interactions identification method (VII) is introduced to decompose the decision variables. It can deal with indirect interactive variables effectively. And then, a reallocate computational resources (RCR) method is proposed to reallocate computational resources among all subpopulations and give more computational resources to the most important subcompacts. Furthermore, a novel ITÖ Algorithm with Cooperation Coevolution strategy (CCITÖ) Algorithm based on above two strategies is designed to solve LSGO problems. The CCITÖ algorithm consists of four factors: particle radius, environmental temperature, drift operator and fluctuate operator. These cooperatively deal with contradictions between the 'exploration' and 'exploitation', experimental results show that the proposed CCITÖ can effectively work, especially its performance is super than or equal to some state-of-art algorithms.

The remainder of this paper is organized as follows. Section 2 presents an overview of ITÖ algorithm. Section 3 introduces our new approach. Section 4 presents the experimental studies. Finally, Sect. 5 is the concluding remarks.

2   ITÖ Algorithm

ITÖ algorithm [13] is first proposed by Dong and Hu in 2007. It mimics the Brown motion of flower power in the surface of the liquid, and can be classified into a novel variant of swarm intelligent optimization algorithm. The Brown motions are first proposed by the English biologist R.Brown in 1827. Einstein gave the mathematical description of this phenomenon in 1905 [13]. Wiener described an accurate notional statement of the Brown motions and built a mathematical model in 1918. In 1940's, a Japanese mathematician Itö Kiyosi extended Weiner's research results and proposed the stochastic differential equation with the Brown motion disturb item $B(t)$ as follows [13]:

Definition 1. *Suppose* $X = \{X(t), t \geq 0\}$ *satisfying Itö integral as: for* $\forall\, 0 \leq t_0 < t < T$,

$$X(t) - X(t_0) = \int_{t_0}^{t} b(s, X(s))ds + \int_{t_0}^{t} \delta(s, X(s))dB(s) \qquad (1)$$

then X *is called Itö (stochastic) process.*

where $\int_{t_0}^{t} b(s, X(s))ds$ is called as drifting rate and denotes the general tendency of stochastic process. The item $\int_{t_0}^{t} \delta(s, X(s))dB(s)$ denotes the tracks fluctuation of variable X and it is called fluctuation rate.

ITÖ algorithm is based on Itö stochastic process. It uses environmental temperature to control the motion ability of the population, and uses particles radius to simulate the characteristics of the particle in Brown motion [14]. It has two operators: drift operator and fluctuate operator. The drift operator corresponding the first item in the Eq. (1), and its controls the particles move toward the global optimum solution. The fluctuate operator corresponding the second item in the Eq. (1), and its controls the random fluctuation of particles in the whole solution space, and ensures the diversity of the population. ITÖ has four parts: particles radius, annealing temperature, drift operator and fluctuate operator.

3   Proposed Approach

3.1   Variable Interactions Identification (VII)

In the cooperative coevolution (CC) framework, group strategy plays a significant role. The accuracy of the grouping method has a direct influence on the performance of the CC algorithm. The proposed variable interactions identification (VII) aims to find the interactions of the decision variables, and try to divide all the variables into different subgroups to make sure that the variables directly or indirectly interactive with each other in the same subgroup and are independent with each other in different subgroups. In VII, we first identify the interactions between separable and nonseparable decision variables, and decompose them into different subgroups. Then, we perform a traversal search for all subgroups and merge the subgroup that contains the same direct or indirect interactive variables into one subgroup. In order to avoid the subgroup is too large, we only merge the subgroup that its size less than $N/10$ for the LSGOs (N is the dimension of the benchmark function). The pseudocode of VII is given in Algorithm 1.

Similar to [15], the interaction between decision variables has three type: (1) two variables has no interaction (called independent), (2) two variables interact directly with each other (called direct interaction) and (3) variables are linked via a third decision variable (called indirect interaction).

For example, we have a function: $f(\mathbf{X}) = x_1 + (x_2 - x_3)^2 + (x_3 - x_4)^2$, $\mathbf{X} \in [-1, 1]^4$. In this function, $\{[x_1, x_2], [x_1, x_3], [x_1, x_4]\}$ are independent variables respectively, $\{[x_2, x_3], [x_3, x_4]\}$ are direct interactive variables respectively, and

---

Algorithm 1. VII

Input: $func, N, lbound, ubound, \varepsilon$
Output: $groups$
1: **for** $i = 1$ to N **do**
2:    $groups \leftarrow \{j\}$;
3:    $X_1 \leftarrow lbound \times ones(1, N)$;
4:    $X_2 \leftarrow X_1$; $X_2(i) \leftarrow ubound$;
5:    $\Delta_1 \leftarrow func(X_1) - func(X_2)$;
6:    **for** $j = i + 1$ to N **do**
7:       $X_1(j) \leftarrow (lbound + ubound)/2$;
8:       $X_2(j) \leftarrow (lbound + ubound)/2$;
9:       $\Delta_2 \leftarrow func(X_1) - func(X_2)$;
10:       **if** $(|\Delta_1 - \Delta_2|) > \varepsilon$ **then**
11:          $groups(i) \leftarrow groups(i) \cup \{j\}$;
12:       **end if**
13:    **end for**
14: **end for**
15: **if** $SIZE(groups) \neq N$ **then**
16:    a=1,b=a+1;
17:    **for** $a < b$ & $a, b \in \{1 : SIZE(groups)\}$ **do**
18:       **if** $groups(a) \cap groups(b) \neq \emptyset$ & $SIZE(groups(a)) < N/10$ & $SIZE(groups(b)) < N/10$ **then**
19:          $groups(a) \leftarrow groups(a) \cup groups(b)$;
20:          delete $groups(b)$;
21:       **end if**
22:    **end for**
23: **end if**
24: **return** $groups$;

$[x_2, x_4]$ are indirect interaction variables. In VII, we first identify the direct interaction between decision variables with differential grouping (DG) [12]. The decision variables of this function $f(\mathbf{X})$ will decompose to three groups: $\{x_1\}, \{x_2, x_3\}, \{x_3, x_4\}$. However, x_2 and x_4 are indirect interaction variables. So, we need to merge this two groups into one group. Finally, The decision variables of this function $f(\mathbf{X})$ will decompose to two groups: $\{x_1\}, \{x_2, x_3, x_4\}$.

In the DG method [12], interactions between decision variables are identified according to the following rule:

Theorem 1. *Let $f(\mathbf{X})$ be an additively separable function. $\forall a, b_1 \neq b_2, \delta \in \mathbb{R}, \delta \neq 0$, if the following condition holds:*

$$\Delta_{x_i} f(\mathbf{X})|_{x_i=a, x_j=b_1} \neq \Delta_{x_i} f(\mathbf{X})|_{x_i=a, x_j=b_2} \tag{2}$$

then x_i and x_j are nonseparable where

$$\Delta_{x_i} f(\mathbf{X}) = f(\cdots, x_i + \delta, \cdots) - f(\cdots, x_i, \cdots) \tag{3}$$

refers to the forward difference of $f(\mathbf{X})$ with respect to x_i with interval δ.

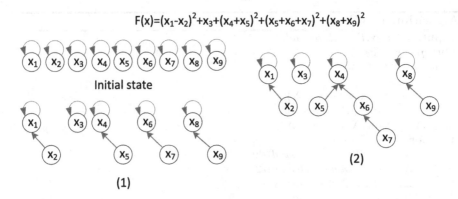

Fig. 1. An examples of disjoint-set

After DG method, the direct interactive variables can be identified. And then, we can find and merge the indirect interactive variables by disjoint-set. The disjoint-set provides a near-constant-time method to keeps track of a set of elements partitioned into a number of disjoint subgroups by the union and find operations. The detail of disjoint-set see Fig. 1.

In the Algorithm 1, the parameter $func$ is the objective function, N is the dimension of the problem, $lbound$ and $ubound$ is the lower bounds and upper bounds of decision variables respectively, ε is the threshold to identify the direct interaction between decision variables. Lines 1–14 identifies the interaction between variables, and lines 15–22 merge the direct and indirect variables into one group.

3.2 ITÖ Algorithm

Algorithm 2 shows a procedure of ITÖ. It has six parts: initialization, selection, update parameters and generate a solution. Line 1 is the initialization step, we set the number of particles and other parameters. Lines 2–17 are the core iteration process of ITÖ. Lines 3–5 sorts the population and save the best solution at the current generation. Lines 6–9 updates the global optimal solution and record the number of iterations. Line 10 update the radius of the particle, the annealing temperature, the motion intensity of the particle and the active strength of the particle respectively. Lines 11–16 updates the particles by the drift and fluctuate, and construct a solution.

Particles Radius. In ITÖ, the best solution has small particles radius, it can maintain the current good solution. The worst solution has big particles radius, in order to find a new better solution around.

$$r(x) = g(f(x)) \tag{4}$$

Algorithm 2. Procedure of ITÖ

Input: $best\_pop, subpop, FE$
Output: $subpop$
1: Initial δ ; T ; ρ ; p;
2: **while** $fitness\_times < FE$ **do**
3:    $particles \leftarrow subpop$;
4:    sort all particles by fitness;
5:    $cbestsolution \leftarrow particles[0]$;
6:    **if** $allbestsolution < cbestsolution$ **then**
7:       $allbestsolution \leftarrow cbestsolution$;
8:    **end if**
9:    $fitness\_times + +$;
10:   update annealing temperature T, all particles radius r, particles motion intensity τ, active strength δ;
11:   **for** $k = 1$ to $popsize$ **do**
12:      $newSolution \leftarrow$ drift and fluctuate by Eq. (7), (8) and (9);
13:      **if** $newSolution < particles[k]$ **then**
14:         $newSolution \leftarrow particles[k]$;
15:      **end if**
16:   **end for**
17: **end while**
18: **return** $newSolution$;

---

where x is a particle, $r(x)$ is the radius of the particle x, $f(x)$ is the fitness of the particle x, $g(\cdot)$ is a inverse scaling function. We can calculate the radius size by the n particles in the population that it is sort of the fitness value, we have:

$$r_i = r_{max} - \frac{r_{max} - r_{min}}{n - 1} * (n_i - 1) \tag{5}$$

where r_{max} and r_{min} are the maximum and minimum values of the radius, respectively. n_i is the i ranked particle in the population.

Annealing Temperature. The environmental temperature controls the mutation ability of the population. In ITÖ, we use the cooling schedule control the temperature.

$$T_i = \rho * T_{i-1} , \; 0 < \rho \leq 1 \tag{6}$$

where T is the cooling schedule, the size L of T is half the number of particles.

Drift Operator. In the diffusion process of information, the hot of information will change over time. The influence of node will be updated by the negative feedback mechanism, we have:

$$\tau^{t+1}(i) = (1 - \rho) * \tau^t(i) \tag{7}$$

where $\tau^{t+1}(i)$ is the motion intensity of particle i at $t+1$ generation. ρ is an attenuation coefficient of the motion intensity, $\rho \in (0, 1)$.

All particles are sorted in descending order according to the fitness value at this at $t+1$ generation. We chose the best solution as the attractive element of all particles, and added into the set of the attractive element σ. Then, the other elements will be attracted by the attractive element and move toward this element.

$$\tau^{t+1}(i) \leftarrow \tau^{t+1}(i) + \delta , \ if \ \tau^t(i) \notin \sigma \tag{8}$$

where δ is drift intensity of the particle, represent the incremental of motion intensity between current solution and the best solution.

Fluctuate Operator. The fluctuation process indicates random perturbations of particles in the search environment and moves to a random position, in order to maintain the diversity of the population, and can effectively jump out of the local optimum, and guarantee the accuracy of the algorithm.

$$\tau^{t+1}(u) \leftarrow \tau^{t+1}(u) + \gamma , rand() < p \tag{9}$$

where γ is the fluctuation intensity, $rand()$ is the random function, $rand() \in [0, 1]$, p is the probability of fluctuate, controls the whole fluctuate capability of ITÖ.

$$\gamma = f(r, T) \tag{10}$$

where r is the particle radius, T is the environmental temperature.

$$f(r, T) = r_{min} + f_1(r) * f_2(T) * (r_{max} - r_{min}) \tag{11}$$

where r_{max}, r_{min} are the maximum and minimum values of the fluctuation intensity, respectively. $f_1(r)$ is the influence of particle radius on fluctuate intensity, $f_2(T)$ is the influence of environmental temperature on fluctuate intensity.

$$f_1(r) = (e^{-\lambda r} - e^{-\lambda r_{max}})/(e^{-\lambda r_{min}} - e^{-\lambda r_{max}}) \tag{12}$$

$$f_2(T) = \exp(-1/T) \tag{13}$$

3.3   Reallocate Computational Resources (RCR)

In the CC framework, every subcomponent has different contributions to the improvement of the global optimum solution. So, we should allocate more computational resources to the most important subcomponents. Intuitively, the subcomponent that has more decision variables is more important, and its need more computational resources to search the best solutions. For example, a function has two subcomponents: S_1, S_2. S_1 have 5 variables and S_2 have 45 variables. In the process of evolution, S_2 should be allocated more computational resources. In this section, we proposed a new reallocate computational resources method on CC framework (CCRCR).

In the Algorithm 3, Line 1 is the initialization step. Lines 2–19 are the core iteration process of CCITÖ. Lines 2 is the grouping step, it decomposes the large scale problem to some subcomponents by VII method. Lines 3–19 are the CC framework. Lines 14–17 allocate computational resources to calculate the most important groups once more in every cycle.

Algorithm 3. CCITÖ

Input: $func, N, lbound, ubound$
Output: $best\_pop, best\_val$
1: Initial $IM\_groups, \theta, \varepsilon$;
2: $groups \leftarrow \text{VII}(func, N, lbound, ubound, \varepsilon)$;
3: $pop \leftarrow rand(popsize, N)$;
4: $(best\_pop, best\_val) \leftarrow min(func(pop))$;
5: **for** $i = 1$ to $cycles$ **do**
6:    **for** $j = 1$ to $SIZE(groups)$ **do**
7:       $rcr_j = true$;
8:       $l \leftarrow sum(SIZE(groups(0 : j - 1)))$;
9:       $u \leftarrow SIZE(groups(j))$;
10:     $subpop \leftarrow pop[:, indicies(l : u)]$;
11:     $subpop \leftarrow \text{ITÖ}(best\_pop, subpop, FE)$;
12:     $pop[:, indicies(l : u)] \leftarrow subpop$;
13:     $(best\_pop, best\_val) \leftarrow min(func(pop))$;
14:     **if** $SIZE(groups(j)) > \theta$ and $rcr_j == true$ **then**
15:       $rcr_j = false$;
16:       GOTO Step 8;
17:     **end if**
18:    **end for**
19: **end for**

3.4 Complexity Analysis

In the Algorithm 1, identify the interaction between variables costs $O(N \log_2 N)$, and the merger step costs $O(SIZE(groups))$. Because $SIZE(groups) < N$, the total computational cost of Algorithm 1 is $O(N \log_2 N)$. In the Algorithm 2, the computational cost of ITÖ is $O(popsize)$. In the Algorithm 3, the computational cost of CC framework is $O(cycles * SIZE(groups))$, so the total computational cost is $O(cycles * SIZE(groups) * popsize) + O(N \log_2 N)$. Because $popsize$, $cycles$ and $SIZE(groups)$ are set to be a small fixed integer value, the total complexity of the CCITÖ is $O(N \log_2 N)$.

4 Experimental Studies

4.1 Experiment Settings

We adopt 20 benchmark functions proposed for the CEC'2010 special session on LSGO, and its detail is given in [16]. In order to test the performance of

Table 1. Comparsions of CEC'2010 LSGO benchmark function

Func	DECC-G	MLCC	DECC-DG	CCPSO2	CCITÖ
F1	2.6E-07±6.9E-08-	**1.5E-27±7.6E-28+**	8.4E+01±2.0E+01-	8.9E-09±1.9E-09-	3.4E-18±5.4E-19
F2	1.3E+03±5.2E+02-	5.5E-01±6.2E-02-	4.4E+03±1.9E+02-	1.1E+02±5.3E+01≈	**9.3E-02±2.1E-02**
F3	9.6E-01±3.8E-02-	**9.8E-13±3.7E-14+**	1.7E+01±3.3E+00-	3.2E-02±6.6E-03-	3.4E-12±6.1E-13
F4	1.2E+13±8.9E+12-	9.6E+12±3.4E+11-	4.1E+12±7.4E+11-	1.4E+12±4.6E+11-	**2.0E+11±7.2E+10**
F5	2.3E+08±7.3E+07≈	3.8E+08±6.9E+07≈	**1.5E+08±6.1E+07≈**	9.5E+08±2.9E+08≈	3.5E+08±4.5E+07
F6	5.1E+06±9.2E+05-	1.6E+07±4.9E+06-	1.6E+01±6.7E+00-	6.3E+07±1.6E+07-	**1.5E+00±5.9E-01**
F7	8.0E+08±2.8E+08-	6.8E+05±1.3E+05-	**4.1E+03±7.4E+02≈**	5.0E+07±4.3E+06-	5.2E+03±5.5E+02
F8	**1.6E+07±6.7E+06≈**	4.3E+07±3.4E+06≈	2.2E+07±8.1E+06≈	4.7E+07±6.9E+06≈	3.3E+07±6.5E+06
F9	4.3E+08±6.1E+07-	1.2E+08±1.3E+07-	4.6E+07±8.1E+06-	7.8E+07±1.0E+07-	**8.4E+06±1.7E+06**
F10	1.0E+04±8.9E+03-	4.4E+03±8.7E+02-	4.2E+03±6.4E+02-	1.1E+04±4.1E+03-	**9.8E+02±2.5E+02**
F11	2.5E+01±4.3E+00-	1.9E+02±6.9E+01-	9.8E+00±1.0E+00-	6.5E+01±1.8E+01-	**9.3E+00±1.6E+00**
F12	9.3E+04±1.8E+04-	3.4E+04±4.2E+03-	1.1E+03±4.8E+02-	9.3E+04±2.3E+04-	**6.3E+02±1.8E+02**
F13	7.5E+03±1.9E+03≈	**2.0E+03±7.2E+02≈**	2.2E+03±8.1E+02≈	9.4E+04±2.1E+04-	5.6E+03±6.8E+02
F14	9.1E+08±2.3E+08≈	3.1E+08±8.7E+07≈	3.2E+08±2.4E+07≈	5.1E+08±8.9E+07≈	**1.9E+08±7.3E+07**
F15	1.1E+04±3.5E+03-	7.1E+03±1.3E+03-	5.8E+03±6.0E+02-	1.3E+04±7.1E+03-	**2.0E+02±5.6E+01**
F16	6.9E+01±2.0E+01≈	3.7E+02±9.7E+01-	**6.2E-13±9.7E-14+**	1.7E+02±9.3E+01-	1.9E+01±7.4E+00
F17	3.0E+05±5.3E+04-	1.5E+05±6.4E+04-	3.6E+04±5.8E+03-	5.5E+05±7.0E+04-	**2.8E+03±5.1E+02**
F18	3.0E+04±5.6E+03-	1.0E+03±4.7E+02≈	1.9E+10±7.4E+09-	8.3E+05±1.6E+05-	**9.1E+02±2.5E+02**
F19	1.1E+06±7.3E+05≈	1.3E+06±7.3E+05≈	1.8E+06±9.5E+04≈	1.4E+07±5.2E+06-	1.3E+06±5.3E+05
F20	4.2E+03±8.9E+02≈	**2.5E+03±8.8E+02≈**	5.9E+10±7.2E+09-	1.0E+04±7.3E+03-	3.7E+03±9.5E+02
-/+/≈	13/0/7	11/2/7	12/1/7	16/0/4	

CCITÖ, we choose four state-of-art algorithms used for comparisons: DECC-G [9], MLCC [11], DECC-DG [12] and CCPSO2 [5]. These algorithms parameters is the default settings suggested by their authors. In CCITÖ, we set $T = 1000, \rho = 0.99, \delta = 0.5, p = 0.6, \theta = 50, \varepsilon = 0.1$ and $popsize = 50$. All experiments are done on an Intel(R) Core(TM) i3 3.60 GHz computer with 4 GB RAM, our codes are implemented in Matlab. To be fair, for all algorithms, the terminated condition is only when a fixed number of function evaluations (FEs, 3×10^6) is reached. Each algorithm is independently performed for 25 times for the propose of statistical comparisons.

4.2   Compare with Other CC Algorithms

In Table 1, the best result is bolded in every problem. Results of CCITÖ are compared with those of DECC-G, MLCC, DECC-DG and CCPSO2, respectively, by Wilcoxon rank sum test at the significance level of 0.05. The marker "−" is worse than the results of CCITÖ, "+" is better than the results of CCITÖ and "≈" is equivalent to the results of CCITÖ.

From Table 1 and Fig. 2, we can see that CCITÖ outperforms the remaining ones on 11 out of total 20 benchmark functions, and 8 out of 10 in the third type (partially-separable functions that consist of multiple independent subcomponents, F_9–F_{18}) problem especially. In the first type Separable functions (F_1–F_3) and third type fully-nonseparable functions (F_{19} and F_{20}), these CC algorithms perform similarly. MLCC perform a litter better than other algorithms. Because MLCC can use its multilevel scheme that it can give a higher probability of being selected in the next co-evolutionary cycle to the better subcomponents. In the second type (partially-separable functions, in which a small number of variables are dependent, F_4–F_8), CCITÖ similar to other CC algorithms. Because, this type only has one group and small dependent variables, CCITÖ can not take advantage of its strategies. In the third type, CCITÖ is significantly superior to the other CC algorithms, due to its VII and RCR strategies. The VII strategy can help CCITÖ find more independent variables and push it to the one group to evaluation, and then RCR strategy can reallocate computational resources that give more computational resources to the most important subcomponents. As a conclusion, CCITÖ performs perfectly on the partially-separable functions and show its robust performance on the LSGO problems.

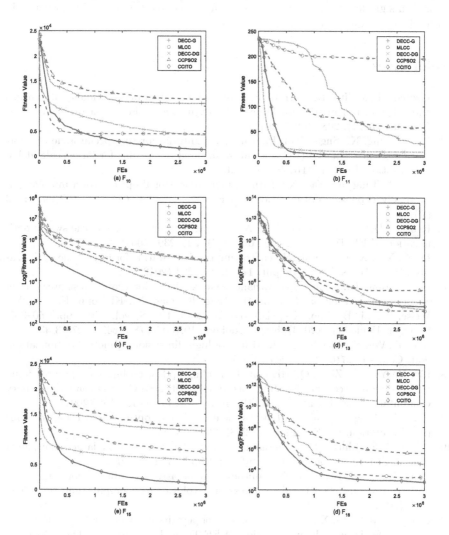

Fig. 2. Evolutionary curves on F10–F13, F15 and F8.

5   Conclusion and Future Work

In this paper, we proposed a novel ITÖ with cooperative coevolution strategy (CCITÖ) Algorithm, which is composed of three components: VII, ITÖ and

RCR. In the decomposition stage, it can recognize the indirect decision variables by a new variable interactions identification (VII) method. In the evaluation stage, it can give more computational resources to the more important subcomponents. Then, we evaluate the performance of our algorithm on the CEC'2010 LSGO benchmark. The experimental results show that our CCITÖ algorithm significantly outperforms other four state-of-the-art CC algorithms in both efficiency and effectiveness.

References

1. Lozano, M., Molina, D., Herrera, F.: Editorial scalability of evolutionary algorithms and other metaheuristics for large-scale continuous optimization problems. Soft. Comput. **15**(11), 2085–2087 (2011)
2. Li, X., Tang, K., Suganthan, P., Yang, Z.: Editorial for the special issue of information sciences journal (ISJ) on nature-inspired algorithms for large scale global optimization. Inf. Sci. **316**, 437–439 (2015)
3. Yang, P., Tang, K., Yao, X.: Turning high-dimensional optimization into computationally expensive optimization. IEEE Trans. Evolut. Comput. **PP**(99), 1–13 (2017)
4. Frans, V.D.B., Engelbrecht, A.P.: A cooperative approach to particle swarm optimization. IEEE Trans. Evol. Comput. **8**(3), 225–239 (2004)
5. Li, X., Yao, X.: Cooperatively coevolving particle swarms for large scale optimization. IEEE Trans. Evol. Comput. **16**(2), 210–224 (2012)
6. Liu, J., Tang, K.: Scaling up covariance matrix adaptation evolution strategy using cooperative coevolution. In: Yin, H., Tang, K., Gao, Y., Klawonn, F., Lee, M., Weise, T., Li, B., Yao, X. (eds.) IDEAL 2013. LNCS, vol. 8206, pp. 350–357. Springer, Heidelberg (2013). https://doi.org/10.1007/978-3-642-41278-3_43
7. Ren, Y., Wu, Y.: An efficient algorithm for high-dimensional function optimization. Soft. Comput. **17**(6), 995–1004 (2013)
8. Liu, Y., Yao, X., Zhao, Q., Higuchi, T.: Scaling up fast evolutionary programming with cooperative coevolution. In: Proceedings of the 2001 Congress on Evolutionary Computation (IEEE Cat. No. 01TH8546), vol. 2, pp. 1101–1108 (2001)
9. Yang, Z., Tang, K., Yao, X.: Large scale evolutionary optimization using cooperative coevolution. Inf. Sci. **178**(15), 2985–2999 (2008)
10. Mahdavi, S., Shiri, M.E., Rahnamayan, S.: Metaheuristics in large-scale global continues optimization: a survey. Inf. Sci. **295**, 407–428 (2015)
11. Yang, Z., Tang, K., Yao, X.: Multilevel cooperative coevolution for large scale optimization. In: 2008 IEEE Congress on Evolutionary Computation, pp. 1663–1670 (2008)
12. Omidvar, M.N., Li, X., Mei, Y., Yao, X.: Cooperative co-evolution with differential grouping for large scale optimization. IEEE Trans. Evol. Comput. **18**(3), 378–393 (2014)
13. Dong, W., Hu, Y.: Time series modeling based on ITO algorithm. In: International Conference on Natural Computation, pp. 671–678 (2007)
14. Nogueras, R., Cotta, C.: Self-healing strategies for memetic algorithms in unstable and ephemeral computational environments. Nat. Comput. 1–12 (2016)

15. Sun, Y., Kirley, M., Halgamuge, S.K.: Extended differential grouping for large scale global optimization with direct and indirect variable interactions. In: Proceedings of the 2015 Annual Conference on Genetic and Evolutionary Computation, GECCO 2015, pp. 313–320. ACM, New York (2015)

16. Tang, K., Li, X., Suganthan, P.N., Yang, Z., Weise, T.: Benchmark functions for the CEC'2010 special session and competition on large-scale global optimization. Technical report, Nature Inspired Computation and Applications Laboratory, USTC, China (2009)

A Conical Area Differential Evolution with Dual Populations for Constrained Optimization

Bin Wu[1], Weiqin Ying[1(✉)], Yu Wu[2(✉)], Yuehong Xie[1], and Zhenyu Wang[1]

[1] School of Software Engineering, South China University of Technology,
Guangzhou 510006, China
yingweiqin@scut.edu.cn
[2] School of Computer Science and Educational Software, Guangzhou University,
Guangzhou 510006, China
wuyu@gzhu.edu.cn

Abstract. During the last decade, multi-objective approaches to dealing with constraints in evolutionary algorithms have drawn more and more attention from researchers. In this paper, a conical area differential evolution algorithm (CADE) with dual populations is proposed for constrained optimization by borrowing the ideas of cone decomposition for bi-objective optimization. In CADE, a conical sub-population and a feasible one are designed to search the global feasible optimum along the Pareto front and the feasible segment, respectively. The conical sub-population aims to construct and utilize the Pareto front by a biased cone decomposition strategy in geometric proportion and a conical area indicator. Afterwards, neighbors in both sub-populations are adequately exploited to help each other. 13 benchmark test instances are used to assess the performance of CADE. The result reveals that CADE is capable of producing significantly competitive solutions for constraint optimization problems compared with the other popular approaches.

Keywords: Constrained optimization · Differential evolution
Multi-objective optimization · Cone decomposition · Dual populations

1 Introduction

Subject to different kinds of constraints, many optimization problems in real-world applications are considered as constrained optimization problems (COPs) [1,2]. Usually, a COP is expressed as follows:

$$minimize\ f(\mathbf{x})$$
$$subject\ to\ g_i(\mathbf{x}) \leq 0, i = 1, 2, \ldots, q$$
$$h_i(\mathbf{x}) = 0, i = q + 1, \ldots m$$
$$\mathbf{x} = \{x_1, x_2, \cdots, x_n\} \in \Omega \qquad (1)$$

© Springer Nature Singapore Pte Ltd. 2018
K. Li et al. (Eds.): ISICA 2017, CCIS 874, pp. 52–64, 2018.
https://doi.org/10.1007/978-981-13-1651-7_5

in which $g_i(\mathbf{x})$ denotes an inequality constraint, $h_i(\mathbf{x})$ is an equality one, and Ω means the decision space. Specifically, a solution satisfying all constraints is said to be feasible.

During the past decades, evolutionary algorithms (EAs) have exhibited significant performance and have been widely used for solving COPs. As unconstrained optimization technique, EAs require additional mechanisms to resolve constraints and a large number of techniques to handle constraints have been developed.

As the most common constraint handling methods, penalty function approaches punish an infeasible solution according to constraint violation so that it has a smaller chance to survive into the next generation. According to the preference of feasible individuals over infeasible ones, Deb [3] suggested a feasibility rule which is the most common method to compare individuals: (1) if one individual is infeasible and the other one is feasible, the feasible individual is selected; (2) between two feasible individuals, the one having a better objective function value is picked; (3) between two infeasible individuals, the one having a better degree of constrain violation is selected. Based on the penalty function and the feasibility rule, a stochastic ranking (SR) method [4] and its variants [5, 6] were proposed to solve COPs. But for the complex test cases, these methods still have difficulties in discovering optimal feasible solutions.

In addition, more and more researchers have used multi-objective evolutionary algorithms (MOEAs) when solving COPs. Suggested by Wang and Cai, an approach of great performance, CMODE [7], combines multi-objective optimization with differential evolution (DE) to deal with COPs. But CMODE doesn't construct the PF systematically to guide the search direction.

Recently, more and more decomposition-based MOEAs have been developed to resolve multi-objective optimization problems (MOPs) by decomposing a MOP into a series of scalar objective optimization subproblems. For instance, a conical area evolutionary algorithm (CAEA), proposed by Ying et al. [8], decomposes a bi-objective optimization problem (BOP) into N conical subproblems. And for each subproblem, an exclusive decision subset Ω_i is assigned. CAEA has exhibited more significant performance and higher efficiency compared with the other popular decomposition-based approaches.

In this paper, a conical area differential evolution algorithm with a dual-population scheme is presented to improve the competitiveness of multi-objective approaches for solving COPs. At first, CADE adopts a dual-population scheme where a conical sub-population and a feasible one are designed to search the optimal feasible solution, respectively, along the Pareto front and the feasible segment. In order to solve COPs, CADE partitions the decision space through a biased cone decomposition strategy different from CAEA, which implies that it needs to maintain more individuals near the feasible region. In addition, CADE can also gain a better diversity of population for COPs.

2   Dual-Population Scheme

In many multi-objective optimization methods to deal with constraints, the constraint violation degree of a solution $\mathbf{x}$ on the i-th constraint is formulated as follows:

$$G_i(\mathbf{x}) = \begin{cases} max\{0, g_i(\mathbf{x})\}, & 1 \le i \le q \\ max\{0, |h_i(\mathbf{x}) - \delta|\}, & q+1 \le i \le m \end{cases} \tag{2}$$

in which the positive tolerance value δ is used to handle equality constraints. Thus, $G(X) = \sum_{i=1}^{m} G_i(\mathbf{x})$ can express the constraint violation degree of $\mathbf{x}$.

To solve COPs by multi-objective optimization methods, a COP is generally converted into a BOP in which two objectives are need to be optimized. The first objective f_1 is to minimize the constraint violation degree $G(\mathbf{x})$, while the second one f_2 is to optimize the primary objective function $f(\mathbf{x})$. Figure 1 demonstrates the BOP converted from a COP. As shown in Fig. 1, all feasible solutions are located on the feasible segment while the non-dominated individuals lie on the Pareto front (PF). In particular, the intersection between the feasible segment and the Pareto front is just the desired global feasible optimum.

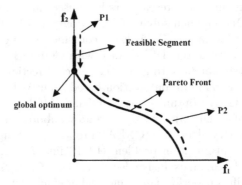

Fig. 1. Dual-population scheme for a BOP converted from a COP.

The advantage of multi-objective approaches such as CMODE is that it can take advantage of the non-dominated solutions to direct the population to search the global optimal feasible individual from infeasible regions to feasible regions. And how to utilize the non-dominated solutions has become the key of multi-objective approaches in the search for global optimum. Based on multi-objective technique, CMODE is able to create competitive results to solve COPs, but it mainly keeps no-dominated solutions and doesn't construct the PF systematically. In the proposed CADE, a biased cone decomposition strategy is employed to construct the PF in a systematic way and a conical area indicator is used to discover a local non-dominated individual in each conical subregion. Figure 1 also illustrates the dual-population scheme in CADE. This scheme consists of two sub-populations: the feasible one and the conical one, written as $P1$ and $P2$, which are designed to search the global feasible optimum, respectively, along the feasible segment and the Pareto front, as shown in Fig. 1.

2.1 Conical Sub-population and Biased Cone Decomposition

To divide the decision space Ω, both a nadir point and an utopian point are acquired in the objective space. Provided a set of individuals $A \in \Omega$, the utopian point and the nadir point over A can be, respectively, formulated as $\mathbf{F}^{min}(A) = (f_1^{min}, f_2^{min})$ and $\mathbf{F}^{max}(A) = (f_1^{max}, f_2^{max})$ where $f_i^{min} = \min_{\mathbf{x} \in A} f_i(\mathbf{x})$ and $f_i^{max} = \max_{\mathbf{x} \in A} f_i(\mathbf{x})$, $i = 1, 2$. For sake of clarity, we convert an objective vector $\mathbf{y}$ by $\overline{\mathbf{y}} = \mathbf{y} - \mathbf{F}^{min}(A)$ so that the utopian point becomes the origin $(0, 0)$.

Definition 1 (Observation vector). *For any transformed point* $\overline{\mathbf{y}} = (\overline{y}_1, \overline{y}_2)$, *its observation vector is* $\mathbf{V}(\overline{\mathbf{y}}) = (v_1, v_2)$ *where* $v_i = \frac{\overline{y}_i}{\overline{y}_1 + \overline{y}_2}, i = 1, 2$.

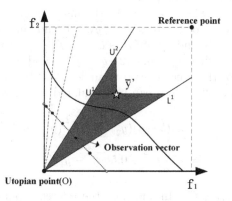

Fig. 2. Biased cone decomposition in geometric proportion.

It could be easily inferred that an observation vector has the following features: $v_i \geq 0$ and $v_1 + v_2 = 1$. Considering a specific number N of partitions, a series of reference observation vectors $\mathbf{V}^k$ in geometric proportion can be defined as follows:

$$\mathbf{V}^k = (\frac{s(1 - q^{k+1})}{1 - q}, 1 - \frac{s(1 - q^{k+1})}{1 - q}), k = 0, 1, \ldots N - 1, \tag{3}$$

where $V^0 = (s, 1 - s)$ is the first reference observation vector in geometric proportion and $q > 1$ denotes the proportion. Furthermore, the region $\mathbf{C} = \{\mathbf{y} = (y_1, y_2) | y_1 \geq 0 \wedge y_2 \geq 0\}$ can be decomposed into N conical subregions and the k-th subregion $\mathbf{C}^k = \{\mathbf{y} \in \mathbf{C} | \frac{s(1 - q^{k-1} + 1 - q^k)}{2(1 - q)} \leq v_1(\mathbf{y}) < \frac{s(1 - q^k + 1 - q^{k+1})}{2(1 - q)}\}$,

$k = 0, \ldots, N - 1$. It is evident form Fig. 1 that the individuals closer to the global optimum on the Pareto front can help the population to generate an offspring near the global optimum in a higher probability, compared with the individuals farther away from the global optimum. In view of this consideration, CADE employs the biased cone decomposition in geometric proportion, as shown in Fig. 2, rather than the uniform cone decomposition in CAEA. Similar to CAEA, the conical area is regarded as a significant indicator and CADE compares the conical areas of individuals in the same subregion and preserves the one with the smallest conical area. However, because of the biased cone decomposition in geometric proportion, the calculation equation of conical area is different from that in CAEA. The definition of conical area is as the following.

Definition 2 (Conical area). *Let $\mathbf{y}^r$ be the reference point dominated by all the other individuals, and $\overline{\mathbf{y}'} \in \mathbf{C}^k (0 \le k \le N - 1)$, then the conical area for $\overline{\mathbf{y}'}$ is the area of the portion $C(\overline{\mathbf{y}'}) = \{\mathbf{y} \in \mathbf{C}^k | \neg(\mathbf{y} \prec \overline{\mathbf{y}'}) \wedge \mathbf{y} \prec \overline{\mathbf{y}^r}\}$, written as $S(\overline{\mathbf{y}'})$.*

It can simply be acquired from Fig. 2 that $C(\overline{\mathbf{y}'})$ denotes the portion non-dominated by $\overline{\mathbf{y}'}$ in $\mathbf{C}^k$ where $\overline{\mathbf{y}'}$ lies. If $1 \le k \le N - 2$, $C(\overline{\mathbf{y}'})$ is a concave quadrilateral $OL^1\overline{\mathbf{y}'}U^2$ where $O = (0, 0)$, the intersection between line $f_2 = \overline{y_2'}$ and the lower boundary of $\mathbf{C}^k$ is $L^1 = (b\overline{y_2'}, \overline{y_2'})$, and the intersection between $f_1 = \overline{y_1'}$ and the upper boundary of $\mathbf{C}^k$ is $U^2 = (\overline{y_1'}, \frac{1}{a}\overline{y_1'})$. Specially, $a = t_a/(1 - t_a)$ and $b = t_b/(1 - t_b)$ where $t_a = 0.5s(2 - q^k - q^{k-1})$ and $t_b = 0.5s(2 - q^k - q^{k+1})$.

Thus, the sum of the areas of two triangles $\triangle OL^1U^1$ and $\triangle yU^2U^1$ is the desired conical area $S(\overline{\mathbf{y}'})$, which can be calculated as follows:

$$S(\overline{\mathbf{y}'}) = 0.5(b - a)(\overline{y_2'})^2 + 0.5\frac{1}{a}(\overline{y_1'} - a\overline{y_2'})^2$$

$$= 0.5\frac{1}{a}\overline{y_1'}^2 + 0.5b\overline{y_2'}^2 - \overline{y_1'}\overline{y_2'} \tag{4}$$

in which $\overline{\mathbf{y}'} = (\overline{y_1'}, \overline{y_2'}) \in \mathbf{C}^k$. In addition, the reference point $\mathbf{y}^r$ is required to calculate $S(\overline{\mathbf{y}'})$ if $\overline{\mathbf{y}'} \in \mathbf{C}^0$ or $\overline{\mathbf{y}'} \in \mathbf{C}^{N-1}$. The nadir point and the utopian point are used to calculate the reference point $(10^3(f_1^{max} - f_1^{min}), 10^3(f_2^{max} - f_2^{min}))$. If $\overline{\mathbf{y}'} \in \mathbf{C}^0$, the portion $C(\overline{\mathbf{y}'})$ is a concave pentagon $OL^1\overline{\mathbf{y}'}T^1T^2$ where T^1 is the intersection between $f_1 = \overline{y_1'}$ and $y_2 = \overline{y_2^r}$ and T^2 is the intersection between $f_2 = \overline{y_2^r}$ and $y_1 = 0$. Therefore, $S(\overline{\mathbf{y}'})$ is the sum of areas of a rectangle $\overline{\mathbf{y}'}T^1T^2U^1$ and a triangle $\triangle OL^1U^1$, i.e., $S(\overline{\mathbf{y}'}) = 0.5b(\overline{y_2'})^2 + (\overline{y_2^r} - \overline{y_2'})\overline{y_1'}$. In similarity, if $\overline{\mathbf{y}'} \in \mathbf{C}^{N-1}$, the conical area $S(\overline{\mathbf{y}'}) = 0.5\frac{1}{a}(\overline{y_1'})^2 + (\overline{y_2^r} - \overline{y_1'})\overline{y_2'}$.

2.2   Feasible Sub-population and Tolerance-Based Sorting

In this paper, a parameter ω represents a constraint tolerance value. Afterwards, the tolerance-based sorting is used for the feasible sub-population $P1$ so that one individual having the lower objective value as well as the constraint violation degree in the range ω precedes the others having higher objective values or the violation degree out of the range ω. The tolerance-based dominance relationship, written as $\leq_\omega$, between a pair of solutions $\mathbf{x}$ and $\mathbf{x}'$ is defined as follows:

$$\mathbf{x} \leq_\omega \mathbf{x}' = \begin{cases} f_2(\mathbf{x}) \leq f_2(\mathbf{x}'), \text{if } f_1(\mathbf{x}) \leq \omega \text{ and } f_1(\mathbf{x}') \leq \omega, \\ f_1(\mathbf{x}) \leq \omega \text{ and } f_1(\mathbf{x}') \geq \omega, \\ f_1(\mathbf{x}) \leq f_1(\mathbf{x}'), \text{if } f_1(\mathbf{x}) > \omega \text{ and } f_1(\mathbf{x}') > \omega. \end{cases} \quad (5)$$

After the individuals are sorted through the tolerance-based dominance relationship, the feasible sub-population $P1$ in CADE is grouped into M levels in sequence. When one new offspring is used to update $P1$, it is easy to find out the level in which it lies. When the offspring has a smaller objective value than the last solution in the k-th level does, the offspring belongs to this level.

Specifically, the constraint tolerance value ω in the tolerance-based sorting needs to be controlled over the function evaluations so that the algorithms can finally obtain high quality solutions with lower constraints violations. In this paper, ω is controlled as follows:

$$\omega(t) = 0.5 \times (1 - \frac{t}{T}) \times f_1(x_\lambda), 0 \leq t \leq T, \quad (6)$$

in which x_λ is the individual in the last conical subregion, T is the permitted maximum number of generations, and t means the current number of generations.

3   Proposed Algorithm: CADE

3.1   Adaptive Hybrid DE Operator

For stability, CADE employs an adaptive hybrid DE operator to generate offsprings. The hybrid operator consists of two classical DE operators, *DE/rand/exp* and *DE/current-to-rand/exp*. Moreover, the hybrid operator uses an adaptive selection parameter *Ada_rate* to control the probability in which the former classical operator is selected. At every generation, *Ada_rate* is updated as follows:

$$Ada\_rate = c \times Ada\_rate + (1 - c) \times \frac{a}{(a + b)}, \quad (7)$$

where a is the number of offsprings generated by the former classical operator updating the population successfully, b denotes the number of those by the latter one, and c is a control parameter, usually set to 0.5. To avoid that only one classical operator is used, *Ada_rate* is set to 0.95 in case of *Ada_rate* > 0.95 and to 0.05 in case of *Ada_rate* < 0.05.

In the hybrid operator, the first parent is chosen according to a tournament selection scheme based on the conical area in probability 0.75 and is picked randomly in probability 0.25 from the whole population consisting of P1 and P2. In addition, it is of great importance for multi-objective optimization to utilize neighbors during generation of offsprings. Therefore, the remainder of parents to be selected are picked from the neighbors of the first parent in probability 0.5 and are selected randomly from the whole population in probability 0.5.

3.2 Update of Sub-populations

After an offspring is generated, both the conical sub-population $P1$ and the feasible one $P2$ need to be updated. Algorithm 1 presents the update procedure for $P2$. When updating P2, the subregion where the offspring lies is firstly located according to the biased cone decomposition and the index k_1 of the corresponding conical subregion is calculated, as described in Line 1 of Algorithm 1. Then, the index k_2 of the conical subregion where the solution $P2.\mathbf{y}^{k_1}$ actually lies is acquired in Line 2. If $k_1 \neq k_2$, then the offspring is preserved and associated with this conical subregion. Otherwise, if $k_1 = k_2$, the conical areas of the offspring and the solution $P2.\mathbf{y}^{k_1}$ are compared and the one with the smaller conical area is preserved.

The procedure of updating $P1$ is listed in Lines 11–15 of Algorithm 2. At the early stage, i.e., when the current number of function evaluations (FES) is less than or equal to a threshold value θ, the offspring $\mathbf{y}$ is utilized to update the first individual worse than $\mathbf{y}$ in the level of $P1$ where $\mathbf{y}$ lies according to the tolerance-based dominance rule for the reason it can help $P1$ keep diversity within the range ω of constraint violation. At the advanced stage, i.e., when FES is bigger than the threshold value θ, CADE picks a solution in a random way from $P1$ and applies the feasibility rule to update it. In such a situation, the feasibility rule can help $P1$ converge faster to a global optimum.

Algorithm 1. Update of the Conical Sub-population

Input: $\mathbf{y}$: an offspring for update; $\mathbf{F}^{min}$: the current utopian point; $P2$: the current conical sub-population.

Output: $P2$: the updated conical sub-population.

1 $k_1 \leftarrow \left\lfloor \log_q \dfrac{2(f_1(\mathbf{y}) - f_1^{min})(1-q)}{s(1+1/q)(f_1(\mathbf{y}) - f_1^{min} + f_2(\mathbf{y}) - f_2^{min})} \right\rfloor$;

2 $k_2 \leftarrow \left\lfloor \log_q \dfrac{2(f_1(P2.\mathbf{y}^{k_1}) - f_1^{min})(1-q)}{s(1+1/q)(f_1(P2.\mathbf{y}^{k_1}) - f_1^{min} + f_2(P2.\mathbf{y}^{k_1}) - f_2^{min})} \right\rfloor$;

3 **if** $k1 \neq k2 \vee S(f(\mathbf{y}) - \mathbf{F}^{min}) < S(f(P2.\mathbf{y}^{k_1}) - \mathbf{F}^{min})$ **then**

4 $\quad | \quad P2.\mathbf{y}^{k_1} \leftarrow \mathbf{y}$;

5 **end**

6 **return** $P2$;

3.3 Procedure of CADE

The procedure of CADE is given in Algorithm 2. In the beginning, N initial individuals $P' = \{\mathbf{y}^1, \mathbf{y}^2, \cdots, \mathbf{y}^N\}$ are generated by sampling from the decision space Ω in a random way. Afterwards, the biased cone decomposition in geometric proportion is utilized to divide the objective space to $size_2$ conical subregions and only one individual is preserved for every conical subregion as described in Line 3. Specifically, each conical subregion for the conical sub-population $P2$ attempts to be associated with the one with the smallest conical area among the individuals belonging to this subregion in P'. Further, each subregion not yet assigned successfully is associated with the one closest to its reference observation vector among the remainder of individuals in P'. Thereafter the remainder of individuals in P' are ranked and grouped through the tolerance-based sorting to form the feasible sub-population $P1$.

Algorithm 2. Procedure of CADE

Input: $size_1$: the size of the conical sub-population $P1$; $size_2$: the size of the feasible sub-population $P2$; MAX_FES: the maximum number of function evaluations.

Output: $\mathbf{x}^\sharp$: the best feasible solution in the final population.

1 Create $N = size_1 + size_2$ initial individuals $P' = \{\mathbf{y}^1, \mathbf{y}^2, \cdots, \mathbf{y}^N\}$ randomly;
2 Initialize the utopian point $\mathbf{F}^{min} = (f_1^{min}, f_2^{min})$ where $f_i^{min} = \min_{\mathbf{y} \in P'} f_i(\mathbf{y})$, $i = 1, 2$;
3 Associate each conical subregion for the conical sub-population $P2$ with one individual from P' according to the biased cone decomposition;
4 Group the remainder of individuals in P' through the tolerance-based sorting to form the feasible sub-population $P1$;
5 **for** $FES \leftarrow 1$ **to** MAX_FES **do**
6 $\quad$ Generate an offspring $\mathbf{y}$ by the adaptive hybrid DE operator and update $\mathbf{F}^{min}$;
7 $\quad$ **if** $\mathbf{F}^{min}$ *is successfully updated and* $FES \leq \theta$ **then**
8 $\quad\quad$ Group the individuals in $P1$ through the tolerance-based sorting;
9 $\quad$ **end**
10 $\quad$ $P2 \leftarrow$ UpdateConicalPop(y,$\mathbf{F}^{min}$,$P2$);
11 $\quad$ **if** $FES \leq \theta$ **then**
12 $\quad\quad$ Update the first individual worse than $\mathbf{y}$ in the level of $P1$ where $\mathbf{y}$ lies according to the tolerance-based rule;
13 $\quad$ **else**
14 $\quad\quad$ Update an individual randomly chosen from $P1$ according to the feasibility rule;
15 $\quad$ **end**
16 $\quad$ **if** $FES \bmod N == 0$ **then**
17 $\quad\quad$ Update Ada_rate and ω;
18 $\quad$ **end**
19 **end**
20 **return** $x^\sharp$;

Subsequently, the adaptive hybrid DE operator is employed to generate an offspring at every iteration. Thereafter, the objective vector of the offspring is used to update the utopian point $\mathbf{F}^{min}$. Then the conical sub-population $P2$ is updated as described in Algorithm 1. Afterwards, the feasible sub-population $P1$ is updated. Besides, the values of ω and Ada_rate are updated after every N iteration. Finally, the best solution in the final population is outputted.

4   Empirical Results and Discussion

The general performance of CADE is first validated on 13 widely used benchmark constrained optimization problems collected in [9] in our experiments. Then, the capability of CADE are compared against with those of the three existing popular algorithms, namely, SR, SaDE [10], and CMODE, on these test instances.

In our experiments, the sizes of the feasible sub-population $P1$ and the conical sub-population $P2$ are set to 120 and 60, respectively, thereby the total population size equals to 180. In CADE, the threshold value θ, the number M of levels for $P1$ and the proportion q for $P2$ are respectively set to 1.0×10^5, 2 and 1.1. In addition, the scaling factor F and the crossover control parameter C_r are picked in a random way, respectively, from [0.5, 0.6] and [0.9, 0.95] for the adaptive hybrid DE operator. The termination criterion is satisfied when the number of function evaluations, FES, reaches 5×10^5 for every algorithm on every test instance. The other parameters of SR, SaDE and CMODE use the corresponding recommended values provided, respectively, by [4,7,10]. All the four algorithms are implemented in C++ and executed on an Intel Core I5-3470 3.20 GHz PC with 4 GB RAM. To evaluate the performances of these techniques, 25 statistically independent runs of each algorithm have been executed for each test case.

4.1   General Performance of CADE

As suggested by Liang et al. [9], if an feasible individual $\mathbf{x}$ has the objective function value $f(\mathbf{x}) - f(\mathbf{x}^*) \leq 0.0001$ in which $\mathbf{x}^*$ represents the global optimal feasible solution, the solution $\mathbf{x}$ could be considered as fulfilling the success condition. Consequently, the difference between $f(\mathbf{x})$ and $f(\mathbf{x}^*)$ is referred as the function error value for the best-so-far individual $\mathbf{x}$ in our experiments.

In our experiment, at least one feasible individual is acquired in each run of CADE on each of 13 instances in 5×10^5 function evaluations. And the convergence curves of function error values (in log scale) obtained by CADE in the median run over FES are displayed in Fig. 3. It is worth noting that all the points whose function error values are less than or equal to 0 are not plotted in Fig. 3. It can be gained from Fig. 3 that CADE is capable of discovering a global optimal individual for g04, g06, g08, g11 and g12 by only using about 1.5×10^5 FES. In addition, Fig. 3 also clearly indicates that CADE requires about 3×10^5 FES for g01, g05 and g09. Furthermore, it can be inferred from Fig. 3 that for 9 out of 13 test instances, i.e., except g02, g06, g09 and g13, CADE can finally find the optimal solutions at 5×10^5 FES, which demonstrates that the feasibility rule of updating $P1$ at the advanced stage help $P1$ rapidly converge to the global optimum.

4.2   Comparison with Some Other Popular DE-Based Methods

In order to clearly discover the advantages of CADE for constrained optimization, CADE is further compared against with three popular DE-based

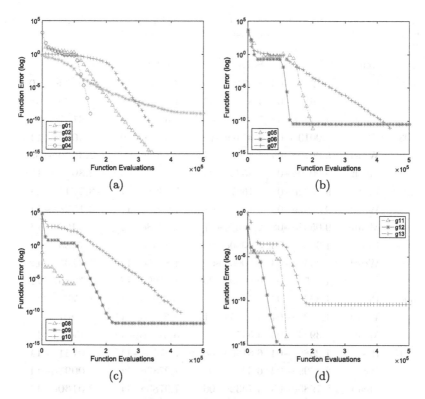

Fig. 3. Convergence curves of function error values obtained by CADE for g01–g04 (a), g05–g07 (b), g08–g10 (c) and g11–g13 (d).

approaches: SR, SaDE and CMODE. Table 1 presents the best, worst and mean of function error values achieved, respectively, by SR, SaDE, CMODE and CADE when FES = 5×10^5 for test instances g02–g10 and g13. Since all four approaches can find the nearly same exact optimal solution in each of their runs on g01, g11 and g12, Table 1 doesn't show the results for these three test instances. In Table 1, the best result among four compared algorithms for each test instance is specially highlighted in boldface. It is evident from Table 1 that CADE performs best, followed by CMODE and SaDE, while SR behaves worst. Specifically, for 8 out of the 10 test cases, CADE obtains obviously lower error values than the other three algorithms. In particular, it is worth noting that, for g02 and g03, only CADE and CMODE can discover the solutions satisfying the success condition and CADE performs better than CMODE there. In conclusion, CADE could produce competitive results due to the dual-population scheme than other three popular approaches.

Table 1. Function error values achieved, respectively, by SR, SaDE, CMODE and CADE when FES $= 5 \times 10^5$ for test instances g02–g10 and g13.

Prob.	Item	SR	SaDE	CMODE	CADE
g02	Best	8.7256e−03	8.0719e−10	4.1726e−09	**3.0904e−10**
	Wrost	9.0653e−02	1.8353e−02	1.1836e−07	**2.7070e−08**
	Mean	4.1432e−02	2.0560e−03	2.0387e−08	**4.8107e−09**
g03	Best	5.0010e−04	1.3749e−10	2.3964e−10	**−1.0002e−11**
	Wrost	5.0010e−04	1.3389e−04	2.5794e−09	**2.3970e−09**
	Mean	5.0010e−04	1.3532e−05	1.1665e−09	**2.3070e−10**
g04	Best	−9.0502e−08	2.1667e−07	7.6398e−11	**−3.2742e−11**
	Wrost	−9.0502e−08	2.1667e−07	7.6398e−11	**−2.1828e−11**
	Mean	9.0502o−08	2.1667e−07	7.6398e−11	**−2.5466e−11**
g05	Best	1.3956e−03	0.0000e+00	−1.8190e−12	**−2.7285e−12**
	Wrost	1.3956e−03	0.0000e+00	−1.8190e−12	**−2.7285e−12**
	Mean	1.3956e−03	0.0000e+00	−1.8190e−12	**−2.7285e−12**
g06	Best	1.3924e+04	4.5475e−11	**3.3651e−11**	3.4561e−11
	Wrost	1.3924e+04	4.5475e−11	**3.3651e−11**	3.4561e−11
	Mean	1.3924e+04	4.5475e−11	**3.3651e−11**	3.4561e−11
g07	Best	1.2586e−04	6.8180e−08	7.9783e−11	**−2.0151e−11**
	Wrost	2.5699e−02	6.3431e−05	7.9783e−11	**−2.0091e−11**
	Mean	5.1985e−03	4.7432e−06	7.9783e−11	**−2.0130e−11**
g08	Best	2.3700e−10	8.1964e−11	8.1964e−11	**−1.8036e−11**
	Wrost	2.3700e−10	8.1964e−11	8.1964e−11	**−1.8036e−11**
	Mean	2.3700e−10	8.1964e−11	8.1964e−11	**−1.8036e−11**
g09	Best	−2.1794e−10	3.7440e−07	**−9.8225e−11**	1.7053e−12
	Wrost	1.7727e−05	3.7440e−07	**−9.8111e−11**	1.7053e−12
	Mean	1.8240e−06	3.7440e−07	**−9.8198e−11**	1.7053e−12
g10	Best	2.4606e−05	6.6393e−11	6.2755e−11	**−3.5470e−11**
	Wrost	4.3716e+00	7.8300e−06	6.3664e−11	**−3.5470e−11**
	Mean	1.4789e+00	1.6907e−06	6.2827e−11	**−3.5470e−11**
g13	Best	8.3338e−06	5.3102e−006	**4.1897e−11**	4.1897e−11
	Wrost	3.8491e−01	3.8491e−001	**4.1897e−11**	4.1897e − 11
	Mean	7.6989e−02	1.0778e−001	**4.1897e−11**	4.1897e−11

5   Conclusion

In this paper, a multi-objective approach, CADE, is proposed to handle constrained optimization problem. CADE employs the dual population scheme so that the information in the Pareto front can be utilized to achieve the global optimum in a more systematic way. Specially, more promising individuals can be preserved in the conical subregions closer to the global optimum by the biased cone decomposition in geometric proportion and the conical area indicator is used to guide the conical sub-population to approximate the front. In addition, the tolerance-based sorting for the feasible sub-population in the early stage can keep a good diversity of population. Experimental results demonstrate that CADE has a great capability for constrained optimization by employing the dual population scheme and biased cone decomposition. Our ongoing research aims on applications of CADE for different types of engineering problems such as the design problem of pressure vessel.

Acknowledgments. This work was supported in part by the Natural Science Foundation of Guangdong Province, China, under Grant 2015A030313204, in part by the Pearl River S&T Nova Program of Guangzhou under Grant 2014J2200052, in part by the National Natural Science Foundation of China under Grant 61203310 and Grant 61503087, in part by the Fundamental Research Funds for the Central Universities, SCUT, under Grant 2017MS043, in part by the Guangdong Province Science and Technology Project under Grant 2015B010131003 and in part by the China Scholarship Council (CSC) under Grant 201406155076 and Grant 201408440193.

References

1. Wang, Y., Cai, Z.: A constrained optimization evolutionary algorithm based on multiobjective optimization techniques. In: IEEE Congress on Evolutionary Computation, pp. 1081–1087. IEEE (2005)
2. Fan, L., Liang, Y., Liu, X., Gu, X.: A hybrid evolutionary algorithm based on dominated clustering for constrained bi-objective optimization problems. In: International Conference on Computational Intelligence and Security, pp. 543–546. IEEE (2017)
3. Deb, K.: An efficient constraint handling method for genetic algorithms. Comput. Methods Appl. Mech. Eng. **186**(2–4), 311–338 (2000)
4. Runarsson, T.P., Yao, X.: Stochastic ranking for constrained evolutionary optimization. IEEE Trans. Evolut. Comput. **4**(3), 284–294 (2000)
5. Liu, J., Fan, Z., Goodman, E.: SRDE: an improved differential evolution based on stochastic ranking. In: ACM/SIGEVO Summit on Genetic and Evolutionary Computation, pp. 345–352. ACM (2009)
6. Ying, W., Peng, D., Xie, Y., Wu, Y.: An annealing stochastic ranking mechanism for constrained evolutionary optimization. In: International Conference on Information System and Artificial Intelligence, pp. 576–580. IEEE (2017)
7. Wang, Y., Cai, Z.: Combining multiobjective optimization with differential evolution to solve constrained optimization problems. IEEE Trans. Evolut. Comput. **16**(1), 117–134 (2012)
8. Ying, W., Xu, X., Feng, Y., Wu, Y.: An efficient conical area evolutionary algorithm for bi-objective optimization. IEICE Trans. Fundam. Electron. Commun. Comput. Sci. **E95–A**(8), 1420–1425 (2012)

9. Liang, J., Runarsson, T.P., Mezura-Montes, E., Clerc, M., Suganthan, P.N., Coello, C.A., Deb, K.: Problem definitions and evaluation criteria for the CEC 2006 special session on constrained real-parameter optimization. Technical report, Nanyang Technological University, Singapore (2006)
10. Huang, V., Qin, A., Suganthan, P.N.: Self-adaptive differential evolution algorithm for constrained real-parameter optimization. In: IEEE Congress on Evolutionary Computation, pp. 17–24. IEEE (2006)

Swarm Intelligence – Swarm Optimization

A Particle Swarm Clustering Algorithm Based on Tree Structure and Neighborhood

Lei Yang[1(✉)], Wensheng Zhang[2], Zhicheng Lai[1], and Ziyu Cheng[1]

[1] College of Mathematics and Informatics, South China Agricultural University,
Guangzhou, China
yanglei_s@scau.edu.cn
[2] Institute of Automation, Chinese Academy of Sciences, Beijing 100190, China

Abstract. Cluster analysis is one of the important research contents in data mining. The basic Particle Swarm Optimization algorithm (PSO) can be combined with the traditional clustering algorithm to achieve clustering analysis. Aiming at the disadvantages of the basic particle swarm optimization algorithm is easy to fall into local extremum, the search accuracy is not high, and the traditional K-means and FCM clustering algorithm are affected by the initial clustering center. This paper proposes a new particle swarm clustering algorithm based on tree structure and neighborhood (TPSO), which designs the structure of the particle group as a tree structure, uses the breadth of traversal, increases the global search ability of the particle, and joins the neighborhood operation to let the particle close to the neighborhood optimal particles and accelerate the convergence speed of the algorithm. Our experiments using Iris, Wine, Seed, Breast-w4 group of UCI public data sets show that the accuracy obtained by the TPSO algorithm implementing the proposed K-means and FCM is statistically significantly higher than the accuracy of the other clustering algorithms, such as K-means algorithm, fuzzy C-means algorithm, the basic particle swarm optimization combined with traditional clustering algorithm, etc., Comparison experiments also indicate that the TPSO algorithm can significantly improve the clustering performance of PSO.

Keywords: Particle swarm optimization · Cluster analysis · TPSO
K-means · FCM

1 Introduction

The explosion of data, extensive applications and huge number make our time become the era of big data. Changes of the time require people to find the tools of data analysis, discover useful knowledge from the big data and serve for the human life. This demand leads to the birth of data mining [1]. Cluster analysis is an effective method to deal with many complicated data. As an important research branch of data mining, it mainly uses the method of similarity measure to classify the identical or similar features into a catalogue, hence achieving clustering [2]. Clustering has been widely used in many areas of life, including business intelligence, image pattern recognition, Web search [3], biology and security [4]. In the field of intelligent business, data clustering can achieve many customer groups, making the group of customers have very similar

© Springer Nature Singapore Pte Ltd. 2018
K. Li et al. (Eds.): ISICA 2017, CCIS 874, pp. 67–85, 2018.
https://doi.org/10.1007/978-981-13-1651-7_6

characteristics, which can be targeted for customer relationship management. In the field of image recognition, clustering can be used to discover "clusters" or "subclasses" in hand-written character recognition systems. As a function of data mining, clustering analysis can also be used as an independent tool for investigating the distribution of data, observing the characteristics of each cluster, and concentrating on further clustering [5]. Clustering analysis has become a very active research topic in the field of data mining [6].

Particle swarm optimization (PSO) is a widely recognized optimization algorithm inspired by social swarm [7], which was proposed by Kennedy and Eberhart in long-term observation birds, fish and other creatures in the nature population in foraging behavior rule. It is a model based on social psychology of social impact and social learning optimization algorithm [8]. Each individual in the particle swarm (PS) follows a simple behavior, that is, to emulate the success of the adjacent individual, and the cumulative behavior will be the best area for searching in a high dimensional space. We can initialize the random number of the solution, and update solution through the iterative, and then cover the previous solution with a better one to find the optimal one. And the fitness is used to evaluate the quality of the solution in particle swarm optimization. Particle swarm optimization algorithm with random search, which does not depend on the objective function, not only has the characteristics of parallelism, fast convergence speed and a wide range of applications, but also has been widely used in function optimization, mechanical fault detection, traffic planning, insurance statistics and decision etc.

Cluster analysis finds hidden patterns through unsupervised learning process and can discover knowledge independently. Many clustering algorithms, such as K-means and FCM clustering, are widely used in data mining and image processing because of their fast speed, simple thought and easy implementation, the method uses iteration to obtain optimal solution, but the clustering quality almost depends on the initial clustering center of random selection. To overcome this shortcoming, many evolutionary clustering algorithms have been proposed, such as genetic algorithm [9], ant colony optimization algorithm, particle swarm optimization algorithm [10], simulated annealing algorithm, etc. The hybrid clustering method with advantages of multiple algorithms has better optimization performance, such as GA + K-means, KCPSO [11], etc. The advantages of clustering method of genetic algorithm are not affected by the initial solution, but the genetic algorithm itself has the problem of slow convergence. The particle swarm optimization algorithm has the advantages of high convergence efficiency and fast speed, but the algorithm also has the problem of falling into the local extreme and the search accuracy is not high. In this paper, a new kind of particle clustering algorithm based on tree structure and neighborhood (TPSO) is proposed by combining particle swarm optimization algorithm and K-means and FCM clustering algorithm. The structure of particle swarm algorithm was designed as a tree, which adopting breadth traversal, increasing the global search ability of particle, adding neighborhood operations at the same time for the particles near to the optimal particle neighborhood, accelerating the algorithm convergence speed in the late. This can divide the related data into the corresponding clusters. After the realization of the algorithm with the MATLAB language, we conducted the experiment by using the four data sets from UCI public test, and compared the result with those obtained from previous some clustering algorithm. The results showed that this TPSO algorithm has certain advantages in clustering.

The remainder of this study is organized as follows. Section 2 describes the traditional particle swarm clustering mining algorithm. Section 3 introduces the new particle swarm clustering algorithm based on the tree structure and neighborhood (TPSO). Section 4 presents the experimental results and discussion. Finally, Sect. 5 summarizes and presents the future research direction.

2 Traditional Particle Swarm Clustering Mining Algorithm

There are many sets of physical or abstract objects in the real world, and clustering is a process of grouping these sets. Each group is a cluster that consists of similar objects and is different from objects in other clusters.

2.1 Cluster Analyses and Algorithm

2.1.1 Euclidean Distance

The sample is the most basic unit in cluster analysis, its structural feature is a data matrix: it measures the characteristics of n samples with d variables (sample attributes), such as job number, name, gender, ethnicity, origin, age, weight and other properties to express the sample "workers". The n × d-dimensional matrix can be used to represent the characteristics of n samples, indicating the structure as follows:

$$
\begin{bmatrix}
x_{11} & x_{12} & L & x_{1d} \\
x_{21} & x_{22} & L & x_{2d} \\
M & M & M & M \\
x_{n1} & x_{n2} & L & x_{nd}
\end{bmatrix}
\tag{1}
$$

There are two samples X_i, X_j, and their eigenvectors are

$$
X_i = \begin{Bmatrix} x_{i1} \\ x_{i2} \\ M \\ x_{in} \end{Bmatrix} = (x_{i1}, x_{i2}, L, x_{in})^T, X_j = \begin{Bmatrix} x_{j1} \\ x_{j2} \\ M \\ x_{jn} \end{Bmatrix} = (x_{j1}, x_{j2}, L, x_{jn})^T
\tag{2}
$$

In general, there are four methods to calculate the distance between samples, respectively, Euclidean distance, Mahalanobis distance, angle cosine distance, binary angle cosine and Tanimoto measure with binary feature. Euclidean distance calculation formula (3) is commonly used to calculate the distance between samples under normal circumstances. The Euclidean distance between the two samples is:

$$
dist_{1,2} = \|X_1 - X_2\| = \sqrt{\sum_{i=1}^{d} (x_{1i} - x_{2i})^2}, X_1 \in \omega_1, X_2 \in \omega_2
\tag{3}
$$

Formula (1) sample X_1, X_2 belong to class ω_1, ω_2.

2.1.2   Cluster Analysis

2.1.2.1 K-means Algorithm

The idea of the K-means clustering algorithm is that the K objects are randomly selected from the samples and used as the initial clustering centers. Then, the Euclidean distance between each object and the cluster centers is given, each object is classified to the nearest clustering center by comparing the distance. Each cluster center and the objects classified to it represent a cluster. The clustering center of each cluster is recalculated based on the objects in the cluster that are categorized into the cluster. Repeating this process until a certain termination condition is met. The termination condition corresponds to the stability of the clustering centers as the number of iterations increases, or the sum of Euclidean distances in each cluster tends to be stable.

K-means algorithm description: define several variables first: N_d represents the input dimension, that is, the number of parameters per vector; N_c represents the number of cluster centers; Z_p represents the p-th vector data; c_j represents the cluster center subscript j point vector; n_j represents the number of data included in the cluster j; C_j represents the data vector subset of data clustering j. The specific process is as follows:

Step 1: Randomly initialize the central vector of N_c cluster.
Step 2: For each vector, the formula (4) is used to calculate the distance between the data to which the data belongs to the jth category and its center:

$$\text{dist}(z_p, c_j) = \sqrt{\sum_{k=1}^{N_d} (z_{pk} - c_{jk})^2} \tag{4}$$

Step 3: Using Eq. (5) is recalculated each cluster center vector.

$$c_j = \frac{1}{n} \sum_{\forall z_p \in c_j} z_p \tag{5}$$

Step 4: Jump to Step 2 until the stop condition is satisfied.

2.1.2.2 Fuzzy C-means Algorithm (FCM)

FCM is a fuzzy partition algorithm, which measures the degree of belonging to a certain cluster by a membership function for the sample processing [12]. The membership function is usually written in $u_A(x)$, it can calculate the degree value of an object x belonging to the set A [13], whose range of values is [0,1], that is, $0 \leq u_A(x) \leq 1$. The number of clusters c and m (the Flexible parameter of the control algorithm) is important parameters of the FCM algorithm.

FCM algorithm has a value function of non-similarity index, the goal of the algorithm is to obtain the smallest function value in the iteration times. The idea of fuzzy division is that each given data point is not explicitly assigned to a group, with the value of 0, 1 between the membership to determine their degree of belonging to each group. The membership matrix U is composed of values between 0 and 1, normalized, and is equal to 1:

$$\sum_{i=1}^{c} u_{ij} = 1, \forall j = 1, \cdots, n \tag{6}$$

Value function (or objective function) is:

$$J(U, c_1, \cdots, c_c) = \sum_{i=1}^{c} J_i = \sum_{i=1}^{c} \sum_{j}^{n} u_{ij}^m d_{ij}^2 \tag{7}$$

Where the value of u_{ij} is between 0 and 1, c_{ij} is the clustering center for fuzzy group, d_{ij} is the Euclidean distance between the ith cluster center and the jth data point, and m is a weighting index.

$$c_i = \frac{\sum_{j=1}^{n} u_{ij}^m x_j}{\sum_{j=1}^{n} u_{ij}^m} \tag{8}$$

The calculation formula of membership function is as shown in Eq. 9.

$$u_{ij} = \frac{1}{\sum_{k=1}^{c} \left(\frac{d_{ij}}{d_{kj}}\right)^{\frac{2}{m-1}}} \tag{9}$$

The specific process is as follows:

Step 1: Initializing the number of iterations t = 0, the number of clustering number c and weight m are determined, and c cluster centers are initialized randomly.
Step 2: According to the formula (9), calculating the membership degree matrix is U.
Step 3: According to formula (8), the clustering center is modified.
Step 4: According to the threshold $\varepsilon > 0$, if the objective function $\|J_t - J_{t-1}\| \le \varepsilon$, then the algorithm terminates, otherwise the iteration number t + 1, jump to Step 2.

2.2   Particle Swarm Optimization

PSO is inspired by the predatory behavior of birds [10]. For an optimization problem, the bird in the model is a viable solution called "particle" in the PSO. Each bird searches near the area of the bird near the food, and in the PSO it is the degree of optimization of the current particle by the fitness value determined by an optimized function. Birds in the flight will continue to change the location and speed, the same particle corresponding to a speed and location information, the speed will determine its direction and location, and then the particles will be closer to the current optimal particle solution space and search [14].

The direction and distance at which each particle is flying in space is determined by the speed of motion, and in general, it is like a bird following a bird that is closest to the food. In each iteration, the particle will update the iteration based on two extremes, one is the particle itself has experienced the best position, but also on their own optimal solution [15]. There is also an optimal position experienced by the whole population, that is, global extremes.

The formula of speed and location is as follows:

$$v_i = \omega v_i + c_1 * rand * (p_i - x_i) + c_2 * rand * (g - x_i) \tag{10}$$

$$x_i = x_i + v_i \tag{11}$$

v is the velocity of the particle. ω is the inertia weight. x is the position of the current particle. p is the optimal solution of particle history. g is the global optimal solution. c_1, c_2 is the learning factor, generally $c_1 = c_2 = 2$.

2.3 PSO Clustering Algorithm

K-mean clustering results will be affected by the initial value, and FCM easily fall into the local optimal. Particle swarm algorithm and K-mean, FCM algorithm can be combined to fill their respective deficiencies.

2.3.1 Particle Swarm Algorithm Combined with K-means Algorithm

Each particle in the particle swarm algorithm is a feasible solution, and then applied to the cluster analysis, each particle is a set of clustering center, and the fitness function is the sum of the distance of each data and its category center Point, through the algorithm of continuous iteration, to obtain the minimum fitness. The following is the algorithmic process.

Suppose the sample is a matrix of n rows and d columns:

Step 1: Random initialization. Determine cluster number M, particle number N, random initialization M cluster center. For the ith particle p (i) is $M * d$ dimensional vector (i.e. M cluster center set). The initial velocity of each particle.

Step 2: According to the initial particle group, the optimal values of each particle and global optimal value are calculated.

Step 3: Update the speed and location of all particles according to formula (10) and formula (11).

Step 4: For each sample, calculate the distance from the cluster center and classify it according to the proximity principle.

Step 5: According to the classification results, calculate the fitness.

Step 6: Update the individual optimal value.

Step 7: Update the global optimal value.

Step 8: If the number of iterations is reached, the algorithm ends, otherwise skip to Step 3.

2.3.2 Particle Swarm Algorithm Combined with FCM Algorithm

The particle swarm optimization combined with FCM algorithm, each particle is a set of clustering centers, fitness function is the value of the FCM function formula (7), and we can obtain the smallest fitness through constant iterative algorithm. The algorithm process is the same as the PSO and k-means in the above, and the parameters are added with some parameters, including the optimal solution change threshold ω and the number of iteration threshold $iter_{stable}$, the following is the algorithm flow.

Suppose the sample is a matrix of n rows and d columns:

Step 1: Random initialization. Determine cluster number M, particle number N, random initialization M cluster center. For the ith particle p (i) is M * d dimensional vector (i.e. M cluster center set). The initial velocity of each particle.

Step 2: According to the initial particle group, the optimal values of each particle and global optimal values are calculated.

Step 3: Update the speed and location of all particles according to formula (10) and formula (11).

Step 4: For each sample, use FCM value function (formula 7) to calculate the fitness and update the fitness of the particle.

Step 5: Update the individual optimal value.

Step 6: Update the global optimal value.

Step 7: If the change in the number of iterations or iteration $iter_{stable}$ is less than ε, then the algorithm is ended, otherwise it will jump to step 3.

3   Particle Swarm Mining Algorithm Based on Tree Structure and Neighborhood

In this section, we discuss in detail our proposed clustering algorithm of particle clustering based on tree structure and neighborhood (TPSO). This section is mainly divided into two parts, namely: particle swarm algorithm based on the tree structure and neighborhood, and improved particle swarm optimization analysis.

3.1   Particle Swarm Optimization Based on the Tree Structure and Neighborhood

The obvious shortcomings of the basic particle swarm algorithm are that the convergence is fast and easy to fall into local optimum. In the case of convergence, the convergence rate of the particles will obviously slow down, because all the particles move closer to the optimal solution and tend to be homogeneous.

Aiming at the deficiency of the basic particle swarm optimization algorithm, we design a particle swarm optimization algorithm with neighbor operation, named TPSO, which is designed as a tree structure based on the basic particle swarm algorithm. The fitness value of the parent node is always better than that of the child node. Before each iteration, the tree structure needs to be compared and adjusted so that the parent node is better than the child node, and the global optimal value is updated by the root node. In the neighborhood of the particle, all nodes have the same number of branches except the bottom node. A neighborhood is defined to contain the particle itself, its parent node, and all its child nodes.

At each iteration, the fitness values for all particles are calculated before the update of the velocity and position. Using the breadth-first traversal tree, each particle is compared with the parent node by the optimal value of the individual, and if the fitness value of the sub node is high, it will swap the position in the network with the parent node. This structure will expand the search space while all particles fly to better particles (not necessarily the global optimal position), while reducing the convergence rate. The following is the speed update formula of the algorithm.

$$
\begin{aligned}
v_i(t) = \omega * v_i(t-1) + c_1 * rand * (pb_i(t-1) - x_i(t-1)) + c_2 * rand \\
* (gb(t-1) - x_i(t-1)) + c_3 * rand \\
* (l_i(t-1) - x_i(t-1))
\end{aligned}
\tag{12}
$$

Where t is the number of iterations, c_3 is the neighborhood learning factor, and l_i is the optimal particle in the domain of the ith particle.

The change formula of ω is:

$$
\omega = \frac{(\omega_{max} - \omega_{min})(h-1)}{H-1} \omega_{min}
\tag{13}
$$

Where $\omega_{max}, \omega_{min}$ is the maximum and minimum of ω, h is the depth of the current particle in the tree, and H is the depth of the whole tree.

Updating formula of particle position:

$$
x_i(t) = v_i(t) + x_i(t-1)
\tag{14}
$$

From the particle velocity update formula (12) and the position update formula (14), it can be seen that the direction of motion of the particle is determined by the four parts: the original velocity v, the distance between the current position of the particle and its optimal position p − x, the distance between the current position of the particle and the optimal position of the population g − x, the distance between the particle's current position and the particle's optimal neighbor's position l − x. The weight of each part is set to represent the relative importance, respectively, by the weighting coefficients ω, c_1, c_2, c_3, and the formula (13) to control a linear change, when the depth of the particle closer to the root node, ω will be bigger, to make good particles can have a strong search ability, and poor search ability of the particles will be weakened.

The algorithm flow is as follows (Fig. 1):

Step 1: Initialize ω, c_1, c_2, c_3, the number of iterations, number of particles, etc., initialize the position and speed of the particle group.
Step 2: Calculate the fitness of each particle, record and select the global optimal fitness.
Step 3: Construct a binary tree according to the rule of the historical optimum of the parent node is better than the best adaptive history of the sub node.
Step 4: The particle swarm with the binary tree structure is traversed in breadth, and the optimal particles in the domain are calculated, update $\omega\ominus$ according to the formula (13), update speed and new particle location according to the formula (12) and formula (14).

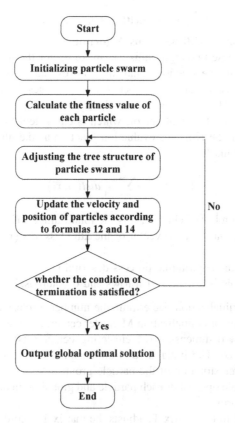

Fig. 1. The algorithm flow chart

Step 5: update the local optimum of the particles.

Step 6: After the completion of the traversal, the local optimum fitness of the root node particles is smaller than the current global optimum, and the global optimum is updated when the particle size is small.

Step 7: If the condition of termination is satisfied, the algorithm ends, or returns step 3.

The tree structure of the particle swarm optimization algorithm is due to the breadth traversal tree, so the assumption of global optimal particle at the far end of the tree leaves, depth of H, then the particles each iteration can only move up a depth, namely need H-1 after iteration to get to the root node, while the global optimal algorithm is updated by the root node of the tree, which is beneficial to slow the progress of the algorithm convergence, increase the global exploring ability of the particles. At the same time, by adding neighborhood operations, by changing the c_3 learning factor, the particle can move closer to the optimal particles in the neighborhood, which is beneficial to accelerate the convergence speed of the algorithm.

3.2   Improved Particle Swarm Algorithm Cluster Analysis

3.2.1   TPSO Combines with K-means Algorithm

The combination of these two algorithms can make up for the deficiency of K-means algorithm according to the previous analysis. The evaluating criterion of K-means algorithm is the degree of dispersion between the classes is minimized, a hybrid algorithm combining TPSO with K-means algorithm, each particle position is composed of m clustering center point vector, which is equivalent to a feasible clustering problem, and then use the K-means evaluation function as the fitness function, which is: the following formula (15).

$$J = \sum_{k=1}^{m} \sum_{i=1}^{n} dist(x_i, \overline{x_k}) \tag{15}$$

Where m is the number of clusters, n is the total number of data, x_i is the ith data point vector, $\overline{x_k}$ is the class-centric vector of the kth class, $dist(x_i, \overline{x_k})$ is the Euclidean distance formula.

The following is the algorithmic process description.

Suppose the sample is a matrix of n rows and d columns:

Step 1: Random initialization. Determine the number of clusters M, the number of particles N, and random initialization M cluster center. For the ith particle p(i) is the vector of the M ∗ d dimension (M clustering center sets), the velocity of each particle is initialized. The initial 1-to-N-dimensional matrix T is used to load the subscript of the tree structure of the particle group.

Step 2: Calculate the optimal of each particle and global optimal values based on the initial particle swarm.

Step 3: According to the matrix T, adjusts the matrix T according to the rule of the historical optimum of the parent node is better than the best adaptive history of the sub node.

Step 4: Update the speed and position of all particles according to formula (12) and formula (14).

Step 5: For each sample, calculate the distance from the cluster center, and classify according to the principle of proximity.

Step 6: Calculate the fitness according to the corresponding clustering, and update the individual optimal value of the particle by the Eq. (15).

Step 7: The global optimal value is updated with the root node particles of the particle tree structure after the overall particle fitness is updated.

Step 8: If the number of iterations is reached, the algorithm ends. Otherwise return Step 3.

3.2.2   TPSO Is Combined with FCM Algorithm

Similarly, a hybrid algorithm combining the improved particle swarm optimization and FCM algorithm, each particle is a set of clustering center, fitness function is the same to FCM value function, and we can obtain the smallest fitness through constant iterative algorithm. The algorithm process is roughly the same as that of the above and K-means, and the parameters are added with some parameters, the optimal solution change threshold ε and the number of iteration thresholds iter$_{stable}$, the following is the algorithm flow.

Suppose the sample is a matrix of n rows and d columns:

Step 1: Random initialization. Determine the clustering number M, the number of particles N, the random initialization M clustering center, for the ith particle p (i) is M * d vector clustering center, the initial velocity of each particle, the initial of a 1 to N for one dimensional matrix T particle swarm breadth traversal of the tree structure of subscript results.

Step 2: Calculate the optimal value of each particle and global optimal value according to the initial particle swarm.

Step 3: According to the matrix T, adjusts the matrix T according to the rule of the historical optimum of the parent node is better than the best adaptive history of the sub node.

Step 4: Update the speed and location of all particles according to formula (12) and formula (14);

Step 5: For each sample, calculate the fitness with FCM value function and update the individual optimal fitness of the particle.

Step 6: The global optimal value is updated with the root node particles of the particle tree structure after the overall particle fitness is updated.

Step 7: If the number of iterations is reached, the algorithm ends. Otherwise return Step 3.

4 Experiment and Result Analysis

4.1 Data Set

The experimental data are four data sets selected from UCI [16], which are Iris, Wine, Seed, and breast-w, respectively. The data set is briefly described as follows.

Iris: a total of 150 data, including four attributes, divided into three categories.
Wine: a total of 178 data, including 13 attributes, divided into three categories.
Seed: a total of 210 data, including seven attributes, divided into three categories, each type of 70 data.
Breast-w: a total of 699 data, including nine attributes, divided into two categories.

Table 1 presents a summary of the data sets used in the experiments. The experimental environment uses a computer MATLAB software programming environment.

Table 1. Data sets used in the experiments

Data set	Number of instances	Number of attributes	Number of categories
Iris	150	4	3
Wine	178	13	3
Seed	210	7	3
Breast-w	699	9	2

4.2 Traditional Clustering Algorithm Experiment

For K-means and FCM algorithm, the four experimental data sets were performed in 10 experiments, and the variance, accuracy and iteration numbers of the 10 results were calculated. The results were as follows.

Table 2. K-means and FCM clustering experiments

Data set	Algorithm	Variance	Accuracy	Number of iterations
Iris	K-means	37.45	86.99	7.6
	FCM	0	89.33	21.5
Wine	K-means	68.17	63.48	12.4
	FCM	0	68.54	49.9
Seed	K-means	0.05	89.33	10.2
	FCM	0	89.52	21
Breast-w	K-means	0	95.85	7
	FCM	0	95.27	13.2

From Table 2, we can see that in terms of accuracy, FCM is better than k-means in the correct classification. In terms of iteration number, the K-means algorithm converges faster, in terms of variance, FCM can get the same result every time because of the calculation formula of the algorithm, the variance is 0, but the k-means result fluctuates a lot. In fact, the K-mean is sensitive to the initialization data or abnormal data, and the initialization is not good, and the effect of clustering will be poor.

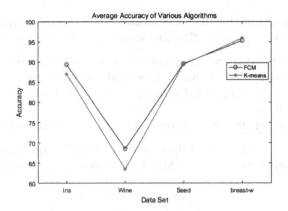

Fig. 2. Average accuracy of two algorithms

Figure 2 illustrates the average accuracy of the two algorithms. For Iris, Seed, and breast-w three data sets, the two algorithms can get better clustering results from the accuracy. For the Wine data set, the clustering effect of the two algorithms is not very good. The reason is that Iris, Seed, and breast-w three data sets have fewer properties, and the Wine data set has more properties and more complexities. This also reflects the poor results of traditional clustering algorithms for high-dimensional data clustering.

4.3 Clustering Experiment Based on Improved Particle Swarm Optimization Algorithm

The algorithm combining the above TPSO with k-means is TPSO_K-means, and the algorithm combined with FCM is called TPSO_FCM. Similarly, we conducted 10 experiments on each of the four data sets, averaging the variance, accuracy, and fitness of TPSO of the 10 results.

4.3.1 The Clustering Experiment of TPSO_K-means Algorithm

The parameters of the above modified PSO algorithm are set to $c_1 = c_2 = 2$, the number of iterations is 100, the number of particles is 10, $\omega_{max} = 0.9$, $\omega_{min} = 0.4$, and the experimental results are as follows:

Table 3. Clustering experiment of TPSO_K-means

Data set	Algorithm	Variance	Accuracy	Number of iterations
Iris	K-means	37.44	86.99	100.43
	PSO_K-means	4.41	91.16	115.11
	TPSO_K-means	4.23	92.13	113.36
Wine	K-means	68.17	63.48	17496.37
	PSO_K-means	0.55	71.73	16697.87
	TPSO_K-means	0.45	72.38	16566.53
Seed	K-means	0.05	89.33	313.45
	PSO_K-means	5.63	87.75	341.64
	TPSO_K-means	2.54	89.04	328.01
Breast-w	K-means	0	95.85	3051.34
	PSO_K-means	2.69	95.48	3367.42
	TPSO_K-means	0.58	96.39	3349.54

As can be seen from Table 3, the TPSO_K-means is better than K-means algorithm and PSO_K-means algorithm in terms of variance and classification accuracy. It can be seen that the improved algorithm is more stable, and the accuracy of clustering is higher. For the fitness value, TPSO_K-means is smaller than PSO_K-means in the fitness of Iris, Seed, and breast-w, but it is larger than that of K-means. In the fitness of the Wine dataset, TPSO_K-means smaller than PSO_K-means and K-means. Where the smaller the fitness, indicating the results of cluster clustering distances and smaller, that is clustered out of the results closer, which is K-means the core evaluation criteria. However, the data of the real life are not necessarily in full compliance with the evaluation criteria of the cluster, so there will be a high value of fitness, but the accuracy is high.

Figure 3 illustrates the average accuracy of the three algorithms, for Iris, Seed, and breast-w, they all belong to the data set with less attribute, and that is, the result of clustering with traditional clustering algorithm is better, so K-means does not easily fall into the local optimal problem. The particle swarm algorithm is clustered with the traditional clustering algorithm. The principle is that the iterations are used to find the feasible solution to find the smaller fitness. And the fitness is extracted from the traditional clustering algorithm. The result is related to the parameter setting of the particle swarm algorithm. Secondly, it is randomness. Therefore, the particle swarm algorithm combined with the traditional clustering algorithm is not necessarily better than K-means the fitness of clustering algorithm is small, but TPSO_K-means is smaller than PSO_K-means, and it can be concluded that the improved particle swarm optimization algorithm is better.

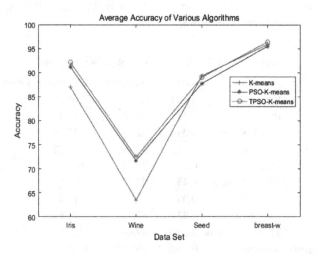

Fig. 3. The accuracy of the three algorithms clustering experiment

For the Wine dataset, it is a high-dimensional data set, from the results of finesse can see TPSO_K-means than PSO_K-means and K-means are small, and the accuracy rate is higher than they are high. The clustering results of the data set and K-means are poor, and the clustering results of the particle swarm algorithm combined with K-means are greatly improved. It can be concluded that the particle swarm optimization algorithm combined with K-means algorithm has some advantages in clustering of high-dimensional datasets. The improved particle swarm algorithm (TPSO_K-means) is smaller than PSO_K-means' fitness, which shows that the improved particle swarm algorithm has higher clustering effect.

The following shows the clustering results of the three algorithms.

As for Fig. 3, we can see that TPSO_K-means has higher clustering accuracy for each data set than the other two algorithms.

For Figs. 4, 5, 6 and 7, you can see TPSO_K-means than PSO_K-means slow convergence, and the fitness value after convergence, TPSO_K-means smaller than

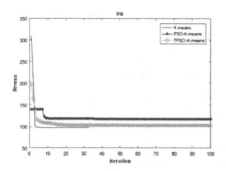

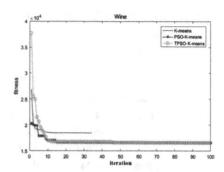

Fig. 4. Three algorithms for Iris clustering convergence graph

Fig. 5. Three algorithms for Wine clustering convergence graph

PSO_K-means, and K-means the fastest convergence speed, can see TPSO_K-means the PSO_K-means preliminary search ability have been improved, the diversity of particles is enhanced. On the other hand, for Fig. 5, the fitness values of TPSO_k-means and PSO_K-means are smaller than that of k-means, for Figs. 4, 6, and 7, TPSO_K -means is not smaller than k-means, but very close.

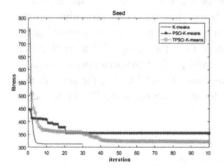

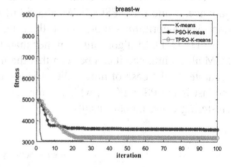

Fig. 6. Three algorithms for Seed clustering convergence graph

Fig. 7. Three algorithms for Breast-w clustering convergence graph

4.3.2 The Clustering Experiment of TPSO and FCM Algorithm

The parameters of TPSO combined with FCM algorithm are set to $c_1 = c_2 = 2$, the number of iterations is 100, the number of particles is 10, $\omega_{max} = 0.9$, $\omega_{min} = 0.4$, and the experimental results are as follows.

As can be seen from Table 4, FCM is the best and most stable in the aspect of variance, while the TPSO_FCM algorithm is better than the PCM algorithm, so the improved algorithm is more stable than the unimproved algorithm. In terms of accuracy, TPSO_FCM algorithm is higher than that of both the PSO_FCM algorithm and FCM algorithm, and it can be seen that the improved particle swarm algorithm is more accurate. In the case of fitness, FCM is the minimum, TPSO_FCM is the second, and the last is the PSO_FCM, which is the reason for the analysis of the TPSO_K-means clustering experimental results.

Table 4. Clustering experiment of TPSO_FCM

Data set	Algorithm	Variance	Accuracy	Number of iterations
Iris	FCM	0	89.33	60.57
	PSO_FCM	7.4	90.68	70.75
	TPSO_FCM	4.62	91.36	67.07
Wine	FCM	0	68.52	1796082.75
	PSO_FCM	0.88	71.85	1953947.32
	TPSO_FCM	0.77	72.24	1895330.21
Seeds	FCM	0	89.33	414.66
	PSO_FCM	0.81	89.27	446.08
	TPSO_FCM	0.48	89.53	439.27
Breast-w	FCM	0	95.27	15202.40
	PSO_FCM	0.04	97	18093.94
	TPSO_FCM	0.04	97.01	17543.16

As can be seen from Table 4, FCM is the best and most stable in the aspect of variance, while the TPSO_FCM algorithm is better than the PCM algorithm, so the improved algorithm is more stable than the unimproved algorithm. In terms of accuracy, TPSO_FCM algorithm is higher than that of both the PSO_FCM algorithm and FCM algorithm, and it can be seen that the improved particle swarm algorithm is more accurate. In the case of fitness, FCM is the minimum, TPSO_FCM is the second, and the last is the PSO_FCM, which is the reason for the analysis of the TPSO_K-means clustering experimental results.

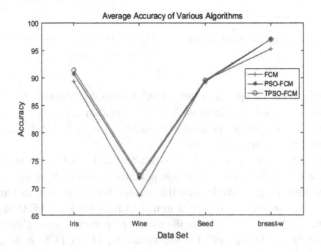

Fig. 8. The accuracy of the three algorithms clustering experiment

Here are three forms of graph representation of clustering results of algorithms. As for Fig. 8, it can be seen that TPSO_FCM is more accurate than that of PSO_FCM and FCM.

In Figs. 9, 10 and 11, we can see the convergence of FCM is fastest, but TPSO_FCM is slower than PSO_FCM convergence, and the convergence of the fitness value of TPSO_FCM is smaller than PSO_FCM, we can see TPSO_FCM preliminary search ability have been improved compared with PSO_FCM, the diversity of particles is enhanced. The fitness of FCM is the smallest, followed by TPSO_FCM, and finally, the PSO_FCM, although TPSO's fitness value is larger than FCM, it is very close.

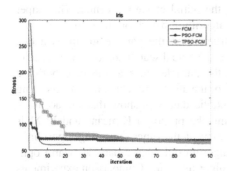

Fig. 9. Three algorithms for Iris clustering convergence graph

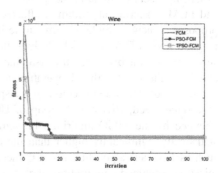

Fig. 10. Three algorithms for Wine clustering convergence graph

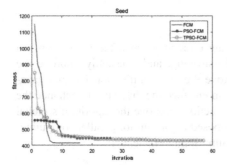

Fig. 11. Three algorithms for Seed clustering convergence graph

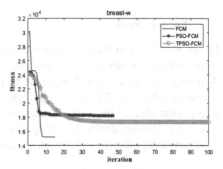

Fig. 12. Three algorithms for Breast-w clustering convergence graph

In Fig. 12, FCM or fastest convergence, TPSO_FCM slower than PSO_FCM convergence, and the fitness value after convergence, TPSO_FCM smaller than PSO_FCM, can see TPSO_FCM PSO_FCM preliminary search ability has been improved, the diversity of particles is enhanced. FCM has the smallest fitness, followed by TPSO_FCM, and finally, the PCM. TPSO the fitness value of PSO and much more, than FCM and comparing the accuracy in Table 3 can be found TPSO and PSO accuracy is higher than FCM, suggesting that the breast - w data set for the evaluation of FCM standard deviation is larger than other three.

5 Conclusion and Future Work

5.1 Conclusion

Aiming at the disadvantages of the basic particle swarm optimization algorithm is easy to fall into local extremum, the search accuracy is not high; the traditional K- means and FCM clustering algorithm is affected by the initial clustering center. This paper proposes a new particle swarm clustering algorithm based on tree structure and neighborhood (TPSO), which designs the structure of the particle group as a tree structure, uses the breadth of traversal, increases the global search ability of the particle, and joins the neighborhood operation to let the particle close to the neighborhood particles and accelerate the convergence speed of the algorithm. Our experiments using Iris, Wine, Seed, Breast-w 4 group of UCI public data sets show that the accuracy obtained by the TPSO algorithm implementing the proposed K-means and FCM is statistically significantly higher than the accuracy of the other clustering algorithms, such as K-means algorithm, fuzzy C-means algorithm, the basic particle swarm optimization combined with traditional clustering algorithm, etc., Comparison experiments also indicate that the TPSO algorithm can significantly improve the clustering performance of PSO. Through the research and experiment of this paper, we can see that the mathematical basis of particle swarm algorithm is relatively weak, such as lack of mathematical derivation analysis in parameter setting, etc.

5.2 Future Work

Future research can focus on several interesting directions. First, the particle swarm algorithm parameters set strict mathematical reasoning, and constantly improve the theoretical research; second, we want to improve the algorithm's global search capabilities step by step. Particle swarm algorithm to enhance the diversity, to enhance the information sharing between particles and conduction, improve the algorithm's global search capabilities; finally, the study of the proposed algorithm for parallel computing, to improve the efficiency of clustering algorithm mining efficiency.

Acknowledgements. This work was partially supported by the Guangdong Provincial Science and Technology Program (Grant Nos. 2017A020224004 and 2016A020212020), the National Natural Science Foundation of China (Grant Nos. 61573157 and 61703170), the Guangdong Provincial Natural Science Foundation Project (Grant Nos. 2015A030313413 and 2016A030313389). The authors also gratefully acknowledge the reviewers for their helpful comments and suggestions that helped to improve the presentation.

References

1. Han, J.W., Kamber, M., Pei, J.: Data Mining: Concepts and Techniques, 3rd edn. Elsevier, Singapore (2012)
2. Raftery, A.E., Dean, N.: Variable selection for model-based clustering. J. Am. Stat. Assoc. **101**(473), 168–178 (2017)
3. Budzik, J., Bradshaw, S., Fu, X.B., et al.: Clustering for opportunistic communication. In: Proceedings of WWW, pp. 726–735. ACM Press (2016)
4. Guha, S., Mishra, N.: Clustering data streams. In: Garofalakis, M., Gehrke, J., Rastogi, R. (eds.) Data Stream Management. DSA, pp. 169–187. Springer, Heidelberg (2016). https://doi.org/10.1007/978-3-540-28608-0_8
5. Wang, S., Wang, D., Caoyuan, L.I., Yan, L.I., Ding, G.: Clustering by fast search and find of density peaks with data field. Chin. J. Electron. **25**(3), 397–402 (2016)
6. Chi, Y., Song, X., Zhou, D., Hino, K., Tseng, B.L., et al.: Evolutionary spectral clustering by incorporating temporal smoothness. In: ACM SIGKDD International Conference on Knowledge Discovery and Data Mining, San Jose, California, USA, August, pp. 153–162. DBLP (2017)
7. Du, W.B., Ying, W., Yan, G., Zhu, Y.B., Cao, X.B.: Heterogeneous strategy particle swarm optimization. IEEE Trans. Circ. Syst. II Exp. Briefs **64**(4), 467–471 (2017)
8. Kennedy, J., Mendes, R.: Neighborhood topologies in fully-informed and best-of-neighborhood particle swarms. In: IEEE International Workshop on Soft Computing in Industrial Applications, vol. 36, pp. 45–50. IEEE (2003)
9. Bandyopadhyay, S., Maulik, U.: Genetic clustering for automatic evolution of clusters and application to image classification. Pattern Recogn. **35**(6), 1197–1208 (2002)
10. Engelbrecht, A.P.: Fundamentals of Computational Swarm Intelligence. Wiley, Hoboken (2009)
11. Kao, Y., Lee, S.Y.: Combining K-means and particle swarm optimization for dynamic data clustering problems. In: IEEE International Conference on Intelligent Computing and Intelligent Systems, pp. 757–761 (2009)
12. Vimali, J.S., Taj, Z.S.: FCM based CF: an efficient approach for consolidating big data applications. In: International Conference on Innovation Information in Computing Technologies, pp. 1–7. IEEE (2016)
13. Kalaiselvi, T., Sriramakrishnan, P., Nirmala, J.: An investigation on suitable distance measure to FCM method for brain tissue segmentation. In: Computational Methods, Communication Techniques and Informatics (2017)
14. Ashraf, E., Mahmood, K., Ahmed, T., Ahmed, S.: Value based PSO test case prioritization algorithm. Int. J. Adv. Comput. Sci. Appl. **8**(1), 389–394 (2017)
15. Jin, Y.X., Shi, Z.B.: An improved algorithm of quantum particle swarm optimization. J. Softw. **9**(11), 2789–2795 (2014)
16. Asuncion, A., Newman, D.J.: UCI machine learning repository. School of Information and Computer Science, University of California, Irvine (2007). http://www.ics.uci.edu/~mlearn/mlrepository.html

Optimization of UWB Antenna Based on Particle Swarm Optimization Algorithm

Mingyuan Yu[1]([✉]), Jing Liang[1], Boyang Qu[2], and Caitong Yue[1]

[1] School of Electrical Engineering, Zhengzhou University, Zhengzhou 450001, China
630032169@qq.com
[2] School of Electronic and Information Engineering,
Zhongyuan University of Technology, Zhengzhou 450007, China

Abstract. In the design and optimization process of ultra-wideband antenna, the fitness function is unknown and the antenna modeling process is complex. To solve these problems, this paper introduces a method to construct evaluation function. The method is based on MATLAB and HFSS joint simulation platform. The antenna modeling and simulation analysis process is packaged as a 'black box' which acts an evaluation function of the optimization algorithm. Real-time data exchange is carried out through the joint simulation platform. The particle swarm optimization algorithm (PSO) is used to optimize the antenna structure parameters automatically. Simulation results show that this method can replace the mathematical fitness function and simplify the antenna modeling process. In addition, the proposed method can reduce the antenna return loss effectively and improve the overall performance of the antenna.

Keywords: Ultra Wide Band (UWB)
Particle Swarm Optimization (PSO) · Evaluation function
Coaxial feed · Return loss

1 Introduction

The performance of the communication system is directly affected by the performance of the antenna. As the application antenna of the UWB system, it is necessary to have fast data transmission speed and wide application frequency band. However, the narrow operating frequency, high return loss and some other problems of current ultra-wideband application antenna restrict its rapid development. To solve these problems, the ultra wideband planar oscillator antenna optimization method is proposed. In [1], with the parametric modeling of the UWB antenna, the automatic optimization of each classification by using the optimization algorithm of HFSS. In [2], Finite-difference time-domain method (FDTD) and micro-genetic algorithm (MGA) are combined to optimize time-domain ultra-wide band (UWB) transverse electromagnetic (TEM) horn antenna

© Springer Nature Singapore Pte Ltd. 2018
K. Li et al. (Eds.): ISICA 2017, CCIS 874, pp. 86–97, 2018.
https://doi.org/10.1007/978-981-13-1651-7_7

array. In [3], the genetic algorithm (GA) and high frequency simulation software (HFSS) was established to optimize the design of the antenna. [4] using Particle Swarm Optimization (PSO) technique to optimize the antenna design, which its application in microwave frequency spectrum from 7.5 GHz to 20 GHz. [5] using Ansoft's High Frequency Structure Simulator (HFSS) and Computer Simulation Tool (CST) softwares to simulate and compare the different radiation characteristics of this array. In [6], the Finite Element Method along with the SFELP formulation was used to try to meet that criteria for the analysis tool. [7] proposed an improved particle swarm optimization (PSO) process to design a high-performance compact ultrawideband (UWB) filter. [8] proposed UWB monopole antenna is also design on the same substrate with DGS, which makes the proposed antenna is a good candidate for portable UWB applications. [9,12] used the particle swarm optimization algorithm to reduce the antenna return loss effectively. In [9], using a modified version of local-best particle swarm optimization (MLPSO) to optimize return loss and dispersion of a planar monopole antenna with microstrip feed line for ultrawideband (UWB). [12] proposed a method of impedance matching for frequency antenna based on chaotic particle swarm optimization algorithm. It realized the impedance matching of single frequency point.

This paper aims to propose a method to avoid the dependence of the optimization algorithm on the evaluation function with algebraic form. The method is based on MATLAB and HFSS joint simulation platform. The antenna modeling and simulation analysis process is packaged as a 'black box' which acts an evaluation function of the optimization algorithm. Real-time data exchange is carried out through the joint simulation platform. The particle swarm optimization algorithm (PSO) is used to optimize the antenna structure parameters automatically. Through the real-time interaction of the data, the particle swarm optimization (PSO) algorithm complete the automatic optimization of the antenna structure parameters.

In the second section, the UWB technology and the particle swarm optimization algorithm are briefly reviewed. The third section presents the antenna theory and the joint simulation platform of MATLAB and HFSS. The fourth part shows the result of this experiment and analyzes the result. Finally, discussions and conclusions are drawn in the fifth section.

2   Ultra-Wideband and Particle Swarm Algorithm

2.1   Ultra-Wideband Technology Features

For a bandwidth-limited signal, we define the highest frequency and the lowest frequency as f_H and f_L at the -10 dB bandwidth respectively. Then the relative bandwidth(B_f) of the signal is defined as follows:

$$B_f = \frac{f_H - f_L}{f_C} = \frac{2(f_H - f_L)}{f_H + f_L} \tag{1}$$

When a signal has an absolute bandwidth (BW1) more than 500 MHz or a relative bandwidth (η) more than 0.2, we define it as a super-bandwidth signal. Absolute bandwidth and relative bandwidth definition formulas are as follows:

$$BW_1 = f_H - f_L \geq \begin{cases} 1.5\,GHz & DARPA \\ 0.5\,GHz & FCC \end{cases} \tag{2}$$

$$\eta = 2\frac{f_H - f_L}{f_H + f_L} \times 100\% \geq \begin{cases} 25\% & DARPA \\ 20\% & FCC \end{cases} \tag{3}$$

In many communication systems, UWB has the significant characteristics that other systems do not have, as following:

- High transmission rate. The communication system can use 7.5 GHz of physical bandwidth, or even support the Gbps level of transmission signals. In addition, the system's space capacity is also large. Per square meter can be achieved 1 Mbps, which will be higher than Bluetooth, WLAN and other narrowband system space capacity.
- Strong ability of spectrum reuse. In theory, ultra-wideband systems can coexist with other wireless services and no noise interference.
- System complexity is low. Its power consumption and manufacturing costs are low. In addition, it is easy to integrate.

2.2   Particle Swarm Optimization Algorithm

Particle Swarm Optimization (PSO) is a swarm intelligence algorithm based on simulated birds foraging behavior on derived. After nearly two decades of continuous development, particle swarm algorithm has been widely used in the electrical, economic, biology and other fields [14,15]. The design of the antenna structure and the optimization of the relevant parameters belong to the standard multi-parameter optimization problem. PSO has wide adaptability in solving the problem of multi-parameter optimization. In PSO, the particle's positional parameters correspond to the antenna structure parameters so that it is not necessary to separate the parameters into binary numbers like genetic algorithms (GA). Therefore, PSO has obvious advantages in solving the optimization problem of UWB parameters.

In the particle swarm optimization algorithm, the concept of "invariance of rotation" is introduced [10,11]. For each particle in each time step, the center of gravity by the following three particles to determine the location: Current particle position $\overrightarrow{p}_i^t$, the closest position to the previous individual optimal particle $\overrightarrow{p}_i^t$, the nearest position of the previous individual optimal particle in the neighborhood $\overrightarrow{l}_i^t$. The formula is as follows:

$$\overrightarrow{p}_i^t = \overrightarrow{X}_i^t + C_1\overrightarrow{U}_i^t \otimes (\overrightarrow{P}_i^t - \overrightarrow{X}_i^t) \tag{4}$$

$$\overrightarrow{l}_i^t = \overrightarrow{X}_i^t + C_2\overrightarrow{U}_i^t \otimes (\overrightarrow{L}_i^t - \overrightarrow{X}_i^t) \tag{5}$$

$$\vec{G}_i^t = \frac{(\vec{X}_i^t + \vec{p}_i^t + \vec{l}_i^t)}{3} \tag{6}$$

x' is defined at any point on the hypersphere $H_i(\vec{G}_i^t, \| \vec{G}_i^t - \vec{X}_i^t \|)$. In the standard particle swarm algorithm, the particle position update formula does not change, the speed update formula is changed as follows:

$$\vec{V}_i^{t+1} = \omega \vec{V}_i^t + H_i(\vec{G}_i^t, \| \vec{G}_i^t - \vec{X}_i^t \|) - \vec{X}_i^t \tag{7}$$

In order to make the algorithm have better global search ability and strong local search ability in the later period, this paper adopts the strategy of linearly decreasing the inertia weight.

$$\omega = \omega_{\max} - (\omega_{\max} - \omega_{\min})\frac{g}{G} \tag{8}$$

where $i = 1, 2, ..., n$, g is the generation index representing the current number of evolutionary generations. G is defined as a maximal number of generations. The maximal and minimal weights $\omega_{\max}$ and $\omega_{\min}$ are usually set to 0.9 and 0.4 [15], respectively.

3   Model Design and Joint Simulation Platform

3.1   Antenna Theory and Model Design

The structure of the coaxial linear rectangular microstrip antenna is shown in Fig. 1. We can seen from the picture, the use of coaxial feeder can make the feed point more match. In addition, the coaxial connector is located under the patch and the presence of the ground plate so that the feeder radiation effect on the antenna is smaller.

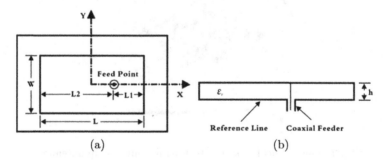

(a)   (b)

Fig. 1. The structure of coaxial feed rectangular microstrip antenna. (a) Top view. (b) Side view

The antenna structure shown in Fig. 1, the thickness of the dielectric substrate of the rectangular microstrip antenna is h, When the operating frequency is f and the speed of light is c, in order to get high efficient radiation patch, width w should meet (if the medium dielectric constant is ε_r):

$$w = \frac{c}{2f}\left(\frac{\varepsilon_r + 1}{2}\right)^{-\frac{1}{2}} \tag{9}$$

The effective permittivity ε_e and the equivalent radiating gap length ΔL are calculated as:

$$\varepsilon_e = \frac{\varepsilon_r + 1}{2} + \frac{\varepsilon_r - 1}{2}\left(1 + 12\frac{h}{w}\right)^{-\frac{1}{2}} \tag{10}$$

$$\Delta L = 0.412h\frac{(\varepsilon_e + 0.3)(^w\!/_h + 0.264)}{(\varepsilon_e - 0.258)(^w\!/_h + 0.8)} \tag{11}$$

Because there is 'Edgeto shorten effect', the actual radiation unit length is:

$$L = \frac{c}{f\sqrt{\varepsilon_e}} - 2\Delta L \tag{12}$$

If the 'edge shortening effect' is ignored, the length L is only related to the wavelength of guided wave λ_e. however, the λ_e is only related to the effective electric constant ε_e, which is

$$L = \frac{\lambda_e}{2} = \frac{c}{2f\sqrt{\varepsilon_e}} \tag{13}$$

The design model of the coaxial feed rectangular microstrip antenna based on the HFSS platform is shown in Fig. 2.

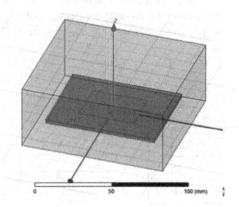

Fig. 2. The physical model of rectangular microstrip antenna.

3.2   Build HFSS and MATLAB Platform

Since we do not know the function that relationship between the geometric parameters of the antenna and the performance of the antenna, this means that the algebraic form and the approximate image of the evaluation function are also unknown. Therefore, the parameter selection is more difficult during the optimization design and isolation assessment. In order to solve this problem, this paper uses the whole modeling process of the antenna as an evaluation function (namely 'black box').

Using HFSS-API as an interface tool for MATLAB and HFSS so that realize real-time data interaction of two software in this paper. High Frequency Simulation Software (HFSS) is a full-wave and three-dimensional electromagnetic simulation software, it is developed by Ansoft Corporation. It can simulate and optimize the design of various types of antennas and accurately calculate the performance of the antenna. Although this paper use the VBScript recording function [1,8], the difference is that changing the VB file to an M file. It is the case through the M file to build the simulation model, not the VB file. The MATLAB simulation platform is used to automatically establish antenna model. The modeling function is stored in the .m file. Through assigning and modifying the variables in the file can be realized the automatically created any model. By MATLAB and HFSS data interaction to complete the entire simulation and optimization process. The algorithm can be summarised as follows:

Algorithm 1. The proposed algorithm

1: Step 1: Build model
2:    To optimize the antenna model
3: Step 2: Evaluation function= model (the 'black box')
4:    Use the HFSS-API toolbox
5: Step 3: Initialization (6)
6:    Randomly generate P in $[X_{\min}, X_{\max}]$
7:    Initialize particles' positions (X_i) and velocities (V_i)
8:    Initialize personal/previous best $\overrightarrow{P}$ and local best $\overrightarrow{G}$
9: **repeat**
10:    **for** $i = 1 : N$ **do**
11:       Update particle's velocity according to (7)
12:       Update particle's position using equation (6)
13:       **if** $f(\overrightarrow{X}_i) < f(\overrightarrow{P}_i)$ **then** minimization of f
14:          Update particle's best-known position $\overrightarrow{P}_i = \overrightarrow{X}_i$
15:       **elseif** $f(\overrightarrow{X}_i) > f(\overrightarrow{P}_i)$ **then** minimization of f
16:          Update the neighbourhoods best-known position $\overrightarrow{L}_i = \overrightarrow{P}_i$
17:       **end if**
18:    **end for**
19: **until** $G + 1 > Gmax$,or $FEs = FEs_{max}$

This 'black box' is used as the evaluation function of the optimize the algorithm. The antenna parameters to be optimized are used as input to this function. The return loss of the antenna is used as the fitness value for this function output. When the fitness value is calculated every time, the program will completely rebuild the antenna model once. Its advantage is that it can release the computer resources that were consumed before the simulation and reduce the resource consumption.

4   Experimental Results and Analysis

In experiment, the objective function of the optimization problem can be expressed as:

$$f(x) = \min_{1 \leq i \leq N} \{S_{11,i(x)}\} \tag{14}$$

In Eq. (14), N is the number of sampling points, $S_{11,i(x)}$ is the reflection coefficient (i.e return less) when the objective function take x at the ith frequency point.

Simulation experiments are carried out on MATLAB 2014a and HFSS 13.0 simulation platforms. In this paper, the variables of the antenna structure parameters are taken as the parameters to be optimized, and the return loss (S_{11}) of the antenna is taken as the objective to be optimized.

Optimization algorithm initialization parameters are: $N = 30$, $FEs = 1500$, $C_1 = C_2 = 2.05$. N is the population size. The FEs is the maximum number of evaluation. C_1 and C_2 is the acceleration factor. Using HFSS design center frequency is 2.45 GHz rectangular microstrip antenna and a fast sweep. The sweep range is set to 1.5 GHz–3.5 GHz. The dielectric substrate is FR4 epoxy board with thickness of 1.6 mm. Choosing 50Ω as the power feeding coaxial antenna feed mode. The initialization parameter values are calculated according to (9–13). Table 1 shows the range of parameters to be optimized. Table 2 shows the antenna initialization and optimization parameter values.

Table 1. Parameter range of values

Parameter	h	L	ω_0	L_1	R_1	R_2
Min	0.80 mm	26.00 mm	36.00 mm	4.00 mm	0.30 mm	0.80 mm
Max	2.40 mm	29.00 mm	38.00 mm	8.00 mm	0.80 mm	3.00 mm

As shown in Table 2, the effect is not obvious after optimization of the antenna in the size of the miniaturization. S_{11} represents the return loss which means is how much energy is reflected back to the source (Port1). The value of the return loss is to be minimized, generally recommended $S_{11} < 0.1$ i.e. -20 dB. Figure 3(a) shows the maximum gain after optimization is 5.9 dB and compared with the initial value the optimization effect is obvious. Figure 3(b) shows the optimized return loss of the antenna is reduced by -69 dB, and the antenna bandwidth about is 65 MHz at -10 dB. The results show the superiority of PSO algorithm on parameter optimization problem.

There are many parameters effect the antenna performance in the antenna optimization design experiment. However, the parameters shown in the Table 1 that impact on the antenna performance is greater. Therefore, this paper put

Table 2. Initialize and optimiza the design parameter

Unit	Structure name	Variable	Initial value	Optimization value
Media substrate	Thickness	h	1.60 mm	1.60 mm
Radiation patch	Length	L	28.00 mm	27.53 mm
	Width	ω_0	37.26 mm	37.07 mm
Feed point	Distance patch center	L_1	7.00 mm	6.61 mm
Coaxial feeder R	Inner radius	R_1	0.60 mm	0.58 mm
	Outer radius	R_2	1.50 mm	1.52 mm
Return less		S_{11}	−26.53 dB	−69.91 dB

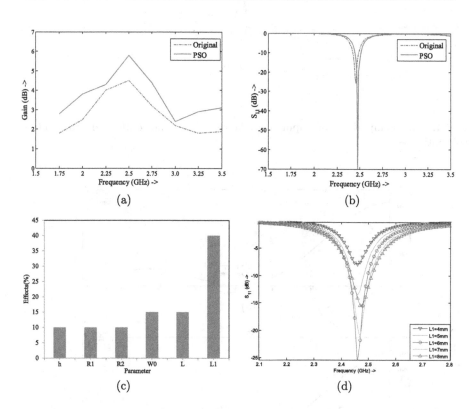

Fig. 3. The comparison of antenna performance index initial and optimized and sensitive parameter analysis. (a) Gain vs frequency. (b) Antenna return loss. (c) Effects for all levels of parameters. (d) L_1 effects return loss

them as this experiment to be optimized sensitive parameters. Based on sensitivity analysis [7], Fig. 3(c) shows the calculated results for effects of all the parameters. The ratio of parameter L_1 is the highest among these sensitivity parameters. In other words, the value of parameter L_1 has a greater effect on antenna performance. Figure 3(d) shows the effect of the most important

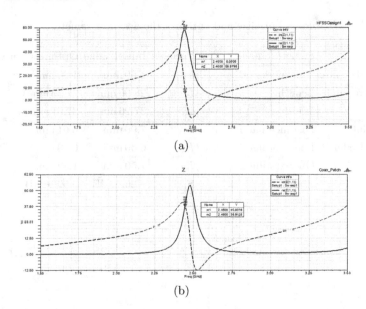

Fig. 4. The comparison of input impedance initial and optimized. (a) Initial. (b) Optimized

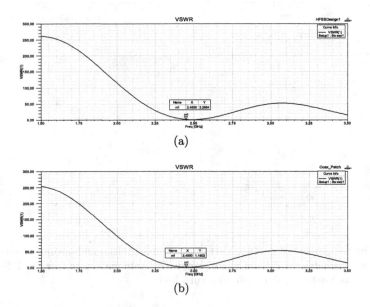

Fig. 5. The comparison of voltage standing wave ratio initial and optimized. (a) Initial. (b) Optimized

sensitivity parameter L_1 on the return loss. The return loss is minimal, and the impedance matching degree is best when L_1 is 6 mm.

The simulation results of the input impedance (Z) and the voltage standing wave ratio (VSWR) of the experimental antenna are shown in Figs. 4 and 5. As shown in Fig. 4, when the center frequency is 2.45 GHz, the calculated input impedance can be optimized closer to 50 Ω. As shown in Fig. 5, the VSWR before the optimization is 2.2684 and the optimized value is 1.1602, this means the optimized antenna feeder and the impedance match are better and the high frequency energy reflection loss is less. VSWR values reflect the efficiency of energy transmission to the transmitter antenna transmission system which closer to 1 that means the higher the conversion efficiency. As shown in Fig. 6, the impedances before and after the antenna optimization are $1.0279 - j0.0981\,\Omega$ and $1.0005 + j0.0010\,\Omega$ according to the Smith char. The analysis shows that the optimized impedance matching is better and the power transmission efficiency is higher.

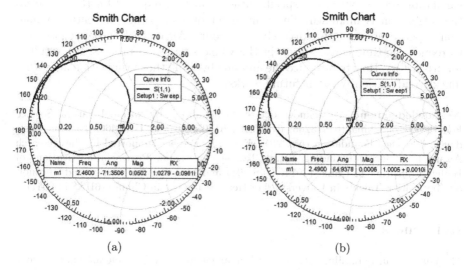

Fig. 6. The comparison of smith chart initial and optimized. (a) Initial. (b) Optimized

5 Conclusions and Future Work

In this paper, the 'black box' method is proposed to solve the difficulty of forming algebraic form of fitness function. Through this numerical experiment, it provides a new idea to solve the unknown problem of fitness function from now on. In the proposed method, HFSS interacts data with MATLAB directly by using HFSS-API as an interface tool. With regard to the joint simulation, it is completed on the two platforms automatically. The optimization program runs on the MATLAB platform and reads HFSS simulation real-time analysis data, which maximize the advantages of the two simulation software. This joint simulation process obviously improves simulation efficiency and saves the occupation of computer resources in terms of computing performance.

In the current study on the optimal design of the antenna, the algorithm used in this paper has some leading advantage in a certain degree. As shown in Table 2, this paper uses particle swarm optimization algorithm can greatly improve the accuracy of the solution. As a consequence, we consider whether the method can be used to solve other related issues. By using the PSO to optimize the parameters of the antenna, the overall performance of the antenna is improved. Compared with the previous optimization, it is worthy noting that the optimized antenna's return loss and voltage standing wave ratio is lower. In other words, the bandwidth of the optimized antenna is wider that it is attributed to the input impedance of the optimized antenna is more matched.

The particle swarm optimization algorithm is adopted to optimize the performance of the antenna, however, optimizing the antenna with the optimization algorithm has disadvantages. It is intensively time-consuming to optimize the simulation process when the specific fitness function required for the optimization algorithm is unknown. The time spent on completing the entire simulation experiment was approximately 47.5 hours. Although this paper also trys to reduce the time consumption in the whole simulation process, the result is unsatisfactory. How to optimize the antenna by using the optimization algorithm quickly and efficiently is our future works.

Acknowledgement. We acknowledge financial support by National Natural Science Foundation of China (61473266, 61673404, 61305080, and U1304602), China Postdoctoral Science Foundation (No. 2014M552013), Project supported by the Research Award Fund for Outstanding Young Teachers in Henan Provincial Institutions of Higher Education of China (2014GGJS-004) and Program for Science and Technology Innovation Talents in Universities of Henan Province in China (16HASTIT041).

References

1. Wu, Q., Su, D.L., Jin, R.H., et al.: Planar monopole and dipole antennas: theory and ultra-wide band applications, pp. 160–178. Beijing Institute of Technology Press (2013)
2. Qin, Y.M., Liao, C., Wei, T., et al.: Combination-optimization method for ultra-wide band TEM horn antenna array using micro-genetic algorithm. In: 2008 IEEE International Conference on Microwave and Millimeter Wave Technology, pp. 739–742 (2014)
3. Sun, S.Y., Lv, Y.H., Zhang, J.L., et al.: Design and optimization of UWB antenna based on genetic algorithm. Chin. J. Radio Sci. **1**, 62–66 (2011)
4. Bhattacharya, A., Roy, B., Islam, M., et al.: An UWB Monopole antenna with hexagonal patch structure designed using particle swarm optimization algorithm for wireless applications. In: 2016 IEEE International Conference on Microelectronics, Computing and Communications, pp. 1–5 (2016)

5. Slimani, A., Bennani, S.D., Alami, A.E., et al.: Conception and optimization of a bidirectional ultra wide band planar array antennas for C-band weather radar applications. In: 2016 IEEE International Conference on Information Technology for Organizations Development, pp. 1–7 (2016)
6. Martinez-Fernandez, J., Gil, J.M., Zapata, J.: Optimization of the profile of a planar ultra wide band monopole antenna in order to minimize return losses. In: 2007 IET European Conference on Antennas and Propagation, pp. 1–5 (2007)
7. Yun, Y.C., Oh, S.H., Lee, J.H., et al.: Optimal design of a compact filter for UWB applications using an improved particle swarm optimization. IEEE Trans. Magn. **52**, 1–4 (2016)
8. Sharma, M.K.: Design and optimization of ultra wide band monopole antenna with DGS for microwave imaging. In: 2016 IEEE Conference (2016)
9. Chamaani, S., Mirtaheri, S.A.: Planar UWB monopole antenna optimization to enhance time-domain characteristics using PSO. AEUE - Int. J. Electron. Commun. **64**, 351–359 (2010)
10. Mauricio, Z.B., Maurice, C., Rodrigo, R.: Standard particle swarm optimisation 2011 at CEC-2013: a baseline for future PSO improvements. In: 2013 IEEE Congress on Evolutionary Computation, pp. 2337–2344 (2013)
11. Maurice, C.: Standard Particle Swarm Optimisation. https://hal.archives-ouvertes.fr/hal-00764996
12. Liu, C.Q., Tan, Y.H., Xiong, Z.T., et al.: Adaptive impedance matching for radio frequency antenna based on dual-aimed chaotic particle swarm optimization algorithm. J. Comput. Appl. **34**, 3624–3627 (2012)
13. Xiao, J.M., Zhou, Q., Qu, B.Y., et al.: Multi-objective evolutionary algorithm and its application in electric power environment economic dispatch. J. Zhengzhou Univ. (Eng. Sci.) **37**, 01–09 (2016)
14. Zhang, C.J., Tan, K.C., Gao, L., et al.: Multi-objective evolutionary algorithm based on decomposition for engineering optimization. J. Zhengzhou Univ. (Eng. Sci.) **36**, 38–46 (2015)
15. Zhan, Z.Z., Zhang, J., Li, Y., et al.: Adaptive particle swarm optimization. IEEE Trans. Syst. Man Cybern. **39**, 1362–1381 (2009)

A Divisive Multi-level Differential Evolution

Huifang Zhang[1,2], Wei Huang[1,2(✉)], and Jinsong Wang[1,2]

[1] Tianjin Key Laboratory of Intelligent and Novel Software Technology,
Tianjin University of Technology, Tianjin 300384, China
[2] School of Computer and Communication Engineering,
Tianjin University of Technology, Tianjin 300384, China
huangwabc@163.com

Abstract. It is generally accepted that the clustering-based differential evolution (CDE) algorithm exhibits better performance in comparison with the standard differential evolution. However, such clustering method mechanism that is only based on input data may lead to some limitations such as premature convergence. In this study, we propose a divisive multi-level differential evolution algorithm (DMDE) to alleviate this drawback. The proposed divisive method is based not only input data but also the output fitness. In particular, DMDE becomes the conventional CDE when the output fitness in not considered in the process of clustering. Several benchmark functions are included to evaluate the performance of the proposed DMDE. Experimental results show that the proposed DMDE exhibits a promising performance when compared with CDE, especially in case of high-dimensional continuous optimization problems.

Keywords: DE · Divisive · Clustering · Parameter adjustment

1 Introduction

Generally, a global optimization problem can be formed as follows:

$$\min f(x)$$
$$s.t. \, x \in S \tag{1}$$

Where S is a feasible region, $S = \{x \in \Re^n : g_i(x) \leq 0, i = 1, 2, \ldots, p\}$, $f(x)$ is the objective function and $f(x) : \Re^n \to \Re$. The general form of global optimization problem contains other types of constraints. Theoretically, it is very difficult to determine that whether or not a local optimum is the same as a global optimum when solving the optimization problems.

Differential evolution (DE) initialized by Storn and Price [1] is a classical evolutionary algorithm (EA) for global optimization, which is a kind of heuristic random search algorithm based on group differences. It has some trail vector generation strategies and three important control parameters, i.e., population size, scaling factor, and crossover rate. But usually by trial and error to find the suitable to solve the current problems of strategy and control parameter require high computational costs [2]. Thus

© Springer Nature Singapore Pte Ltd. 2018
K. Li et al. (Eds.): ISICA 2017, CCIS 874, pp. 98–110, 2018.
https://doi.org/10.1007/978-981-13-1651-7_8

the algorithm CODE [2] proposed the random combination of various strategies and parameters to enhanced search capabilities of DE. Furthermore, the adaptive algorithm [3] proposed a parameter adaptive strategy, which makes the parameters need to be adapted to the optimization of the population to solve the problem of high computation costs. Recently CDE algorithm [4] use one-step k-means clustering acts as several multi-parent crossover operators to utilize the information of the population efficiently. But there also have a series of shortcomings to be solved. The clustering of the population belonging to the pure random process and the control parameters is set in advance, which can't change with the optimization process of population.

Here we proposed a new clustering based differential evolution, namely Guide Clustering-based differential evolution (DMDE). The population is first divided into two categories according to the fitness values of individual, so as to provide a guide to the second classification of the population, and to avoid the blind and random of the second clustering. According to the relationship between individuals of the two categories to analyze population state, then fine tune the control parameters, thus overcoming the shortcoming of control parameters can't be adaptive to changes. It makes the population in each generation to be able to match the current status of the control parameters. And, the clustering method of CDE algorithm is used to carry out the second classification to achieve the further optimization of the population.

The rest of this paper is organized as follows. Section 2 introduces the related study DE and CDE. Section 3 describes the detail descriptions of DMDE algorithm. In Sect. 4, the CDE algorithm is compared with DMDE algorithm through 20 benchmark functions. Experimental results are also discussed. At last, Sect. 5 mainly focuses on the summary and outlook.

2   Related Study

2.1   Differential Evolution

Differential Evolution (DE) [5] is a stochastic optimization algorithm based on swarm intelligence. It has the unique ability of memory, so that it can dynamically track the current search situation, to adjust its search strategy [6]. Therefore it has a strong global convergence and robustness. Like other evolutionary algorithms, DE consists of mutation, crossover and selection operators [7–9].

Mutation Operator: Assume that NP represents population size, N stands for the solution dimension, $P \subset R^N$ and the individual i can be formed as a vector of G in the space of a real solution in the following way $P_i^G = \left(P_{i1}^G, P_{i2}^G, P_{iN}^G, \ldots, \right)$, $\forall i \in \{1, 2, \ldots, N\}$.

DE mutation operator generates a corresponding temporary individual u_i^G. As shown in the variation formula (2). DE randomly selects three individuals from the contemporary population as the parent to generate the variant u_i^G.

$$u_{ij}^G = p_{r1,j}^G + F \times (p_{r2,j}^G - p_{r3,j}^G) \tag{2}$$

Where u_{ij}^G is the jth dimension of variable individual $u_i^G, p_{r1}^G, p_{r2}^G, p_{r3}^G$ are different individuals of generation G and $r_1 \neq r_2 \neq r_3$, F is a constant value, known as the "scaling factor", its range general $F \in [0, 1]$.

Crossover Operator: In the DE algorithm, new candidate solutions v_i^G are generated using cross operator, as shown in the reference [10]. Binomial crossover is a common way of DE, as shown in formula (3).

$$v_{ij}^G = \begin{cases} u_{ij}^G, & if (rand() \leq CR) \vee (j = Ir) \\ p_{ij}^G, & otherwise \end{cases} \tag{3}$$

Where v_{ij} is the jth dimension of candidate solutions, $rand()$ is a uniform random number between [0,1), CR is a constant number in [0,1], known as "crossover rate". Ir is a random integer between [1, N), which keep the only for the different dimensions of the same individual.

It follows from formula (3) that variation individual u_i^G at least one element is inherited to the new candidate solution. Hence, DE can maintain the diversity of the population [11].

Selection Operator: Greedy selection strategy is used to update the population of DE, and through the comparison of the newly generated trial individuals and the corresponding individuals of current generation to choose the better value of the individuals as the offspring into new species, as shown in formula (4).

$$p_i^{G+1} = \begin{cases} v_i^G, & if\ f(v_i^G) < f(p_i^G) \\ p_i^G, & otherwise \end{cases} \tag{4}$$

Where $f(x)$ is the objective function for the minimization problem.

2.2   CDE

Clustering is not given in designated classification case, according to the similarity of information packets of information, is a kind of unsupervised learning. There is no direct information, classified completely according to the distribution of data. K-means is one of the most widely used clustering algorithms based on partition [12].

And the k-means clustering is employed in the design of CDE. Unlike several iterations used in the classical K-means algorithm, the CDE algorithm only use a step clustering. CDE algorithm references the k-means algorithm, but did not inherit the k-means algorithm completely, in which there is no repeated iteration to cluster the population, so also in certain intensity reduced the computational time [4].

3 DMDE Algorithm

To further balance the convergence speed of DE and explore the optimal solution, we proposed a divisive multi-level differential evolution. The algorithm we use a division technique, the so-called division we adopt a top-down strategy, the beginning of the population of all individuals in the same cluster. Then we use the guide-clustering method we proposed to divide the population into two basic large clusters. On this basis, the population is subdivided into smaller and smaller clusters until each individual becomes a cluster or reaches the end condition of our population optimization.

In addition, the algorithm also references the k-means algorithm, which is a classic algorithm to solve the clustering problem. The algorithm is scalable and efficient for dealing with large data sets. Especially when clustering is intensive and the distinction between classes is significant.

Therefore, in order to make full use of the advantages of k-means algorithm, a multi-level differential evolution is used after the first division of the population, that is, according to the density of individuals in the two clusters with different update rules. If the difference between individuals in the cluster is large, the population is updated by the CDE, so that the clusters are gradually divided into smaller clusters. If not, update the individuals with classical differential evolution, so as to improve the search capability and scalability of the entire algorithm.

And last in order to further enhance the effect of the algorithm, we made some changes in parameters of the algorithm, thus to ensure our algorithm can efficiently and quickly find the optimal solution.

3.1 Division

Clustering is a statistical analysis method for the classification of the sample and the index [13, 14]. Clustering analysis is based on the taxonomy. In the ancient taxonomy, people mainly rely on experience and expertise to achieve classification [15]. With the development of human science and technology, the requirements for classification are higher and higher, so that sometimes it is difficult to classify by experience and expertise. So people gradually introduced the mathematical tool to the taxonomy, formed the numerical taxonomy, and then introduced the multivariate analysis technique to numerical taxonomy to form the cluster analysis [16].

However, clustering algorithm has certain blindness. In order to overcome the blindness and random of CDE, we proposed a new development for clustering algorithm to divide the population. It takes into account the fitness value of individuals to divide the population into two basic large clusters. On this basis, the population is subdivided into smaller and smaller clusters until each individual becomes a cluster or reaches the end condition of our population optimization. And it not only breaks the previous DE algorithm of imprisonment, and provides a significant guidance for the multi-level differential evolution algorithm.

The divisive method in DMDE is an excellent progress for clustering. The distribution of fitness values of individuals in the same cluster may be irregular because the traditional clustering only calculates the Euclidean metric between them. While DMDE classify the population in accordance with the fitness value of individuals aiming at the target of method (find the individual which has the best fitness value). Thereby the individuals who have the same characteristics of individual fitness value will be gathered in the same cluster. And ultimately makes the algorithm faster and more targeted search for the optimal solution nearby.

And here we take 10 points in space as a population sample to show the difference between them and the superiority of guide-clustering clearly in Fig. 1.

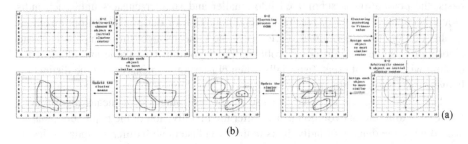

(b)

(a)

Fig. 1. Example of clustering (a) k-means clustering process and (b) k-means process with guide-clustering

Algorithm 1: Divisive Multi - level DE: DMDE	
1: Initialize the population P randomly	18: **if** case1 is greater than ((k1* k1)/5) or
2: Evaluate the fitness for each individual in **P**	case2 is greater than((k2* k2)/5)
3: generation counter g=1	**then**
4: **while** The halting criterion is not satisfied **do**	19: F=0.5+0.1*tempd
5: Use DE to update the population	CR=0.9-0.05*tempd
6: **if** g%m==0 **then**	20: **else**
7: Adopt the **guide clustering** to divide the popu-	21: F=0.5-0.1*tempd
lation into two groups (group 1 and group 2)	CR=0.9+0.05*tempd
(Parameter adjustment)	**(Mutil-level)**
8: **in group 1(the length is k1)**	22: **if** case1 is greater than case2
9: **if** the distance of fitness value between two	23: Use CDE to update the individuals in
individuals is greater than (2* k1/3) then	group 1(the length is k1) and use DE
10: case1++;	to update the individuals in group 2(the
11: Assignment d1 is the biggest distance of them	length is k2)
12: **in group 2(the length is k2)**	24: **else**
13: **if** the distance of fitness value between two	Use CDE to update the individuals in
individuals is greater than (2* k2/3) **then**	group 2(the length is k2) (see lines 7-
14: case2++;	10 in Algorithm 2). and use DE to up-
15: Assignment d2 is the biggest distance of them	date the individuals in group 1(the
and d is the biggest distance of population	length is k1)
16: Initialize tempd=0;	25: **end if**
17: **if** d1 is greater than d2, tempd=d1/d,	26: g=g+1
else tempd=d2/d	27: **end while**

Fig. 2. Pseudo-code of DMDE

3.2   Multi-level

The multi-level refers to according to the distribution of the individual to adopt a different update method in the process of population division. It can not only make full use of k-means algorithm in the data with significant differences to identify dense and sparse regions, and found the global distribution pattern. But also for some individuals continue to use differential evolution algorithm to dynamically track the current search situation. The pseudo code of the algorithm is shown in Fig. 2

Table 1. An instance of clustering

O	Original data			Guide one-step (k means)			One–step (k means)		
	y	x_1	x_2	y	x_1	x_2	y	x_1	x_2
1	2	0	2	5/3	1/3	4/3	2	0	2
2	0	0	0	0	0	0	0	0	0
3	1.5	1.5	0	1.5	1.5	0	1.5	1.5	0
4	5	5	0	5	5	0	5	5	0
5	7	5	2	7	5	2	7	5	2
6	3	1	2	3	1	2	3	1	2
7	4	1	3	4	1	3	4	1	3
8	3.5	2.5	1	2.5	2	1/2	3.5	2.5	1
9	4.5	1.5	3	4.5	1.5	3	4.5	1.5	3
10	6	3	3	29/6	11/6	3	5.2	3.4	1.8

First of all, the population is divided into two basic clusters by using the guide clustering method. The length of cluster1 and cluster2 are respectively k_1 and k_2. If $case_1$ is greater than $case_2$. It is proved that the difference between individuals in cluster1 is larger than cluster2, then k-means method can play better advantage to get a better result in cluster1. On the contrary, the conventional differential evolution algorithm is used to track the current search situation.

Besides, there is an example in Table 1 to further prove the optimization of our algorithm. The original data represents data that has not been manipulated. The function is $y = x_1 + x_2$. In order to facilitate the example, we set the value of k is 2.

Guide one-step k means

(1) Guide clustering
   (A) Choose $O_1(0, 2)$, $O_{10}(3, 3)$ as the centers of Guide clustering, and cluster the data according to the fitness values(y). $G_1 = O_1(0, 2)$, $G_2 = O_{10}(3, 3)$.
   (B) For each remaining object, assign it to the nearest cluster according to its distance from the center of each cluster. For example $O_2(0, 0)$
   $$d(G_1, o_2) = \sqrt{(2 - 0)^2} = 2, \quad d(G_2, o_2) = \sqrt{(6 - 0)^2} = 6,$$ where
   $d(G_1, o2) \leq d(G_2, o_2)$, so assign o_2 to cluster GC_1. Remaining data are clustered in turn respectively. Finally new centers are obtained $GC_1 = \{o_1, o_2, o_3, o_6, o_8\}$ and $GC_2 = \{o_4, o_5, o_7, o_9, o_{10}\}$.

(2) One-step k means clustering (k = 2)

    (A) Choose $O_1(0, 2)$, $O_8(2.5, 1)$ as the centers. $M_1 = o_1(0, 2)$, $M_2 = O_8(2.5, 1)$.

    (B) Assign each remaining object to the nearest cluster according to its distance from the center of GC_1. $o_2(0,0)$: $d(M_1, o_2) = \sqrt{(0-0)^2 + (2-0)^2} = 2$ $d(M_2, O_2) = \sqrt{(2.5-0)^2 + (1-0)^2} = \sqrt{7.25}$. Where $d(M_1, o_2) \leq d(M_2, o_2)$, Assign o_2 to cluster C_1. Finally $C_1 = \{o_1, o_2, o_6\}$, $C_2 = \{o_3, o_8\}$. New centers: $M_1' = (\frac{1}{3}, \frac{4}{3})$, $M_2' = ((1.5+2.5)/2, (0+1)/2) = (2, \frac{1}{2})$. The M_1' is better than M_1 and M_2' is better than M_2. Replace $M_1 = o_1(0, 2)$ with M_1'. Replace $M_2 = o_8(2.5, 1)$ with M_2'.

    (C) The same way in GC_2, Choose $O_4(5, 0)$, $o_{10}(3, 3)$ as the centers of clustering. $M_1 = v_1(0, 2)$, $M_2 = o_{10}(3, 3)$. New clusters: $C_1 = \{o_4, o_5\}$ and $C_2 = \{o_7, o_9, o_{10}\}$. New centers: $M_1' = ((5+5)/2, (0+2)/2) = (5, 1)$, $M_2' = (\frac{11}{6}, 3)$. Replace $M_2 = O_{10}(3, 3)$ with M_2'.

    (D) Update the dataset.

One-step k means

(1) Choose $O_1(0, 2)$, $O_{10}(3, 3)$ as centers of clustering. $M_1 = O_1(0, 2)$, $M_2 = O_{10}(3, 3)$.

(2) Assign remaining objects to the nearest cluster. $C_1 = \{o_1, o_2, o_3, o_6, o_7\}$, $C_2 = \{o_4, o_5, o_8, o_9, o_{10}\}$. New centers $M_1' = ((1.5+1+1)/5, (2+2+3)/5) = (0.7, 1.4)$, $M_2' = (3.4, 1.8)$. Replace $M_2 = O_{10}(3,3)$ with M_2'.

(3) Update the dataset.

3.3 Parameter Tuning

Because of two control parameters of DE algorithm: scaling factor (F) and crossover factor (CR) have different functions. To generate a vector mutation with a higher value of F, this makes the difference with the basis vectors larger, thus helping to maintain the diversity of the population [18, 19]. In contrast, a small F is more helpful to the convergence of the population. A large value of CR will lead to the new generation vector converge to locally optimal vector, and a small value of CR makes the whole population is more diverse [20, 21].

Based on the above considerations, the strategies for adjusting the population parameter values F and CR can be defined as follows. The population is divided into two categories according to the Guide-Clustering. If the distance of fitness value between individuals in each category is far, but they are in the same category, individuals of the population in the situation as shown in Fig. 3(a). It can be seen that the population should increase the search space to increase population diversity, reducing the speed of convergence to further optimize population, namely, in this case, should be appropriate to increase scaling factor F, reduce the crossover rate CR. On the other hand, in the same class while fitness values are relatively close to the individual, as shown in Fig. 3(b). We should accelerate speed of convergence, which is increasing the crossover factor CR, reducing scaling factor F to make the solution of population closer to the optimum [22], so that the population to achieve a better state of optimization.

Table 2. Best error values of DE, CDE and DMDE on 20 test functions (f1–f20) [17]

F	D	Max_NFFEs	DE		DMDE		CDE		CDE-DMDE
			Mean	Std Dev	Mean	Std Dev	Mean	Std Dev	T-test
f01	30	150000	2.01E−17	1.14E−17	**1.88E−38**	**1.47E−38**	1.07E−28	7.65E−29	9.79
f02	30	200000	3.86E−14	9.28E−15	**1.58E−28**	**7.21E−29**	4.21E−21	1.85E−21	15.9
f03	30	500000	5.04E−11	2.46E−11	**9.82E−37**	**2.82E−36**	1.64E−34	9.18E−34	1.20
f04	30	500000	8.81E−08	2.39E−08	1.69E+00	1.25E+00	6.48E−22	1.18E−21	−9.01
f05	30	500000	5.15E−22	1.21E−21	2.43E+01	1.81E+00	0	0	−10.8
f06	30	150000	0	0	0	0	0	0	0
f07	30	300000	0.0078	0.0017	**7.15E−04**	**3.48E−04**	0.0013	7.37E−04	5.15
f08	30	300000	0	0	0	0	0	0	12.1
f09	30	300000	0	0	0	0	0	0	0
f10	30	150000	1.21E−09	3.14E−10	**4.57E−15**	**1.30E−15**	5.28E−15	1.67E−15	2.41
f11	30	200000	0	0	0	0	0	0	0
f12	30	150000	1.46E−18	7.33E−19	3.02E−17	1.85E−32	1.79E−30	1.50E−30	−14.1
f13	30	150000	1.59E−16	6.79E−17	**2.88E−17**	**2.47E−32**	9.42E−29	8.40E−29	24.0
f14	2	10000	0	0	0	0	0	0	0
f15	4	40000	1.85E−19	4.00E−19	1.31E−03	2.60E−18	1.03E−19	3.12E−19	−1.18
f16	2	10000	1.28E−14	4.71E−14	**0**	**0**	7.99E−16	4.44E−15	16.2
f17	2	10000	1.74E−11	6.58E−11	**0**	**0**	4.33E−13	1.46E−12	9.1
f18	2	10000	7.08E−15	1.43E−14	2.48E+05	3.55E+04	4.39E−15	4.93E−15	−8.9
f19	3	10000	0	0	0	0	0	0	0
f20	6	20000	2.92E−12	2.04E−11	1.90E−01	1.78E−15	1.40E−14	7.04E−14	−4.9

Here the update of population algorithm is similar to those in G3 model [23, 24]. The update method of individuals in the population according to the updated manner of CDE algorithm [4] because of the use of the K-means algorithm. In addition, in order to the effective use of search space, here used the idea of clustering period, which is similar to [25]. The use of clustering period for the reason of the population also need some time to detect the search space from clustering. A large clustering periodic value allows the algorithm to achieve a more stable clustering effect. However the smaller value of clustering period will lead to the clustering errors of judgment. The specific impact of clustering period on the algorithm will introduce later.

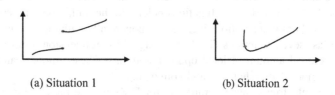

(a) Situation 1   (b) Situation 2

Fig. 3. Most of the individuals in a population.

Table 3. Comparison of CDE and DMDE for different values of K

F	D	CDE	DMDE	CDE	DMDE
		K=4	K=4(G=2,C=2)	K=8	K=8 (G=2,C=4)
f01	30	3.73E-30±2.51E-30	**4.02E-36⊕3.29E-36**	5.59E-31±2.40E-31	**1.03E-37⊕1.08E-37**
f02	30	3.12E-22±1.17E-22	**5.14E-27⊕1.38E-27**	7.22E-23±2.95E-23	**3.32E-28⊕1.90E-28**
f03	30	5.49E-32⊕1.34E-32	**5.24E-42⊕1.42E-41**	4.40E-33⊕5.65E-33	**4.88E-40⊕6.78E-40**
f04	30	4.08E-20⊕1.28E-19	8.04E-01⊕3.05E-01	4.74E-20±1.40E-19	1.36E+00⊕5.56E-01
f05	30	4.60E+00⊕7.68E+03	2.29E+01⊕2.72E+00	4.65E+05±1.4E+04	2.35E+01⊕1.68E+00
f06	30	1.76E+00⊕8.47E+02	**0.00E⊕00**	1.74E+04±1.33E+03	**0.00E⊕00**
f07	30	1.05E-03⊕3.18E-04	**6.43E-04⊕3.44E-04**	8.56E-04±4.04E-04	**7.26E-04⊕3.54E-04**
f08	30	1.20E+00⊕7.34E+03	**0.00E⊕00**	1.18E+05±3.66E+03	**0.00E⊕00**
f09	30	2.13E+00⊕4.36E+03	**0.00E⊕00**	2.07E+05±5.62E+03	**0.00E⊕00**
f10	30	4.14E-1⊕0.00E+00	**4.35E-15⊕1.07E-15**	4.14E-15±0.00E+00	4.64E-15⊕1.36E-15
f11	30	9.04E+00⊕1.32E+03	**0.00E⊕00**	8.99E+04±1.99E+03	**0.00E⊕00**
f12	30	8.26E-32⊕5.98E-32	3.02E-17⊕1.85E-32	3.07E-32±2.17E-32	3.02E-17⊕1.85E-32
f13	30	9.89E-30⊕1.33E-29	2.88E-17⊕2.47E-32	1.29E-30±8.18E-31	2.88E-17⊕2.47E-32
f14	2	8.61E+00⊕6.08E+02	**7.60E-22⊕2.28E-21**	8.77E+03±4.51E+02	**1.96E-05⊕5.88E-05**
f15	4	2.00E-19⊕4.24E-19	**1.38E-03⊕2.60E-18**	2.06E-19±4.34E-19	**1.38E-03⊕2.60E-18**
f16	2	6.02E-15⊕1.58E-14	**0.00E⊕00**	5.00E-15±9.71E-15	**0.00E⊕00**
f17	2	8.87E-12⊕2.71E-11	**0.00E⊕00**	3.12E-10±9.82E-10	**0.00E⊕00**
f18	2	3.91E-15⊕5.05E-15	2.43E+05⊕4.96E+04	7.82E-15±4.12E-15	2.53E+05⊕4.14E-02
f19	3	7.74E+00⊕1.05E+02	**0.00E⊕00**	7.70E+03±3.15E+02	**0.00E⊕00**
f20	6	1.73E+04⊕1.07E+03	**1.90E-01⊕1.78E-15**	1.77E+04±9.02E+02	**1.90E-01⊕1.78E-15**

4   Experimental Results

4.1   Benchmark Functions and Experimental Setup

In this section, we will test the performance of our algorithm with 20 benchmark functions [26, 27]. These functions can be divided into three categories [4]. The former 6 functions are unmoral functions. These functions can be used to test the convergence speed of algorithm. The next seven functions belong to the multimodal functions [28], the number of local optimal solutions increases exponentially with the dimension of the problem. The last seven functions f14–f20 belong to low dimension multimodal [29] function with a small amount of local optimal solutions. These unique features make our experiment more comprehensive and convincing.

There are mainly four control parameters for DMDE. The three belonging to the original DE algorithm are the population size of NP, the scaling factor F, and the crossover rate CR and a parameter is clustering period, but our algorithm DMDE is to fine tune the parameters F and CR. The control parameters that we use outside of the cluster period: $NP = 100$. $F = 0.5$. $CR = 0.9$. The value of Clustering period: $m = 10$. The scheme of DE: DE/rand/1/exp.

And in the clustering period, in addition to the scaling factor F and the crossover rate CR need to fine tune according to the results of classification and other parameters were held constant in all experiments, unless other explanations of the changes.

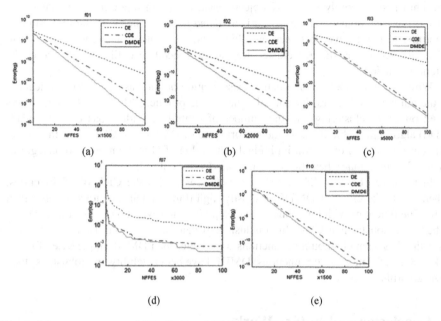

(a)   (b)   (c)

(d)   (e)

Fig. 4. Mean error curves of DE, CDE and DMDE. (a) f01. (b) f02. (c) f03. (d) f07. (e) f10.

4.2 Comparison Between CDE and DMDE

Here briefly introduce the performance evaluation error to the performance of algorithm. Error of a solution X is defined as $f(X) - f(X^*)$, where is the local optimal solution. In 50 times run separately when it reaches the value NFFES maximum then the minimum error value will be recorded. The corresponding error average value and standard difference also be calculated [13].

In this part, we compared DMDE and CDE to demonstrate the superiority of the DMDE algorithm. In the experiment, all functions are run independently 50 times. The best *error* results of all functions with DE, CDE and DMDE algorithm in all experiments as show in Table 2. And Fig. 4 has shown some representative graph of convergence with these algorithms.

It can be seen from the Table 2 that the DMDE algorithm is superior to the CDE algorithm in 8 functions. There are 6 functions DMDE and CDE algorithm performs quite well. CDE and DMDE algorithm can obtain the global optimal solution in the 50 independent runs. And from Fig. 4 we can see the convergence rate of DMDE to the optimal solution is faster than CDE and DE algorithm. In conclusion, CDE and DMDE all have ability to solve these problems, either for high-dimensional or low dimensional functions. And the DMDE algorithm we proposed performs better than the CDE algorithm slightly.

4.3 Influence of Cluster Centers

DMDE can be regarded as an extended CDE. The K-means algorithm used in CDE, which is an indirect clustering method based on the similarity of the samples. This algorithm first randomly selects K objects, each of which represents the center of a cluster. For each object and the rest, it is assigned to the cluster with the most similar, according to the distance between the object and the centers of a cluster. Then, calculate the new center of each cluster. Repeat the procedure until convergence of the criterion function.

While we have made an alternative classification on the basis of this classification, so the number k of central individual equal to the product of first number of clusters based on Guide-clustering and the number of central individual based on K-means (CDE) algorithm. Table 3 shows the performance of DMDE and CDE algorithm under different number (k) of central individual. The value of G is the number of categories for the first time that when the population is divided into clusters.

In the case of k = 4, DMDE is better obviously than the CDE of 15 functions. When k = 8, the effect of DMDE in mostly high dimensional functions is better than CDE. The functions with low dimension except f18 and several other functions DMDE is better obviously than CDE. In conclusion, It can be seen from Table 3 that CDE algorithm has a small amount of functions were better than DMDE in the case of k = 4 or k = 8. But for most of the functions, DMDE have better stability and robustness than CDE algorithm.

5 Conclusion and Future Work

In this paper, we proposed a DMDE that makes corresponding adjustment on population control parameters according to clustering results. The characteristic of DMDE algorithm is to get the fitness value information of the individual to cluster the population firstly. The state of the population is analyzed through the result of cluster analysis. And the population is classified secondly with the idea of CDE algorithm after adjusting the parameters.

And we test the algorithm based on series of benchmark functions to evaluate the performance. Experimental results show that the proposed DMDE algorithm have appropriate search capability both in multimodal and unmoral problems.

For future work, we will add more comprehensive adaptive parameter strategy and more suitable clustering strategy to enhance the algorithm ability to solve the problem, and extend the algorithm DMDE to solve other more complex problems such as dynamic and multi-objective optimization problem.

Acknowledgments. This work was supported by the National Natural Science Foundation of China (Grant no. 61673295), and supported by the Tianjin Science and Technology Major Project (Grant no.15ZXZNCX00050).

References

1. Liang, J.J., Qin, A.K., Suganthan, P.N., Baskar, S.: Comprehensive learning particle swarm optimizer for global optimization of multimodal functions. IEEE Trans. Evol. Comput. **10** (3), 281–295 (2006)
2. Wang, Y., Cai, Z., Zhang, Q.: Differential evolution with composite trial vector generation strategies and control parameters. IEEE Trans. Evol. Comput. **15**(1), 55–66 (2011)
3. Yu, W.J., et al.: Differential evolution with two-level parameter adaptation. IEEE Trans. Cybern. **44**(7), 1080–1099 (2014)
4. Cai, Z., Gong, W., Ling, C.X., Zhang, H.: A clustering-based differential evolution for global optimization. Appl. Soft Comput. **11**(1), 1363–1379 (2011)
5. Storn, R., Price, K.: Differential evolution-a simple and efficient heuristic for global optimization over continuous spaces. J. Glob. Optim. **11**(4), 341–359 (1997)
6. Srinivas, M., Patnaik, L.: Adaptive probabilities of crossover and mutation in genetic algorithms. IEEE Trans. Syst. Man Cybern. **24**(4), 656–667 (1994)
7. Mallipeddi, R., Suganthan, P.N.: Differential evolution algorithm with ensemble of parameters and mutation and crossover strategies. In: Panigrahi, B.K., Das, S., Suganthan, P.N., Dash, S.S. (eds.) SEMCCO 2010. LNCS, vol. 6466, pp. 71–78. Springer, Heidelberg (2010). https://doi.org/10.1007/978-3-642-17563-3_9
8. Eiben, A.E., Hinterding, R., Michalewicz, Z.: Parameter control in evolutionary algorithms. IEEE Trans. Evol. Comput. **3**(2), 124–141 (1999)
9. Price, K.V.: An introduction to differential evolution. In: New Ideas Optimization, pp. 293–298. McGraw-Hill, London (1999)
10. Gamperle, R., Muller, S.D., Koumoutsakos, P.: A parameter study for differential evolution. In: Proceedings of Advances in Intelligent Systems, Fuzzy Systems, Evolutionary Computation, Crete, Greece, pp. 293–298 (2002)
11. Saidi, K., Allad, M.: Fuzzy controller parameters optimization by using genetic algorithm for the control of inverted pendulum. In: International Conference on Control, Engineering & Information Technology, pp. 1–6. IEEE (2015)
12. Brest, J., Greiner, S., Boskovic, B., Mernik, M., Zumer, V.: Self adapting control parameters in differential evolution: a comparative study on numerical benchmark problems. IEEE Trans. Evol. Comput **10**(6), 646–657 (2006)
13. Noman, N., Iba, H.: Accelerating differential evolution using an adaptive local search. IEEE Trans. Evol. Comput. **12**(1), 107–125 (2008)
14. Jain, A., Murty, M., Flynn, P.: Data clustering: a review. ACM Comput. Surv. **31**(3), 264–323 (1999)
15. Zhang, J., Chung, H.S., Lo, W.L.: Clustering-based adaptive crossover and mutation probabilities for genetic algorithms. IEEE Trans. Evol. Comput. **11**(3), 326–335 (2007)
16. Wang, Y., Zhang, J., Zhang, C.: A dynamic clustering based differential evolution algorithm for global optimization. Eur. J. Oper. Res. **183**(1), 56–73 (2007)
17. Xue, L.I., Cui, D.W., Hua, J., et al.: Research on optimization of control parameters for genetic algorithm based on fitness landscape. J. Xian Univ. Technol. (2010)
18. Basak, A., Das, S., Tan, K.C.: Multimodal optimization using a biobjective differential evolution algorithm enhanced with mean distance-based selection. IEEE Trans. Evol. Comput. **17**(5), 666–685 (2013)
19. Zhang, J., Sanderson, A.C.: Adaptive differential evolution with optional external archive. IEEE Trans. Evol. Comput. **3**(5), 948–952 (2009)
20. Wang, Y., Cai, Z.X.: Combining multi objective optimization with differential evolution to solve constrained optimization problems. IEEE Trans. Evol. Comput. **16**(1), 117–134 (2012)

21. Zaharie, D.: Control of population diversity and adaptation in differential evolution algorithms. In: Matousek, R., Osmera, P. (eds.) Proceedings of Mendel 9th International Conference on Soft Computing, Brno, Czech Republic, pp. 41–46 (2003)
22. Damavandi, N., Safavi-Naeini, S.: A hybrid evolutionary programming method for circuit optimization. IEEE Trans. Circuits Syst.-I $52(5)$, 902–910 (2005)
23. Suganthan, P.N., et al.: Problem definition and evaluation criteria for the CEC 2005 special session on real-parameter optimization. Nanyang Technology University, Singapore, IIT Kanpur, Kanpur, India, Technical report, KanGAL#2005005, pp. 341–357 (2005)
24. Olorunda, O., Engelbrecht, A.P.: Differential evolution in high dimensional search spaces. In: Proceedings of 2007 IEEE Congress on Evolutionary Computation, Singapore, pp. 1934–1941 (2007)
25. Qin, A.K., Huang, V.L., Suganthan, P.N.: Differential evolution algorithm with strategy adaptation for global numerical optimization. IEEE Trans. Evol. Comput. $13(2)$, 398–417 (2009)
26. Storn, R., Price, K.: Home page of differential evolution. http://www.ICSI.Berkeley.edu/~storn/code.html
27. Noman, N., Iba, H.: Accelerating differential evolution using an adaptive local search. IEEE Trans. Evol. Comput. $12(1)$, 107–125 (2008)
28. Liu, J., Lampinen, J.: A fuzzy adaptive differential evolution algorithm. Soft Comput. $9(6)$, 448–462 (2005)
29. Ping, J., Peiguang, W.: Parameters optimization of active disturbance rejection controller with genetic algorithm for cascade speed control system. In: Fourth International Conference on Intelligent Computation Technology and Automation, vol. 1, pp. 464–467. IEEE Computer Society (2011)

Complex Systems Modeling – System Dynamic

A Comparative Summary of the Latest Version of MapReduce Parallel and Old Version from the Perspective of Framework

Xinze Li[✉] and Qi Liu

Yunnan Open University, Kunming 650223, Yunnan, China
xinzeli2018@126.com

Abstract. After MapReduce Parallel refactoring, there occurred an enormous difference between the latest version of MapReduce parallel and the old version on framework. At first, this paper introduces the framework of the old version of MapReduce parallel, the procedure of implementing tasks and how to schedule tasks and resources allocations, etc. Also, it points out the limitation of the old parallel. Then, it explained a framework of the new version of MapReduce parallel, task scheduling and resource allocation and so on. From the perspective of the framework, task scheduling, and resource allocation, it compared these two different versions, putting forward the advantages of the renewed framework. New-generation of MapReduce has a framework YARN, which is sharing model, in other words, it's an application program that enables various compiling of computing structures to run on the same cluster. Meanwhile, it makes operation and maintenance more smooth and makes full use of cluster resource.

Keywords: MapReduce · YARN · Task scheduling · Resource allocation

1 Introduction

In August 2006, Google first put forward the concept of "Cloud Computing" at the search engine conference. After several years of rapid development, the industry has practiced a number of cloud computing examples, such as Amazon's EC2, Microsoft's Azure, IBM's "Blue Cloud", Alibaba's "Flying Apsaras". Google's cloud computing platform is most famous. From a technical perspective, Google's cloud computing infrastructure consists of 3 separate and tightly integrated systems: Distribute File System GFS; Parallel Programming Model and Task Scheduling Model MapReduce; Large-scale Distribute Database Bigtable capable of handling structured and semi-structured data [1].

Apache Hadoop is an open-source cloud computing platform developed based on batch processing technology, written in the Java language, running on Linux operating systems, composed of two core components including HDFS (Hadoop Distributed File System) and Hadoop MapReduce, which are open-source implementations of GFS and GMR (Google MapReduce) [2]. In just a few years, Hadoop has grown into a de facto standard for big data and has developed its own complete ecosystem. Figure 1 shows

© Springer Nature Singapore Pte Ltd. 2018
K. Li et al. (Eds.): ISICA 2017, CCIS 874, pp. 113–124, 2018.
https://doi.org/10.1007/978-981-13-1651-7_9

the most important product in the Hadoop system, but fundamentally, the core of Hadoop is still MapReduce.

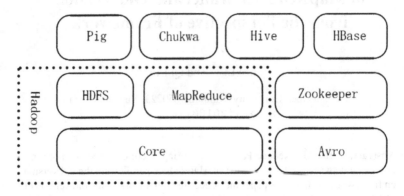

Fig. 1. The main products of Hadoop ecosystem

According to the source of data, big data can be roughly divided into two categories: one from the physical world and the other from human society. The former is mostly scientific experiment data or sensing data, while the latter is related to human activities, especially the Internet [3]. According to statistics by Digital Universe project, the global new data volume was 1.2 ZB and 1.8 ZB in 2010 and 2011 respectively, but the number soared up to 2.8 ZB in 2012. As a powerful tool to deal with BigData, distribute systems and parallel computing technology are becoming increasingly important.

MapReduce [4] was introduced by Google in 2004, which is a programming model which can process big concurrent data set (over 1 TB). Its name originates from two core operations in the functional programming model: Map and Reduce. Sure enough, both of them are based on the same concept. Map is a process of separating data, while Reduce combines the separated data. Its characteristics are that it's simple and easy to learn and also used widely because it lowers the level of difficulty in concurrent programming so that programmers get rid of the concurrent programming, writing concurrent programs more effectively and with ease [5].

With the development of big data in recent years, the larger the data size is, the greater the value of data excavation. It's a better way to utilize MapReduce to complete the work. Hadoop MapReduce is an open source edition of MapReduce developed by Google. As a programming model, MapReduce provides a high-level programming language, while it avoids low-level implementation details, lowering the level of programming difficulty so as to improves programming efficiency. Though it is deployed on a cluster of cheap machines, it can also obtain excellent performance. However, as a compute engine, it still has many disadvantages, such as single point of failure. In addition, because it has its own functions of resource allocation and task scheduling, even with High Availability of MapReduce, it's extremely difficult for failure recovery as the master node is connected with a large number of task information and resource information. Therefore, a new tool for task scheduling is needed to complete scheduling tasks and resource allocation for MapReduce.

Precisely because of commonalities between functional expressions and vector programming languages of MapReduce, this programming pattern is especially suitable for massive data search, mining, and analysis, as well as intelligent learning of machine. MapReduce can handle TB and PB volume-level data while enjoying a distinct advantage in handling TB volume-level mass data. MapReduce is not a panacea. It has shortcomings in the following 3 areas:

- MapReduce is mainly applied to loosely coupled data processing, and it is very inefficient for the tightly coupled computing tasks which are hard to be decomposed into many independent subtasks.
- MapReduce cannot explicitly support iterative computations;
- MapReduce is an off-line computing framework, which is not suitable for streaming computation and real-time analysis.

2   Researches on MapReduceV1

MapReduce utilizes the idea of "To divide and process, then integrate". Hadoop divides a big task into smaller ones and then executes them in parallel. Each MapReduce task is initialized to a Job, which then can be divided into two stages, Map stage and Reduce stage [6]. These two stages are represented by two functions. Map function receives a set of <key, value> input, and then generates the same set of <key, value> intermediate output. The machine which implements tasks of Reduce excavates <Key, Value> data with the same Key, and then delivers it to the Reduce function, which receives such a set of input as <key, (the list of value)>, followed by set processing and result output of the value. The output of Reduce is in form of <key, value> [7]. Figure 2 is the flow chart about how MapReduce processing data change.

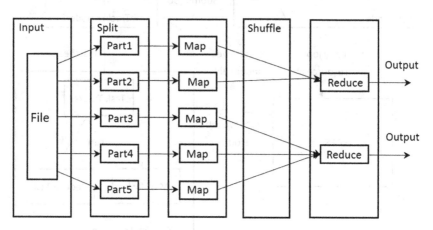

Fig. 2. Changes in MapReduce data flow

MapReduce utilizes master/slave framework, where the master node is JobTracker (job server), and the slave node is TaskTracker (task server). They communicate through a timed heartbeat.

The JobTracker is mainly responsible for communicating with clients, accepting commands from clients, such as submitting jobs, killing tasks, etc., and sending these commands to TaskTracker to perform. The communication between JobTracker and TaskTracker goes on in the form of the timed heartbeat. Jobtracker assigns tasks to Tasktracker, computing resources (CPU, internal storage, bandwidth, etc.) for Task-Tracker, and records the task status. It stores all information about tasks.

TaskTracker is a bridge between JobTracker and tasks: on one side, it receives and executes commands sent by JobTracker. On the other side, the status of each task and progress of executing tasks on this node are sent to JobTracker [8] in the form of a timed heartbeat.

Figure 3 shows the specific process of MapReduce job execution:

2.1 Submit Jobs

After a MapReduce job is submitted to Hadoop, it will enter the fully automated execution process, in which users are not allowed to interfere the job except monitoring the execution and forcing to terminate jobs. Therefore, before submitting jobs, users need to make sure that all parameters configuration that jobs need is complete.

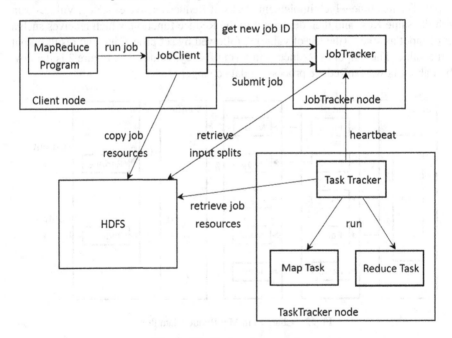

Fig. 3. MapReduce framework

2.2   Initialize Jobs

In the client, after the user job calls the submitJob() method on JobTracker, JobTracker puts the job into a job queue and performs scheduling with default schedule FIFO. When the job is scheduled to execute, JobTracker initializes the job and creates a Map task and Reduce task according to the data processing data.

2.3   Assign Tasks

TaskTracker is a software that runs in a separate loop and can send a heartbeat message (heartbeat) to JobTracker periodically, for example, it tells JobTracker whether the TaskTracker is available and ready to execute a new task. As soon as JobTracker receives the heartbeat message, it will assign a task to TaskTracker which is ready if there is a task which is prepared to be assigned in the task queue. Then JobTracker encapsulates distribution information (including information about which task to execute and the resources allocation) in the return value of the heartbeat message and sends it back to TaskTracker. According to the return information, TaskTracker executes the task. When JobTracker assigns tasks to TaskTracker, it will consider distributing tasks to the Task-Tracker where the data exits in order to minimize I/O overhead of data movement.

2.4   Execute Tasks

TaskTracker performs tasks based on the information of the heartbeat return value and also tracks the status of the task execution, reporting the status of the task to JobTracker periodically until the task execution is completed [9].

- Fault handling mechanism.
  In the process of task scheduling, fault handling is an important aspect, which includes two major faults: hardware fault and task execution fault.
- Hardware fault, namely, the server outage.
  From the perspective of MapReduce task execution, the hardware involved is JobTracker and TaskTracker. Obviously, the hardware fault refers to the machine error of JobTracker and TaskTracker.

In the Hadoop cluster, there is only one JobTracker at any time. So, single point of failure of JobTracker is the worst of all failures. So far, there is still no relevant solutions in the Hadoop. It's can be solved simply by creating multiple standby JobTracker nodes. After the failure of the master JobTracker, a new JobTracker node can be created by means of election algorithm (a kind of algorithm commonly used in the Hadoop to determine the master node) to determine. When some enterprises provide Hadoop services, they use this method to solve such problems [10].

TaskTracker failures are relatively common, but MapReduce has a corresponding solution, mainly to re-perform tasks. In the actual tasks, MapReduce jobs also encounter the task's failure caused by user code debts or process crashes. The user code debts result in casting failure during execution, causing the failure of task execution. In this case, TaskTracker reports to JobTracker, which will reboot the task.

3   Researches on MapReduceV2

The latest version of MapReduceV2, also known as YARN, was designed to solve the problem of the old version of MapReduce. The limitations of the old MapReduce are:

- Poor reliability.
  The old version used the master-slave framework, which resulted in the single point of failure of the master node. Once there is something wrong with the master node, the whole cluster will not be able to run. Once the single point of failure occurs, recovery is quite difficult.
- Poor extensibility.
  JobTracker in the old MapReduce not only plays the role as a resource manager and also the job scheduler. When the cluster is gradually expanding, JobTracker pressure will be more and more intense, which become the bottleneck of the Hadoop system, and restricted scalability of the Hadoop cluster.
- It is only limited to MapReduce programming framework.
  With the rapid development of the Internet, it's difficult for such simple and offline batch processing framework to meet all requirements. Now, there is a real-time stream processing framework such as storm and iterative computing framework such as Spark and so on, while the old version of MapReduce can't support these frameworks.
- Low resource utilization.
  In order to solve these limitations of the old MapReduce and promote the long-term development of Hadoop, Hadoop carried out complete refactoring of the MapReduce framework and it changed fundamentally as shown in Fig. 4.

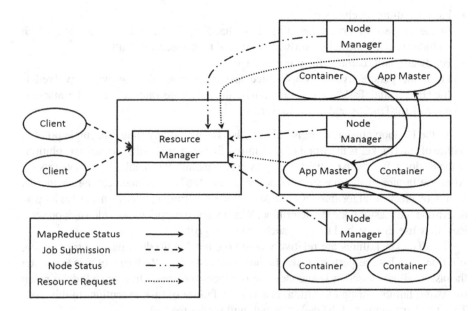

Fig. 4. MapReduce2 framework

The fundamental idea of refactoring is to separate the two main functions of JobTracker, resource management, and task scheduling, which are performed by two separate modules. The new resource manager manages the allocation of all computing resources on the entire cluster; YARN assigns an ApplicationMaster to each application, which is responsible for corresponding jobs scheduling and resource coordination.

ResourceManager is a master node that is responsible for managing the unified management and allocation of all resources in the cluster and receiving resource reporting information from NodeManager on each node [11]. It mainly composes of two modules: Resource Scheduler and Application-Master. Based on the situation of the job, the computing resources in the cluster, Resource Scheduler assigns the resources in the system to the running application. The unit of resources that Resource Scheduler allocated for each application is a Container, which is an abstract unit, a kind of dynamic resource allocation unit, encapsulating resource such as CPU, internal storage, disk, and network, etc. [12]. The Application-Manager is responsible for generating and managing all of the Application-Managers that run in the entire cluster and monitoring the running condition of the Application-Masters, and restarting when they fail.

The master node ResourceManager of YARN is only responsible for simple task scheduling and resource allocation. It no longer monitors and records the execution status of each task and it's no longer responsible for the restart of failed missions. Instead, it's only responsible for monitoring and restarting ApplicationMaster, which greatly reduced its own resource overhead. Moreover, single point of failure has little effects on the whole cluster so that it is much easier to recover.

NodeManager is slave node, the task and resource manager on a single node, which is deployed in each machine of each cluster. It manages the resources of a single node in the cluster, monitoring Container's life cycle and how resources are used. Also, it is responsible for communicating with ResourceManager, reporting the uses of computing resources on nodes, and health condition of nodes, etc.

ApplicationMaster is the manager of a single job, also responsible for task monitoring and each job has a corresponding one, which should apply for resource management to the ResourceManager and needs to monitor the status of all tasks [13].

YARN is an elastic computing platform that is no longer limited to MapReduce programming model. Instead, it supports multiple programming models simultaneously such as Spark. In other words, it is a model of sharing cluster. Below are advantages of YARN:

- High resource utilization.
  One framework for one cluster can cause unbalanced resource utilization, while excessive application of a framework leads to intense resources shortage in a cluster or less application of a framework causes cluster resources waste. Through sharing a cluster by a variety of computing frameworks, YARN is beneficial to high resource utilization of a cluster.
- Data sharing.
  As the data volume increases, the cost of mobile data is growing, while YARN model can reduce the cost in an effective way.
- Costs of operation and maintenance are low.

Changing from operation and maintenance of multiple clusters to that of a cluster reduces its cost greatly.

4   Comparison of the Latest Version and the Old Version

First of all, their similarities show that the client doesn't change, the API interfaces that the application calls mostly maintain compatible so that the previous code can be easily ported to the new framework.

4.1   Differences of the Two Version

The master nodes in the original framework, JobTracker, and TaskTracker, disappeared, which are replaced by ResourceManager, NodeManager, and ApplicationMaster.

Among them, ResourceManager took place of JobTracker and has played a role in resources allocation. About job scheduling, it only needs to start, monitor Application-Master that contains a task, and restart after a fault occurs. ApplicationMaster is no longer responsible for monitoring and restarting each task like JobTracker in the old framework, which make the single point of failure has little impact and it's easy to recover. ApplicationMaster is responsible for the entire lifecycle of a job, having the function of task scheduling as JobTracker in the original framework. However, each job owns an ApplicationMaster, which can not only run on ResourceManager but also on NodeManager, so there is no problem of single point of failure for ApplicationMaster.

4.2   Advantages of the New Framework

- The new framework separated two functions of JobTracker, which reduces its resource consumption so as to make the system recover from a single point of failure more easily. Moreover, it monitors the subtask status of each job. The programming is distributed, which becomes more secure.
- In the new framework, ApplicationMaster can change, which makes it possible to write its own ApplicationMaster for different computing frameworks. It enables more computing frameworks to run on the Hadoop cluster.
- In the old framework, the biggest burden of JobTracker is to monitor the operation of tasks under each job, in addition to rebooting when tasks faults occur. However, in a new framework, the latter function is performed by ApplicationMaster, and ResourceManager only needs to monitor the status of ApplicationMaster.
- The container has a great effect on resource isolation.

5   Experiment

5.1   Experimental Environment and Setup

We compared and evaluated the performance of MapReduce and YARM in terms of execution efficiency and data scalability.

The experimental cluster we used consisted of 1 JobTracker and 16 TaskTrackers, and the node configuration of the cluster is shown in Table 1.

Table 1. Configuration information for computing node

Item	Configuration information
CPU	4 Core Intel Xeon 2.4 GHz × 2
Memory	24 GB
Disk	2 TB SAS × 2
Network bandwidth	1 Gbps
OS	Red Hat Enterprise Linux Server 6.0
JVM version	Java 1.6.0
Hadoop version	Hadoop 1.0.3

The experimental data set was composed of a synthesized data set-LUBM (Lehigh University Benchmark), two real data sets-WordNet electronic dictionary semantic data and DBpedia dataset. All data was encoded and compressed and then stored in HDFS in the format of Hadoop SequenceFile. The default block size was 64 MB and the blocks were evenly distributed at computing nodes in the cluster.

The LUBM dataset is currently recognized as the test benchmark of Semantic Web knowledge base system oriented towards large-scale ontology application environment. In the experiment, five groups of test data were generated by fitting tools: LUBM-100, LUBM-250, LUBM-500, LUBM-750, and LUBM-1000, respectively, which respectively contained semantic data of 100, 250, 500, 750 and 1000 universities. The scale of their triples was respectively 13 million, 33 million, 66 million, 100 million and 133 million.

WordNet is a real semantic data set, which is a large English vocabulary database developed by the Cognitive Science Laboratory of Princeton University in 1985. In WordNet, words are organized according to semantics and contain 1,942,887 triples in total.

DBpedia is a widely used RDF dataset, whose data is structured information extracted from Wikipedia pages and contains a lot of entity information in many fields. It is widely used as a standard data set for RDF data query. Four groups of DBpedia data sets of different sizes were selected in the experiment: DBpedia-1 (35 Million triples), DBpedia-2 (85 Million triples), DBpedia-3 (140 Million triples) and DBpedia-4 (210 Million triples).

5.2 Execution Performance Testing

We conducted five tests on 10 sets of data (5 sets of LUBM data, 1 set of WordNet, and 4 sets of DBpedia data) under MapReduce and YARM respectively, and finally took the average running time. The experimental results are shown in Table 2.

Table 2. Reasoning execution time

Data set	Triples/M	MapReduce/s	YARM/s	Speed lifting multiples
LUBM-100	13.0	630.066	46.547	13.5
LUBM-250	33.0	742.164	54.486	13.6
LUBM-500	66.0	834.114	58.512	14.2
LUBM-750	100.0	911.106	64.760	14.0
LUBM-1000	133.0	964.143	72.732	13.2
WordNet	1.9	373.548	70.718	5.3
DBpedia-1	35.0	146.651	22.548	6.5
DBpedia-2	85.0	383.295	33.639	11.4
DBpedia-3	140.0	662.131	40.642	16.3
DBpedia-4	210.0	734.424	43.646	16.8

The results indicated that compared with MapReduce, YARM is about 10 times faster at execution time. In addition, although the dataset size of WordNet is smaller than LUBM-100, yet the reasoning time of YARM in WordNet is greater than the reasoning time in LUBM-100, which is because the WordNet reasoning is much more complex than LUBM. The experimental results showed that the YARM efficiency is influenced by the reasoning complexity, while the factors that affect the MapReduce are more the data set size. So in large-scale RDF reasoning tasks, YARM has better performance than the MapReduce.

5.3 Data Scalability Testing

In order to observe the change of the actual running time of YARM when the data scale increases and try to avoid the influence on the result caused by using different datasets with different complexity, we adopted five groups of LUBM datasets and four groups of DBpedia datasets with the same complexity but different sizes for the test. The scale of LUBM increased from 100 universities to 1,000, and five tests were carried out on each level of scale, finally, the average running time was taken.

DBpedia used four sets of data with different sizes, being respectively 35M records, 85M records, 140M records and 210M records. Five tests were made for each set of data to finally take the average running time. The experimental results are shown in Figs. 5 and 6. The vertical coordinate is the execution time and the horizontal coordinate is the data scale. It can be seen that with the increase of the data scale, the reasoning time of YARM shows a trend of approximate linear increase, hence compared with MapReduce. YARM has better data scalability.

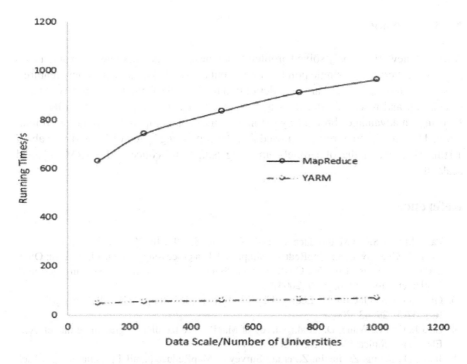

Fig. 5. Comparison of running time when LUBM data scale changes

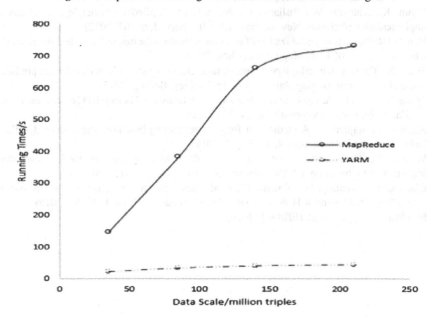

Fig. 6. Comparison of running time when DBpedia data scale changes

6 Conclusion

A new framework not only solved problems that the old framework was confronted with such as recovery of the single point of failure but also added a number of new features. This paper introduces the characteristics of two frameworks from the perspective of task scheduling and resource allocation at first. After that, it compares their differences, focusing on advantages brought by the new framework compared with the old framework. The experimental results showed that the computing speed of YARM was about 10 times higher than that of parallel computing using MapReduce, and YARM had good scalability.

References

1. Pace, M.F.: BSP vs MapReduce. Procedia Comput. Sci. **9**, 246–255 (2012)
2. Dean, J., Ghemawat, S.: MapReduce: Simplified data processing on large clusters. In: OSDI 2004 Proceedings of the 6th Conference on Symposium on Operating Systems Design & Implementation, vol. 6, p. 10 (2004)
3. Ghemawat, S., Gobioff, H., Leung, S.T.: The Google file system. ACM SIGOPS Oper. Syst. Rev. **37**(5), 29–43 (2003)
4. Li, J.J., Cui, J., Wang, D., et al.: Survey of MapReduce parallel programming model. Acta Electronica Sinica **39**(11), 2635–2642 (2011)
5. Jiang, D., Zheng, Z., Jiexin, Z., et al.: Survey of MapReduce parallel programming model. Comput. Sci. **42**(S1), 537–541, 564 (2015)
6. Liyun, K., Xiaoyue, W., RuJiang, B.: Analysis of MapReduce principle and its main implementation platforms. New Technol. Libr. Inf. Serv. **2**, 60–67 (2012)
7. Han, H.: Researches on optimization of resource allocation for mapreduce scheduling. South China University of Technology, Guangzhou (2013)
8. Shuqi, X.: The research on high performance task scheduling technology based on mapreduce in cloud computing. Beijing University of Technology, Beijing (2013)
9. Wang, B., Chen, L.: Analysis and application of mechanism of Hadoop RPC communication. J. Xi'an Univ. Posts Telecommun. **6**, 74–77 (2012)
10. Xiang, R., Lingjuan, L.: A method for frequent set mining based on MapReduce. J. Xi'an Univ. Posts and Telecommun. **4**, 37–39, 43 (2011)
11. Wentao, Z., Zhongliang, A., Zhonglin, L., et al.: A kind of high-priority job scheduler implementation based on YARN. Comput. Eng. Softw. **3**, 84–88 (2016)
12. Chuntao, D., Wenting, L., Qingxia, C., et al.: Research on the framework and resource scheduling mechanisms of Hadoop YARN. Inf. Commun. Technol. **1**, 77–84 (2015)
13. http://hadoop.apache.org/. [EB/OL] (2016)

A Third-Order Meminductor Chaos Circuit with Complicated Dynamics

Zhiping Tan[1] and Shanni Li[2(✉)]

[1] Guangdong University of Science and Technology, Dongguan 532000, Guandong, China
Tzp2008ok@163.com
[2] Capital Management Co., Ltd., Shenzhen 518000, Guandong, China
lishanni@sina.com

Abstract. A novel meminductor circuit model is showed in this paper, and the mathematical model is also derived. By utilizing this circuit model, a simple series chaotic circuit is constructed by resistor, capacitor, and meminductor. Stability and dynamical behaviors of the third-order meminductor chaos circuit are studied by a variety of nonlinear analyzing tools. The result of the research shows that the system has complicated nonlinear phenomena; it is very suitable for confidential communication.

Keywords: Meminductor · Chaos · Series chaotic circuit

1 Introduction

According to the symmetry rules of the circuit, the memristor was considered as the fourth fundamental circuit element by Chua in 1971 [1], but it has not been confirmed until in 2008, the memristor was proved by the researchers in Hewlett–Packard (HP) laboratory [2]. Then, the memristor in the nanometer state are shown. Soon after the memristor have gradually become a focus, Because of its potential application in engineering, which attracted a lot of interest in both industry and academia. All kinds of equivalent circuits are developed to explore the memristors potentials applications. Besides, on account of the particular nonlinear dynamics, memristors are suitable for constructing a chaotic system. By using theoretical analysis, the memristor chaotic circuits are proposed by scholars, and multiple attractors are found in memristor based chaotic circuit and complex dynamics are related to initial conditions [3–5]. In 2009, Di Ventra and Chua postulated the memcapacitor and the meminductor, which in order to extend the concept of the memory device [6], its potential value has attracted the attention of the researchers [7–10]. The meminductor was proposed to be a fundamental circuit memdevice parallel with the memristor, in comparison to the memristor, the study of the meminductor-based chaos system just begins. A few simple memristive oscillators which are consist of fundamental circuit elements were reported in Ref. [11–13]. So as to explore meminductor-based chaos system [14–18]. In this paper a third-order chaos system based on meminductor is showed, which consists of a resistor connected in series with a capacitor and a flux-controlled meminductor. Compared with other chaos circuits,

© Springer Nature Singapore Pte Ltd. 2018
K. Li et al. (Eds.): ISICA 2017, CCIS 874, pp. 125–132, 2018.
https://doi.org/10.1007/978-981-13-1651-7_10

the proposed circuit has complex dynamical behaviors, and the circuit consist of only three circuit components.

In this paper, a novel meminductor circuit model is showed, and a mathematic mode of smooth curve based on meminductor is proposed. Furthermore, the meminductor is applied to construct a simple chaos circuit. The rest of this paper is organized as follows. In Sect. 2, the mathematic model of meminductor is described. In Sect. 3, a chaos circuit based on the meminductor is showed. In Sect. 4, complex dynamical behaviors of the meminductor-based oscillator are analyzed. Finally, some conclusions are drawn from the present study in Sect. 5.

2   The Mathematic Model of Meminductor

Following the definition given by Chua, a memristor is a passive two terminal circuit element. The circuit symbol of memristor is shown in Fig. 1.

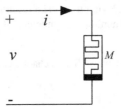

Fig. 1. The circuit symbol for a memristor

A common meminductor can be defined as providing variable inductance related to current or flux by Chua in 2009. The circuit symbol of memristor is shown in Fig. 2.

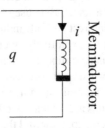

Fig. 2. The circuit symbol for a meminductor

A flux-controlled meminductor can be described by the equations:

$$I(t) = L^{-1}(\rho, \phi, t) \cdot \phi(t) \tag{1}$$

$$d\rho/dt = f(\rho, \phi, t) \tag{2}$$

Where L^{-1} is the inverse meminductance and ρ is the integral of flux ϕ, and a current-controlled meminductor can be described by the equations:

$$\phi(t) = L(q, I, t) \cdot I(t) \tag{3}$$

$$dq/dt = f(q, I, t) \tag{4}$$

Where L is the meminductance, $I(t)$ and q are the internal current and the internal charge through the meminductor.

In this paper, the inverse meminductance is assumed as $L^{-1} = a + b\rho^2$. Then the nonlinear current-flux relationship of the meminductor can be defined as follows,

$$I(t) = (a + b\rho^2)\phi(t) \tag{5}$$

$$d\rho/dt = \phi(t) \tag{6}$$

3   Meminductor Based Chaotic Circuit

The third-order chaos circuit based on meminductor is showed in Fig. 1, which is consist of resistor, capacitor and meminudctor, According to Kirchhoff's circuit laws, the equations for the chaos circuit in Fig. 3 are:

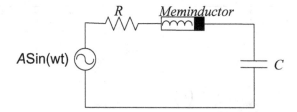

Fig. 3.   Meminductor based chaotic circuit

$$\begin{cases} dv_c/dt = i_m/C \\ d\phi/dt = -R \cdot i_m - v_c + A\sin(wt) \\ d\rho/dt = \phi \end{cases} \tag{7}$$

Where v_C, i_L, and ϕ are the voltage across capacitor C, the current through inductor, the current through meminductor, and the magnetic flux which has passed through the meminductor, and the $A\sin(wt)$ is the incentive signal source.

The normalized form of the circuit Eq. (7), which are convenient for numerical analysis are obtained as

$$\begin{cases} dx/dt = \alpha \cdot f \\ dy/dt = -\beta \cdot f - x + A\sin(wt) \\ dz/dt = y \end{cases} \tag{8}$$

The state variables used are $v_c = x, \phi = y, \rho = z, i_m = f$.

The rescaling parameters used for this normalization are

$$\alpha = 1/C, \beta = R, f = (a + 3bz^2)y \qquad (9)$$

So the proposed meminductor based chaotic circuit can be deemed as a three-dimensional system, its dynamic characteristics are determined by Eq. (8). Let $\alpha = 1000$, $A = 7.9$, $\beta = 10$, $w = 6.74$, $a = -0.1$, $b = 6.6$, for initial conditions (0.01, 0, 0.1), the system (8) is chaotic and a double-scroll is shown in Fig. 4.

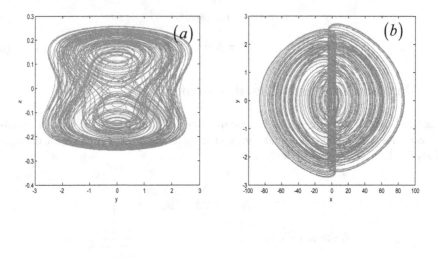

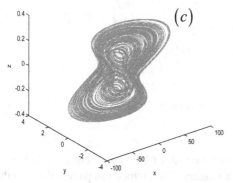

Fig. 4. Some typical chaotic attractors in the phrase plane (a) y-z, (b) x-y (c) x-y-z

4   Dynamical Properties of the Chaotic System

4.1   Dissipativity and Equilibrium Point

The discrepancy of Eq. (8) is regarded as

$$\nabla v = \frac{\partial \dot{x}}{\partial x} + \frac{\partial \dot{y}}{\partial y} + \frac{\partial \dot{z}}{\partial z} = -\beta \cdot (a + bz^2) \tag{10}$$

According to, $\nabla v < 0$, for $-\beta \cdot (a + bz^2) < 0$, revealing that the system is dissipative, and the circuit might be a chaos circuit.

The equilibrium state of system (8) is given by set $B = \{(x, y, z)|x = y = 0, z = c\}$, which corresponds to the u-axis. Here c is a real constant. The Jacobian matrix J_A at this equilibrium set is given by:

$$J_A = \begin{bmatrix} 0 & \alpha \cdot (a + 3bz^2) & 6\alpha byz \\ -1 & -\beta \cdot (a + 3bz^2) & -6\beta byz \\ 0 & 1 & 0 \end{bmatrix} \tag{11}$$

Its characteristic equation is given by

$$|\lambda I - J_A| = 0 \tag{12}$$

And its characteristic equation at the equilibrium point B consequently is written as:

$$\lambda^3 + k_1 \lambda^2 + k_2 \lambda = 0 \tag{13}$$

In particularly, $k_1 = \beta(a + 3bc^2), k_2 = \alpha(a + 3bc^2)$, form Eq. (12), it can be resumed, the system has a zero characteristic root and two nonzero characteristic roots, according to Routh–Hurwitz condition, the system is stable on condition that the real parts of the roots of Eq. (13), except the zero eigenvalue, are negative if and only if

$$E_k = \begin{vmatrix} k_1 & 1 \\ 0 & k_2 \end{vmatrix} > 0 \tag{14}$$

Where $k = 1, 2$, namely

$$\begin{cases} E_1 = k_1 > 0 \\ E_2 = k_1 k_2 > 0 \end{cases} \tag{15}$$

So, if $\beta(a + 3bc^2) > 0, \alpha(a + 3bc^2) > 0$, the system is stable. On the contrary, the system is unstable, if set parameters $\alpha = 1000, \beta = 10, a = -0.1, b = 6.6$, we can obtain:

$$|c| > 0.2 \tag{16}$$

4.2 Lyapunov Spectra and Bifurcation Diagrams

Lyapunov exponents and bifurcation diagrams are the effective methods to judge whether or not a system is chaotic. A system with only one positive Lyapunov exponent is in chaotic, while a system with more than one positive Lyapunov exponent is in hyperchaotic. In sake of to confirm the dynamical behaviors in system (8), the Lyapunov exponent spectra and the corresponding bifurcation diagram are given in Figs. 5 and 6, respectively. In the numerical simulations, $\alpha = 1000$, $A = 7.9$, $w = 6.74$, $a = -0.1$, $b = 6.6$, and initial condition $(0.01, 0, 0.1)$ are fixed, while β acts as the control parameter and satisfies $9.5 \leq \beta \leq 12.5$.

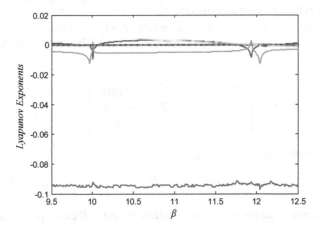

Fig. 5. The Lyapunov exponents with respect to parameter β.

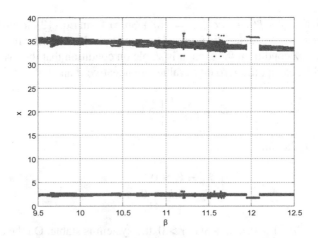

Fig. 6. The bifurcation diagram with respect to parameter β.

Obviously, the corresponding Lyapunov exponent spectrums and bifurcation diagrams, it is easy to see that the system has different kinds of transient chaos with different parameters.

5   Conclusion

In this paper, a novel meminductor circuit model is showed, and the mathematical model is also derived. Based on the meminductor, a simple chaotic circuit is proposed. By theoretical analysis, we demonstrate that this circuit can exhibit complex dynamical behaviors, including the coexisting bifurcation modes, the coexisting attractors and the chaotic transients. Numerical simulation results can be well verified the theoretical analysis, due to its complex chaotic behavior, it is important potential applications in the engineering application.

Acknowledgements. The authors wish to thank the reviewers for their fruitful advice to improving this paper. The work was supported by youth special innovation project of guangdong province (Grant No 61176032).

References

1. Chua, L.O.: Memristor—the missing circuit element. IEEE Trans. Circuit Theory **18**, 507–513 (1971)
2. Strukov, D.B., Snider, G.S., Stewart, D.R., Williams, R.S.: The missing memristor found. Nature **453**(7191), 80–83 (2008)
3. Wei, W., Deng, N.: Memristor interpretations based on constitutive relations. J. Semicond. **38**(10), 79–82 (2017)
4. Cao, J., Li, R.: Fixed-time synchronization of delayed memristor-based recurrent neural networks. Sci. Chin. Inf. Sci. **60**(03), 108–122 (2017)
5. Ho, P.W.C., Almurib, H.A.F., Kumar, T.N.: Memristive SRAM cell of seven transistors and one memristor. J. Semicond. **37**(10), 60–63 (2016)
6. Di Ventra, M., Pershin, Y.V., Chua, L.O.: Circuit elements with memory: memristors, memcapacitors, and meminductors. Proc. IEEE **97**, 1717–1724 (2009)
7. Fouda, M.E., Radwan, A.G.: Meminductor response under periodic current excitations. Circuits Syst. Signal Process. **33**(5), 1573–1583 (2014)
8. Biolek, D., Biolek, Z., Biolková, V.: PSPICE modeling of meminductor. Analog Integr. Circuits Signal Process. **66**(1), 129–137 (2011)
9. Sah, M.P., Budhathoki, R.K., Yang, C., Kim, H.: Mutator-based meminductor emulator for circuit applications. Circuits Syst. Signal Process. **33**(8), 2363–2383 (2014)
10. Ci-Yan, Z., Dong-Sheng, Y., Yan, L., Meng-Ke, C.: Composite behaviors of dual meminductor circuits. Chin. Phys. B **24**(11) (2015). Project supported by the Fundamental Research Funds for the Central Universities, China (Grant No. 2013QNB28)
11. Wang, S.F.: The gyrator for transforming nano memristor into meminductor. Circuit World **42**(4), 197–200 (2016)
12. Yu, Q., Bao, B.C., Xu, Q., Chen, M., Hu, W.: Inductorless chaotic circuit based on active generalized memristors. Acta Phys. Sin. **64**, 170503 (2015)

13. Bao, B.C., Liu, Z., Xu, J.P.: Steady periodic memristor oscillator with transient chaotic behaviours. Electron. Lett. **46**, 237–238 (2010)
14. Iu, H.H.C., Yu, D.S., Fitch, A.L., et al.: Controlling chaos in a memristor based circuit using a Twin-T notch filter. IEEE Trans. Circuits Syst. I **58**(6), 1337–1338 (2011)
15. EI-Sayed, A.M.A., Elsaid, A., Nour, H.M., Elsonbaty, A.: Dynamical behavior, chaos control and synchronization of a memristor-based ADVP circuit. Commun. Nonlinear Sci. Numer. Simul. **18**, 148–170 (2013)
16. Fitch, A.L., Yu, D., Iu, H.H.: Hyperchaos in a memristor-based modified canonical chua's circuit. Int. J. Bifurc. Chaos **22**, 1250133–1250138 (2012)
17. Wang, X.Y., Iu, H.H.C., Yi, G., Liu, W.W.: Study on time domain characteristics of memristive RLC series circuits. Circuits Syst. Signal Process. **35**(11), 4129–4138 (2016)
18. Riaza, R.: Second order mem-circuits. Int. J. Circuit Theory Appl. **43**(11), 1719–1742 (2015)

Mathematical Model of Cellular Automata in Urban Taxi Network – Take GanZhou as an Example

Zhaosheng Wang[1(✉)] and Shiyu Li[2]

[1] Ganzhou Teachers College, No. 35 Development Zone Road, Ganzhou 341000, Jiangxi, China
1239331840@qq.com
[2] Jiangxi University of Science and Technology, Ganzhou 341000, China

Abstract. Urban traffic is an extremely complex dynamic system. Urban traffic modeling and forecasting is still a challenge, the main difficulty is how to determine supply and demand and how to parameterize the model. This paper tries to solve these problems with the help of a large number of floating taxi data. We describe the first solution to the challenge of finding a taxi destination. The tasks included at the beginning of its trajectory prediction of a taxi destination, it is expressed as the GPS point of variable length sequences, and related information, such as the departure time, the driver id and customer information. We use a neural network based approach that is almost completely automated. The architecture we are trying to use is a multi-layer perception, bidirectional recursive neural network, and a model inspired by the recently introduced memory network. Our approach can be easily adapted to other applications, with the goal of predicting the fixed-length output of a variable length sequence.

Keywords: Mathematical model · Artificial neural network engineering
Traffic prediction · Taxi destination prediction

1 Introduction

Traffic congestion has become a serious problem limiting the operating of many big cites. In order to avoid or easy the traffic prediction is becoming more and more important for trip planning and traffic management. For quite a long time the lack of high quality traffic data has been impeding the development of traffic modeling and prediction on large scale road networks. Recently, large amount of floating taxi data and mobile phone data covering the whole urban area with high spatial and temporal resolution have become available. These data provide new sources of information, allowing us to push the limit of traffic modeling and prediction on Urban-wide scale. In this paper we consider the problem of short-term prediction of the traffic speed on Urban-wide scale. Existing traffic prediction approaches can be roughly divided into statistical methods and model-based methods. Statistical approaches usually use time series analysis we

Foundation item: Supported by Jiangxi provincial education department science and technology research project (GJJ161333).

K. Li et al. (Eds.): ISICA 2017, CCIS 874, pp. 133–141, 2018.
https://doi.org/10.1007/978-981-13-1651-7_11

propose a new and relative simple approach for Urban traffic modeling and simulation. Our method is model-based, but also requires heavily use of statistical tools in setting up the model. Because of the limitation of the computation power and data, we do not aim to faithfully model the real traffic system in fully detail. Instead, we just want to capture the effective traffic dynamics. In particular, we simulates a large number of vehicles moving on uniform grids. The grid system has a spatial resolution of 100 m _ 100 m, which is much large compared with the microscopic cellular automata models. Now that one grid can hold more than one taxis, so we call it the coarse-grained cellular automata model.

This model is different than continuous traffic models as it keeps track of the individual vehicles. It is also different than the discrete models in that there is no explicit taxi interaction. Instead, vehicles are coupled through the coarse-grained state variables [1]. Because of these differences, we cannot directly import the parameters of other traffic models into our case. We obtain an effective traffic flow system for the taxis of the entire Urban. Because taxis share the same road with other vehicles, the simulated speed of this effective system can be used to predict Urban-wide traffic speed. To evaluate the model, we extract the Origin-Destination demands and driving trajectories of taxis from historical floating taxi data and feed it to the model. Promising results are obtained: on the Urban-wide level, the simulated traffic flow pattern is similar to that observed in the historical data; on the road-segments level, the model predicted traffic speed scores a reasonable accuracy. This work can serve as a starting point a practical Urban traffic forecasting system.

2   Data Preprocessing

The floating taxi data is collected from GPS devices mounted on every taxis in Ganzhou, China. This kind of device will repeatedly transmit vehicle information to the server, including the current time, GPS coordinate and operating status. The frequency of data transmission varies between 20 s to 60 s, depending on the standard that the taxi companies use. The total number of taxis in Ganzhou is about 1,400, continuously generating records each day from 0:00 to 24:00. In our example, the input layer receives a representation of the taxi prefix, and its associated metadata and output layer predicts the destination (latitude and longitude) of the taxi. We use the standard hidden layer, which consists of matrix products, followed by bias and nonlinearity [2]. We choose to use a nonlinear rectifier is linear units, it's just calculate maximum value compared with the traditional s activation function, disappear RLU limits the gradient problem, because when x for timing, its derivative is always 1. For our winning approach, we used a hidden RLU neuron.

We only considered the first k point of the trajectory prefix and the last k point, which gives us the 2k points of our input vectors, or the 4k values. We take k = 5 for the winning model. These GPS points are standardized (zero mean, unit variance). The first and last k points overlap when the path's prefix contains less than 2 k. When the path prefix contains four forecast of artificial neural network is applied to taxi destination, less than k, we by repeating the first or the first way to populate the input GPS point finally [3].

To deal with discrete metadata, including customer ID, taxi ID, date and time information, we learn the model of embedding this information with the model. The embedded table of the word is included in the model parameters of our study and is represented as a conventional parameter matrix: the embedding is initialized first and then modified by the training algorithm.

3   Coarse-Grained Cellular Automata Model

Generally speaking, to build a traffic model, one need to answer three questions: (i) demands: what are the Origin-Destination demands and the routes they would take? (ii) supply: what is the road cap Urban and free-flow speed of each road segment? (iii) driving model: how the taxis move on the roads? In this section we will address these questions. We will introduce in this section of the model in the specific aim of competition test set didn't show very well, but we believe that they can be for other problems involving fixed length output and input variable length provides interesting insights.

3.1   Demands

The model simulates a group of taxis moving on a grid. To do this, we need to know the history and destination of each trip and the route of each car. For fully integrated traffic simulation and prediction systems, it is ideal to generate this information [4]. Predicting the needs of the original destination and route selection itself is a very difficult issue and will not be discussed in this job. However, we just want to do a concept verification test to see if the model can capture traffic flow mechanics. Therefore, we extract all the original destination and trace information from historical data.

3.2   Supply

For the Urban, we currently use an implicit road system, and the entire urban road network is represented by a unified grid. Each grid contains four valid sections pointing to the right, top, left and bottom. Each segment can be occupied by multiple cabs at the same time. There are no turning restrictions, no traffic lights. From historical data, we can estimate that the caper cities of each road may exceed: more taxis means more roads, and no taxis means no roads. This article will discuss how to evaluate the details of road cap Urban. This grid system can be considered an approximation of the actual road network.

3.3   Model

To let the traffic flow, we need to do two things:

1. Given the position of all vehicles on the grids, determine the driving speed on each road segment V_i.
2. Given V_i, update the position of all vehicles on road.

To address the issue, we try to learn a V-N relation from the historical data. In particular, we compute the historical V_i and N_i at every ten-minute time windows (non-overlapping) according to Eqs:

$$V_i(t_n) = \frac{1}{n} \sum_{i=1}^{n} V_i(t_{n-s}) \qquad (1)$$

$$N_i(t_n) = \frac{1}{n-1} \sum_{i=1}^{n} N_i(t_{n-s}) \qquad (2)$$

For each grid, these historical values generates. In the figure the x-axis is N_i and the y-axis is V_i. The average ux. The pattern looks similar to the classical fundamental diagram for small occupancy, the increases linearly with road occupancy; When the occupancy exceeds a threshold (vertical dashed line), traffic congestion may occur. A small difference is that, when the occupancy is small, the speed can sometimes be very low. This is probably due to the punctuations of the ratio of taxis among the whole traffic. For each road link, V-N relation from the data points. First we determine the two points P and Q, which corresponds to the center of the green and blue points.

Currently we adopt a naive way of selecting the green and blue points: first we rule out the top 5% data points as outliers [5]. Then we select the left-most 20% from the top-most 20% data points as the green ones. Lastly we select all the data points on the right-hand-side of the green dots as the blue ones. Then we assume that $V = V_f$ is constant when $N < N_c$ and takes the form $V = a/(N-b)$ when $N > N_c$, where a and b are two constants whose value are to be determined. To fit this type of V-N function we need four parameters, namely V_f, V_s, N_c, N_m. V_f is the free-flow speed of the road segment, which equals to the slope of the line-segment OP. N_c is the effective cap Urban for taxis on the road segment, corresponds to the x-value of point P. The other two parameters, V_s and N_m corresponds to the slope of line-segment OQ and the x-value of Q, respectively. In terms of these four parameters, the function V(N) can be written as

$$V(N) = \begin{cases} V_f; & \text{if } N \leq N_c \\ \dfrac{N_f V_s(N_m - N_c)}{(V_f - V_s)N - (V_f N_c - V_s N_m)}, & \text{if } N > N_c \end{cases} \qquad (3)$$

From this V(N) function we can predict a target speed for a given occupancy. V_f, V_s, N_c, N_m estimated this way give a quantitative description of the intrinsic properties of the road network. Since there are four directions in each grid, the total number of parameters we need to learn from the data is $256 \times 256 \times 4 \times 4 = 1,048,576$. There are many grids that are seldom visited by taxis, and we simply assign a default value to them because these regions do not affect the over all traffic much (we use $V_f = 20$ km/h, $V_s = 5$ km/h, $N_c = 1$, $N_m = 0.5$). All these parameters are pre-computed and stored. Now we address the second issue, which is to update vehicle positions when the speed V_i is given. As we mentioned earlier, from the historical data we know when and where a taxi enters into the grids. We also know the route it will take. Denote $l_j(t)$ as the position.

Estimating model parameters from historical data. For each grid and each of the four directions, one scatter plot is generated from historical data. Four of these plots are shown in the figure, with their grid locations marked by red circles and directions indicated by arrows in (B). Data points in (A) correspond to the ten-minute average occupancy (x-value) and flux (y-value). Green dots and blue dots are two selected groups of data points whose centers determines the shape of the fundamental diagram. The model parameters V_f; V_s; N_c; N_m can then be derived from the fundamental diagram according to the procedure. The heat-map in right-lower panel shows the estimated V_f. Note that the estimated speeds on oblique roads tend to be larger than those on horizontal and vertical roads because of the zigzagged representation in the gridded road system [6]. Of the j-th taxi on its pre-determined route at time t. $l_j = 0$ corresponds to the starting point and l_j is increased by 100 m every time it passes a grid. During the simulation, at each time step we update l_j by

$$l_j(t + dt) = l_j(t) + V_j(t)dt \qquad (4)$$

where dt is the simulation time step size (we choose dt to be one minute), and $V_j(t)$ is the speed of the taxi. $V_j(t)$ is determined by

$$V_j(t) = \lambda V_{current} + (1 - \lambda)V_{next} + \eta \qquad (5)$$

Here $V_{current}$ and V_{next} is the average speed of the grid where the taxi is currently at and the next grid that the taxi is heading to, respectively. It is a constant controlling the look-ahead effect, the impact of the speed of the upstream taxis on the downstream taxis, and it is a random noise. Note that a taxi may travel multiple grids within one time step. Once the taxi arrives its destination, it is removed from the system.

3.4 Parameter Turning

The parameters V_f, V_s, N_c, N_m at each grid learned directly from the data provide a good starting point for the model. However, they may subject to errors during the fitting processes and hence are not optimal. As we shall see later, the simulation results are quite sensitive to the choice of parameters. In order to improve the accuracy of the model, we need to fine tune the parameters. To do so, we simply iteratively adjust V_f, V_s, N_c and N_m in the following way. First, we define V regular simulate and V regular real to be the average of speed during regular hours (from 12:00 to 15:00) obtained from the simulated and historical data at each grid, respectively. If $V_{regular\ simulate} > V_{regular\ real}$.

We reduce V_f at this grid by a small amount. Vice versa, if $V_{regular\ simulate} < V_{regular\ real}$ then V_f is increased by a small amount. After many times of iterations, V_f should approximate the mean speed during the selected regular hour. Next we define $V_{rush\ simulate}$ and $V_{rush\ real}$ to be the minimum speed recorded during the morning rush hour (7:00–9:00) at each grid. Similarly, if $V_{rush\ simulate} > V_{rush\ real}$ we reduce N_c at this grid by a small amount, and if $V_{rush\ simulate} < V_{rush\ real}$ we increase N_c. When changing N_c, we also change Nm accordingly to keep the ratio $N_c = N_m$ unchanged. As we shall see in the numerical results, after using this simply intuitive way to optimize the parameters, the accuracy of the model is greatly improved. We will investigate other parameter tuning methods in future work.

3.5 Incorporating Real Time Traffic Data

Mistakes will accumulate over time. Using real-time traffic data to adjust the simulated traffic condition can improve the prediction accuracy. In numerical weather forecasting, this program is called data assimilation and can be used in a method [7]. In our example, we use recently estimated that the speed of each grid to modify the speed of simulation (for example, in before 9 o 'clock to predict flow before 10 o 'clock, we will use in 9 speed to update the current history). For the sake of simplicity, we assume that the noise under the estimated speed compared with the simulation of the noise is negligible, so as long as we have a new set of the estimated speed, we can simply replace the simulation speed. However, the update footprint is more complex and is not implemented here.

4 Results

To demonstrate the potential of our model in traffic forecasting, we use it to predict road speed one hour into the feature in Ganzhou. We choose a day and compute the average speed in every 10 min from the historical data. The historical speed information is used for two purposes: one is to serve as the ground truth for comparison with the model predicted results, and the other is to be used as the real time input data for runtime speed correction. We construct two sets of O-D and route information, one come from the day to be simulated, and the other comes from another weekday. We use this alternative route-set to show how different route choice would affect the simulation results. Additionally, in order to demonstrate how sensitive the simulation results depend on model parameters, we consider two parameter sets. One is directly learned from the historical data, and the other is the result of parameter tuning. The former is referred as parameter in the following. With different route set and parameter set, we run the simulation and predict the speed one hour into the future during time period 6:20 to 18:00 of the selected day.

First we test if our traffic simulator can generate traffic state that is consistent with that observed in the historical data on a city-wide scale. Here consistent is in the sense that the predicted congested locations agree with the historical data in a qualitative manner. We plot the speed predicted for 7:00 and 13:00. In each plot we show the combined right and up speed. The combination is done via a weighted sum, with weights equal to the volumes in the corresponding directions. We can see that, the model predicted traffic speed is in general consistent with the historical data. On the other hand, there appears to be a large discrepancy when using parameter-set 2, indicating that the model results is sensitive to the choice of parameters. For route-set2, the overall pattern remains similar with that of the route-set 1, suggesting that the simulation results are not very sensitive to travel demands. This is an important point which makes it possible for us to replace the exact route-set with simulated ones in future. Local road segment speed Next we take a closer look at the predicted traffic speed on road segments. To do so, we select a number of road segments, each of which is made of ten connected grids (total length is 1 km). We select eight road segments as shown. Four of them are part of arterial roads, while the other four are part of local roads. The speed of a road segment is defined as the average speed of its compositing grids. To avoid bias in selection we let the starting grid of all road segments to be either 128-th row or 128-th column. Figure 8 shows the predicted road segment speed using different implementations. The model predicted speed agrees well with the real value on most of the selected roads, but there are significant discrepancy for road segment 3 and 6. In these two cases, there are sudden drops in speed during regular traffic hours. Our current model fails to catch these irregular behaviors in urban traffic, probably because of insufficient data. If using parameter-set 2, the predicted speed is quite different from the real one, and in road segments 7 and 8 there is even severely deviation in the average of the speed. This explains why parameter tuning is important.

Next we take a closer look at the predicted traffic speed on road segments. To do so, we select a number of road segments, each of which is made of ten connected grids (total length is 1 km). We select eight road segments as shown. Four of them are part of arterial roads, while the other four are part of local roads. The speed of a road segment is defined as the average speed of its compositing grids. To avoid bias in selection we let the starting grid of all road segments to be either 128-th row or 128-th column. It shows the predicted road segment speed using different implementations. The model predicted speed (red square curve) agrees well with the real value on most of the selected roads, but there are significant discrepancy for road segment 3 and 6. In these two cases, there are sudden drops in speed during regular traffic hours. Our current model fails to catch these irregular behaviors in urban traffic, probably because of insufficient data. If using parameter-set 2, the predicted speed (green circle curve) is quite different from the real one, and in road segments 7 and 8 there is even severely deviation in the average of the speed. This explains why parameter tuning is important. On the other hand, using route-set 2 does not change the results much (Table 1).

Table 1. One hour prediction error in road segments speed. For each method, both RMSE (the smaller the better) and Accuracy (the larger the better) are computed. See main text for detailed information.

	RMSE	Accuracy
Model prediction	9.55	0.72
Extrapolation	11.77	0.70
Parameter-set 2	13.15	0.55
Route-set 2	9.60	0.70

To further quantify the prediction accuracy, we compute the Rooted Mean Square Error (RMSE) and Accuracy (one minus the Mean Absolute Percentage Error, or MAPE):

$$\text{RMSE} = \sqrt{\frac{\sum_{i=1}^{T} (V_{real} - V_{predict})}{T}}$$

$$\text{Accuracy} = 1 - \text{MAPE} = 1 - \sum_{i=1}^{T} \frac{|V_{real} - V_{predict}|}{V_{real}}$$

In both equations the summation is taken for the whole simulation time period and then averaged over all the selected road segments. As a control experiment, we also consider an extrapolation method of prediction, which simply uses the current speed as the prediction. In the Table, the RMSE and Accuracy between the model prediction and real value are given. We can see that, the model prediction scores the best result in both RMSE and Accuracy compared with others. Also, when using route-set 2 the prediction accuracy do not degrade much. Note that we do not attempt to compare the current model with other machine-learning based prediction here. In fact, for now the state-of-art machine learning prediction program can do better reported an accuracy of 0.84, and reported an accuracy nearly 0.94 for one hour prediction). We can only conclude that the current model gives a reasonable prediction, Travel time estimation One benefit of the model-based method is that the travel time of each trip naturally comes out from it. It shows the histogram of the relative divergence between the model predicted and historical travel time. The samples that produces this histogram include all the trips whose real trip time falls between 20 min to 30 min. For most trips, the estimated travel time has an error smaller than 50%. However, there are a number of trips exhibits large errors. The reasons may due to the model or the data itself (it is possible that some taxi may stop during the trip). While the detailed analysis will be left for future work, the current result shows that the model provide a roughly unbiased travel time estimation.

5   Conclusions

In this paper we propose a coarse-grained cellular automata model for Urban traffic forecasting. We do not aim to faithfully model the real traffic system with fully details. Instead, limited by the computation power and data, we design a model to describe the effective dynamics of the taxis. By leaving many details in the traffic system to be unresolved, the model is simple enough to be implemented on a Urban-wide scale. It is interesting to mention the similarity between our method and numerical weather prediction. In the later, it is also impractical (due to limited computation power and unpredictable effects) to solve the real physical system in the nest scale. Instead, weather models are build on grids. Histogram of the relative error of estimated travel time. real travel time corresponds to the recorded time lapse between the destination and origin in the floating taxi data. Pried travel time for each trip is estimated based on the one hour ahead predicted traffic state. The red-dashed vertical line separates the positive (underestimation) and negative (over-estimation) values. Relatively low spatial resolution, leaving many detailed physical processes in the sub-grid level to be unresolved. As a result, the dynamics on the resolved coarse grid is an effective system [8], whose model parameters also need to be adjusted using the observation data. As part of an on-going work, our model can be considered as a prototype of Urban traffic forecasting system. We show that our model gives reasonable prediction to the Urban traffic. Currently we are using only the taxi GPS data. With more data in various form available in the future, there is ample room for further improvements. Some thoughts are discussed below. Incorporating the information of other vehicles The traffic flow of the taxis constitute a relative small portion of the total traffic in Ganzhou (about 10% based on rough estimation). To make the problem even more difficult, the ratio of taxis among all the vehicles changes over time and space. For example, in suburbs the taxi ratio tends to be high in the morning, while in urban regions tends to increase in the evening. For now, we only fitted the overall proportional change of taxis by rescaling the capacity everywhere using the same function (t). This is far from ideal. To solve this problem we need more information, say the total traffic volume on a number of representative roads. From them we should be able to estimate the proportion of taxis on these road, and then extrapolate the result to other roads that do not have the data. If we can correctly estimate the ratio of taxis on every roads, the accuracy of the model will likely to be greatly improved.

References

1. Doetsch, P., Kozielski, M., Ney, H.: Fast and robust training of recurrent neural networks for offline handwriting recognition. In: 2014 14th International Conference on Frontiers in Handwriting Recognition (ICFHR), pp. 279–284. IEEE (2014)
2. Graves, A., Mohamed, A.-R., Hinton, G.: Speech recognition with deep recurrent neural networks. In: 2013 IEEE International Conference on Acoustics, Speech and Signal Processing (ICASSP), pp. 6645–6649. IEEE (2013)
3. Glorot, X., Bordes, A., Bengio, Y.: Deep sparse rectifier neural networks. In: International Conference on Artificial Intelligence and Statistics, pp. 315–323 (2011)
4. Zhou, X., Taylor, J.: DTALite: a queue-based macroscopic traffic simulator for fast model evaluation and calibration. Cogent Eng. 1(1), 961345 (2014)
5. Sak, H., Senior, A., Beaufays, F.: Long short-term memory based recurrent neural network architectures for large vocabulary speech recognition. arXiv preprint arXiv:1402.1128 (2014)
6. Xu, K., et al.: Show, attend and tell: neural image caption generation with visual attention. arXiv preprint arXiv:1502.03044 (2015)
7. Bahdanau, D., Cho, K., Bengio, Y.: Neural machine translation by jointly learning to align and translate. arXiv preprint arXiv:1409.0473,2014.12
8. Li, P., Mirchandani, P., Zhou, X.: Hierarchical multiresolution traffic simulator for metropolitan areas: architecture, challenges, and solutions. Transp. Res. Rec.: J. Transp. Res. Board 2497, 63–72 (2015)

Hybrid Colliding Bodies Optimization for Solving Emergency Materials Transshipment Model with Time Window

Xiaopeng Wu[1], Yongquan Zhou[1,2], and Qifang Luo[1,2(✉)]

[1] College of Information Science and Engineering,
Guangxi University for Nationalities, Nanning 530006, Guangxi, China
l.qf@163.com
[2] Key Laboratory of Guangxi High Schools Complex System
and Computational Intelligence, Nanning 530006, China

Abstract. This paper introduces a time satisfaction function to build the emergency materials transshipment model, combining the traditional point to point transport model and hub-and-spoke distribution mode. The proposed model, emergency materials transshipment model with time window constraints, has two vital factors that are quantity and time of emergency material transportation. The quantity of materials is considered as the weight of time satisfaction. The total time satisfaction is the sum of product of the quantity and time satisfaction. A hybrid of colliding body's optimization (CBO) and genetic algorithm (GA) imbedding with linear programming algorithm is proposed to solve the problem through analyzing the trait of the model. The hybrid algorithm improves the performance of CBO algorithm in the discrete field. Experimental results demonstrate the efficiency of the model and algorithm.

Keywords: Emergency materials transshipment model
Time window constraints · Linear programming algorithm
Colliding body's optimization

1 Introduction

In recent years, natural disasters happened all over the world frequently. It caused a huge economic loss to human society every time. Especially, the earthquake disaster has destructive, difficult to predict and other characteristics. But how to dispatch emergency materials timely and efficiently is the key to ensure the effectiveness of emergency rescue work.

Many scholars have made some successful study about emergency material dispatch problem. Chang proposed greedy-search-based multi-objective genetic algorithm for emergency logistics scheduling [1]. This paper cans efficient dispatch relief to victims of disaster. Zhang uses discrete Particle Swarm Optimization (PSO) [2] to deal with emergency logistics center location problems [3]. Liu and Xie proposed dynamic programming and ant colony optimization and apply it to Emergency materials transportation model [4]. It helps government to make a transport plan of vehicles to rescue the disasters. Jiang and Wang use random process theory to establish emergency

© Springer Nature Singapore Pte Ltd. 2018
K. Li et al. (Eds.): ISICA 2017, CCIS 874, pp. 142–151, 2018.
https://doi.org/10.1007/978-981-13-1651-7_12

supplies distribution and decision model for multi-level network [5]. They proposed a modified discrete particle swarm optimization (MBPSO) to solve the problem. Many studies about emergency materials scheduling problem aim to minimize the unsatisfied demand, cost or total time as the objective function. Most model design is based on this criterion. In fact, minimizing the total time is similar to maximizing the time satisfaction. Both of them need to make a reaction after the disasters in a short time. But there is something different. Minimizing the total time does not consider the difference between the different demand points for the same arrival time of the emergency material. In addition, the different time points of the disaster area are not a linear relationship in terms of its effect. Based on these two points of view, this paper proposes an emergency materials transshipment model with time window constraints which introduces the time satisfaction function for emergency materials by considering the characteristics of the survivability probability of trapped people and combines with the two modes that are point-to-point transportation and transit transportation. The rest of this paper introduces the basic CBO algorithm [6] and a hybrid approach of colliding bodies Optimization (CBO) and Genetic Algorithm (GA) [7] imbedding with linear programming algorithm, and using the hybrid approach to solve the model. Finally, the experimental data are given to illustrate the effectiveness of the model and algorithm.

2   Emergency Materials Transshipment Model with Time Window Constraints

2.1   Model Description

Emergency materials transshipment model has m supply stations (supply points) and n disaster area (demand points), and there are l emergency materials transit stations (transit points) between the supply points and the demand points. The transportation time is known between the supply points, the transfer points and the demand points. The emergency material can be transported through the point-to-point direct transport mode from the supply point to the demand point, or through the transit transport mode by selecting the appropriate transit candidate points as a transit station. Two question need to be determined: One is which transit candidate points should be selected? The other is the transport quantity of goods through point-to-point transport mode and transit transport mode. The goal is to maximize the satisfaction of the affected area for emergency materials.

Hypothesis:

(1) The transport time is given among the supply points, demand points and transfer points.
(2) The maximum supply quantity of materials at each supply point is known.
(3) The demand quantity of materials at each demand point is known.
(4) The urgency of each demand points for materials is known.
(5) The total supply quantity of materials of each supply point is greater or equal to total demand quantity of materials of each demand point.

(6) The materials through the transit transport mode must be concentrated in the transit point, and then transported to the demand points.

(7) Each demand point up to only one transit point to service it.

(8) Based on the assumption that Fiedrich's [8] survivability probability function of the trapped people. Time satisfaction function is shown as:

$$g_j = e^{-t^2/\theta_j} \tag{1}$$

Where the j represents the jth demand point. t is the arrival time of materials. θ_j denotes the urgency of the jth demand point for arrival time.

2.2 Parameters and Variable Settings

The notations used in the formulation are shown as follows: M denotes point set of supply points, where the number of supply points is m. L denotes point set of transit points, where the number of transit points is l. D denotes point set of demand points, where the number of demand points is d. a_i is maximum supply quantity of emergency materials supply point i, $i \in M$. b_j is the demand quantity of emergency materials demand point j, $j \in D$. c_k is capacity of emergency materials transit point k, $k \in L$. t_{ij}^d denotes the time required for the material supply point i to the demand point j through the point to point transport mode. t_{ik}^h denotes the time required for the material supply point i to the transit point k. t_{kj}^h denotes the time required for the transit point k to the demand point j. θ_j denotes the urgency of the jth demand point for arrival time. x_{ij}^d denotes the quantity of the supply point i to be transported to the demand point j through the point to point transport mode. x_{ik}^h denotes the quantity of the supply point i to be transported to the transit point k. x_{kj}^h the quantity of the transit point k to be transported to the demand point j. If the value of y_{kj} is 1, it denotes that the demand point j is severed by transit point k. If its value is 0, it denotes that they do not have a relationship.

2.3 Mathematical Model

The objective function aims to calculate the maximum time satisfaction. The equality is shown as:

$$\max z = \sum_{j \in D} \sum_{i \in M} x_{ij}^d g_j \left(t_{ij}^d \right) + \sum_{j \in D} \sum_{i \in M} x_{kj}^h g_j \left(\max_{i \in M} \{ t_{ik}^h \delta (x_{ik}^h) \} + t_{kj}^h \right) \tag{2}$$

Subject to:

$$\sum_{j \in N} x_{ij}^d + \sum_{k \in L} x_{ik}^h \leq a_i, i \in M \tag{3}$$

$$\sum_{i\in M} x_{ij}^d + \sum_{k\in L} x_{ik}^h \le b_i, j \in D \tag{4}$$

$$\sum_{i\in M} x_{ik}^h = \sum_{j\in N} x_{kj}^h, j \in l \tag{5}$$

$$\sum_{j\in N} x_{kj}^h \le c_k \delta \left(\sum_{j\in N} y_{kj} \right), k \in l \tag{6}$$

$$x_{kj}^h \le y_{kj} c_k, k \in L, j \in D \tag{7}$$

$$\sum_{k\in L} y_{kj} \le 1, j \in D \tag{8}$$

$$x_{ij}^d, x_{ik}^h, x_{kj}^h \ge 0, \; y_{kj} \in \{0,1\}, i \in M, j \in D, k \in L \tag{9}$$

Where $\delta(x)$ is instruction function. When $x > 0$, the value of $\delta(x)$ is 1. Otherwise, its value is 0. In this model, the first item denotes the product of the quantity and time satisfaction of the goods using the point to point method. The second item has two steps. The first step is a collection of emergency materials from all the supply point to the transit point. The next step is transporting the collection to the demand point.

In these constraints, constraint Eq. (3) indicates that the total amount of the delivery of each supply point through the point-to-point and transit modes is limited by the maximum quantity of supply point. Constraint Eq. (4) indicates that the demand amount for each demand point is satisfied through two modes of transportation. Constraint Eq. (5) indicates the number of materials that the transit point receives is equal to the number that the transit point delivery. Constraint Eq. (6) indicates the number of materials that the transit point receives is limited by the maximum capacity of the transit point. Constraint Eq. (7) ensure that the transport quantity x_{kj}^h is limited by y_{kj} where is 0–1 variable. Constraint Eq. (8) indicates that there is no more than one transit point provides services for each demand point. Constraint Eq. (9) indicates that all the transport quantity meets the nonnegative condition. The relationship between the transit point and demand point is a 0–1 variable.

3 A Hybrid of Colliding Bodies Optimization and Genetic Algorithm

3.1 Colliding Bodies Optimization

Metaheuristic optimization, such as ant colony optimization (ACO) [9], artificial bee colony (ABC) [10], cuckoo search (CS) [11], bat algorithm (BA) [12]. Colliding body's optimization (CBO) is an efficient metaheuristic optimization algorithm. It has been applied to many problems. CBO algorithm is used to solve truss structures [13]. The combination of CBO and k-median method is applied to optimize domain decomposition [14]. The improved algorithm Improved Colliding Bodies Optimization is used to optimal power flow [15]. The basic principle of CBO is one-dimensional collisions. The agent solution is seen as the body. Two bodies move and collide. After the collision, the two bodies with mass will have new velocities and then both of them will get a new position. This collision improves the bodies move to a better place; consequently improve the agent get a better solution. Genetic Algorithm (GA) is an evolutionary algorithm which simulates the process of natural selection. Combining the characteristic of CBO and GA, this paper proposed a hybrid algorithm Genetic Algorithm-Colliding Bodies Optimization GACBO.

3.2   Genetic Algorithm

The most important process of Genetic Algorithm is genetic operator which includes selection operator, crossover operator and mutation operator. Selection operator usually uses the elite individual preservation strategy and roulette method. The highest fitness function value of individuals must be selected. According to the crossing probability p_c, a number of parents and pairing are selected to start cross operator. Mutation operator is necessary to randomly determine the mutated individuals according to a certain mutation probability p_m and to carry out the corresponding mutation operation.

3.3   CB Updating Mechanism and Corresponding Genetic Operations

This paper proposed updating mechanism is different from the Kaveh proposed the updating mechanism in construction site layout planning problem [16]. In emergency materials transshipment model, the equation $x_i^{new} = x_i + rand \circ v_i'$ is infeasibility. Because emergency materials transshipment model is a permutation problem. In this permutation problem, velocity of CBO algorithm is not a vector. It is considered as an exchange sequence between the stationary group and moving group. The body in stationary group has a corresponding body of moving group. Versa case is right. The value of the current variable is updated according to its reference. Each CB make crossover with its reference. Then each CB gets a new solution for the problem. The mutation process is that the variable of CB is generated random in the range [0–L]. It can increase the diversity of the solution.

3.4   Hybrid CBO and GA Algorithm for Solving Emergency Materials Transshipment with Time Window

According to the CB updating mechanism and corresponding genetic operations, the proposed hybrid algorithm (GACBO) is applied as follows. The objective function is for maximum value.

Step 1: The hybrid algorithm initialization. Utilizing the CB encoding scheme to initialize the possible combination. 0 denotes demand point get emergency materials through point to point transport mode. Other numbers denotes demand point get emergency materials through transit transport mode. The value of the number indicates which transit point is selected. Several notations are predefined:

$n = 50$. The number of CB;

$P_m = 0.25$. Mutation probability

$Iter_{max} = 500$. The maximum iteration of the hybrid algorithm

best. The best fitness function value of current iteration

Step 2: The second item of objective function is non-linear component. So, if given a y_{kj} which is satisfied Eq. (2), it can certain the value of $\max_{i \in M}\{t_{ik}^h \delta(x_{ik}^h)\}$. The nonlinear component of objective function is transformed to linear component. Utilizing the linear programming (linprog) algorithm to solve the model get the optimal solution and best value.

Step 3: Selection operator. The best CB with best fitness function value *best* is selected absolutely. And other CBs are selected by adopting roulette wheel selection approach. The probability of each CB is calculated by the Eq. (10).

$$P_i = \frac{fit(i)}{\sum\limits_{i=1}^{n} fit(i)} \tag{10}$$

Where P_i indicates the probability of selection. i indicates ith CB. $fit(i)$ denotes the fitness function value of the agent i. If the random number $r < P_i$, ith CB is selected.

Step 4: Creating groups. All the CBs are sorted ascendingly according to the satisfaction of transport plan. And then the CBs are divided into two groups. The bigger value of total time satisfaction is a group which named stationary group. The smaller value of total time satisfaction is a group which named moving group. Let the moving group go to collide with the corresponding stationary group.

Step 5: Determine the exchange sequence. The exchange sequence of stationary group is shown as:

$$v_i = x_i - x_{i+\frac{n}{2}} \ i = 1, 2, \ldots, \frac{n}{2} \tag{11}$$

The exchange sequence of moving group is shown as:

$$v_i = x_{i+\frac{n}{2}} - x_i \ i = \frac{n}{2} + 1, \frac{n}{2} + 2, \ldots, n \tag{12}$$

where v_i is the exchange sequence of ith CB. x_i is the ith transport plan before the collision.

Step 6: Update transport plan. New transport plan is determined by exchange sequence and transport plan before the collision. Choose *random* ($1 < random < D, D$ is the set of the demand points) values of position from the transport plan x_i before the collision and choose $(D - random)$ values of position from the exchange sequence v_i. Combine two group values, the new transport plan x_i^{new} is generated.

Step 7: Mutation operator. If random number r is smaller the mutation probability P_m. The CB is selected to mutate. One value or several value is regenerated random in the range of $[0, l]$ (l is the total number of transit points).

Step 8: If the current iteration is equal to $Iter_{max}$. Stop the algorithm. Otherwise, go to Step 2.

4   Simulation Experiments and Result Analysis

Assuming transport emergency materials from 6 supply points to 8 demand points. There are 4 transit points between supply points and demand points. Generate 10 sets of data random for the model. Explanation of data values: Material supply quantity of supply point is a random integer in the range of $[100, 300]$. Material demand quantity of demand point is a random integer in the range of $[80, 100]$. The capacity of demand point is a random integer in the range of $[30, 80]$. The transport time from supply point to demand point is a random integer in the range of $[50, 150]$. The transport time from supply point to transit point is a random integer in the range of $[10, 50]$. The transport time from supply point to transit point is a random integer in the range of $[10, 50]$. The transport time from transit point to demand point is a random integer in the range of $[10, 30]$. The parameter θ_j of objective function is a random integer in the range of $[1000, 3000]$. $m = 6$ is the number of the supply points. $d = 8$ is the number of the demand points. $l = 4$ is the number of the transit points. $popsize = 50$ is the number of CBs $P_m = 0.25$ is the mutation probability. $Iter_{max} = 500$ is maximum iteration. 10 group random data calculation results are shown as Table 1.

Table 1. 10 group random data calculation results

Group	Best	Worst	Mean	Std
1	276.2656	258.5969	267.5028	6.059032
2	221.6576	206.6108	214.3216	4.827426
3	232.184	207.2623	221.1921	7.852979
4	254.1106	227.6731	238.0442	8.566014
5	179.0179	164.3387	171.171	4.560152
6	209.3995	197.7349	204.0964	3.067934
7	220.9763	203.4297	213.9519	5.704231
8	219.2715	206.8644	212.7244	3.734486
9	272.8501	249.7595	262.3692	7.613957
10	201.165	177.5906	188.9624	5.3315

Every group data runs twenty times independently. Each group data including best, worst, mean, std is different from others. Because the parameters of initialization are generated random. The results have large values and small values. The values of std are small indicating the good stability of the hybrid algorithm. Here, The tenth group data is selected to detail. Table 2 is the transport schedule. The value of the table is the specific transport time.

Table 2. Transport schedule

		Demand point j								Transit point k			
		1	2	3	4	5	6	7	8	1	2	3	4
Supply point i	1	81	102	62	133	70	58	110	147	22	18	39	16
	2	81	95	89	97	134	69	118	138	35	47	11	43
	3	142	133	102	88	107	145	74	115	11	19	32	40
	4	56	97	120	93	87	131	143	59	16	24	19	24
	5	120	123	54	106	60	62	103	114	39	16	47	35
	6	100	90	73	97	88	147	51	57	31	25	31	34
Transit point k	1	22	29	24	14	19	23	11	24				
	2	22	23	19	17	14	11	25	12				
	3	11	29	26	21	22	24	11	17				
	4	18	28	27	14	26	23	28	13				

Where θ = [1224, 1863, 1598, 2278, 1165, 1065, 2605, 2481]; a = [177, 297, 221, 172, 252, 244]; b = [90, 82, 82, 84, 94, 94, 97, 92]; c = [69, 61, 36, 42].

Table 3. Best solution

		Demand point j								Transit point k			
		1	2	3	4	5	6	7	8	1	2	3	4
Supply point i	1						54				61		
	2											36	
	3				15					69			
	4	54											42
	5			82		94							
	6		40					97	71				
Transit point k	1				69								
	2						40		21				
	3	36											
	4		42										

Table 3 is the best solution. The value of the table is the transport quantity from one point to other point. The blank box shows there are no materials transported between the relative points. Convergence curve of fitness function value is shown as Fig. 1.

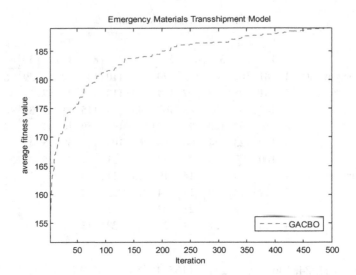

Fig. 1. Convergence curve of fitness function value

5 Conclusions

In this paper, the hybrid algorithm of Genetic Algorithm and Colliding Bodies Optimization (GACBO) has been applied to the emergency material transshipment model with time window constraints. The proposed hybrid algorithm shows good performance. GACBO improves the ability of CBO algorithm in solving discrete problem. Mutation operator improves ability of the algorithm to prevent premature convergence and escape local optimum. In CB encoding scheme, successfully solving the problem of CB encoding and the problem of bring CB into the objective function. The experiment data proves the hybrid algorithm can deal with different scale problem.

Acknowledgements. This work is supported by the National Science Foundation of China under Grant Nos. 61563008 and 61463007, and the Project of Guangxi Natural Science Foundation under Grant No. 2016GXNSFAA380264.

References

1. Chang, F.-S., Wu, J.-S., Lee, C.-N., Shen, H.-C.: Greedy-search-based multi-objective genetic algorithm for emergency logistics scheduling. Expert Syst. Appl. **41**, 2947–2956 (2014)
2. Kennedy, J., Eberhart, R.: Particle swarm optimization. In: Proceedings of the IEEE International Conference on Neural Networks, Perth, Australia, vol. IV, pp. 1942–1948 (1995)
3. Zhang, Y.-Y.: Choice of emergency logistics center location based on particle swarm optimization. Comput. Model. New Technol. **18**(12A), 392–395 (2014)

4. Liu, J., Xie, K.: Emergency materials transportation model in disasters based on dynamic programming and ant colony optimization. Kybernetes **46**(4), 656–671 (2017)
5. Qi, L., Jiang, D., Wang, Z.: A modified discreet particle swarm optimization for a multi-level emergency supplies distribution network. Int. J. Eng. (IJE) Trans. C: Aspects **29**(3), 359–367 (2016)
6. Kaveh, A., Mahdavai, V.R.: Colliding bodies optimization: a novel meta-heuristic method. Comput. Struct. **139**, 18–27 (2014)
7. Goldberg, D.E.: Genetic Algorithms in Search Optimization and Machine Learning. Addison-Wesley, Boston (1989)
8. Fiedrich, F., Gehbauer, F., Rickers, U.: Optimized resource allocation for emergency response after earthquake disasters. Saf. Sci. **35**, 41–57 (2000)
9. Socha, K., Dorigo, M.: Ant colony optimization for continuous domains. Eur. J. Oper. Res. **185**(3), 1155–1173 (2008)
10. Karaboga, D., Basturk, B.: A powerful and efficient algorithm for numerical function optimization: artificial bee colony (ABC) algorithm. J. Glob. Optim. **39**(3), 459–471 (2007)
11. Yang, X.S., Deb, S.: Cuckoo search via levy flights. In: World Congress on Nature and Biologically Inspired Computing (NaBIC 2009), pp. 210–214. IEEE, USA (2009)
12. Yang, X.S.: A new metaheuristic bat-inspired algorithm. In: Gonzalez, J.R., Pelta, D.A., Cruz, C. (eds.) Nature Inspired Cooperative Strategies for Optimization. SCI, vol. 284, pp. 65–74. Springer, Berlin (2010). https://doi.org/10.1007/978-3-642-12538-6_6
13. Kaveh, A., Mahdavai, V.R.: Colliding bodies optimization method for optimum design of truss structures with continuous variables. Adv. Eng. Softw. **70**, 1–12 (2014)
14. Kaveh, A., Mahdavi, V.R.: Optimal domain decomposition using colliding bodies optimization and k-median method. Finite Elem. Anal. Des. **98**, 41–49 (2015)
15. Bouchekara, H.: Optimal power flow using an improved colliding bodies optimization algorithm. Appl. Soft Comput. **42**, 119–131 (2016)
16. Kaveh, A.: Construction site layout planning problem using two new meta-heuristic algorithms. Iran. J. Sci. Technol. Trans. Civ. Eng. **40**, 263–275 (2016)

A Dual Internal Point Filter Algorithm Based on Orthogonal Design

Yijin Yang, Tianyu Huo, Bin Lan, and Sanyou Zeng[⊠]

School of Mechanical Engineering and Electronic Information,
China University of Geosciences, Wuhan 430074, China
1064935578@qq.com, 1837889013@qq.com, 874432599@qq.com,
sanyouzeng@gmail.com

Abstract. The Primal-dual interior-point methods with a filter are one of hot issues in optimization with both equality and inequality constraints. Extensive attention has been paid and great progress has been made for a long time. Interior-point methods not only have polynomial complexity, but are also highly efficient in practice.

In this paper, we first generalize the dual interior filter algorithm for referred paper. Then a dual interior filter algorithm based on the orthogonal design is proposed. The orthogonal design is used to evenly sample the solution space, then the points with small constraint violations are chosen as the initial points of the algorithm. The dual interior point filter algorithm based on orthogonal design (DIPFA-OD) choosing the initial points is compared with the dual interior point filter algorithm based on random initial points (DIPFA-RND). The experimental results show that DIPFA-OD has a faster convergence speed and obtains better objective values than DIPFA-RND.

Keywords: Dual interior filter algorithm · Orthogonal design
Nonlinear programming

1 Introduction

The growing interest in efficient optimization has led to the development of in-point or barrier methods. The early method is proposed by Gould and Toint [1] that focused on solving equality constrained problems. The latter method has the additional capability of being able to solve problems with both equality and inequality constraints [2,3].

In recent years, more efficient algorithms with global convergence have been proposed [4,5]. The use of filter algorithms to improve poor initial points provides an alternative function to ensure that the entire algorithm has global convergence. The global convergence of the interior point method with filter algorithm is analyzed in [6].

The paper we present a more efficient method that it increases the convergence speed by combining the orthogonal experiment design with the original

---

Y. Yang and T. Huo—Contributed equally to this work and should be considered co-first authors.

interior-point method. The orthogonal design is used to generate the initial point of the algorithm, which will get several initial points with little constraint violation, so that the algorithm has better global search ability. Compared with the random initial point, the initial points generated by the orthogonal array are uniformly distributed, which can effectively avoid the local optimization and accelerate the convergence speed of the algorithm.

The paper is organized as follows. Section 2 presents the related work, which introduce the overall interior-point algorithm and orthogonal design. In Sect. 3, we generalize the new problem with equality and inequality constraints. In Sect. 4, we proved the improved method is more efficient through a series of test questions.

2   Related Work

2.1   Orthogonal Experimental Design

Research on orthogonal experimental design can be found in [7].

2.2   Existing Dual Internal Point Filter Algorithm

Nonlinear programming is a mathematical programming with nonlinear constraints or objectives. One form is as in Eq. (1)

$$
\begin{aligned}
min \quad & f(x) \\
s.t. \quad & \mathbf{c}(x) = 0 \\
& x \geq 0
\end{aligned}
\tag{1}
$$

A more generalized form has with both equality and inequality constraints in Eq. (2).

$$
\begin{aligned}
min \quad & f(\mathbf{x}) \\
s.t \quad & \mathbf{h}(\mathbf{x}) = 0 \\
& \mathbf{g}(\mathbf{x}) \leq 0
\end{aligned}
\tag{2}
$$

Among them, $\mathbf{x} = (x_1, x_2, \ldots, x_n)^T$ represents the decision variable, $f(\mathbf{x})$ represent the target space, $\mathbf{h}(\mathbf{x}) = (h_1(\mathbf{x}), h_2(\mathbf{x}), \ldots, h_l(\mathbf{x}))^T$ represents the equality constraints, $\mathbf{g}(\mathbf{x}) = (g_1(\mathbf{x}), g_2(\mathbf{x}), \ldots, g_m(\mathbf{x}))^T$ represent the inequality constraints.

Wachter [6] proposed a dual internal point filter algorithm to solve the problem Eq. (1). For the details of the algorithm, please see [6]. This paper will generalize the dual internal point filter algorithm to solve the general problem Eq. (2) in Sect. 3.

3   Generalizing the Dual Internal Point Filter Algorithm

Dual interior point filter algorithm, when dealing with inequality constraints, generally transforms it into an equality constraint, and then processes it.

Here, the relaxation variables $\mathbf{s} = (s_1, s_2, \ldots, s_m)^T$ are defined, subject to $\mathbf{g}(\mathbf{x}) + \mathbf{s} = 0$. So that the original problem is transformed into:

$$
\begin{aligned}
min \quad & f(\mathbf{x}) \\
s.t \quad & h(\mathbf{x}) = 0 \\
& \mathbf{g}(\mathbf{x}) + \mathbf{s} = 0, \quad s \geq 0.
\end{aligned} \tag{3}
$$

According to the barrier method, the above problem can be transformed into the following problem:

$$
\begin{aligned}
min \quad & \varphi_\mu(\mathbf{x}) = f(\mathbf{x}) - \mu \sum_{i=1}^{m} \ln s_i \\
s.t \quad & h(\mathbf{x}) = 0 \\
& g(\mathbf{x}) + s = 0, \quad s \geq 0.
\end{aligned} \tag{4}
$$

Here, the barrier parameter μ is defined to control the speed of relaxation variable s trend to 0. Then the Lagrange function is:

$$
L(\mathbf{x}, s, \lambda_h, \lambda_g) = f(\mathbf{x}) - \mu \sum_{i=1}^{m} \ln s_i + \lambda_h h(\mathbf{x})^T + \lambda_g (g(\mathbf{x}) + \mathbf{s})^T. \tag{5}
$$

$\lambda_h = (\lambda_{h1}, \lambda_{h2}, \ldots, \lambda_{hl})^T$, $\lambda_g = (\lambda_{g1}, \lambda_{g2}, \ldots, \lambda_{gm})^T$ are the Lagrange multipliers. The proposed method is to compute the approximate solution (4) of the barrier problem with a fixed value of μ, then reduce the barrier parameters and continue to solve the approximate solution to the previous barrier problem.

The **KKT** condition for the above barrier problem is:

$$
\begin{aligned}
\nabla f(\mathbf{x}) + \lambda_h \nabla h(\mathbf{x})^T + \lambda_g \nabla g(\mathbf{x})^T &= 0 \\
h(\mathbf{x}) &= 0 \\
g(\mathbf{x}) + \mathbf{s} &= 0 \\
s\lambda_g - \mu &= 0.
\end{aligned} \tag{6}
$$

We define the optimal error for the barrier problem as follows:

$$
\begin{aligned}
E_\mu(\mathbf{x}, s, \lambda_h, \lambda_g) := max \Big\{ & \frac{\|\nabla f(\mathbf{x}) + \lambda_h \nabla h(\mathbf{x})^T + \lambda_g \nabla g(\mathbf{x})^T\|_\infty}{s_d}, \\
& \|h(\mathbf{x})\|_\infty, \|g(\mathbf{x}) + s\|_\infty, \frac{\|s\lambda_g - \mu\|_\infty}{s_c} \Big\}
\end{aligned} \tag{7}
$$

Solution of the Barrier Problem. In order to solve the barrier problem, we applied the damping Newton method to the original dual equations when dealing with the formula (4). Let the formula expand at the initial point according to Taylor's formula, and select the first term and the second term as the approximate expression of the formula (6).

$$\begin{pmatrix} W_k & 0 & H_k & G_k \\ 0 & A_k & 0 & I \\ H_k^T & 0 & 0 & 0 \\ G_k^T & I & 0 & 0 \end{pmatrix} \begin{pmatrix} d_k^x \\ d_k^s \\ d_k^{\lambda_h} \\ d_k^{\lambda_g} \end{pmatrix} = - \begin{pmatrix} \nabla f(x_k) + H_k \lambda_{h,k} + G_k \lambda_{g,k} \\ -\frac{\mu}{s} + \lambda_{g,k} \\ h(x_k) \\ g(x_k) + s. \end{pmatrix} \tag{8}$$

Here, W_k donate the Hessian $\nabla_{xx}^2 L(x_k, s_k, \lambda_h, \lambda_g)$ of the Lagrangian function (for the original problem (1)),

$$L(x_k, s_k, \lambda_h, \lambda_g) = f(x) + h(\mathbf{x})^T \lambda_h + g(\mathbf{x})^T \lambda_g. \tag{9}$$

H_k is the gradient of the equality constraint; G_k is the gradient of the inequality constraint; $\nabla_{xx}^2 f(x_k)$ is the Hessian matrix of the objective function with respect to x; $\nabla_{xx}^2 h(x_k)$ is the Hessian matrix of the equality Constraint with respect to x; $\nabla_{xx}^2 g(x_k)$ is the Hessian matrix of the inequality constraint with respect to x; otherwise, $A_k = \frac{\lambda_g}{s}$.

Instead of solving the nonsymmetric linear system (8) directly, the proposed method computes the solution equally by first solving the smaller problem:

$$\begin{pmatrix} W_k + G_k G_k^T & H_k \\ H_k^T & 0 \end{pmatrix} \begin{pmatrix} d_k^x \\ d_k^{\lambda_h} \end{pmatrix} = - \begin{pmatrix} \nabla f(x_k) + H_k \lambda_{h,k} + G_k \frac{\mu}{s} + G_k A_k (g(x_k) + s) \\ h(x_k). \end{pmatrix} \tag{10}$$

Then the vector $d_k^s; d_k^{\lambda_g}$ can be obtained from:

$$\begin{aligned} d_k^s &= -g(x_k) - s - G_k^T d_k^x \\ d_k^{\lambda_g} &= \lambda_g (g(x_k) + G_k d_k^x)/s. \end{aligned} \tag{11}$$

In order to be able to calculate the search direction, we need to ensure that the iterative matrix is non-singular, and we must ensure that the matrix in the upper-left corner of the matrix (10) is positive definite in the null space of H_k^T. In our implementation, we modified the matrix:

$$\begin{pmatrix} W_k + G_k G_k^T + \delta_w I & H_k \\ H_k^T & -\delta_c I \end{pmatrix} \begin{pmatrix} d_k^x \\ d_k^{\lambda_h} \end{pmatrix}$$
$$= - \begin{pmatrix} \nabla f(x_k) + H_k \lambda_{h,k} + G_k \frac{\mu}{s} + G_k A_k (g(x_k) + s) \\ h(x_k) \end{pmatrix} \tag{12}$$

for $\delta_w, \delta_c \geq 0$.

After calculating the search directions for (8) and (10), we also need to find the step size $\alpha_k, \alpha_k^{lambda}$ in (0, 1) to get the next iteration is:

$$\begin{aligned} x_{k+1} &:= x_k + \alpha_k d_k^x \\ \lambda_{k+1} &:= \lambda_k + \alpha_k d_k^\lambda \\ \lambda_{h,k+1} &:= \lambda_{h,k} + \alpha_k^\lambda d_k^{\lambda_h} \\ \lambda_{g,k+1} &:= \lambda_{g,k} + \alpha_k^\lambda d_k^{\lambda_g}. \end{aligned} \tag{13}$$

Since x and λ_g are both positive at an optimal solution of the barrier problem (4), this property is maintained for all iterates. It is attained using the *fraction-to-the-boundary*

$$\alpha_k^{max} := max\{\alpha \in (0,1] : x_k + \alpha d_k^x \geq (1 - \tau_j)x_k\}$$
$$\alpha_k^{\lambda} := max\{\alpha \in (0,1] : \lambda_k + \alpha d_k^{\lambda} \geq (1 - \tau_j)\lambda_k\}. \tag{14}$$

To ensure global convergence, the step size $\alpha_k \in (0, \alpha_k^{max}]$ is determined by a backtracking search procedure that reduces the trial step size $\alpha_{k,l} = 2^{-l}\alpha_k^{max}(with l = 0, 1, 2, \ldots)$.

Before the next section examines this procedure, in order to satisfy a requirement for the convergence, we set:

$$\lambda_{k+1} = max\left\{ min\left\{\lambda_{k+1}, \frac{\kappa_{\Sigma}\mu_j}{s_{k+1}}\right\}, \frac{\mu_j}{\kappa_{\Sigma}s_{k+1}}\right\} \tag{15}$$

for fixed $\kappa_{\Sigma} \geq 1$ after each procedure.

A Line-Search Filter Method. The filtering method was originally proposed by Fletcher and Leyffer fletcher2002Nonlinear. The basic idea behind this approach is to interpret (4) as a two-objective optimization problem in the context of solving the hurdle of μ_j (4), minimizing the objective function $\varphi_{\mu_j}(x)$ and constraint violation $\theta(x) = max\{h(x), g(x) + s\}$. Following this paradigm, it is acceptable if $x_k(\alpha_{k,l}d_k^x)$ during the retrograde row search is sufficient if it leads to sufficient progress compared to the current iteration if

$$\theta(x_k(\alpha_{k,l})) \leq (1 - \gamma_{\theta})\theta(x_k)$$
$$or \quad \varphi_{\mu_j}(x_k(\alpha_{k,l})) \leq \varphi_{\mu_j}(x_k) - \gamma_{\varphi}\theta(x_k) \tag{16}$$

holds for fixed constants $\gamma_{\theta}, \gamma_{\varphi} \in (0,1)$. However, the above criterion is replaced by requiring sufficient progress in the barrier objective function, whenever for the current iterate we have $\theta(x_k) \leq \theta^{min}$, for some constant $\theta^{min} \in (0, \propto]$, and the following "switching condition"

$$\nabla\varphi_{\mu_j}(x_k)^T d_k^x < 0 \quad and \quad \alpha_{k,l}[-\nabla\varphi_{\mu_j}(x_k)^T d_k^x]^{s_{\varphi}} > \delta[\theta(x_k)]^{s_{\theta}} \tag{17}$$

with constants $\delta > 0, s_{\theta} > 1, s_{\varphi} \geq 1$ holds. If $\theta(x_k) \leq \theta^{min}$ and (19) is true for the current step size $\alpha_{k,l}$, the trial point has to satisfy the Armijo condition

$$\varphi_{\mu_j}(x_k(\alpha_{k,l})) \leq \varphi_{\mu_j}(x_k) + \eta_{\varphi}\alpha_{k,l}\nabla\varphi_{\mu_j}(x_k)^T d_k^x \tag{18}$$

instead of (16), in order to be acceptable. Here, $\eta_{\varphi} \in (0, \frac{1}{2})$ is a constant.

For each iteration, the algorithm includes a filter, defined as $F_k := \left\{(\theta, \varphi) \in \mathbb{R}^2 : \theta \geq 0\right\}$, The filter F_k contains the default value θ and the target function φ. During a linear search, $\varphi_{\mu_j}(x_k(\alpha_a t F_k K, L)))$, then the point is rejected by the filter. At the beginning of the optimization, the filter is initialized to

$$F_0 := \left\{ (\theta, \varphi) \in \mathbb{R}^2 : \theta \geq \theta^{max} \right\} \tag{19}$$

for some θ^{max}, so that the algorithm will never allow trial points to be accepted that have a constraint violation larger than θ^{max}. Later, the filter is augmented after every iteration, where the accepted test step does not satisfy the switching condition (17), or the Armijo condition (18) does not hold. This ensures that the iteration can not return to the neighborhood of x_k. On the other hand, the filter remains unchanged if both (17) and (18) hold an acceptable step size. The update formula is as follows:

$$F_{k+1} := F_k \cup \left\{ (\theta, \varphi) \in \mathbb{R}^2 : \theta \geq (1 - \gamma_\theta)\theta(x_k) \quad and \quad \varphi \geq \varphi_{\mu_j}(x_k) - \gamma_\varphi \theta(x_k) \right\}, \tag{20}$$

In general, this process ensures that the algorithm can not loop, for example, between two points that alternately reduce the constraint violation and the barrier objective function.

Finally, in some cases, it may not be possible to find the trial step size $\alpha_{k,l}$ that satisfies the above criteria, so that the algorithm has to start again. So we use the linear model of the correlation function to approximate the minimum expected step size. On the other hand, if $\nabla \varphi_{\mu_j}(x_k)^T d_k^x < 0$, from (17) the trial step size $\alpha_{k,l}$ have to satisfy the following condition:

$$\alpha_{k,l} > \frac{\delta[\theta(x_k)^{s_\theta}]}{[-\nabla \varphi_{\mu_j}(x_k)^T d_k^x]^{s_\varphi}}. \tag{21}$$

However, when the trial step size $\alpha_{k,l}$ does not satisfy condition (17), we will refer to the following to decide whether to move to the *feasibility restoration phase*:

$$\theta(x_k + \alpha d_k^x) = \theta(x_k) - \alpha\theta(x_k)$$
$$\varphi_{\mu_j}(x_k + \alpha d_k^x) = \varphi_{\mu_j}(x_k) + \nabla \varphi_{\mu_j}(x_k)^T d_k^x. \tag{22}$$

Obtained from (16), when $\alpha_{k,l} \geq \gamma_\theta$ and $\nabla \varphi_{\mu_j}(x_k)^T d_k^x < 0$, we have $\alpha_{k,l} \geq \frac{\gamma_\varphi \theta(x_k)}{-\nabla \varphi_{\mu_j}(x_k)^T d_k^x}$. For this, we define

$$\alpha_k^{min} := \begin{cases} min\left\{ \gamma_\theta, \alpha_{k,l} \geq \frac{\gamma_\varphi \theta(x_k)}{-\nabla \varphi_{\mu_j}(x_k)^T d_k^x}, \alpha_{k,l} > \frac{\delta[\theta(x_k)^{s_\theta}]}{[-\nabla \varphi_{\mu_j}(x_k)^T d_k^x]^{s_\varphi}} \right\} \\ \qquad if \quad \nabla \varphi_{\mu_j}(x_k)^T d_k^x < 0 \quad and \quad \theta(x_k) \leq \theta^{min} \\ min\left\{ \gamma_\theta, \alpha_{k,l} \geq \frac{\gamma_\varphi \theta(x_k)}{-\nabla \varphi_{\mu_j}(x_k)^T d_k^x} \right\} \\ \qquad if \quad \nabla \varphi_{\mu_j}(x_k)^T d_k^x < 0 \quad and \quad \theta(x_k) \leq \theta^{min} \\ \gamma_\theta \\ \qquad other \quad conditions \end{cases} \tag{23}$$

If the backtrack search encounters a trial step of $\alpha_{k,l} \leq \alpha_k^{min}$, then the algorithm starts again. Here, the algorithm will try to find a new iteration $x_{k+1} > 0$ if the trial step size is acceptable for the current filter and satisfies (16).

Second-Order Corrections (SOC). Many non-linear optimization methods use second-order corrections to improve the situation when the test point is rejected. A second-order correction (SOC) aims to reduce the infeasibility. If the initial trial step size $\alpha_{k,0}$ is rejected and once $\theta\left(x_k\left(\alpha_{k,0}\right)\right) \in \theta\left(x_k\right)$, a second-order correction $d_k^{x,soc}$ is worked out:

$$H_k^T d_x^{x,soc} + h\left(x_k + \alpha_{k,0}d_k^x\right) = 0. \tag{24}$$

The corrected search direction is obtain from

$$d_k^{x,cor} = \alpha_{k,0}d_k^x + d_k^{x,soc}. \tag{25}$$

To prevent extra matrix factorization, we use (10) to calculate the entire calibration step $d_k^{x,cor}$ from

$$\begin{pmatrix} W_k + \Sigma_k + \delta_w I & H_k \\ H_k^T & -\delta_c I \end{pmatrix} \begin{pmatrix} d_k^{x,cor} \\ d_k^{\lambda_h} \end{pmatrix}$$
$$= -\begin{pmatrix} \nabla f\left(x_k\right) + K_k \lambda_{h,k} + G_k \lambda_{g,k} + G_k\left(\frac{\mu}{s} + A_k g_k^{soc}\right) \\ h_k^{soc} \end{pmatrix}. \tag{26}$$

We choose

$$\begin{aligned} h_k^{soc} &= \alpha_{k,0}h\left(x_k\right) + h\left(x_k + \alpha_{k,0}d_k^x\right) \\ g_k^{soc} + s_k^{soc} &= \alpha_{k,0}\left(g\left(x_k\right) + s_k\right) + g\left(x_k + \alpha_{k,0}d_k^x\right) + s_{k+1}, \end{aligned} \tag{27}$$

which will be get from (26), (24) and (25).

When the corrected search direction is computed, we adapt the fraction-to-the-boundary rule

$$\alpha_k^{soc} := max\left\{\alpha \in (0,1] : x_k + \alpha d_k^{x,cor} \geq (1 - \tau_j)x_k\right\}, \tag{28}$$

and then check if the trial point $x_k^{soc} := x_k + \alpha_k^{soc}d_k^{x,cor}$ is acceptable to the filter.

If the new trial point meets the above criteria, it will be accepted as the next iteration point. If not, we apply the second-order correction again until the correction step does not reduce the constraint violation of a fraction $k_{soc} \in (0,1)$ or has performed a second-order correction. At this time, the original search direction d_k^x is saved and a regular backtrack search is performed using $\alpha_{k,1} = \frac{1}{2}\alpha_{k,0}$.

The Algorithm. We give the concrete steps of the Algorithm 1 to solve the barrier problem (4) in the following part.

In our implementation, the l_1 norm is used to determine if it is infeasible, $\theta(x) = \|c(x)\|_1$. The values of the constants in our implementation are as follows: $\theta_\mu = 1.5; \tau_{min} = 0.99; \gamma_\theta = 10^{-5}; \gamma_\varphi = 10^{-5}; \delta = 1; \kappa_\epsilon = 10; \kappa_\mu = 0.2; \gamma_\alpha = 0.05; s_\theta = 1.1; \kappa_{soc} = 0.99; p^{max} = 4; s_\varphi = 2.3; \eta_\varphi = 10^{-4}; \mu_0 = 0.1; \epsilon_{tol} = 10^{-8}; \theta^{max} = 10^4, max\{1, \theta(x_0)\}$ and $\theta^{min} = 10^{-4}, max\{1, \theta(x_0)\}$.

Algorithm 1. LINE-SEARCH FILTER BARRIER METHOD

Input: Starting point $(x_0, s_0, \lambda_{h,0}, \lambda_{g,0})^T$ with $s_0 > 0$; Initialize the barrier parameters $\mu_0 > 0$; constants $\epsilon_{tol} > 0$; $\lambda_\epsilon > 0$; $\gamma_\theta, \gamma_\varphi \in (0,1)$; $\delta > 0$; $s_{max} \geq 1$, $\kappa_\mu \in (0,1)$; $\kappa_{soc} \in (0,1)$; $\theta_\mu \in (1,2)$; $\tau_{min} \in (0,1)$; $\kappa_\Sigma > 1$; $\gamma_\alpha \in (0,1]$; $s_\theta > 1$; $s_\varphi \geq 1$; $\theta^{max} \in (\theta(x_0), \infty)$; $\theta^{min} > 0$; $\eta_\varphi(0, \frac{1}{2})$; $p^{max} \in \{0, 1, 2, 3, ...\}$.

Step 1. Initialize the iteration calculator $j \leftarrow 0$ and $k \leftarrow 0$, and the filter F_0 can be initialized from the previous formula (19).

Step 2. Check whether global convergence meets the requirements, which refer to $E_0(x_k, s_k, \lambda_{h,k}, \lambda_{g,k}) \leq \epsilon_{tol}$ and then stop.

Step 3. Check whether convergence for the barrier problem meets the requirements. If $E_0(x_k, s_k, \lambda_{h,k}, \lambda_{g,k}) \leq \kappa_\epsilon \mu_j$, then:

    Step 3.1. Compute μ_{j+1} and τ_{j+1}, and set $j \leftarrow j + 1$;

    Step 3.2. Initialize the filter $F_k \leftarrow \{(\theta, \varphi) \in \mathbb{R}^2 : \theta \geq \theta^{max}\}$;

Step 4. Compute the search direction. Compute $(d_k^x, d_k^s, d_k^{\lambda_h}, d_k^{\lambda_g})^T$ from (12), and δ_w and δ_c are determined by Inertia Correction in [6].

Step 5. Backtrack line search.

    Step 5.1. Initialize the line search, whose key is α_k^{max} from (14).

    Step 5.2. Compute a new trial point.Set $x_k(\alpha_k, l)$ from (13).

    Step 5.3. Check whether θ and φ meet the requirements with the filter. If satisfy $\left(\theta(x_k(\alpha_{k,l})), \varphi_{\mu_j}(x_k(\alpha_{k,l}))\right) \in F_k$, filter rejects the trial step and go to Step 5.5.

    Step 5.4. Check whether the current iteration is declining enough. We have to judge whether θ meet the Two Accelerating Heuristics in [6]. If satisfy the conditions, then enter Step 6. Otherwise, go to the next step.

    Step 5.5. Initialize the SOC. If $\theta(x_{k,0} < \theta_k)$ or $l > 0$, skip the SOC and go to Step 5.10. Otherwise, initialize the SOC counter $p \leftarrow 1$, $g_k^{soc} + s_k^{soc}$ and h_k^{soc} from (27). Initialize $\theta_{old}^{soc} \leftarrow \theta(x_k)$.

    Step 5.6. Compute the SOC direction. Compute $d_k^{x,cor}$ and d_k^λ from (26), α_k^{soc} from (28), and x_k^{soc} from (13).

    Step 5.7. Check whether θ and φ meet the requirements with the filter (in SOC). If satisfy $\left(\theta(x_k^{soc}), \varphi_{\mu_j}(x_k^{soc})\right) \in F_k$, filter rejects the trial step and go to Step 5.10.

    Step 5.8. Check whether the current iteration is declining enough(in SOC). We have to judge whether θ whether meet the Two Accelerating Heuristics in [6]. If satisfy the conditions, then enter Step 6. Otherwise, go to the next step.

    Step 5.9. Next SOC. If $\theta(x_k^{soc}) > \kappa_{soc}\theta_{old}^{soc}$ or $p = p^{max}$, stop and enter next step. Otherwise, increase the SOC counter $p \leftarrow p+1$, and set $h_k^{soc} = \alpha_k^{soc}h_k^{soc} + h(x_k^{soc})$, $g_k^{soc} + s_k^{soc} = \alpha_k^{soc}(g_k^{soc} + s_k^{soc}) + g(x_k^{soc}) + s_k^{soc}$ and $\theta_{old}^{soc} \leftarrow \theta(x_k^{soc})$. Go back to Step 5.6.

    Step 5.10. Choose the new trial step size. Set $\alpha_{k,l+1} = \frac{1}{2}\alpha_{k,l}$, $l \leftarrow l+1$.If the trial step size is regarded too small, $\alpha_{k,l} < \alpha_k^{min}$ with α_k^{min} defined in (23), restart. Otherwise, go back to Step 5.2.

Step 6. Accept the trial point that meets requirement. Set $\alpha_k := \alpha_{k,l}$, and update $s_{k+1}, \lambda_{h,k+1}, \lambda_{g,k+1}$ and α_k^λ by applying the formula (14). Apply (15) to correct $\lambda_{g,k+1}$ if necessary.

Step 7. Update the filter if necessary. If α_k can't meet (17) or (18), update the filter using (20). Otherwise leave the filter remained, set $F_{k+1} := F_k$. Go to Step 8.

Step 8. Continue at the next iteration. Increase the iteration counter $k \leftarrow k+1$ and return to Step 2.

4 DIPFA-OD

4.1 The Algorithm

Based on the original dual interior point filter algorithm, DIPFA-OD Algorithm 2 is raised.

The value of the constants in our implementation is level $m = 29$.

Algorithm 2. DIPFA-OD

Input: Identify the number of factors n and level m.

Step 1. Evenly sample the solution space by the orthogonal design.

Step 2. Order the violation values from small to large and choose the first three as the initial point.

Step 3. Dual internal point filter algorithm. Execute Algorithm 1. If the last violation value is not near zero, we go to Step 4, Otherwise, go to Step 5.

Step 4. Randomly generate the initial point and iterations accumulate. Go back to Step 3.

Step 5. Print the result.

4.2 Numerical Experiments

In order to test the dual internal point filter algorithm, 18 test problems are selected to test the algorithm. The definition of these functions is described in the literature [10]. Each function is experimentally run on the PC 25 times. The statistical results are shown in Table 1, where f denotes the optimal solution of the problem, and mean denotes the algorithm to solve the average of the final result, best denotes the minimum value of the experiment, worst denotes the maximum value of the experiment, and std denotes the standard deviation. And tbest denotes the minimum of iterations, tworst denotes the maximum of iterations, tmean denotes the mean of iterations, tstd denotes the std of iterations.

In order to test the orthogonal design method, we can improve the search performance of the dual internal point filter algorithm. We compare and analyze the test results of Table 1, from the result of precision, the convergence rate two aspects to compare the performance of DIPFA-OD with DIPFA-RND.

In terms of accuracy, it can be seen from Table 1 that DIPFA-OD is more accurate in the functions of P4, P5, P7, P13, P14, P15 and P17, except P8, and the other results are similar.

In addition, iterations of DIPFA-OD in functions P1, P3, P4, P5, P6, P7, P8, P9, P10, P11, P12, P13, P15, P16, P17 and P18 is better than DIPFA-RND, except P14and iterations of P2 is similar.

From the above research, whether it is the optimal accuracy or the number of iterations, DIPFA-OD overall better than DIPFA-RND, showing better optimization performance.

Table 1. Comparison of inequality constraints results with DIPFA-OD and DIPFA-RND.

	pro	f	best	worst	mean	std	tbest	tworst	tmean	tstd
OD	p1	1	1.000000	1.000000	1.000000	4e−16	12	24	**17**	2.902872
RND	p1	1	1.000000	1.000000	1.000000	1.43e−14	8	31	17.84	5.865242
OD	p2	−99.96	−99.959973	−99.959973	−99.959973	1.42e−14	6	6	6	0.0
RND	p2	−99.96	−99.959973	−99.959973	−99.959973	1.42e−14	6	6	6	0.0
OD	p3	5.0	4.999988	4.999988	4.999988	1.63e−13	18	32	**23.04**	5.198564
RND	p3	5.0	4.999988	4.999988	4.999988	1.36e−13	11	67	24.8	11.949895
OD	p4	2	2.000211	2.000214	**2.000213**	1.38e−6	8	8	**8**	0.0
RND	p4	2	1.999895	9.472542	6.284151	3.695	6	30	14.29	5.880529
OD	p5	306.50	306.500000	306.500000	**306.500000**	5.69e−14	17	17	**17**	0.0
RND	p5	306,50	306.500000	360.379767	307.936793	8.680422	16	61	22.46	6.637938
OD	p6	0	−0.0000016	−0.0000016	−0.0000016	0.0	14	14	**14**	0.0
RND	p6	0	−0.0000016	−0.0000016	−0.0000016	5e−15	6	38	14.37	4.955195
OD	p7	0.25	0.250029	0.250029	**0.250029**	2.29e−9	9	13	**10.70**	1.116821
RND	p7	0.25	0.250029	23.144729	3.913181	8.393335	8	30	12.70	4.661254
OD	p8	38.198	38.198799	40.198802	39.532134	0.942810	9	15	**11.22**	1.438271
RND	p8	38.198	38.198799	40.198802	**39.212134**	0.973333	9	83	22.26	15.457756
OD	p9	1.0	1.000061	1.000093	1.000067	6.95e−6	10	14	**11.72**	0.775629
RND	p9	1.0	1.000061	1.000089	1.000065	6.79e−6	10	46	15.54	6.421409
OD	p10	100.99	100.990099	100.990099	100.990099	1.42e−14	10	138	**29.88**	22.315890
RND	p10	100.99	100.990099	100.990099	100.990099	1.42e−14	10	119	31.00	22.225511
OD	p11	1.0	1.000000	1.000000	1.000000	3.17e−14	11	11	**11**	0.0
RND	p11	1.0	1.000000	1.000000	1.000000	4.86e−14	11	22	12.73	1.871423
OD	p12	6.0	6.000030	6.000030	6.000030	8.12e−14	7	8	**7.33**	0.471405
RND	p12	6.0	6.000030	6.000030	6.000030	5.39e−12	7	46	14.96	8.675544
OD	p13	1	0.999780	0.999780	**0.999939**	1.84e−4	42	141	**73.02**	19.824215
RND	p13	1	0.999770	1.000242	0.999944	1.92e−4	47	165	75.37	26.157887
OD	p14	1.9259	1.925921	1.925931	**1.925921**	1.92e−6	7	12	8.66	1.619328
RND	p14	1.9259	1.925921	1.925931	1.925923	4.04e−6	6	12	**8.45**	1.745801
OD	p15	−1	−0.999953	−0.881399	**−0.998340**	1.35e−2	7	34	**13.33**	0.942809
RND	p15	−1	−0.999953	−0.871550	−0.996580	2.03e−2	5	33	13.45	9.898543
OD	p16	−3456	−3455.999999	−3455.999999	−3455.999999	7.62e−13	15	51	**26.65**	8.422974
RND	p16	−3456	−3455.999999	−3455.999999	−3455.999999	8.06e−13	12	52	28.00	8.648699
OD	p17	−44	−43.907239	−43.866558	**−43.905595**	6.29e−3	27	106	**52.33**	13.811428
RND	p17	−44	−43.907015	−43.866000	−43.904554	8.74e−3	20	129	55.64	19.494026
OD	p18	−6961.8	−6961.813875	−6961.813875	−6961.813875	7.24e−12	12	74	**22.76**	13.976018
RND	p18	−6961.8	−6961.813875	−6961.813875	−6961.813875	8.08e−12	16	72	29.90	11.160852

5   Conclusion

The initial point of DIPFA-RND is random initial, with a large blindness, results in a certain degree of ineffective search, so the convergence rate is relatively slow. In this paper, we obtains evenly sample the solution space by the orthogonal design with small constraint violation as the initial points to accelerate the convergence speed of the algorithm. We first generalize the dual interior filter algorithm for referred paper [6] from only equality to both equality and inequality constraints. Then a dual interior filter algorithm based on the orthogonal design is proposed. The orthogonal design is used to evenly sample the solution space, then the points with small constraint violation is chosen as the initial points of the algorithm. The DIPFA-OD choosing the initial points is compared with DIPFA-RND. We run on Algorithm 2 25 times, while we randomly choose 3 initial points every time and run on 25 times. The experimental results show that DIPFA-OD has a faster convergence speed and obtains better objective values than DIPFA-RND.

Acknowledgment. The authors are very grateful to the anonymous reviewers for their constructive comments to this paper. This work is supported by the National Science Foundation of China under Grant 61673355, 61271140 and 61203306.

References

1. Gould, N.I.M., Toint, P.L.: Nonlinear programming without a penalty function or a filter. Math. Program. **91**(2), 239–269 (2010)
2. Curtis, F.E., Gould, N.I.M., Robinson, D.P., Toint, P.L.: An interior-point trust-funnel algorithm for nonlinear optimization. Math. Program. **161**(22), 73–134 (2017)
3. Bohme, T.J., Frank, B.: Introduction to nonlinear programming, pp. 3–18. Springer, Potsdam (2017)
4. Darvay, Z., Takács, P.R.: New method for determining search directions for interior-point algorithms in linear optimization. Optim. Lett. **12**, 1–18 (2017)
5. Yang, X., Zhang, Y., Liu, H., Shen, P.: A new second-order infeasible primal-dual path-following algorithm for symmetric optimization. Numer. Funct. Anal. Optim. **37**(4), 499–519 (2016)
6. Wachter, A., Biegler, L.T.: On the implementation of an interior-point filter line-search algorithm for large-scale nonlinear programming. Math. Program. **106**(1), 25–57 (2006)
7. Zhang, Q., Zeng, S., Wu, C.: Orthogonal design method for optimizing roughly designed antenna. Int. J. Antennas Propag. **2014**(6), 1–9 (2014)
8. Byrd, R.H., Liu, G., Nocedal, J.: On the local behavior of an interior point method for nonlinear programming. Numer. Anal. **19**(2), 37–56 (1998)
9. Fletcher, R., Leyffer, S.: Nonlinear programming without a penalty function. Math. Program. **91**(2), 239–269 (2002)
10. Betts, J.T.: An accelerated multiplier method for nonlinear programming. J. Optim. Theory Appl. **21**(2), 137–174 (1977)

Complex Systems Modeling – Multimedia Simulation

A Beam Search Approach Based on Action Space for the 2D Rectangular Packing Problem

Aihua Yin[1], Lei Wang[2,3], Dongping Hu[1(✉)], Hao Rao[1],
and Song Deng[1]

[1] School of Software and Internet of Things Engineering,
Jiangxi University of Finance and Economics, Nanchang 330023, Jiangxi, China
hdp337@126.com
[2] College of Computer Science and Technology,
Wuhan University of Science and Technology, Wuhan 430081, Hubei, China
[3] Hubei Province Key Laboratory of Intelligent Information Processing and
Real-Time Industrial System, Wuhan 430081, Hubei, China

Abstract. A beam search algorithm is presented to solve the 2D rectangular packing problem. The basic algorithms work according to 7 rule vectors of heuristic selection rules designed to select a corner-sticking action. Furthermore, the trade-off scheme of breadth first search (BFS) and the depth first search (DFS) increases the algorithm's effectiveness and efficiency. The improved version of the algorithm adopts a rough phase to get a height for the stripe and a refine phase to obtain better solution for the problem. Computational experiments run on two sets of well-known benchmark instances and the computational results show that the algorithm outperforms the current best algorithms. Especially, for two benchmark instances ZDF6 and ZDF7, our algorithm finds the best packing configurations so far.

Keywords: Rectangular packing · Action space · NP hard · Heuristic

1 Introduction

Rectangular packing and cutting problem arises in a number of industrial areas with different constraints and alternative objectives. In container transportation industry, objects with different sizes are planned to be packed into a standard container as many as possible. In space industry, the instrumentations may have to be packed into a very limited space. In furniture factory, cutting board or plate usually needs to take into accounts the texture of a board or plate [1].

The two dimensional rectangular packing problem (2D-RPP) considers packing n rectangular blocks $B_i (i = 1, 2, \ldots, n)$ into a rectangular box with given height H and width W, the objective is keep the occupied height to be as low as possible. In the literature, some authors study one subtypes of this problem, i.e., fixing the pieces orientation without guillotine cutting (OF). In this paper, a further discuss on it is presented.

The 2D-RPP is classic NP-hard problem. To solve this sort problem, Alvarez-Valdes et al. presented a branch and bound algorithm for 2D-RPP [1]. However, most

© Springer Nature Singapore Pte Ltd. 2018
K. Li et al. (Eds.): ISICA 2017, CCIS 874, pp. 165–174, 2018.
https://doi.org/10.1007/978-981-13-1651-7_14

of the researchers focus on heuristics algorithms or metaheuristics approaches. Such as, the largest caving degree first rule [2], quasi-human algorithm [3, 4], genetic algorithm [5] and particle swarm optimization [6].

In the latest literature, several state-of-art metaheuristic algorithms are presented. Wei et al. present a least-wasted-first algorithm LWF [7]. Leung et al. show a two-stage intelligent search approach by combining a local search strategy and a simulated annealing method [8]. Leung and Zhang get a very fast layer-based deterministic algorithm which is inspired by the wall-building-rule of brick layers [9].

The rest of the paper is organized as following. Section 2 presents the scheme of the beam search. Section 3 describes the improved version of it. Section 4 reports the computational result on two instance sets. Conclusions are presented in Sect. 5.

2   The Schemes of Beam Search

Given a box of height H and width W, and a number of rectangular blocks $B_i(i = 1, 2, \ldots, n)$ of height h_i and width w_i. Set up a Cartesian coordinate system by coinciding a vertex of the box, its width and height with $(0, 0)$, the origin of the coordinate, X-axis and Y-axis, respectively. The 2D-RPP is to maximize the area usage of the box, i.e., to maximize the total area of the blocks packed into the box. The two constraints are as follows: (1) Any edge of the block in the box should be coincided with or parallel to a side of the box. (2) Any two blocks in the box should not overlap for each other.

2.1   Basic Conceptions

In this paper, to pack all the blocks into the box, a kind of construction algorithms are adapted to pack the blocks one by one. Initially, all the blocks are outside of the box, and then blocks are selected one-by-one to pack into the box. By the kth state, k blocks are already packed while another $n-k$ blocks are still waiting in the block queue. The process terminates when all the blocks are packed into the box or no more blocks can be packed into the box. To facilitate the model the algorithm, some essential definitions are listed below.

Definition 1 (*Configuration, CON*). With some blocks remained outside, the others have been packed into the box. There is a configuration with 4 blocks in the box and the others are outside it (see Fig. 1).

At the beginning, the initial configuration $CON_{initial}$ just has an empty box with all the blocks being outside. On the other side, at the end of the packing process, the terminal configuration $CON_{terminal}$ has a filled box with none block outside or no more block outside can be set into the box.

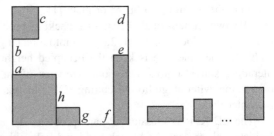

Fig. 1. A configuration with 4 blocks in the box

Definition 2 (*Corner*). Under a configuration, a right angle in the box, whose sides belong to two blocks or the box or one side belongs to a block and another one to the box, is called a corner. Figure 1 shows 7 corners, a, b, c, d, e, f, g and h.

Definition 3 (*Corner-sticking action, CSA*). A corner-sticking action is to stick one block in a corner. In Fig. 1, each of the 4 blocks in the box are packed into the box by a *CSA*.

To solve the packing problem, the *CSA* is natural and important. In fact, this packing problem is equal to the cutting problem which means to cut down all required the rectangular blocks from the given rectangular block (the box). After the four blocks are cut down in Fig. 1, there is a unregular board remaining (see Fig. 2). To cut another rectangular block from this remaining block, human life experience is to cut a corner from any one of a to h, which is the most likely way to save material.

In this paper, every block will be packed into the box by a *CSA*. In an old configuration, a *CSA* is performed and a new configuration is yielded. In other word, the new configuration is the result of a *CSA* acting on the old configuration.

Definition 4 (*Action space, AS*). At any configuration, if one corner is occupied by a proper dummy box whose other two sides just touch one side of a fixed block or the box, then this dummy box is called an action space [10].

Any *CSA* is performed in some *AS*. For example, in Fig. 1, a and b share one *AS*, c, d and e share another *AS*, f and g share the third one, and just h determines its own *AS*. In the 0th configuration, the only one *AS* is the original box.

Definition 5 (*Planeness of a Configuration, PoC*). The planeness of a configuration is $\pi^{4-\alpha-\beta}$, where α and β refers to the numbers of the corners and the *AS*s of this configuration, respectively.

The sum of the corners and *AS*s indicate the *PoC* for a certain configuration. The smaller the sum is, the more flatness the configuration is. However, the corner and the *AS* are not one-to-one correspondence.

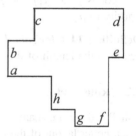

Fig. 2. A configuration with 4 blocks in the box

Definition 6 (*Planeness increment of an action, PIoA*). The planeness increment of an action is the difference of *PoC* between the new configuration and the old configuration after a *CSA* is performed.

A *CSA* may change the number of corner and *AS* of a configuration, which means that some *CSA*s increase or decrease these numbers but others do not.

Definition 7 (*Filling pit action, FPA*). A *CSA* is called a Filling pit action if the number of either corners or *AS*s does not increase once this action is performed.

Definition 8 (*Coincidence edge degree of an action, CEDoA*). For a *CSA*, its Coincidence degree is defined as the number of the block's sides Coinciding with the *SA*. In the absence of confusion, we call the *CEDoA* of a *CSA* is that of the corresponding being packed block.

It is clear that one or two sub-ASs are created after a CSA is carried. For a certain sub-AS, if none of the blocks outside can be put into it, then it is called a fade zone. In future, part of the fade zone can not be used which is called the true fade zone.

Definition 9 (Lose rate of an action, $LRoA$). When a CSA is performed in some AS, the lose rate of this CSA is defined as the ratio of the true fade zone area to the AS area.

Let the bottom-left and top-right coordinates of block B_i be (x_{li}, y_{li}) and (x_{ri}, y_{ri}). Let the width and height of AS j be W_j and H_j. For the sake of clearness, let the length, the width, the perimeter and the area of the block B_i be l_i, w_i, p_i and a_i, separately.

Definition 10 (Usage ratio of an action, $URoA$). The usage ratio of a given CSA is $l_i w_i / W_j H_j$.

Definition 11 (Caving degree of an action, $CDoA$). The caving degree (a simplified version of the original definition proposed by Huang et al.) of a given CSA (see Fig. 3) is $1 - \min(d_x, d_y) / (l_i w_i)^{1/2}$, where $d_x = W_j - (x_{ri} - x_{li}), d_y = H_j - (y_{ri} - y_{li})$.

Definition 12 (*Perimeter priority of block, PPoB*). For the blocks B_i and B_j, the $PPoB$ of B_i is priority to the $PPoB$ of B_j if any constraint of the following is satisfied: (1) $p_i > p_j$; (2) $p_i = p_j$ and $l_i > l_j$; (3) $p_i = p_j$, $l_i = l_j$ and $i > j$.

Definition 13 (*Priority of block left-bottom vertex, PoBV*). Given two points $p_1(x_1, y_1)$ and $p_2(x_2, y_2)$ in the coordinates 0-xy, p_1 is priority to p_2 if $y_1 < y_2$, or if $y = y_2$ and $x_1 < x_2$. For two CSAs, i.e., CSA_1 and CSA_2, if the block left-bottom vertexes are the two points $p_1(x_1, y_1)$ and $p_2(x_2, y_2)$, then CSA_1 is priority to CSA_2 is equal to p_1 is priority to p_2.

Definition 14 (*Horizontal degree of an action, HDoA*). The horizontal degree of an action is the length of the corresponding block's horizontal edge.

2.2   Rule Vector

Usually, there are many CSA candidates to be selected in each configuration. To get much better layout of the blocks, we should design enough good approach to select a block each time. In this paper, we eclectically combine the breadth-first search (BFS) and the depth-first search (DFS) in the search tree. Those characters of CSA above can be used to obtain the effective approaches.

Definition 15 (*Rule vector, RV*). The rule vector is a sequence of some of the characters above. Such as, $\pi^{(1)} = <\pi_{PIoA}, \pi_{CEDoA}, \pi_{PPoB}, \pi_{CDoA}, \pi_{LRoA}, \pi_{PoBV}, \pi_{HDoA}>$ is a seven elements rule vector and the seven descending order of the priority of the changes is: $\pi_{PIoA}, \pi_{CEDoA}, \pi_{PPoB}, \pi_{CDoA}, \pi_{LRoA}, \pi_{PoBV}, \pi_{HDoA}$.

In this paper, we stipulate the rule vector can not only act on a set of actions but also act on a configuration, as well.

First, when the rule vector $\pi^{(1)}$ acts on a set of actions it yields a descending sequence of the actions based on the priority of the elements in $\pi^{(1)}$. Let Ω be the set of all the actions of a configuration, $|\Omega|$ be the number of the actions and $\Omega\downarrow^{(1)}$ be the descending sequence of $\pi^{(1)}$ acts on Ω, i.e.,

$$\pi^{(1)}(\Omega) = \Omega \downarrow^{(1)} \tag{1}$$

Second, when the rule vector $\pi^{(1)}$ can also act on a configuration $CON_{current}$ with all actions being in the set $\Omega_{current}$ it yields a new configuration CON_{new}. Which means that the CON_{new} is the result of the first element of $\Omega_{current} \downarrow^{(1)}$ acting on the $CON_{current}$. So,

$$\pi^{(1)}(CON_{current}) = CON_{new} \tag{2}$$

and there are $\left(\pi^{(1)}\right)^2(CON_{current}) = \pi^{(1)}(CON_{new})$, $\left(\pi^{(1)}\right)^3(CON_{current}) = \left(\pi^{(1)}\right)^2 (CON_{new})$, and so on. Specially, for an instance with n blocks, there is

$$\left(\pi^{(1)}\right)^n(CON_{initial}) = CON_{terminal} \tag{3}$$

Definition 16 (Value of a configuration, *VC*). A configuration's value is defined as the sum of the areas of the blocks inside the box.

Definition 17 (Length priority of block, *LPoB*). For the blocks B_i and B_j, the *LPoB* of B_i is priority to the *LPoB* of B_j if any one of following constraints is satisfied: (1) $l_i > l_j$; (2) $l_i = l_j$ and $w_i > w_j$; (3) $l_i = l_j, w_i = w_j$ and $i > j$.

Definition 18 (Area priority of block, *APoB*). For the blocks B_i and B_j, the *APoB* of B_i is priority to the *APoB* of B_j if any one of the following constraints is satisfied: (1) $a_i > a_j$; (2) $a_i = a_j$ and $l_i > l_j$; (3) $a_i = a_j, l_i = l_j$ and $i > j$.

For a rule vector, it shows quite limited characters of the actions or the configuration and the descending order is very rigid. Next, we can create several other rule vectors to show the characters of the actions or the configuration as many as possible. Furthermore, the

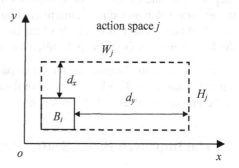

Fig. 3. Caving degree of an action

combination of these rule vectors can make the descending order elastic in practice. The other eleven rule vectors are listed as following:

$\pi^{(2)} = \, <\pi_{PIoA}, \pi_{CEDoA}, \pi_{LPoB}, \pi_{CDoA}, \pi_{LRoA}, \pi_{PoBV}, \pi_{HDoA}>$, replacing π_{PPoB} with π_{LPoB} in $\pi^{(1)}$.

$\pi^{(3)} = \, <\pi_{PIoA}, \pi_{CEDoA}, \pi_{APoB}, \pi_{CDoA}, \pi_{LRoA}, \pi_{PoBV}, \pi_{HDoA}>$, replacing π_{PPoB} with π_{APoB} in $\pi^{(1)}$.

$\pi^{(4)} = \, <\pi_{PIoA}, \pi_{CEDoA}, \pi_{CDoA}, \pi_{PPoB}, \pi_{LRoA}, \pi_{PoBV}, \pi_{HDoA}>$, swapping π_{PPoB} and π_{CDoA} in $\pi^{(1)}$.

$\pi^{(5)} = \, <\pi_{PIoA}, \pi_{CEDoA}, \pi_{LRoA}, \pi_{CDoA}, \pi_{PPoB}, \pi_{PoBV}, \pi_{HDoA}>$, swapping π_{PPoB} and π_{LRoA} in $\pi^{(1)}$.

$\pi^{(6)} = <\pi_{PIoA}, \pi_{CEDoA}, \pi_{PPoB}, \pi_{URoA}, \pi_{LRoA}, \pi_{PoBV}, \pi_{HDoA}>$, replacing π_{CDoA} with π_{URoA} in $\pi^{(1)}$.

$\pi^{(7)} = <\pi_{CDoA}, \pi_{CEDoA}, \pi_{PPoB}, \pi_{PIoA}, \pi_{LRoA}, \pi_{PoBV}, \pi_{HDoA}>$, swapping π_{PIoA} and π_{CDoA} in $\pi^{(1)}$.

2.3   The Base Beam Search Algorithms

Based on the seven rule vectors, the basic methods, $BSA\text{-}I_i(i = 1, 2, \ldots, 7)$, are to perform an alternation of BFS and DFS. The details of the schemes are described as follows.

Beam Search Algorithms $BSA\text{-}I_i$

Step 0. Let $CON_{current} = CON_{initial}$;

Step 1. Act $\pi^{(i)}$ on all the actions of $CON_{current}$ to get a descending sequence of $CSAs$ in $CON_{current}, \Omega_{current} \downarrow^{(i)}$;

Step 2. Act the first Φ $CSAs$ of $\Omega_{current} \downarrow^{(i)}$ on the $CON_{current}$ and yield Φ candidate configurations;

Step 3. Act $\pi^{(i)}$ on each of these candidate configurations Θ times one by one to get Φ corresponding new configurations. Compute the values of the Φ new configurations;

Step 4. Select the candidate configuration whose corresponding new configuration has the greatest value as the new current configuration;

Step 5. If there is no block outside of the box or the blocks outside can not be put into the box, then stop. Go to Step 1. otherwise.

It is not easy to Compromise these two parameters Φ and Θ. Here, we adopt Φ with variable value and Θ with an alternative fixed value, i.e., $\Theta = max(60, n/100)$ where is the number of blocks of the instance.

3   The Improved Beam Search Algorithm

The seven versions of algorithm $BSA\text{-}I$ can apply for solving the 2D-RSPP. To solve this problem more efficiently, some improvement of $BSA\text{-}I$ should be made. Denote this new algorithm $BSA\text{-}II$. For the RSPP, there is no height for the "box", an approach to create a height for the box is presented in $BSA\text{-}II$.

The outline of $BSA\text{-}II$ is as following: (1) *Rough search*, the seven versions of algorithm, $BSA\text{-}I_i(i = 1, 2, \ldots, 7)$ are used to get a box height boundary such that all the blocks can be put into the box (line 3 to 19). (2) *Refined search*, these seven algorithms are adopted to generate better configurations (line 20 to 35).

Let $sum = \sum AreaB_i$ $(i = 1, 2, \ldots, n)$, then an obvious height boundary of the box is sum/W. The details of the computation steps are described in pseudo C language as follows. Let $LB = [sum/W]$, $\delta = min(3, LB/100)$, the algorithm $BSA\text{-}II$ is described in details as follow.

Algorithm: Pseudo-Code of the *BSA-II*

1: input: The box and all blocks, LB, δ.
2: output: UB and the terminal configuration
3: $h = LB$
4: **repeat**
5: $flag = 0$; Set $h = $ the box height;
6: **for** $\Phi = \{5, \ldots, 305\}$ **do**
7: **for** $i = \{1, \ldots, 7\}$ **do**
8: Run *BSA-I$_i$* up to T_UB seconds;
9: **if** all the blocks are put into the box **then**
10: $UB = h$; $flag = 1$; break;
11: **end for**
12: **if** $flag == 1$ **then** break
13: **end if**
14: **if** $flag == 1$ **then** break;
15: **end if**
16: Φ += 150
17: **end for**
18: h += δ
19: **until** the stop criterion is met
20: $UB^+ = UB + \delta/2$; $UB^- = UB - 2$;
21: **repeat**
22: $flag = 0$; $mid = (UB^+ + UB^-)/2$; Set $mid = $ the box height;
23: **for** $\Phi = \{5, \ldots, 305\}$
24: **for** $i = \{1, \ldots, 7\}$ **do**
25: Adopt algorithm *BSA-I$_i$* to computation up to T_UB seconds;
26: **if** all the blocks are placed into the box **then** set $fg = 1$, break;
27: **end if**
28: **end for**
29: **if** $flag == 1$ **then** break;
30: **end if**
31: **if** $flag == 1$ **then** $UB^- = mid - \delta/2$; $UB = mid$;
32: **end if**
33: **else** $UB^+ = mid + \delta/2$;
34: Φ += 150
35: **until** $UB^- \geq UB^+$

To reduce the time complexity of the algorithm, a time upper bound is set to limit the execution time of *BSA-I$_1$*, ..., and *BSA-I$_7$*. Naturally, the size of the benchmark instance is larger, its T_UB is bigger. The number of the blocks is denoted as n.

(1) Small size benchmark ($n \leq 200$): $T\_UB = 400$.
(2) Medium size benchmark ($200 < n \leq 500$): $T\_UB = 800$.
(3) Large size instance benchmark ($500 < n \leq 10000$): $T\_UB = 2000$.
(4) Very Large size benchmark ($n > 10000$): $T\_UB = 20000$.

The algorithm *BSA-II* is implement in *C* Language and run it on a personal computer with a 2.8 GHz CPU.

4   Computational Results

In this paper, two sets of 37 well-known benchmarks are used test out algorithm *BSA-II*. For each set of these instances, we compare *BSA-II* with the best algorithms reported in the literature up to now. All these benchmarks are available at http://paginas.fe.up.pt/esicup/tiki-index.php. The two set are listed below.

(1) C: 21 instances provided by Hopper and Turton [11].
(2) ZDF: 16 instances provided by Leung and Zhang [8].

Where, C is zero-waste instances. ZDF is non-zero-waste instances.

Table 1 shows the effect of *T_UB* on ZDF 6–7. The optimal configurations of ZDF 6–7 are still unknown. In this algorithm, if *T_UB* is set to be higher, the better solution is obtained. Table 2 shows the effect of Φ on C16, C21 and ZDF7. Φ is an important parameter for the breadth-first search scheme. A smaller Φ means that the less candidate *CSA*s are considered, however, some good *CSA* candidates may be discarded. A larger Φ means that the more *CSA*s candidate are considered and longer time is costed. Here, a dynamic scheme is adopted for Φ, i.e., first set $\Phi = 5$, and increase it by 150 up to 305.

In the following tables, let *n*, *LB* and *W* denote the number of the blocks, the lower bound of the box height and the box width, respectively. For all instances, the best results are shown in bold.

In Table 3, we compare *BSA-II* with other two well-known and well-performed algorithms in the literatures. These algorithms are QH [4], *ISA* [8], *HA* [12], *GRASP* [13], *IDBS* [14] and *SRA* [15]. We list the average gap (*AveGap*) between the best solutions and the *LBs* of the instances in the two sets. For the randomized algorithm, we just take account into their best solutions and average running time of 10 runs. Furthermore, *BSA-II* gets four the smallest *AveGap*s on the two sets.

Table 1. Effect of parameter *T_UB* on two nonzero-waste instances ZDF 6–7 (OF subtype)

Instance				T_UB					
				800		2000		20000	
Name	n	W	LB	H	Time	H	Time	H	Time
ZDF 6	1532	3000	4872	5005	9381	4980	26043	**4944**	312425
ZDF 7	2432	3000	4852	5035	26832	4962	32965	**4946**	302069
Average				5020		4974		**4934**	

It is not difficult to know that *QH* is the second algorithm in Table 3, which means that *BSA-II* is quite better than *QH* on the four sets of instances.

Table 2. Effect of parameter Φ on four instances (OF subtype)

Instance				$\Phi = 5$		$\Phi = 155$		$\Phi = 305$	
Name	n	W	LB	H	Time	H	Time	H	Time
C16	97	80	120	127	4019	**120**	4982	**120**	5965
C21	196	160	240	256	8760	243	9951	**241**	11065
ZDF 7	2432	3000	4852	5001	216455	4951	285732	**4946**	302069

Table 3. Comparison of the average gap of six algorithms on the instances (OF subtype)

Instances	GRASP	ISA	IDBS	SRA	HA	QH	BSA-II
C (21)	0.95	0.76	0.20	0.69	0.51	0.36	**0.10**
ZDF (16)	4.22	2.67	2.42	1.5	1.44	0.28	**0.25**

5 Conclusions

Although the beam search scheme in the search tree is effective, it is necessary to construct the proper actions and action rules. Inspired by human being's social life experience of bricklaying, packing of goods, and chess-playing, this paper presents a deterministic heuristic algorithm to solve the rectangular packing problem. The trade-off idea of taking advantage of BFS and DFS has made great contributions on the algorithm performance. The computational results have shown that our algorithm is efficient enough to solve OF subtype of the problem. For the future study work, it maybe to improve the algorithms by combining this beam search with local search.

Acknowledgments. This research was supported by the National Natural Science Foundation of China (Grant Nos. 61702238, 61262011, 61462037), the Natural Science Foundation of Hubei Province under Grant No. 2014CFC1121.

References

1. Alvarez-Valdes, A., Parreno, F., Tamarit, J.M.: A branch and bound algorithm for the strip packing problem. OR Spect. **31**(2), 431–459 (2009)
2. Huang, W.Q., Chen, D.B., Xu, R.C.: A new heuristic algorithm for rectangle packing. Comput. Oper. Res. **34**(11), 3270–3280 (2007)
3. Huang, W.Q., He, K.: A carving degree approach for the single container loading problem. Eur. J. Oper. Res. **196**(1), 93–101 (2009)
4. Wang, L., Yin, A.H.: A quasi-human algorithm for the two dimensional rectangular strip packing problem: in memory of Prof. Wenqi Huang. J. Comb. Optim. **32**(2), 416–444 (2016)
5. Bortfeldt, A.: A genetic algorithm for the two-dimensional strip packing problem with rectangular pieces. Eur. J. Oper. Res. **172**(3), 814–837 (2006)
6. Jiang, J.Q., Liang, Y.C., Shi, X.H., Lee, H.P.: A hybrid algorithm based on PSO and SA and its application for two-dimensional non-guillotine cutting stock problem. In: Bubak, M., van Albada, G.D., Sloot, Peter M.A., Dongarra, J. (eds.) ICCS 2004. LNCS, vol. 3037, pp. 666–669. Springer, Heidelberg (2004). https://doi.org/10.1007/978-3-540-24687-9_98

7. Wei, L.J., Zhang, D.F., Chen, Q.S.: A least wasted first heuristic algorithm for the rectangular packing problem. Comput. Oper. Res. **36**(5), 1608–1614 (2009)
8. Leung, S.C.H., Zhang, D.F., Sim, K.M.: A two-stage intelligent search algorithm for the two-dimensional strip packing problem. Eur. J. Oper. Res. **215**(1), 57–69 (2011)
9. Leung, S.C.H., Zhang, D.F.: A fast layer-based heuristic for non-guillotine strip packing. Expert Syst. Appl. **38**, 13032–13042 (2011)
10. He, K., Jin, Y., Huang, W.Q.: Heuristic for two-dimensional strip packing problem with 90 rotations. Expert Syst. Appl. **40**, 5542–5550 (2013)
11. Hopper, E., Turton, B.C.H.: An empirical investigation of meta-heuristic and heuristic algorithm for a 2D packing problem. Eur. J. Oper. Res. **128**(1), 34–57 (2001)
12. Defu, Z., Yuxin, C., Furong, Y., Yain-Whar, S., Leung, S.C.H.: A hybrid algorithm based on variable neighborhood for the strip packing problem. J. Comb. Optim. **32**(2), 513–530 (2016)
13. Alvarez-Valdes, R., Parreno, F., Tammrit, J.M.: Reactive GRASP for the strip-packing problem. Comput. Oper. Res. **35**(4), 1065–1083 (2008)
14. Wei, L.J., Oon, W.C., Zhu, W.B., Lim, A.: A skyline heuristic for the 2D rectangular packing and strip packing problems. Eur. J. Oper. Res. **215**(2), 337–346 (2011)
15. Yang, S.Y., Han, S.H., Ye, W.G.: A simple randomized algorithm for two-dimensional strip packing. Comput. Oper. Res. **40**(1), 1–8 (2013)

On the Innovation of Multimedia Technology to the Management Model of College Students

Yuanbing Wang[✉]

Department of Finance and Economics, Guangdong University of Science and Technology, Dongguan, China
315724104@qq.com

Abstract. If the university management is going to keep up with the pace of development of the times. In this process, we should give full play to the dominant position of students. Because in the present social environment, college students on the application of advanced scientific and technological achievements have a high sensitivity. This reality is the interest of college students. Therefore, colleges and universities is to implement effective management of students. If it still use the traditional management model, it is difficult to play the timeliness of student management. We are now in the information age, multimedia technology in all areas of society are very popular. College students lead the forefront of network technology, so the use of traditional management model is a clear lag of management. Information environment, colleges and universities must use multimedia technology to carry out student management, in order to achieve student management innovation and improve the efficiency of student management. In this article, we first make a simple price introduction to the research background of university student management and bring out the key word of multimedia technology. Then, we will show the current mode of university student management and the research status quo. In view of this status quo, the main problems that exist in the management of college students are pointed out. Finally, according to the problems encountered in the actual process, put forward specific solutions to the management of colleges and universities.

Keywords: Innovation · Multimedia technology · Management model

1 Introduction

With the advent of the multimedia era, large data also attracted more and more attention. In fact, large data is not a technology, but because of the growing amount of data and data types are gradually derived from the social phenomenon. At present, the scale of college enrollment has been expanded. The number of students has been increasing rapidly, and the quality of students has been declining. These have made the traditional management mode of traditional students unable to adapt to the trend of popularization of higher education in the new period [1]. Student managers are also facing increasing pressure on labor intensity. Therefore, the major colleges and universities to further strengthen the modern management tools used in student management work, especially

© Springer Nature Singapore Pte Ltd. 2018
K. Li et al. (Eds.): ISICA 2017, CCIS 874, pp. 175–182, 2018.
https://doi.org/10.1007/978-981-13-1651-7_15

in college student management in the data management is particularly urgent. This article is in this context, to discuss the impact of multimedia technology on college students management mode based. At the same time, it is also actively looking for ways to innovate management.

In 2016, Jiang Xuan published an article at the Forum on Industry and Technology entitled "The Current Situation and Countermeasures of Student Management in Institutions of Higher Learning in the Information Age". In the article, she explained the status quo and countermeasures of university student management in the information age, which provided a lot of reference value for the writing of this article. In the same year, Wu Yongsheng also published an article in Chifeng University (Natural Science Edition). Article entitled "The development of online media to strengthen the management of university students analysis." The difference from the previous article is that this article focuses on the online media for college students inspired by the management. To a certain extent, this article also provides some help for the writing of this article.

2   Research Background

In short, multimedia is the Internet company through the search engine, access records, App tracking and other technical means can get a lot of users to browse information, thus establishing a large database, so as to carry out data analysis. It is such a reality for the college student management has brought opportunities and challenges. As an important department of educational informationization in colleges and universities, how to improve the data storage and processing capacity of mass growth in the multimedia age, and to find a new way of information service is the information management center must think and study. Only by rationally understanding the "multimedia" and the changes brought to our environment can we better tap the internal relations of massive information and draw new conclusions in order to provide better suggestions for the development of information construction in colleges and universities. In the context of multimedia, the information technology is strongly cited to the university teaching, research, management and information release and other schools daily work and life of all links, to achieve the historical transformation of college student management is the trend.

3   The Tradition Patterns and Research Status of College Student Management

3.1   The Traditional Model of Student Management

In our country, student management generally adopts the control type management pattern. The work system is a typical two-level management of the school. School set up a student work office is a party and government of the specialized agencies, it and the Office of Academic Affairs, Finance Department, the Communist Youth League and other institutions together to complete the goal of student work management. Under the student office, but also the establishment of student archives, ideological education room, student management room, national defense education room and other organizations. They subdivide their duties, bear the students' daily thinking education and student management, and educational affairs. Students rely on the rights granted by the school, the use of the introduction of documents, the implementation of the notice, such as the management of students. The focus of its work is to emphasize uniformity and obedience. In the college, the establishment of the party committee, the Communist Youth League, to accept the guidance and delegation of academic office. Among them, in the college, the Communist Youth League in charge of daily student activities to carry out approval. Communist Youth League not only by the school, the school Communist Youth League and other school-level functional departments of the guidance, but also accept the leadership of the party and government organizations.

In these sectors, each department has different responsibilities and different divisions, but there are overlapping functions in management practice. For example, student dormitory management involves school logistics group, school Communist Youth League, academic department, security office and so on [2]. In the case of cross-management, inevitably there will be people's prevarication and other issues. In addition, due to the implementation of the system of party and government, in the actual major affairs not only to obtain the consent of the Communist Youth League, but also with the consent of the party committee, which will lead to inefficiency. In many schools, this phenomenon is not uncommon.

3.2 China's Information Technology Teaching Management Status Quo

3.2.1 Student Management Information Construction Overview

At present, the most cutting-edge research is education and scientific research network (CERNET) and satellite video system. These network systems are clearly not mature enough, but the current stage of information technology has begun to take shape. Due to the high cost of high-speed Internet and the speed is not stable enough, so the proportion of put into use is not big enough [3]. At present, many colleges and universities use information means to strengthen the university student management work, they through the purchase of high hardware equipment and management system, the information technology used in college student management. Although this initiative to improve the efficiency, but there are still many problems to be solved. Some college students manage the organizational structure, management model, business process can not meet the needs of information background. Some colleges and universities students management information technology practice lack of effective performance evaluation system. So many school students management work time is poor, poor accuracy, poor performance.

3.2.2 China's Student Management Information Model

Below, I will use a model to illustrate the status of China's information management student management. On the one hand, the school through the student health management system, student status management system, students outside the school practice three systems to collect data, the data into the electronic growth file. Students electronic growth files include physical and mental health, academic progress, growth experience, personality skills 4 dimensions [4]. On the other hand, parents through the IPTV platform can be concerned about the status of students. That is, under the supervision of both, all the information will form a strong database (Fig. 1).

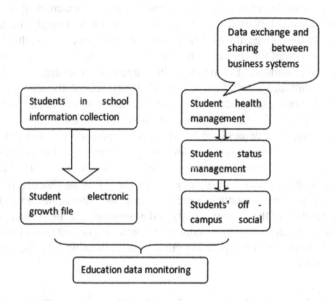

Fig. 1. China's student management information model

In short, the multimedia "navigation" can be a comprehensive and objective record of student growth trajectory, precipitation and accumulation of multi-dimensional student growth data, so that students reflect the state of development data complete display, promote the quality of education changes, The scientific development of management.

4 The Main Problems in the Management of College Students

With the digitalization of the campus, the gradual deepening of the information construction, the campus of various information resources integration has entered a comprehensive planning and implementation stage. Take any of our undergraduate universities as an example to simulate the construction of the campus card system and set up a network of the whole university campus. Departments to strengthen their own network construction, under limited resources, the development of resource sharing platform, so that students and teachers can be very much in the LAN to share and release resources. That is, to improve students' self-education, self-management, self-service ability, but also to share the task of student management tasks [5]. At the same time, funds can also learn from abroad, through the market mechanism, cited enterprises as technology shares, to ensure the technical problems of information technology.

4.1 Multi-card Multi-purpose, Time-Consuming Trouble

Now students in the hands of student cards, rice cards, bank cards, etc., ranging from three or four, as many as five or six, to bring a lot of inconvenience to life. To a college student, from the beginning of their entry into the school there are two bank cards for their daily life, there are other campus cards, water cards, student cards and other card. These cards have their own use, in different occasions need to produce different documents. For example, in the library by the umbrella must produce a student card, and the general students if there is no special needs, go out most of the cases will not bring student cards and other commonly used documents.

4.2 System Independent, Resource Separation

Existing systems have independent certification methods and consumption; schools can not be unified management, resources can not be a reasonable application and sharing. Such as the current school of the bedroom electricity, net fee to achieve credit card system, but the supermarket, access control, etc. are not fully realized.

4.3 A Large Number of Students, Management "Failure"

Student management is also faced with the information processing capacity becomes larger, processing accuracy requirements more and more fine plight. Student managers are also facing increasing pressure on labor intensity. There are a large number of students, each class, each professional, college needs statistical management of the number of business, the school's hardware equipment and management system obviously can not adapt to our huge amount of information, while the school student management information level is low, There is no good use of multimedia technology, most of the work also rely on manual operation, low efficiency.

4.4 Monitoring Weakness, "Vulnerability" Frequency Out

Student management informatization practice lacks effective performance evaluation system. Student work management staff lack of effective supervision, resulting in complex institutional system not only did not play a highly efficient and efficient role, but caused the delay in time, waste of resources, managers loose and other undesirable phenomenon [6]. The emergence of these issues called for a more comprehensive emergence of the campus management system, later demonstrated smart card system and multimedia library, information platform maintenance update through the existing system to further improve, can solve this problem [7].

5   The Specific Solution to the Management of Colleges and Universities

5.1   Smart Card System

For the emergence of "multi-card multi-purpose, time-consuming trouble" phenomenon, design a "smart card." As shown above, the campus card to replace the previous various documents (including student ID, work permit, library card, medical certificate, access card, etc.) all or part of the function, and with the bank card to achieve self-service trap. In addition to this card can achieve a variety of school spending, but also can achieve personal identity and campus access control attendance management. In the attendance management subsystem, to achieve a credit card attendance, but also can reduce or even solve the phenomenon of absenteeism late. In addition, the campus card to improve the attendance, daily life consumption management subsystem, enrich its function in student life [8]. In the access control subsystem, to achieve access card system, can prevent theft or other security issues. And between the various subsystems, through the server terminal database, any two subsystems can establish contact between. For example, in the access control and attendance management of the two subsystems combined, according to the students out of the bedroom and classroom situation can try to avoid the phenomenon of generation and so on.

5.2   Construction of Off-Campus Information Platform

In order to overcome the "system independence, resource separation" defects, schools can launch mobile client and other interactive platform. Important information about the school and related announcements, such as the school in the examination, competitions, competitions and other aspects of the major achievements, the school received higher recognition, the school to carry out important activities such as sports, speech contest, essay contest and other relevant information, As well as parental notice, school leave time, examination arrangements, payment notice and so on. This school, parents and students to provide effective management of the three parties, the daily life of the students also provides a great convenience [9].

5.3   Establish a Database System

Constructing an Effective Performance Evaluation System to Supervise Students' Work Managers. Supervision and implementation of all-round multi-angle supervision, so that a strong monitoring, "loopholes" fill vacancies, and the supervision results as an important basis for the performance evaluation of managers, so that students work managers have a sense of crisis, strengthen self-moral cultivation, and strive to complete the work. Evaluation of the channel: blog, microblogging, paste it, WeChat, Fetion, QQ, and so on. The evaluation system is as follows: self-evaluation, evaluation, evaluation of teachers and students; evaluation methods: direct review and proxy review;

In addition, the use of the portal, blog, microblogging, online media, webcast, on the other hand to the phone to receive the terminal form of media, such as mobile phone, mobile phone text messages, mobile TV and mobile Internet. With the campus card intelligent management system platform to promote the school units, departments of information technology, standardized management process. Table 1 below summarizes the problems and solutions of the student management system in colleges and universities [10].

Table 1. Problems and solutions to the management system of college students

The management of college students	Problem	Solutions
Four questions, three countermeasures	Multi-card multi-purpose, time-consuming trouble	Smart card system
	System independence, resource separation	Construction of off - campus information platform
	Many students, management "failure"	Create a database system
	Monitoring weakness, "vulnerability" frequency out	/

6   Conclusion

The arrival of the multimedia era provides a new opportunity for students to manage the work, student management information construction will also become the only way for college students to work. Student work managers to adapt to the requirements of the data age, bold innovation and change, so that students work management more scientific, standardized, information, unified, to achieve data sharing; maximize the conservation of resources, reduce the strength of managers, The Characteristics of the Student Information Management [11]. Although the idea of the card and the database can solve many management problems, but there are still many deficiencies. Such as the card once in custody inadvertently or lost, the student's personal information may be a lot of leakage, at the same time, a short period of time there is no good remedy, re-card procedures need to pass a variety of binding is very complicated; The construction of the database inevitably requires a lot of manpower and material resources. But the idea of such a student management, the purpose is to make readers more profound understanding of student management information is a very practical system engineering, student management in the cause of more attention and research, so that college students management information Towards a more scientific and systematic direction [12].

References

1. J. Changchun Inst. Educ. (Soc. Sci. Ed.) (2015). 【Co-citations】 1 Hits
2. School of Economics and Management, Changsha University of Science and Technology, Changsha, China: On the management of college students in the network information age. J. Changsha Railway Inst. (Soc. Sci. Ed.) (2016)

3. Nie, C.: The use of multimedia technology to optimize mathematics classroom teaching. In: Chinese Education Theory and Practice Research Papers Selected Results, vol. IV (2013)
4. School of Computer Science and Technology, Tsinghua University, Beijing, China: Application of multimedia technology in computer basic teaching. Chinese Education Theory and Practice
5. Wang, W.: Application of multimedia technology in display space. Shandong University of Architecture (2015)
6. Popova, M.: St. Petersburg Water cosmos museum of multimedia technology application. Harbin Institute of Technology (2015)
7. J. Shanxi Univ. (Philos. Soc. Sci. Ed.) (2009). A new idea of student management in colleges and universities based on network community. Shanxi Youth (2016)
8. Wu, Y., School of Economics and Management, Zhejiang University, Hangzhou, China: Analysis on the development of network media. J. Chifeng Univ. (Nat. Sci. Ed.) (2016)
9. Application of fine management in financial budget management of colleges and universities. Mod. Econ. Inf. (2016)
10. Tian, Y., Chen, W.: Analysis of multimedia technology in college students in the management of the application and practical significance (2016). Reading Abstract
11. School of Economics and Management, Zhejiang University, Hangzhou, China: Discussion on the management of college students in the context of new internet. Dev. Hum. Resour.
12. Fei, X.: Application of new network media in student management in colleges and universities. New Media Res. (2015)

Convenient Top-k Location-Text Publish/Subscribe Scheme

Hong Zhu[1], Hongbo Li[1,2], Zongmin Cui[3(✉)], Zhongsheng Cao[1],
and Meiyi Xie[1]

[1] School of Computer Science and Technology, Huazhong University of Science
and Technology, Wuhan, Hubei, China
[2] School of Computer Science and Information Technology,
Daqing Normal University, Daqing, Heilongjiang, China
[3] School of Information Science and Technology, Jiujiang University,
Jiujiang, Jiangxi, China
cuizm01@gmail.com

Abstract. With the popularity of social media and GPS equipment, a large number of the location-text data have been produced in the form of stream. The popularity leads to a variety of applications, such as based-location advice and location information transmission. Existing top-k location-text publish/subscribe schemes need subscriber to set a threshold, k value (the number of returned top results) and preference parameter δ (the decision of which one is more important about location or text). The threshold brings lots of disadvantages to subscribers and publishers. Therefore in this paper, we propose an efficient top-k location-text publish/subscribe scheme without threshold which is named as TGT. Our scheme only needs subscriber to input k value and preference parameter δ. Then TGT returns top-k results to the subscriber based on k and δ without any threshold. Therefore, our scheme can reduce redundant computation, improve the recall ratio and facilitate the subscriber. Extensive experiments prove the efficiency and effectiveness of the proposed scheme.

Keywords: Top-k · Location-text · Publish/subscribe · Without threshold
Data query

1 Introduction

Recently, with the ubiquity of GPS mobile devices and social media, a large number of location data have been produced in the form of flow text [1–3]. The ubiquity leads to the popularity of the location-text publish/subscribe system [4–6]. Hence, the location-text publish/subscribe has been applied in many scenarios, such as advice and social networks based on location [7–9].

In such a system, each individual user can register his own interest (like food and exercise) and his own location as location-text keyword subscription [10–12]. A location-text information (such as electronic coupons promotion and microblog with location information, etc.) was produced by a publisher (such as local businesses) continuously and was quickly provided to relevant subscribers (i.e. users). As the

© Springer Nature Singapore Pte Ltd. 2018
K. Li et al. (Eds.): ISICA 2017, CCIS 874, pp. 183–191, 2018.
https://doi.org/10.1007/978-981-13-1651-7_16

user's energy is limited, only returning to the user top-k results has become the most common way of data query [13].

Publish/subscribe system [14] consists of two roles: the publisher and the subscriber [15]. The subscriber accesses the published data, such as the buyer. The publisher publishes data, such as the seller. The subscriber registers a subscription to capture his interests. For example, a set of sellers publish a set of commodity information, while another set of buyers subscribe the publication to find the needed commodity.

When dealing with top-k issues, most of latest location-text publish/subscribe methods [5, 16, 17] require users to set a threshold τ, which is used to combine location and text. It is difficult that the final calculated number of publications is exactly equal to the number of required publications (i.e. k). Meanwhile, it is also difficult for a user to set a threshold τ obviously.

For example, there are two users who set the same threshold. Therefore the system calculates 6 relevant results according to the threshold. However, the first user wants only one result (i.e. k = 1). The second user wants 8 results (i.e. k = 8).

Obviously, there are three problems in this process as follows.

(1) It is a waste of time to set τ for the two users.
(2) For the first user, five other results waste the precious computational resources of the system.
(3) For the second user, the system misses two results. Meanwhile, as the calculated number of results is less than k (i.e. 6 < 8), the calculated results do not have to be scored. So in this case, the system has wasted computing resources.

In conclusion, the efficiency and effectiveness are not good enough in lots of existing methods [5, 16, 17]. Moreover, the methods are not very convenient to users. To remove the problems, we propose a top-k location-text publish/subscribe scheme without threshold (called as TGT).

This scheme allows users to only input k (i.e., the number of returned top results) and preference parameter δ (the decision of which one is more important about location or text). Then the publish/subscribe system only returns k results based on k and δ. Therefore, TGT can reduce the redundant calculation, improve the recall ratio and facilitate the user.

The rest of the paper is organized as follows. Section 2 presents the related concepts. Section 3 describes our system. Section 4 shows the experimental results. Finally, we conclude the paper in Sect. 5.

2 Basic Concepts

In order to accurately define the facing problems of this paper, we show the related definitions as follows.

Definition 1 (Subscription). A subscription s includes a text s_T, a location s_L, a preference parameter δ (used to coordinate text similarity and location similarity, if $\delta > 0.5$, text similarity is more important than location similarity) and a k value (the number of the returned results). That is, a subscription is expressed as $s = (s_T, s_L, \delta, k)$. s_T is a set of keywords $\{t_1, t_2, \ldots, t_{|s_T|}\}$. Each keyword t_i is associated with a weight $w(t_i)$. Our keyword weight is set as inversed document frequency (IDF [18]) of keyword. s_L is a location composed of latitude and longitude.

Definition 2 (Publication). A publication p includes a text p_T and a location p_L. That is, a publication is expressed as $p = (p_T, p_T)$. Like Definition 1, p_T is a set of keywords $\{t_1, t_2, \ldots, t_{|p_T|}\}$ and p_L is a location composed of latitude and longitude.

Example 1. Table 1 shows the examples of 8 subscriptions and 8 publications.

Table 1. The examples of subscriptions and publications

Subscription		Publication	
Name	Content	Name	Content
s_0	$(s_{T_0}, s_{L_0}, 0.7, 3)$	p_0	(p_{T_0}, p_{L_0})
s_1	$(s_{T_1}, s_{L_1}, 0.2, 2)$	p_1	(p_{T_1}, p_{L_1})
s_2	$(s_{T_2}, s_{L_2}, 0.9, 1)$	p_2	(p_{T_2}, p_{L_2})
s_3	$(s_{T_3}, s_{L_3}, 0.5, 3)$	p_3	(p_{T_3}, p_{L_3})
s_4	$(s_{T_4}, s_{L_4}, 0.6, 5)$	p_4	(p_{T_4}, p_{L_4})
s_5	$(s_{T_5}, s_{L_5}, 0.8, 4)$	p_5	(p_{T_5}, p_{L_5})
s_6	$(s_{T_6}, s_{L_6}, 0.6, 1)$	p_6	(p_{T_6}, p_{L_6})
s_7	$(s_{T_7}, s_{L_7}, 0.5, 2)$	p_7	(p_{T_7}, p_{L_7})

A location-text publish/subscribe system sends relevant publications to each subscriber. To quantify the correlation between a publication and a subscription, we adopt a similarity method as follows.

Definition 3 (Location Similarity). The Location Similarity LSIM(s, p) between a subscription s and a publication p is computed as Formula (1).

$$\text{LSIM}(s, p) = \max(0, 1 - \frac{DIST(s_L, p_L)}{MAXDIST}) \tag{1}$$

In Formula (1), $DIST(s_L, p_L)$ is the Euclidian distance between s_L and p_L. MAXDIST is the maximum user-tolerated Euclidian distance between subscriptions and publications [4].

Definition 4 (Text Similarity). The Text Similarity TSIM(s, P) between a subscription s and a publication p is computed as Formula (2).

$$\text{TSIM}(s, p) = \frac{\sum_{t \in s_T \cap p_T} w(t)}{\sum_{t \in s_T} w(t)} \tag{2}$$

In Formula (2), $w(t)$ is the weight of keyword t.

Based on above definitions, we show our core definition as follow.

Definition 5 (Location-Text Similarity). The Location-Text Similarity SIM(s, P) between a subscription s and a publication p is computed as Formula (3).

$$SIM(s, p) = \delta \cdot TSIM(s, p) + (1 - \delta) \cdot LSIM(s, p) \tag{3}$$

In Formula (3), $\delta \in [0, 1]$ is a preference parameter. If $\delta > 0.5$, text similarity is more important than location similarity.

To facilitate the description, we use SIM(s, P) to denote the set of Location-Text Similarities between a subscription s and a set of publications P in the next sections.

Example 2. Table 2 shows the set of Location-Text Similarities $SIM(s_1, P)$ between a subscription s_1 and all publications.

Table 2. The examples of location-text similarities based on Table 1

Location-text similarity	Value	Location-text similarity	Value
$SIM(s_1, p_0)$	0.1	$SIM(s_1, p_1)$	0.3
$SIM(s_1, p_2)$	0.5	$SIM(s_1, p_3)$	0.7
$SIM(s_1, p_4)$	0.2	$SIM(s_1, p_5)$	0.6
$SIM(s_1, p_6)$	0.4	$SIM(s_1, p_7)$	0.9

3   Publish/Subscribe System

The publish/subscribe system structure of TGT is shown in Fig. 1. Publishers upload their own publications to the system (mobile phone, ipad, cloud, Internet, etc.). Subscribers upload their own subscriptions to capture their interests. The system returns top-k publications to the subscriber according to location-text similarities.

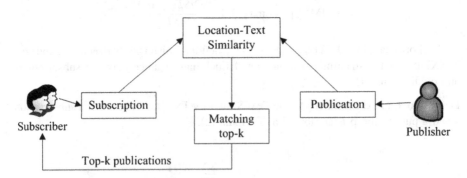

Fig. 1. System structure of TGT

Algorithm 1 Match regulates the publish/subscribe matching process of TGT. The algorithm takes a subscription s and the location-text similarity set SIM(s, P) as input. Meanwhile, the algorithm takes the top-k matching results M_k as output.

First, if the calculated number of results |SIM(s, P)| are less than k, the calculated results do not have to be ranked. So in this case, Algorithm 1 directly returns SIM(s, P) to the subscriber s without any operation (Steps 2, 15 and 16).

Otherwise, the algorithm inserts the publication M (who has the maximum location-text similarity with s) into M_k (Steps 4–11). Then, Algorithm 1 removes M from SIM(s, P) (Step 12).

By the same way, the algorithm next finds out the maximum value among the rest of SIM(s, P). Finally, the algorithm inserts top-k publications (who have top-k highest scores of location-text similarities) into M_k.

Algorithm 1: Match	
Input: s, SIM(s, P)	
Output: M_k	
1:	$M_k := \emptyset$
2:	**If** |SIM(s, P)|>k **then**
3:	**For all** $i \in [1, k]$ **do**
4:	M:= SIM(s, p_i)
5:	**For all** $j \in [2, |SIM(s, P)|]$ **do**
6:	**If** SIM(s, p_j)> SIM(s, p_i) **then**
7:	M:= SIM(s, p_j)
8:	**End if**
9:	**End for**
10:	Let the publication of M is p_m
11:	$M_k := M_k \; Y \; p_m$
12:	SIM(s, P):= SIM(s, P)/M
13:	**End for**
14:	**Else**
15:	$M_k :=$ SIM(s, P)
16:	SIM(s, P) := $\emptyset$
17:	**End if**
18:	**Return**(M_k)

Example 3. Base on Tables 1 and 2, we run Match (s_1, SIM(s_1, P)). The top-k result is $M_2 = \{p_7, p_3\}$.

First, from Table 1, we find s_1's k = 2 and |SIM(s_1, P)| = 8 > 2. Second, the maximum location-text similarity among Table 2 is SIM(s_1, p_7) = 0.9. Thus, $M_2 = \{p_7\}$. Third, we remove SIM(s_1, p_7) = 0.9 from Table 2. Fourth, the maximum location-text similarity among rest of $SIM(s_1, P)$ is $SIM(s_1, p_3) = 0.7$. Thus, $M_2 = \{p_7, p_3\}$. That is, the final matching results of s_1 is $M_2 = \{p_7, p_3\}$.

4 Experiments

IGPT [5, 16, 17] is the most classic and close related work to our method. Therefore, we compare our scheme TGT to IGPT to verify efficiency and effectiveness.

4.1 Experimental Setup

We use Visual c++ 6.0 to implement our experiments. The experimental scenario includes only one computer with 3.4 GHz dual-core CPU and 32 GB memory. Following the tradition settings (for example, [7, 16]) of publish/subscribe system, we assume that the index is stored in the memory to support real-time response.

The data sets of publication and subscription are randomly generated by the system. For each subscription, preference parameter δ is randomly selected in the interval [0, 1]. The comparison between TGT and IGPT is always under the same data set.

In our experiments, the keyword number is from 10 to 50. The keyword weight is computed according to IDF [18]. Our k value (i.e. the number of returned results) is from 2 to 10. The subscription number is from 10M to 50M (1M = a million). The detailed experimental parameters are shown in Table 3.

Table 3. Experimental parameters

Keyword number	10, 20, 30, 40, 50
k	2, 4, 6, 8, 10
Subscription number	10M, 20M, 30M, 40M,50M

4.2 Memory Usage

As index is stored in the memory, the first set of experiments test the memory usage of TGT and IGPT. There are two parameters affecting memory usage: subscription number and keyword number. Therefore, the experimental results are shown in Fig. 2, where Y-axis denotes the memory usage, whose unit is GB. Meanwhile, X-axes denote subscription number and keyword number respectively.

(1) Figure 2(a). When subscription number increases from 10M to 50M, the memory usage of our scheme TGT increases slowly. Meanwhile, as TGT is less than IGPT a parameters: the threshold, IGPT has almost 1.3-fold memory usage of TGT. That is, TGT significantly decreases memory usage.

(2) Figure 2(b). When keyword number increases from 10 to 50, the memory usage of IGPT increases more quickly than TGT. This is because that the TGT is less than IGPT a parameters: the threshold. Thus there is no threshold index in the memory, which results in lower memory consumption.

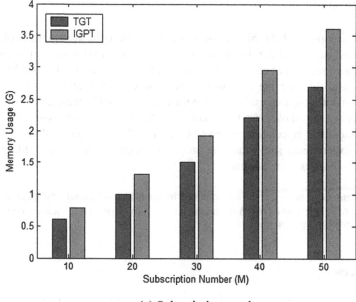

(a) Subscription number

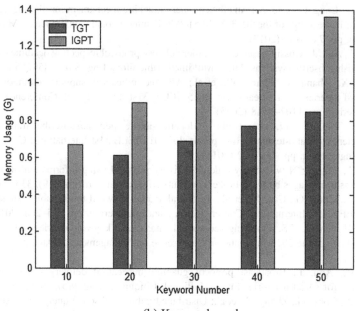

(b) Keyword number

Fig. 2. The comparison of memory usage

5 Conclusions

In solving the top-k problem of location-text publish/subscribe system, most of existing methods need user to set the threshold. Setting threshold mostly makes the number of final calculated results not equal to the number of required results. On one hand, these methods waste time for the user to set parameters. On the other hand, they waste valuable computing resources. To remove the issues, we put forward a scheme which is a parameter (i.e. threshold) less than existing methods. Without threshold, our scheme reduces redundant calculation, improves the recall ratio and facilitates the user. Extensive experiments show that the proposed technique has better efficiency and effectiveness than existing methods.

Acknowledgment. This research was supported by the National Natural Science Foundation of China [Nos. 61772215 and 61762055]; and the Jiangxi Provincial Natural Science Foundation of China [No. 20161BAB202036].

References

1. Li, G., Wang, Y., Wang, T., et al.: Location-aware publish/subscribe. In: Proceedings of the 19th ACM SIGKDD International Conference on Knowledge Discovery and Data Mining, pp. 802–810 (2013)
2. Chen, L., Cong, G., Cao, X.: An efficient query indexing mechanism for filtering geo-textual data. In: Proceedings of the 2013 ACM SIGMOD International Conference on Management of Data, pp. 749–760 (2013)
3. Kim, C., Ahn, J.: Gossip-based causal order delivery protocol respecting deadline constraints in publish/subscribe systems. Int. J. Multimed. Ubiquitous Eng. **8**(6), 245–256 (2013)
4. Wang, X., Zhang, Y., Zhang, W., et al.: AP-Tree: efficiently support continuous spatial-keyword queries over stream. In: 2015 IEEE 31st International Conference on Data Engineering, pp. 1107–1118 (2015)
5. Hu, H., Liu, Y., Li, G., et al.: A location-aware publish/subscribe framework for parameterized spatio-textual subscriptions. In: 2015 IEEE 31st International Conference on Data Engineering, pp. 711–722 (2015)
6. Kim, C., Ahn, J.: New causally ordered delivery protocol using information of immediate predecessor messages between brokers and subscribers. Data Process. **9**(9), 335–350 (2014)
7. Chen, L., Cong, G., Cao, X., et al.: Temporal spatial-keyword top-k publish/subscribe. In: 2015 IEEE 31st International Conference on Data Engineering, pp. 255–266 (2015)
8. Cong, G., Jensen, C.S.: Querying geo-textual data: spatial keyword queries and beyond. In: Proceedings of the 2016 International Conference on Management of Data, pp. 2207–2212 (2016)
9. Kim, C., Ahn, J.: Broadcast protocol guaranteeing causal delivery order consistency condition for social networks. Int. J. Multimed. Ubiquitous Eng. **9**(5), 49–62 (2014)
10. Chen, Z., Cong, G., Zhang, Z., et al.: Distributed publish/subscribe query processing on the spatio-textual data stream. arXiv preprint arXiv:1612.02564 (2016)
11. Zhao, K., Liu, Y., Yuan, Q., et al.: Towards personalized maps: mining user preferences from geo-textual data. Proc. VLDB Endow. **9**(13), 1545–1548 (2016)
12. Kim, C., Ahn, J.: Causal order protocol based on virtual synchronous group membership in wireless sensor networks. Int. J. Control Autom. **8**(2), 9–20 (2015)

13. Jiang, H., Zhao, P., Sheng, V.S., et al.: An efficient location-aware top-k subscription matching for publish/subscribe with Boolean expressions. In: International Conference on Database Systems for Advanced Applications, pp. 335–350 (2016)

14. Liu, S., Ma, X., Wang, X.: An improved model of general data publish/subscribe based on data distribution service. Int. J. Database Theory Appl. **9**(11), 83–94 (2016)

15. Cui, Z., Wu, Z., Zhou, C., et al.: An efficient subscription index for publication matching in the cloud. Knowl.-Based Syst. **110**, 110–120 (2016)

16. Wang, X., Zhang, Y., Zhang, W., et al.: SKYPE: top-k spatial-keyword publish/subscribe over sliding window. Proc. VLDB Endow. **9**(7), 588–599 (2016)

17. Wang, X., Zhang, Y., Zhang, W., et al.: Top-k spatial-keyword publish/subscribe over sliding window. arXiv preprint arXiv:1611.03204 (2016)

18. Broder, A.Z., Carmel, D., Herscovici, M., et al.: Efficient query evaluation using a two-level retrieval process. In: Proceedings of the Twelfth International Conference on Information and Knowledge Management, pp. 426–434 (2003)

Effects of Foliar Selenium Fertilizer on Agronomical Traits and Selenium, Cadmium Contents of Different Rape Varieties

Bin Du[1,2,4], HuoYun Chen[1,2,3], and DanYing Xing[1,2,3(✉)]

[1] Institution of Crops Selenium-Enrichment Application Technology,
Yangtze University, Jingzhou 434025, China
97573133@qq.com, xingdy_2006@126.com
[2] School of Agriculture Jingzhou Hubei, Yangtze University,
Jingzhou 434025, China
[3] Hubei Collaborative Innovation Center for Grain Industry,
Jingzhou 434025, Hubei, China
[4] Department of Crop Science and Technology, College of Agriculture,
South China Agricultural University Guangzhou, Guangdong 510642, China

Abstract. To elucidate the relationship between agronomic traits and Se/Cd contents of rapeseeds, we conducted randomized block tests with 6 varieties of rapeseed (30 g/hm^2 foliar Se fertilizer). With the response relationship between agronomic traits and Se fertilizer of rapeseed as a template, the effects of foliar Se fertilization on the primary branch number, secondary branch number, rapeseed yield, rapeseed Se content, and grain Cd content were analyzed to find out the differences among cultivars in response to Se application. This study was aimed to actively promote the search for Se-sensitive rape, screen Se advantage in the future, and provide reference for rational and effective use of Se resources into rapeseed varieties. Least significant difference analysis of variance model was used to scrape out the average Se content of 518 seeds of Deye Oil 0.248 mg/kg, The potential for Se enrichment was strong, and it was expected to become a typical variety of Cd-tolerant rapeseeds. Seeds of Huayouza 9 made maximum use of exogenous Se under relatively low Se content (0.298 mg/kg). The soil Se content was lower for the main varieties of planted areas. The Cd accumulation of Huiyouzaoxian 6815 was weak and did not change much when the leaves were sprayed with 30 g/hm^2 Se, which could be used as a favorable material for Se-rich and low-Cd production.

Keywords: Rape · Selenium · Agronomical traits · Cadmium

1 Introduction

Heavy metal cadmium (Cd) was classified as a human carcinogen by International Agency for Research on Cancer (IARC 1993) and US National Toxicology Program (NTP 2000). Cd is a pollutant in soils and poses risk for human health through the food

---

B. Du—Contributes equally.

chain due to its plant availability in soils (He et al. 2004; Belkhadi et al. 2010; Filek et al. 2010; Gallego et al. 2012; Lin et al. 2012; Saidi et al. 2014). Cd in soils and groundwater is easily taken up by plant roots and efficiently transported to the aboveground organs; this efficient Cd transfer from soils to plants poses a potential threat to human health through the food chain (Kuboi et al. 1986; Grant et al. 1998; Clemens et al. 2013).

Selenium (Se) is an essential micronutrient with important benefits for human health (Allan et al. 1999). While no evidence proves the essentiality of Se for higher plants, many beneficial effects of Se on plant physiology have been recognized in several plant species (Pilon-Smits et al. 2009). In plants, Se protects against Cd and other toxic elements such as copper, lead, antimony, mercury and arsenic (Gotsis 1982; He et al. 2004; Srivastava et al. 2009; Taspinar et al. 2009; Feng et al. 2013; Zhou et al. 2013). The protective effect of Se against heavy metals also affects the accumulation and distribution of toxic metals in plant organs (Zembala et al. 2010; Duan et al. 2013).

Oilseed rape (*Brassica napus* L.) is a very important oil crop grown in China and the world and has been confirmed with some characteristics representative of a Cd hyperaccumulator, such as high Cd-accumulating ability, fast growth rate, large biomass, and ease of harvest (Carrier et al. 2003; Meng et al. 2009). Many studies concentrate on the growth (Novo et al. 2014) and Cd absorption capacity (Su and Wong 2004; Ru et al. 2004) of oilseed rape in contaminated soil or a simulated environment. Similarly, we need many rapeseed and oil-free crops not polluted by Cd as a necessity for human daily life. Thus, this study was conducted.

2 Materials and Methods

2.1 Test Materials

Six varieties were used here: Dexuan 518, Huayouza 9, Dexinyou 53, Huiyouza 6815, Chuyouza 79, and Dingyouza 4.

2.2 Experimental Design

Experiments were carried out at a base of College of Agriculture, Yangtze University (111°150' E, 29°260' N, 32.6 M), Jingzhou, Hubei Province from 2015 to 2016 and from 2016 to 2017. The single-factor randomized block design of Se fertilizer (30 g/hm^2 Se) was applied at the early flowering stage of rapeseeds. Meanwhile, a blank control experiment was carried out. The soil was featured by high fertility and contained 0.350 mg/kg Se and 0.298 mg/kg Cd. In the seeding periods of October 25, 2015 and October 28, 2016, sowing was planted in the plot area of 2 × 6 m^2. The experiments were repeated three times. Before sowing, 40 kg/667 m^2 Shihe compound fertilizer (N: P$_2$O$_5$: K$_2$O = 15:15:15) and 15 kg/667 m^2 urea were used as basal fertilizers. Field management was consistent and fertilization level was medium. Other management was normally high.

2.3   Climate Characteristics at Experimental Area

The experimental base belongs to the subtropical monsoon humid climate zone. The climate characteristics were total annual solar radiation = 104–110 kcal • cm^{-2}, annual average temperature = 15.9–16.6 °C, ≥ 10 °C accumulated temperature = 5000–5350 °C, annual frost-free period = 242 to 263 days, annual sunshine hours = 1800–2000 h, and average annual rainfall = 1200 mm. From April to October, the precipitation, solar radiation and accumulated temperature ≥ 10 °C accounted for 80%, 75% and 80% of the whole year, respectively. The climatic conditions were the same as the season of agricultural production, and suitable for the growth and development of many crops.

2.4   Indoor Test Species

After maturing, the rapes were harvested in small plots. Each plot was sampled at 5 o'clock. The main indexes were plant height, rhizome thick, high branching point, height of main axis, number of effective branches (primary and secondary), and production.

2.5   Determination of Se, Cd Contents

After the completion of tests, the test materials of rape stems and rapeseeds were smashed with a high-speed grinder, crushed by a 1 mm sieve, digested in a mixed acid (nitric acid: perchloric acid = 4: 1) at 170 °C to be colorless and transparent, and determined on an AFS-810 atomic fluorescence spectrometer for Se and Cd contents according to *Determination of Selenium in Foods* (GB/T5009.93-2003) and *Determination of Cadmium in Foods* (GB 5009.15-2014), respectively.

2.6   Data Processing

Experimental data were analyzed on EXCEL and SPSS 20.

3   Results

3.1   Agronomic Traits

Effects of Se fertilizer on the agronomic traits of rapeseeds are showed in Table 1. The data of two years show the rape plant height, root rhizome, branch height, spindle height, primary and secondary branch numbers, number of spindle pectin, numbers of primary crested and secondary crested fruits, and yield after application of Se formula fertilizer to each rapeseed cultivar compared with the control (Table 1). Further data analysis shows the application of Se fertilizer compared with the control in 2015 did not significantly affect the number of primary pachylope or the number of primary pachyrhizi (F = 0.028 and 2.224 < F$_{0.05}$). The application of Se fertilizer can significantly increase the number of main rapeseed pods in Huayouza 9, Huiyouza 6815 and Chuyouza 79, and the average numbers of spindle pods are 83.00, 62.28, 52.33

respectively, or increase by 23%, 48% and 25% respectively compared with the controls. The application of Se fertilizer also increased the number of primary shoots of rapeseeds. The average number of branchlets across cultivars is 168.53, an increase of 16% compared with the controls. Results show significant difference in the number of secondary branchlets of rapeseeds (F = 28.517 > $F_{0.05}$). The application of Se fertilizer effectively promoted the secondary branchlet number of rapeseeds. The cultivar Dingyouza 4 showed obvious performance, as the average number of secondary branchlets reached 30.17, 0.44 and 4.56, respectively. The Se application did not significantly affect the yield of rapeseeds (F = 0.984) (199.64 kg vs. 12.78 kg in the control) with an increase of 6.8%. In 2016, the number of primary pods or the number of primary branchlets did not significantly change after application of Se fertilizer compared with the control (F = 0.497 and 0.753 < $F_{0.05}$), Compared with 9%, 26%, 4%, 2% increase of The numbers of fruit branches of Huayouza 9, Huiyaoza 6815, Chuyouza 79 and Dingyouza 4 after application of Se fertilizer increased 33%, 81%, 45% and −10% compared with the controls, respectively, which are basically consistent with the results of 2015 (F = 4.913 > $F_{0.05}$). The application of Se fertilizer could effectively promote the secondary branching of rapeseeds. The average number of fruit pods on secondary branches across cultivars was 8.48, which increased by 6.09 compared with the control. The average yield across cultivars was 148.38 kg, with an increase of 20.92 kg or 16.4% compared with the controls.

3.2   Grain Se Contents

The effects of Se fertilizer on Se contents of rapeseeds are shown in Table 2. Application of Se foliar fertilizer to rapes in 2015 significantly increased the Se contents of rapeseed kernels (Table 2). The average Se content of rapeseed kernels after application of Se fertilizer was 0.2605 mg/kg, with an increase of 0.1875 mg/kg compared with the control (0.073 mg/kg), indicating an obvious effect of Se enrichment. Under blank treatment, the absorption and utilization effects of Se were different among the rapeseed varieties. The absorption and utilization rate of natural Se by Dexinyou 53 were insensitive to the other five varieties, and the Se content was only 0.047 mg/kg. Compared with other five varieties, the oil absorption and utilization rate of Chuyouza 79 on natural Se was sensitive, and the grain Se content was 0.093 mg/kg, an increase of 97.87% compared with Dexinyou 53 (0.046 mg/kg). The Se contents of Chuyouza 79 and Dingyouza 4 were 0.089 and 0.083 mg/kg, respectively, indicating strong selenium absorption and utilization ability. In the application of Se foliar fertilizer, the Se accumulation of varieties of all rapeseed was obvious, as the grain Se contents exceeded 0.2 mg/kg. In particular, Chuyouza 79 and Dexuan 518 had higher Se absorption and utilization ability, as the grain Se contents reached 0.297 and 0.287 mg/kg, respectively. Huiyouza 6815 and Dexinyou 53 showed low Se absorption and utilization ability, as grain Se contents were 0.207 and 0.236 mg/kg, respectively.

Annual data in 2016 shows that application of foliar Se fertilizer increased Se content in rapeseed kernels. The average Se content of rapeseed kernels after fertilization was obviously higher compared with the control (0.151 vs. 0.136 mg/kg). Under the blank treatment, no significant difference among cultivars was found in the

Table 1. Effects of selenium formula fertilizer on agronomic properties of rapes

Treatment	Variety	Height/cm	Stem diameter/mm	Branch height/cm	Spindle height/cm	Branch		Pod number of branch			Yield (kg/667 m^2)
						First	Second	Spindle	First	Second	
2015Se	Dexuan518	82.82	18.63	46.33	59.93	5.22	3.67	17.33	148.83	30.17	185.56
	Huayouza9	98.26	14.44	63.66	57.21	5.67	0.44	83.00	116.89	0.44	220.01
	Dexinyou53	91.82	15.75	48.86	54.64	6.00	0.22	61.67	209.22	2.89	215.57
	Huiyouza6815	103.27	15.34	72.66	55.01	7.00	0.00	62.28	210.89	0.00	234.46
	Chuyouza79	98.19	17.18	55.09	45.71	6.67	1.11	52.33	185.00	4.56	177.79
	Dingyouza4	99.71	15.29	58.68	48.63	4.78	0.00	46.44	140.33	0.00	164.45
2015CK	Dexuan518	90.78	15.72	46.74	64.61	4.83	9.00	49.00	191.89	0.00	195.57
	Huayouza9	93.59	18.20	57.39	57.22	4.78	0.00	67.33	136.17	0.00	220.01
	Dexuan53	79.37	13.57	55.18	57.18	4.67	0.33	61.83	105.83	4.00	185.56
	Huiyouza6815	94.31	15.56	60.21	54.01	6.44	0.22	42.17	165.78	1.22	183.34
	Chuyouza79	91.94	13.25	58.5	55.5	5.89	2.00	41.78	150.11	1.5	155.56
	Dingyouza4	84.78	13.10	58.18	55.48	5.00	0.00	68.5	121.00	0.00	181.12
2016Se	Dexuan518	130.58	32.44	44.74	59.8	3.78	0.44	57.67	83.78	1.00	149.77
	Huayouza9	138.97	38.61	43.92	62.14	4.67	0.61	67.00	144.00	2.22	178.45
	Dexinyou53	135.34	33.67	48.9	55.24	4.78	1.50	56.00	134.44	6.78	153.94
	Huiyouza6815	143.56	34.00	53.83	60.18	4.56	0.33	67.67	132.89	0.44	198.52
	Chuyouza79	122.10	29.67	41.00	59.09	3.67	1.89	57.00	127.67	20.33	103.63
	Dingyouza4	127.52	32.89	36.00	52.36	4.56	2.33	55.33	124.50	20.11	105.97
2016CK	Dexuan518	144.02	39.67	45.49	67.42	4.44	0.56	68.67	148.61	0.00	147.18
	Huayouza9	133.79	36.67	47.41	60.64	4.11	0.78	61.33	108.33	2.00	149.41
	Dexinyou53	129.11	34.89	43.54	57.86	4.33	1.78	57.33	139.22	4.00	133.96
	Huiyouza6815	128.67	31.96	49.63	58.49	3.67	0.59	53.5	73.50	0.00	138.35
	Chuyouza79	114.36	30.44	40.22	53.94	3.33	1.11	55.00	88.17	0.17	101.77
	Dingyouza4	124.96	38.44	38.6	50.33	4.94	2.78	54.00	138.17	8.17	94.1

Table 2. Effects of selenium formula fertilizer on selenium, cadmium content in rape seeds

Year	Content/mg/kg	Dispose	Variety					
			Deuan518	Huaouza9hao	Deinyou53	Huyouza6815	Chuouza79	Digyouza4hao
2015	Se	Se	0.29	0.26	0.24	0.21	0.30	0.27
		CK	0.06	0.09	0.05	0.07	0.09	0.08
2016		Se	0.21	0.17	0.10	0.19	0.12	0.12
		CK	0.02	0.02	0.01	0.018	0.01	0.01
2015	Cd	Se	0.13	0.11	0.13	0.10	0.12	0.11
		CK	0.14	0.11	0.13	0.10	0.12	0.11
2016		Se	0.07	0.10	0.06	0.08	0.08	0.08
		CK	0.07	0.10	0.06	0.08	0.08	0.07

absorption or utilization of Se. The absorption and utilization of natural Se in Dingyou 4 was insensitive to 5 other cultivars, and the grain Se content was only 0.011 mg/kg; Dexuan 518, Huiyouza 6815 and Huayouza 9 had similar abilities to absorb and utilize natural Se, and the average grain Se content was 0.018 mg/kg, and the rapeseed Se content increased by 0.007 mg/kg or 63.64% compared with Dingyouza 4. After the application of Se formula fertilizer, all varieties showed obvious accumulative effect on Se, and the grain Se contents all exceeded 0.1 mg/kg. Among them, the highest Se content was 0.208 mg/kg, while Dexinyou 53, Chuyouza 79 and Dingyouza 4 were less able to absorb and utilize Se, and the grain Se contents were 0.104, 0.117 and 0.123 mg/kg, respectively.

3.3 Agronomic Traits

The effects of Se fertilizer on the rapeseed grain Cd contents are shown in Table 2. In 2015, the average Cd content of rapeseed leaves was 0.116 mg/kg, which was 2.82% lower than the control (Table 2), indicating Se fertilizer could reduce the Cd accumulation in rapeseeds. Under the blank treatment, no significant differences in the absorption and utilization of Se were found among the rape cultivars. The Cd accumulation in Huayouza 9, Huiyouza 6815 and Dingyouza 4 was less than the other three rape cultivars, as Ca accumulation was about 0.1 mg/kg. After the application of Se fertilizer, the Cd accumulation in the seeds of Dexuan 518, Dexinyou 53, Chuyuza 79 and Dingyouza 4 was lower than the controls. Among them, the Cd accumulation in Dexuan 518, Dexinyou 53 and Dingyouza 4 varieties decreased by 6.9%, 5.1% and 2.7%, respectively, compared with the control, indicating Se fertilizer could effectively reduce the Cd content in rapeseed kernels. The results of Cd accumulation in two treatments show that the use of 30 g/hm² foliar Se fertilizer could effectively control Cd accumulation in rapeseed kernels in Huiyouza 6815 and Dingyouza 4.

In 2016, the average Cd content in rapeseed leaves was 0.077 mg/kg, which was 1.63% less than that of the control, indicating Se fertilizer could reduce the Cd accumulation in rapeseeds. Under the blank treatment, significant difference in the absorption and utilization of Se was found among rape varieties. In particular, the Cd accumulation in the seeds of Dexuan518, Huiyouza 6815 and Dingyouza 4 was less than the other three cultivars. After the application of Se fertilizer, the Cd accumulation in seeds of Dexuan 518, Huayouza 9, Dexinyou53, Huiyouza 6815, Chuyouza 79 was significantly higher than these results of 2015. Compared with the controls, the Cd accumulation in Dexuan518, Huayouza 9 and Dexinyou 53 decreased by 5.7%, 6.7% and 9.4% respectively, indicating which the use of Se fertilizer could effectively reduce the Cd contents of rapeseeds. Two years of data show that the application of 30 g/hm^2 foliar Se fertilizer can effectively reduce the Cd accumulation in rapeseeds of Dexuan 518 and Dexinyou 53.

3.4   Effect of Se Fertilizer on Se Contents of Rape Seeds

The effects of Se fertilizer on Se contents in rapeseed are shown in Fig. 1 and Fig. 1 (2015). In 2015, the application of foliar Se fertilizer led to differences in Se absorption and utilization among the rape varieties (Fig. 1 (2015)). Under the blank treatment, the Se content in rapeseed was significantly different from Dexinyou 53, and the Se enrichment potential of Dexinyou 53 was weaker. No significant difference in rapeseed grain Se content of Dexuan 518, HuiYouza 6815, Dingyouza 4 was found from other varieties, which indicates the general potential of selenium. Further analysis showed the rapeseed Se contents significantly differed among different treatments (F = 350.737 > F$_{0.01}$). After application of foliar Se fertilizer, the average Se content of rapeseeds was 0.2605 mg/kg, which increased significantly by 0.1875 mg/kg from 0.073 mg/kg.

Results of the 2016 trial also showed the application of the foliar Se fertilizer had different effects on the Se absorption and utilization of the rapeseed varieties (Fig. 1 (2016)). Under blank treatment, the rapeseed Se contents were not significantly different among all the rapeseed varieties, but Dexuan 518 showed a seed Se content of 0.208 mg/kg and relatively strong potential of Se enrichment, which is expected to be a good selenium-enriched material. Further analysis showed the rapeseed Se contents significantly differed among different treatments (F = 150.667 > F$_{0.01}$). After application of Se fertilizer, the average Se content of rapeseeds increased from 0.015 mg/kg by 0.136 mg/kg to 0.151 mg/kg.

By comparing the Se contents in different cultivars under two years of different treatments, it was found rape cultivars with higher Se-enrichment ability under blank conditions was significantly different in rapeseed Se contents after the use of exogenous Se, which effectively improved the rapeseed grain Se content. In addition, an inherent difference in the Se content in the field was found, which can absorb and utilize the Se source well and increase the Se content after exogenous selenium was supplied to Dexuan 518. Therefore, Dexuan 518 can be the predominant source of selenium-enriched production of materials; Dexinyou 53 had poor utilization ability of exogenous selenium, especially in relatively low-Se soils compared to other varieties. The rapeseed kernel Se content of Dexinyou 53 was only 0.104 mg/kg, which failed to achieve the state provision of food Se level. When soil Se content was 0.298 mg/kg, the Se content of rapeseed leaves treated with foliar fertilizer increased by 0.154 mg/kg, which increased by 905.8% than the blank control. When the content was 0.35 mg/kg, the Se content in rapeseed leaves treated with foliar fertilizer increased by 0.164 mg/kg or 176.3% than the blank control. This is relatively low when the use of exogenous Se can be maximized in low-Se areas as a dominant species to promote planting.

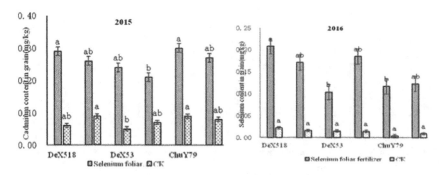

Fig. 1. Effect of Se fertilizer on rape seed Cd content

3.5 Effect of Se Fertilizer on Rapeseed Se Content

The effect of Se fertilizer on rapeseed kernel Cd content is shown in Figs. 2 (2015) and (2016). In 2015, after application of foliar Se, the Cd absorption and accumulation were both different among rapeseed varieties (Fig. 2 (2015)). Under the blank treatment, the Cd accumulations in rapeseed kernels were significantly different between Dexuan 518 and other cultivars. Among them, the Cd accumulation in Dexuan 518, Dexinyou 53 and Dingyouza 4 decreased by 6.9%, 5.1% and 2.7%, respectively compared with the control, indicating Se fertilizer could effectively reduce the Cd content in rapeseed kernels. Further analysis showed the rapeseed Se contents were significantly different ($F = 0.927 < F_{0.05}$), but the average Cd content in rapeseed was 0.116 mg/kg, 2.82% less than the control, indicating Se fertilizer reduced the Cd accumulation in rapeseed grains.

The 2016 annual trial also showed the rape cultivars differed in the absorption and utilization of Cd after the application of foliar Se fertilizer (Fig. 2 (2016)). Under blank treatment, the Cd accumulation in Huayouza 9 was significantly different from other cultivars, and was reduced by 6.7% compared with the control. Further analysis showed the rapeseed Se contents significantly differed among different treatments ($F = 0.073 < F_{0.05}$). The average Cd content in rapeseed was 0.077 mg/kg, which was 1.63% less than the control, indicating Se fertilizer reduced the Cd accumulation in rapeseed grains.

By comparing the trends of Se contents in different rape varieties under two years of different treatments, it was also found rapeseed varieties with lower Cd-enriching ability under blank condition had less Cd accumulation in rapeseed with the presence of exogenous Se. In addition, the Se contents of Huiyouza 6815 inherently differed in the field under the supply of exogenous Se. Compared with the control, the Cd accumulated amount in the grain was almost not changed, and the Cd accumulation ability of this variety was weak, which was advantage of Se and low Cd production for the material selection. The variation of grain Cd accumulation of Dingyouzao 2 differed in the two years, and the specific reasons are to be verified next.

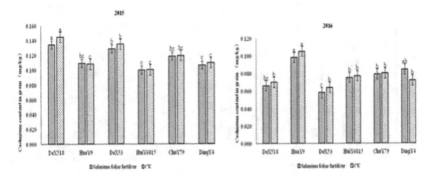

Fig. 2. Effects of Se fertilizer on Se content in rape seeds. Short captions are centered, while long ones are justified. The macro button chooses the correct format automatically.

4   Discussion

No evidence proves Se is essential for higher plants, but many beneficial effects of Se on plant biochemistry have been recognized in some plant species, as reviewed by Pilon-Smits et al. (2009). Owing to the similarity between Se and S, the uptake, movement and metabolism of Se in plants are generally thought to mimic those of S (Anderson 1993). Compared with the control, the Se fertilizer significantly affected the number of secondary branches of rapeseeds (F = 4.913 > $F_{0.05}$), indicating the application of Se fertilizer could effectively promote the secondary branching of rapeseeds. The average pod number on secondary branches was 8.48, which was 6.09 higher than the control. The average yield of rapeseed after Se application was 148.38 kg, increased by 20.92 kg or 16.4% compared with the control in the same period of last year. These results are consistent with Malik et al. (2011) and square up that Se supplementation might improve growth and defense ability against Cd-induced toxicity due to increased antioxidant potential. As reported, As treatment alone significantly reduced dry matter yield of grain and root, indicating its toxicity (Tripathi et al. 2012, 2013; Kumar et al. 2014). Supplementation of Se benefited plant growth and yield by alleviating both biotic and abiotic stresses and nutrient imbalance (Filek et al. 2008; Kumar et al. 2013).

Se fertilizer application and other biological strengthening measures have become the main ways to produce Se-enriched agricultural products of variety screening oil crop. Studies show the amount of exogenous Se can increase plant Se content and reduce the accumulation of heavy metals in crops. Francis (1973) found increasing Se concentration in plants can significantly reduce the Cd concentration in plants. The appropriate Se concentration can modestly mitigate Cd poisoning. Our rape experiments in JingZhou in 2015–2016 show application of exogenous Se can increase the Se accumulation in rape and reduce the Cd content of rapeseed cultivar by 10%–30%. Moreover, the use of 30 g/hm^2 Se foliar fertilizer to Huiyouza 6815 and Dingyouza 4 could control the Cd accumulation in rapeseed grains, which is consistent with Elguera et al. (2017).

5 Conclusions

Effects of foliar Se application on the number of secondary shoots in rapeseeds were studied. Application of Se fertilizer significantly increased the Se content of rapeseed kernels. Application of foliar Se fertilizer into rapeseeds is an effective measure to improve Se content in rapeseeds. The uptake and utilization of exogenous Se both differed among rapeseed cultivars, as rapeseed cultivars with lower Cd accumulation ability under blank conditions and less Cd accumulation in rapeseed were interfered with exogenous Se. The Se content of Huiyouza 6815 in the field inherently differed under the supply with exogenous Sd. The accumulated amount of Cd in the grain was almost not changed from the control, and the Cd accumulation ability of this variety was weak, which indicate the advantages of Se-rich low-Cd production for material selection.

References

Belkhadi, A., Hediji, H., Abbes, Z., Nouairi, I., Barhoumi, Z., Zarrouk, M., et al.: Effects of exogenous salicylic acid pretreatment on cadmium toxicity and leaf lipid content in Linum Usitatissimum L. Ecotoxicol. Environ. Saf. **73**(5), 1004 (2010)

Carrier, P., Baryla, A., Havaux, M.: Cadmium distribution and microlocalization in oilseed rape (Brassica napus) after long-term growth on cadmium-contaminated soil. Planta **216**(6), 939–950 (2003)

Clemens, S., Aarts, M.G., Thomine, S., Verbruggen, N.: Plant science: the key to preventing slow cadmium poisoning. Trends Plant Sci. **18**(2), 92–99 (2013)

Duan, G., Liu, W., Chen, X., Hu, Y., Zhu, Y.: Association of arsenic with nutrient elements in rice plants. Metallomics Integr. Biometal. Sci. **5**(7), 784 (2013)

Elguera, J.C.T., Barrientos, E.Y., Wrobel, K., Wrobel, K.: Effect of cadmium (Cd(II)), selenium (Se(IV)) and their mixtures on phenolic compounds and antioxidant capacity in lepidium sativum. Acta Physiol. Plant. **35**(2), 431–441 (2013)

Francis, S.H.: Acoustic-gravity modes and large-scale traveling ionospheric disturbances of a realistic, dissipative atmosphere. J. Geophys. Res. Atmos. **78**(13), 2278–2301 (1973)

Filek, M., Keskinen, R., Hartikainen, H., Szarejko, I., Janiak, A., Miszalski, Z., et al.: The protective role of selenium in rape seedlings subjected to cadmium stress. J. Plant Physiol. **165**(8), 833 (2008)

Feng, R., Wei, C., Tu, S., Liu, Z.: Interactive effects of selenium and antimony on the uptake of selenium, antimony and essential elements in paddy-rice. Plant Soil **365**(1–2), 375–386 (2013)

Gallego, S.M., Pena, L.B., Barcia, R.A., Azpilicueta, C.E., Iannone, M.F., Rosales, E.P., et al.: Unravelling cadmium toxicity and tolerance in plants: insight into regulatory mechanisms. Environ. Exp. Bot. **83**(5), 33–46 (2012)

Grant, C.A., Buckley, W.T., Bailey, L.D., Selles, F.: Cadmium accumulation in crops. Can. J. Plant Sci. **78**(1), 1–17 (1998)

Gotsis, O.: Combined effects of selenium/mercury and selenium/copper on the cell population of the alga *Dunaliella minuta*. Mar. Biol. **71**(3), 217–222 (1982)

He, P.P., Lv, X.Z., Wang, G.Y.: Effects of se and Zn supplementation on the antagonism against Pb and Cd in vegetables. Environ. Int. **30**(2), 167 (2004)

Kumar, N., Mallick, S., Yadava, R.N., Singh, A.P., Sinha, S.: Co-application of selenite and phosphate reduces arsenite uptake in hydroponically grown rice seedlings: toxicity and defence mechanism. Ecotoxicol. Environ. Saf. **91**(2), 171–179 (2013)

Kuboi, T., Noguchi, A., Yazaki, J.: Family-dependent cadmium accumulation characteristics in higher plants. Plant Soil **92**(3), 405–415 (1986)

Kumar, A., Singh, R.P., Singh, P.K., Awasthi, S., Chakrabarty, D., Trivedi, P.K., et al.: Selenium ameliorates arsenic induced oxidative stress through modulation of antioxidant enzymes and thiols in rice (Oryza sativa L.). Ecotoxicology **23**(7), 1153–1163 (2014)

Listed, N.: Beryllium, cadmium, mercury, and exposures in the glass manufacturing industry. Working group views and expert opinions, Lyon, 9–16 February 1993. IARC Monogr. Eval. Carcinog. Risks Hum. 58, 1 (1993)

Lin, L., Zhou, W., Dai, H., Cao, F., Zhang, G., Wu, F.: Selenium reduces cadmium uptake and mitigates cadmium toxicity in rice. J. Hazard. Mater. **235–236**(2), 343 (2012)

Meng, H., Hua, S., Shamsi, I.H., Jilani, G., Li, Y., Jiang, L.: Cadmium-induced stress on the seed germination and seedling growth of Brassica napus, L., and its alleviation through exogenous plant growth regulators. Plant Growth Regul. **58**(1), 47–59 (2009)

Malik, J.A., Kumar, S., Thakur, P., Sharma, S., Kaur, N., Kaur, R., et al.: Promotion of growth in mungbean (Phaseolus aureus Roxb.) by selenium is associated with stimulation of carbohydrate metabolism. Biol. Trace Elem. Res. **143**(1), 530–539 (2011)

Novo, L.A.B., Manousaki, E., Kalogerakis, N., González, L.: The effect of cadmium and salinity on germination and early growth of Brassica juncea (L.) var. Juncea. Fresenius Environ. Bull. **22**(12a), 3709–3717 (2014)

National Toxicology Program: NTP toxicology and carcinogenesis studies of methyleugenol (CAS No. 93-15-2) in F344/N rats and B6C3F1 mice (gavage studies). Natl. Toxicol. Program Tech. Rep. **491**(7), 1 (2000)

Pilon-Smits, E.A., Leduc, D.L.: Phytoremediation of selenium using transgenic plants. Curr. Opin. Biotechnol. **20**(2), 207–212 (2009)

Pilon-Smits, E.A., Quinn, C.F., Tapken, W., Malagoli, M., Schiavon, M.: Physiological functions of beneficial elements. Curr. Opin. Plant Biol. **12**(3), 267 (2009)

Ru, S., Wang, J., Su, D.: Characteristics of Cd uptake and accumulation in two Cd accumulator oilseed rape species. J. Environ. Sci. **16**(4), 594 (2004)

Saidi, I., Chtourou, Y., Djebali, W.: Selenium alleviates cadmium toxicity by preventing oxidative stress in sunflower (Helianthus annuus) seedlings. J. Plant Physiol. **171**(5), 85–91 (2014)

Srivastava, M., Ma, L., Rathinasabapathi, B., Srivastava, P.: Effects of selenium on arsenic uptake in arsenic hyperaccumulator Pteris vittata L. Bioresour. Technol. **100**(3), 1115–1121 (2009)

Su, D.C., Wong, J.W.: Selection of mustard oilseed rape (Brassica juncea L.) for phytoremediation of cadmium contaminated soil. Bull. Environ. Contam. Toxicol. **72**(5), 991 (2004)

Stulen, I., De Kok, L.J., Rennenberg, H., Brunold, C., Rauser, W.E.: Sulfur nutrition and assimilation in higher plants; regulatory, agricultural and environmental aspects. Scand. J. Clin. Lab. Invest. Suppl. (1993)

Taspinar, M.S., Agar, G., Yildirim, N., Sunar, S., Aksakal, O., Bozari, S.: Evaluation of selenium effect on cadmium genotoxicity in Vicia faba using RAPD. J. Food Agric. Environ. **7**(3–4), 857–860 (2009)

Tripathi, P., Mishra, A., Dwivedi, S., Chakrabarty, D., Trivedi, P.K., Singh, R.P., et al.: Differential response of oxidative stress and thiol metabolism in contrasting rice genotypes for arsenic tolerance. Ecotoxicol. Environ. Saf. **79**(3), 189–198 (2012)

Tripathi, P., Tripathi, R.D., Singh, R.P., Dwivedi, S., Goutam, D., Shri, M., et al.: Silicon mediates arsenic tolerance in rice (*Oryza sativa*, L.) through lowering of arsenic uptake and improved antioxidant defence system. Ecol. Eng. **52**(2), 96–103 (2013)

Zhou, X., Wang, W., Yu, S., Zhou, Y.: Interactive effects of selenium and mercury on their uptake by rice sedlings. Res. J. Appl. Sci. Eng. Technol. **5**(19), 4733–4739 (2013)

Zembala, M., Filek, M., Walas, S., Mrowiec, H., Kornaś, A., Miszalski, Z., et al.: Effect of selenium on macro- and microelement distribution and physiological parameters of rape and wheat seedlings exposed to cadmium stress. Plant Soil **329**(1–2), 457–468 (2010)

Fresh-Water Fish Quality Traceability System Based on NFC Technology

Longqing Zhang[1(✉)], Lei Yang[1], Liping Bai[2], Yanghong Zhang[1],
and Kaiming You[1]

[1] Guangdong University of Science and Technology,
Dongguan, Guangdong, China
bjzlq@qq.com
[2] Macau University of Science and Technology, Weilong Road, Macau, China

Abstract. Because of the insufficiency of such traceability systems as conventional bar code, and two-dimensional code, fresh-water fish traceability system based on NFC technology is designed, which consists of NFC cellphone, and Mifare electronic tag. As for Mifare electronic tag, disposable wristband Mifare tag is adopted due to its features of usage convenience, water and high temperature resistance and anti-copy by special modes. Three identities are contained in the system, providing different users with different operation authorities. Encrypted key traceability information is written into Mifare sectors through APP of NFC cellphone, which, by consumers, can be placed near tags, for the system to automatically analyze and read the traceability information, thus achieving the purpose of traceability.

Keywords: NFC technology · Traceability system · Mifare RFID tag

1 Introduction

Food security issues took place in recent years have far raised people's attention to the transparency of food production and transportation process [1], especially nutritional value, freshness and quality safety of water products. On account of the particularity in breed aquatics, transport and consumption of fresh-water fish products, product information for product traceability might be unable to be fully transmitted and difficult to be planned, thus enhancing the difficulty tracing defective products precisely. This paper applies NFC technology to efficiently solves these problems. Near Field Communication (NFC), also known as short distance wireless communication, is the kind of short-distance high-frequency communication technology [2], which grants mobile devices, consumer electronics products, and PC to conduct non-contact point-to-point data transmission or exchange with intelligent control instruments. With the fresh-water fish products quality traceability system designed with NFC technology application in

Supported by FDCT NO.102/2013/A, this research also Supported by Innovation and entrepreneurship training program for college students in Guangdong No. 201713719017.

this paper, users are enabled to acquire key information about quality and safety of fresh-water fish products conveniently via cellphone, which pays special attention to product name, product number, manufacturers, responsible person, contact information, logistics code, quality certificate, date of production and expiration date.

2 System Structure

General structure of primary function modules of fresh-water fish quality traceability system based on NFC is shown as Fig. 1:

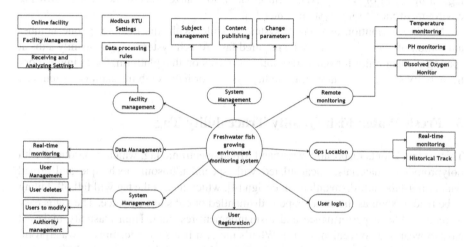

Fig. 1. Functional structure of fish quality traceability system based on NFC

Main function modules of fresh-water fish quality traceability system based on NFC consists of: manufacturers, consumers, administrators, users log-in, express check, growing environment monitoring system, number attribution and attach system.

Manufacturers module: for manufacturers, it is quite necessary to maintain a set of complete and full information management system to trace various information and logistics process of fresh-water fish products, and acquire relevant quality safety level presented by qualified institutions for fresh-water fish quality-safety certification, thus laying solid foundation for quality safety and product traceability.

Consumers module: for consumers, the major concern is the safety of fresh-water fish products and convenience in acquiring information. What concerns them most are whether the fresh-water fishes are safe, whether the raw materials applied in breeding fresh-water fish are safe, which means most attention are paid to understanding of growing environment by consumers.

Administrators module: administrators conduct management over sectors of Mifare electronic tags, and ensure that traceability information be written into sectors. Administrators utilize both password management and user identity management to secure the safety of traceability system.

Users log-in module: users log-in falls into three identities of manufacturer, consumers, administrators, each with different operation authorities.

Express check module: manufacturers and consumers are allowed to skip to WEB system via express check module, and inquire into logistics links through logistics codes.

Growing environment monitoring system module: manufacturers and administrators record web-pages of growing environment monitoring data centers into electronic tags in the form of writing NDEF information to realize consumers' visit to growing environment monitoring system center [3].

Number attribution and attach system module: with this module, phone number attribution of manufacturers can be inquired fast. Attach system module mainly aims at sparing compatibility for other traceability systems on this platform, for example, pork traceability system, vegetable traceability system, or hairy crab traceability system etc.

3 Fresh-Water Fish Quality Traceability Tag

This system adopts disposable wristband Mifare electronic tag, which is produced with polypropylene materials, encapsulated with special ultrasonic technique and armed with armor-tape anti-dismantlement design [4], which means the tag will fall off, failing to be read as soon as the armor-tape is dismantled or cut short by force. The tag is easy to read, stable in performance and dismantlement-resistant. High elasticity makes it easy to wear and convenient to use. What's more, it bears such features as waterproof, moistureproof, quakeproof, high-temperature resistance and anti-copy. In the aspect of the technique, RFID electronic tag, with the frequency of 13.56 MHz and S50 packaging of Mifare series produced by NXP company is adopted [5], whose reading interval is 3 to 10 cm, the perimeter is 20 cm, and width of band is 16 m. More importantly, it supports ISO 14443A/15693 International Standard Protocol.

The globally-only uid sequence number is stored within Mifare electronic tag, within which there are 16 sectors. Each sector contains four parts, and each part contains 16 bytes [6]. According to system requirements, traceability information (for example, product name, product number, manufacturers, responsible person, contact information, logistics code, quality certificate, date of production and expiration date) can be written into each sector of wristband chip. Specific practice is that on fishing for each batch of fresh-water fish products, each one that is up to standards after examination is equipped with a wristband Mifare electronic tag by quality-safety certification notary party, whose traceability information is written into electronic tag through NFC cellphone APP by manufacturers and quality-safety certification notary party, and consumers are accessible to such information via cellphone with NFC later.

4   Production of NFC Traceability Tag

4.1   Interface Interchange of NFC Cellphone

Prior to visiting NFC hardware and correctly dealing with NFC Intent container, it shall be asserted to <uses-permission> element in AndroidManifest.XML about visiting NFC hardware, which is: android: name = "android.permission.NFC".

However, since Android API Level 9 version merely supports limited tag dispatch via ACTION_TAG_DISCOVERED before, API Level 10 contains wide read-and-write support [9], and provide Android Beam element to create NDEF record conveniently, and method to push NDEF outward to other methods. Hence, application program in this paper is necessary to assert that the supportable minimum Android SDK version is Level 10. That is android: minSdkVersion = "10".

Furthermore, in Google Play, it shall be asserted that the application program only displays in devices with NFC hardware. That is android:name = "android.hardware. nfc" android:required = "true". NFC hardware [10], when operating, must call getDe-faultAdapter() method to check whether NFC is effective. In the end, configuration for filtering MIME of text/plain type is ACTION_NDEF_DISCOVERED [11]:

1. android:name = "android.nfc.action.NDEF_DISCOVERED"
2. android:name = "android.intent.category.DEFAULT"
3. android: mimeType = "text/plain" (Fig. 2).

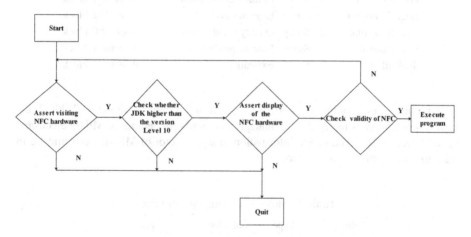

Fig. 2. Interface interchange process of NFC traceability process

4.2   Content of Traceability Tag

Based on features of different users, functions and needs of different users are analyzed in this section for further design of content of traceability tag. There are three kinds of users and identities in this system:

Manufacturers: manufacturers arm each fresh-water fish that is up to standards through examination by quality-safety certification notary party (taking soft-shelled turtle as an example) in each batch of fishing fresh-water fish with a Mifare electronic tag, and the manufacturers and quality-safety certification notary party will write traceability information into electronic tag through NFC function of cellphone (for instance, tag code, product name, product number, manufacturers, responsible person, contact information, logistics code, quality certificate, date of production and expiration date) (Table 1).

Consumers: all that consumers need to do is place cellphone with NFC function close to electronic tag and initiate the APP to inquire and browse traceability information within the electronic tag, primarily including tag code, product name, product number, manufacturers, responsible person, contact information, logistics code, quality certificate, date of production and expiration date, etc.

Table 1. Structure of manufacturers and consumers

Name	Type	Function	Location
Rfid uid	Char	Sole mark of electronic tag	0 sector 0 block
Product name ET	String	Product name	2 sector 0 block
Product id ET	Int	Product code	2 sector 1 block
Factory name ET	String	Manufacturer name	2 sector 2 block
Duty Man.ET	String	Person in charge	3 sector 0 block
Phone no. ET	String	Contact information	3 sector 1 block
Logistic no. ET	String	Logistics code	3 sector 2 block
Quality certificate ET	String	Quality certification	4 sector 0 block
Production date ET	String	Date of production	4 sector 1 block
Shelf-life ET	String	Expiration date	4 sector 2 block

Administrators: as administrators of this software system, as shown in Table 2, the main responsibility is to check and manage passwords for sectors of Mifare electronic tag, and keep sellers' and users' information in some sector of Mifare electronic tag in order to ensure information security.

Table 2. Structure of administrator table

Name	Type	Remarks
Id	int	Administrator id
Name	Varchar	Users' name of administrator
Password	Varchar	Password of administrator
Password past	Varchar	Password for tag identification

4.3 Read-Write of Electronic Tag

First element for reading and writing of Mifare Tag is correct Key value, while as three control bits of data block 0, 1 and 2 of all sectors exist in data block 3, the access control byte in both positive and negative forms [12], data block 3 determines its access control permission. Initial access control default value of block 3 is (C1 $\times$ 0, C2 $\times$ 0, C3 $\times$ 0 = 000; C1 $\times$ 1, C2 $\times$ 1, C3 $\times$ 1 = 000; C1 $\times$ 2, C2 $\times$ 2, C3 $\times$ 2 = 000; C1 $\times$ 3, C2 $\times$ 3, C3 $\times$ 3 = 001) and KeyA as well as KeyB default values. Access control value in this paper is set to be FF 07 80 69. As shown in Table 3, through comparing with manufacturer data block access control permission table, when its access control value is C1 $\times$ 1, C2 $\times$ 1, C3 $\times$ 1 = 000, correct calibration of KeyA or KeyB is fundamentally necessary to permit reading of content in block 1 [13], otherwise, reader-writer will fail to read and transmit data due to password mistake in some sector.

Executive procedure of reading and writing electronic tag is shown in Fig. 3. Step 1, Mifare card calls: when electricity in the card restores, by sending answer-back code to request, reader-writer is able to respond to all request orders sent-out within its antenna. Step 2 is to conduct anti-collision cycle, in which sequence number of every card is read. Supposed that there are several cards within working range of reader-writer, they can be discriminated by their unique sequence numbers, and will be selected for further transaction. Step 3, the certain card is selected. Reader-writer will choose a certain card via select card order to conduct identification and relevant operation of storer. The selected card then returns designated answer-back code (ATS = 08 h),in order to clarify type of the selected card [14]. Step 4, three rounds of secret key identification over KeyA and KeyB is conducted. Once the card is selected, NFC cellphone will assign memory location for following reading and writing [15], and conduct three rounds identification with corresponding secret keys. After identification being achieved, all operation of storer will be encrypted.

Table 3. Value FF078069 access permission table

Bit	7	6	5	4	3	2	1	0
Bit 6	1-	1-	1-	1-	1-	1-	1-	1-
Bit 7	0	0	0	0	0-	1-	1-	1-
Bit 8	1	0	0	0	0	0	0	0
	CX $\times$ 3	CX $\times$ 2	CX $\times$ 1	CX $\times$ 0	CX $\times$ 3	CX $\times$ 2	CX $\times$ 1	CX $\times$ 0

Remarks: X Stands for block, "-" means negation

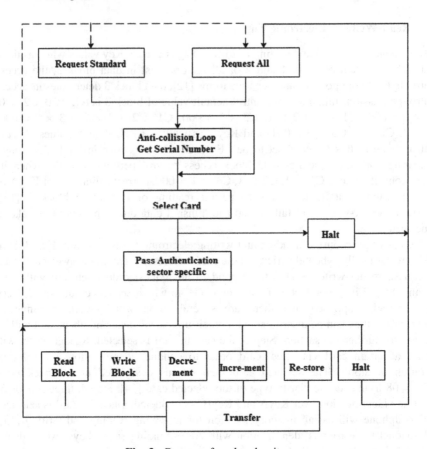

Fig. 3. Process of read and write tag

5   Realization of Traceability System

In this system, fresh-water fish is first rated by the quality supervision bureau, and brand a safety certificate, then manufacturer writes the key information of the product to the Mifare electronic tag which is read by the consumer with NFC Mobile phone [5]. Mifare tag including fish traceability and growth of environmental monitoring website.

5.1   Form of Realization

In this system, we write such information as product name, manufacturers' names, persons in charge, quality certification of water products into assigned sectors and blocks in the form of Chinese character of ISN. Tag code, product code, contact information, logistics code, date of production, and expiration date are stored in relevant sectors and blocks in the form of hexadecimal (Hex) specific operational rules. Specific storage contents and forms are shown in Table 4:

Table 4. Storage form of traceability information in Mifare Tag

Name	Location	Content	ISN or HEX
Tag code	0 sector 0 block	Tag code	4D030906
Product name	2 sector 0 block	Soft-shelled turtle	BCD7 D3E3
Product code	2 sector 1 block	20150613	3230313530363133
Manufacturer name	2 sector 2 block	Some-aquatic products company	C4B3 CBAE B2FA B9AB CBBE
Person in charge	3 sector 0 block	Zhang San	D5C5 C8FD
Contact information	3 sector 1 block	18618181234	3138363138313831323334
Logistics code	3 sector 2 block	3100059755143	333130303035393735353131433
Quality certification	4 sector 0 block	Aquatic products inspection Bureau of Hunan Province	BAFE C4CF CAA1 CBAE B2FA BCEC B2E2 BED6
Date of production	4 sector 1 block	20150614	3230313530363134
Expiration date	4 sector 2 block	20150714	3230313530373134

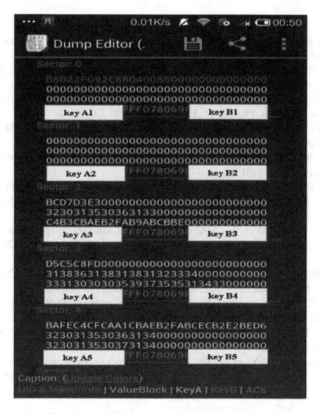

Fig. 4. Dump file with traceable electronic tags

5.2 Experimental Verification

After writing traceability information into this system, it is stored in Mifare Tag after certain encryption processing. To verify the operation condition of this system, by certain access permission set and handling, it is found out that with specific software, stored ISN and data in form of Hex can be accessed [16]. The internal storage is shown in Fig. 4.

6 Conclusions

Fresh-water fish traceability system based on cellphone NFC technology is able to trace and identify quality of fresh-water fish products, and with which consumers are enabled to read actual data information of manufacturing locations rather conveniently. Establishment of the traceability system provides a new thought for realizing effective tracking and tracing of fresh-water fish, and lays a foundation for realizing real-time quality monitoring over fresh-water fish products and cracking down fake or defected goods.

References

1. Zhixi, L.: Food safety risk analysis and its application in food quality management. Food Saf. Guide (18) (2016)
2. Liu, Z.: Key technology study of short-distance wireless communication. Sci. Technol. Econ. Market **1**, 6–7 (2016)
3. Mcafee, I.: System and method for profile based filtering of outgoing information in a mobile environment. Biochem. Pharmacol. **34**(34), 1035–1047 (2015)
4. Guo, F., Li, B.: Design of integrated pressure sensor applicable for RFID tags. Chin. J. Electr. Dev. **39**(4) (2016)
5. Kim, B.Y.: Method of controlling NFC-related service provision and apparatus performing the same: EP2833284 (2015)
6. Dai, B.: The product authentication application design based on NFC (2015)
7. Minjun, Y.: Study on smart card access control system with integration technology. China Secur. Prot. **8**, 55–59 (2014)
8. Chan, D.C., Benjapolakul, W.: NFC-enabled android smartphone application development to hide 4 digits passcode for access control system. Procedia Comput. Sci. **2016**(86), 429–432 (2016)
9. Tech A.: NFC android applications development guidelines. Card Manuf. (2016)
10. Valerio, P.: NXP releases free android dev kit for MiFare & NTAG
11. Dhanabal, S.: NFC connection handover protocol: an application prototype. Geochim. Cosmochim. Acta **41**(8), 1180 (2015)
12. Yang, T.C., Liu, Z.U.: Method for reading data from block of flash memory and associated memory device (2015)
13. He, L., Cheng, P., Gu, Y., et al.: Mobile-to-mobile energy replenishment in mission-critical robotic sensor networks. In: Proceedings IEEE, INFOCOM 2014, pp. 1195–1203. IEEE (2014)

14. Farrugia, A.J., Chevallier-Mames, B., Kindarji, B., et al.: Securing implementation of cryptographic algorithms using additional rounds: US20130067212, US (2013)
15. Lee, S.H.: Apparatus and method for controlling functions of a mobile phone using near field communication (NFC) technology: US 9172786 B2 (2015)
16. Kunitz, H., Mettken, W.: Encryption of data to be stored in an information processing system: US8024582, US (2011)

Intelligent Information Systems – Information Retrieval

An Information Filtering Model Based on Neural Network

Rongrong Li[✉]

Department of Computer Science,
Guangdong University of Science and Technology, Dongguan, China
408976182@qq.com

Abstract. Thorough analysis of the traditional linear model of information filtering, an improved model is proposed based on neural network, which reflects the user's expectation. Taking 200 Email as the test object, the advantages and disadvantages of the linear model and the improved model are compared. The improved information filtering model has strong self-learning ability and adaptive ability, and improves the recognition rate.

Keywords: Information filtering · Neural network
Perceptual machine network model · Information value

1 Introduction

In the 1960s, the West began to study the automatic classification of information; in the early 1990s, a redundant information processing system based on probability theory was produced, but the effect was not ideal. Domestic information automatic classification technology is not mature, redundant information processing system has yet to be developed. Faced with a lot of information on the WWW, congestion full of mail boxes, people often helpless [1]. This model will provide people with an effective tool to replace manual sorting information, so that people from the heavy information sorting labor freed.

(1) **Vector Space Model Algorithm.** The vector space model is a model of text representation proposed by Salton, which uses the vectors (w1, w2, ..., wn) formed by the feature items to represent the document information and the filtering information. Where wi is the weight of the i-th feature. At present, most filtering algorithms use the word frequency vector space model, in which the weights of the key words are generated directly according to the word frequency statistics. No matter what matching algorithm is used, the parameter plays a decisive role in filtering the final result. The LFM vector space model has some obvious shortcomings, such as simple and easy to understand, easy statistics and so on. However, in the absence of reference context and lack of prior knowledge, the filter often regards the high frequency interference information in local text as Feature is extracted, while ignoring the part of the theme-oriented information, such misjudgment is the filtering results deviate from the expected theme of the main reason.

K. Li et al. (Eds.): ISICA 2017, CCIS 874, pp. 217–227, 2018.
https://doi.org/10.1007/978-981-13-1651-7_19

(2) **Vector Space Model Difficulties and Defects.** In the vector space model, the threshold needs to be determined for each information category. When a text is above the threshold of the information classification, the text belongs to the category. However, the determination of the threshold is very difficult, and now generally adopts the predetermined initial value. This method has two disadvantages. First, the initial value is not easily determined based on experience or simple tests. Second, the magnitude of the adjustment Unable to determine, when the initial value is too high or too low need to increase or decrease, the range of increase or decrease can't be well determined, only repeated testing, repeated adjustments, thus greatly increasing the workload.

(3) **Vector Space Model Improvement Measures.** In order to get a more reasonable template of illegal document, the traditional information filtering algorithm should be improved in terms of filtering speed and precision. In this case, we need to match the feature items with the whole context. But deeper research on information filtering, such as semantic analysis, is also needed.

2   Linear Model of Information Filtering

To filter out useful information from a large amount of information, first of all, how to measure the value of a message. There are many schemes for measuring the value of information. Traditionally, the average information is used (the entropy of the information source) to measure, as follows:

$$H(x) = \sum_{i=1}^{n} P_l J_l - \sum_{k=1}^{m} Q_k H_k \tag{1}$$

Where x_i represents the information element that composes the source information, n represents the number of information units, and $p(x_i)$ represents the probability that each information element appears in the information unit. Obviously, this program one-sided emphasis on the information itself, the organizational form of the impact of the value of information, the value of information and information equalized [2]. Some scholars in Japan believe that "the information that meets the user's purpose is the valuable information." Accordingly, he proposed a representative information value formula:

$$H(x) = - \sum_{i=1}^{n} P(x_i) log_2 P(x_i) \tag{2}$$

Where P_i denotes the probability of using the information, Q_k denotes the probability that the information is not used, J_i denotes the proceeds when the information is used, and H_k denotes the proceeds obtained when the information is not used. Compared with the traditional scheme, this scheme more truly reflects the value of information, but this measure of the value of information is based on the use of information on the basis of the information value formula can only be used for information value Posterior assessment.

Both of these schemes have some limitations, and it is difficult to apply the information filtering model to measure the value of unknown information according to the user's requirement. As the information is ultimately for the user service, in line with user expectations (collection purposes) information is valuable to the user.

Because the user expectations of the information is not specific, but only some of the characteristics of information requirements, such as the date of issue, therefore, it is necessary to extract feature information from each unit, with the characteristics of information unit vector to represent the information unit [3]. On the other hand, in order to reflect the degree of matching between source information and user expectation information, we introduce the concept of information redundancy (D) to measure the value of information and define: $D = 0$ when the source information exactly matches the expected information; Conversely, $D = 1$.

Based on the above considerations, we define the information as follows:

Source information: $I = \{X_1, X_2, \ldots, X_n\}$
Information element eigenvector: $X_n = (x_1, x_2, \ldots, x_i)$
Information element eigenvalue: X_i

For simple information value calculation, the redundancy calculation of information unit X can be expressed as follows:

$$D = \sum_{i=1}^{n} (x_i \oplus x_{0i}) \circ k_i \qquad (3)$$

$$k_i = (x_{0i} \circ w_i) / \sum_{j=1}^{n} (x_{0i} \circ w_i) \qquad (4)$$

Where X_i denotes the characteristic value of the information element X_m, X_{0i} denotes the characteristic value of the expected information element X_{0m}, ω_i denotes the weight of the i-th information unit, w_j denotes the priority level of the j-th information unit, and $\oplus$ denotes the exclusive OR logic Operation.

For example, set the source information element: $X = (1, 0, 1, 1, 0)$, expectation information unit: $X_0 = (1, 1, 0, 1, 1)$, priority level: $W = (2, 2, 1, 3, 1)$, the weights are: $(0.25, 0.25, 0, 0.38, 0.12)$, Source Information Element redundancy: $D = 0 * 0.25 + 1 * 0.25 + 1 * 0.38 + 0 * 0.12 + 1 * 0.2 = 0.6$.

This scheme takes full account of the form of information itself and the impact of information user's requirements on the value of information, and it is easy to implement. The information filtering model based on this scheme is shown in Fig. 1, where D0 is the maximum tolerated redundancy determined by the user.

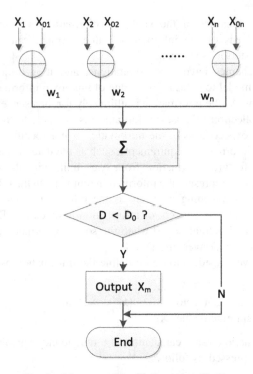

Fig. 1. Linear model of information filtering

In the above scheme, the information redundancy is obtained by the linear processing (XOR) of the eigenvector of the information element. However, the eigenvalues of the eigenvectors of the information unit often have complicated relation, and only the linear processing is not enough [4]. To this end, we propose an improved model based on neural networks.

3   A Model of Information Filtering Based on Neural Network

The basic idea of using neural network to filter information is to divide the redundancy of information units into K grades [5]. If there is an unknown redundant information unit, the redundancy level should be recognized. For this reason, the characteristics of the information element extracted first. In the eigenvector, the vector value indicates the degree of prominence of the feature, 1 indicates that there is this feature, 0 means that there is no such feature at all; after proper processing, To the I input of the neural network. If the neural network has been trained before, then the output layer can be an output vector, the vector value that the information unit redundancy corresponds to the level of the degree, if the maximum vector value corresponding to the level of k, that

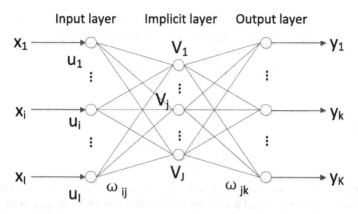

Fig. 2. MLP neural network structure

this The redundancy level of the information element D = k. Here, we have chosen the most widely used multi-layer perceptron network model (MLP), the structure shown in Fig. 2:

In the figure, a three-layer MLP model is shown in which each small circle represents an artificial neuron, and its typical input-output relationship is as follows:

$$v_j = f\left(\sum_{i=1}^{I} k_{ji} u_i - T_j\right) \quad j = 1, \ldots, J \tag{5}$$

$$f(x) = \frac{1}{1+e^{-\lambda x}} \tag{6}$$

In the actual network, λ is always taken 1, v_j is the output of the j-th neuron of the hidden layer, u_i is the output of the i-th neuron of the input layer, that is, the i-th input of the j-th neuron of the hidden layer, ω_{ij} is the connection right (i.e., the connection strength) of the i-th neuron connected to the input layer and the j-th neuron of the hidden layer, and T_j is the threshold value (correction value) of the jth neuron of the hidden layer. The theory proves that MLPs with a layer number greater than (or equal to) 3 can be used for any complex pattern recognition [6]. In this model, the number of input neurons in the perceptron is determined by the dimension of the eigenvector of the information element, and the number of output neurons is determined by the number of redundancy levels. $(x_{n1}, \ldots, x_{ni}, \ldots, x_{nl})$ of the nth input information element is denoted by X_n, Y_n is the nth input of the nth input information element, and the nn input of the nth input information element is represented by the famous BP algorithm (Back Propagation) $(y_{n1}, \ldots, y_{nk}, \ldots, y_{nK})$ corresponding to the information unit. (X_n, Y_n), n = 1, ..., N (i.e., N samples are given to train the MLP) while training the network is always given both input and reference output vector pairs. For each X_n, we find v_j, j = 1, ..., J for the formula (7) (8), then find the actual output of the network: $Y_n' = \left(y_1'^n, ..., y_k'^n, ..., y_K'^n\right)$, for all inputs, the squared error function is:

$$E = \frac{1}{2}\sum_{n=1}^{N}\left(Y_n - Y_n'\right)^2 = \sum_{n=1}^{N}E_n \tag{7}$$

Among them,

$$E_n = \left(Y_n - Y_n'\right)^2 \tag{8}$$

The error function for a single sample. According to the gradient algorithm, the training process can be adjusted according to the following formula, so that the error E gradually reduced:

$$k_{ji}(t+1) = k_{ji}(t) + \frac{\partial E}{\partial k_{ji}(t)} + T\Delta k_{ji}(t) > 0 \tag{9}$$

$$\frac{\partial E}{\partial k_{ji}} = \sum_{k=1}^{N}\frac{\partial E_k}{\partial k_{ji}} \tag{10}$$

Where $\omega_{ji}(t)$ is the value of the connection of the neurons j and i at time t, and $\mu > 0$ is the adjustment factor that can be selected in the experiment. Compared with the general training algorithm, the algorithm adopted in this paper adds an item after the weight adjustment formula, which is called the inertia term. Experiments show that this is conducive to reducing the oscillation when training. When the scale factor α is large, the convergence speed of the network can be accelerated, and the convergence precision of the network can be improved when the scale factor α is large. Before the network training to allow the error value of ε, when E $\leq$ ε, that the network has been trained; if E > ε, you also need to train the network (that is, adjust the connection weight).

When the dimension of the eigenvector is large, the neural network is larger and the training speed is slower. At this time, the neural network can be trained by the proportional training method. The basic idea is to reduce the large-scale BP neural network to a certain proportion Small BP neural network, the use of BP learning algorithm to train this small BP neural network [7]. And then to the same proportion to the original large-scale BP neural network, then use the BP algorithm to train the large-scale neural network, which will greatly speed up the network training speed [8]. Based on the above analysis, the model of the information processing process shown in Fig. 3:

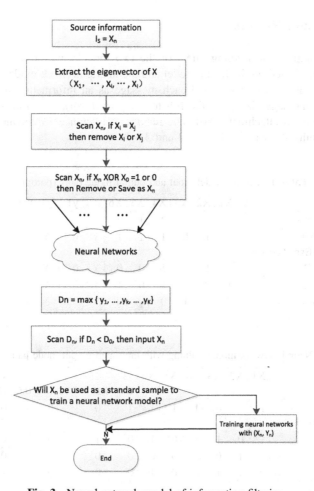

Fig. 3. Neural network model of information filtering

Wherein there are two pre-filtering steps before the source information unit inputs the neural network. The step is to scan X_n, if $X_i = X_j$, then remove X_i or X_j, the purpose of this step is to remove the source information set of duplicate information unit; the other step is to scan X_n, if $X_p \oplus X_0 = 0$ or 1 (X_0 Table expectation information vector), then remove or save X_p. This two-step "denoising" process greatly reduces the amount of information processing of the neural network, speeding up the speed of information filtering [9]. In addition, the model finally provides a user interaction process, asking the user whether the input vector and neural network processing output vector as a standard input and output. This step allows the user according to actual needs, select the standard sample to train the neural network, to provide users with great flexibility [10]. This is one of the characteristics of this improved model - self-learning and adaptive ability.

4 Experiment Results

In order to compare the filtering effect of the two models, we developed two mail filtering systems based on the linear model and the neural network model. We treat all messages as source information, and each message as an information unit. (x1), the date (x2), the message size (x3), the full text keyword (x4), the priority level (x5), whether there is an attachment (x6). The maximum tolerance redundancy D0 = 0.5, part of the results shown in Tables 1, 2 and 3.

Table 1. Linear model input and output mode pairs (portion)

	X1	X2	X3	X4	X5	X6		y1	Output?
W	3	1	1	2	1	2			
X0 (Expected value)	1	1	0	1	1	1	Y0	0	Y
X1	1	1	1	1	0	0	Y1	0.20	Y
X2	1	0	1	1	0	1	Y2	0.33	Y
X3	0	1	1	0	0	0	Y3	0.83	N

Table 2. Neural network model training with input and output mode pairs (portion)

	X1	X2	X3	X4	X5	X6	y1 D = 0	y2 D = 0.33	y3 D = 0.67	y4 D = 1
X0 (Expected value)	1	1	0	1	1	1	1	0	0	0
X1	1	0	1	1	0	1	0	1	0	0
X2	1	1	0	1	0	0	0	0	1	0
X3	0	1	1	0	0	1	0	0	0	1

Table 3. Neural network model input and output mode pairs (portion)

	X1	X2	X3	X4	X5	X6		y1 D = 0	y2 D = 0.33	y3 D = 0.67	y4 D = 1	Output?
X1	0	0	0	0	1	1	Y0	0.2	0	0	0.8	N
X2	1	0	1	0	1	0	Y1	0	0.9	0	0	Y
X3	1	1	1	0	1	1	Y2	0.8	0	0	0.2	Y
X4	1	1	1	0	0	0	Y3	0	0	0.8	0.2	N

Neural network training network performance function using MSE (Mean-Square-Error) between the network output and the target output, After training and testing. The number of network cycles is limited to 5000000 times, the mail sample loops 128630 times to reach convergence, Fig. 4 depicts a graph of the average error of mail documents trained on the network to 20000 times in a neural network.

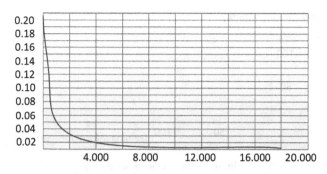

Fig. 4. Neural network training to 20 000 average error curves

After the neural network is fully trained and tested, a large number of connection weights remained stable, then can be used for information filtering. The basic parameters before training and after training are shown in Table 4, the two models of filtration efficiency shown in Figs. 5, 6 and 7.

Table 4. Pre-training and training after the basic parameters

	Linear filter model	Neural network model
Filtering speed comparison	Not sure, sometimes slower	Faster
Filtering accuracy comparison	80%	95%
After a period of comparison of recognition rate	Decline faster	95%–96% after the steady
Effectiveness comparison	Lower	Higher
Training time used to compare	Less	Many
Overall performance comparison	General	Better

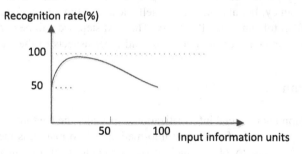

Fig. 5. Linear model

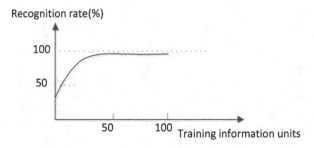

Fig. 6. Neural network model

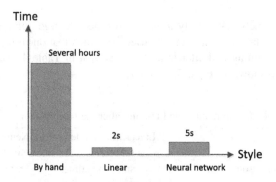

Fig. 7. Time cost ratio (200 E-mail)

5   Features and Directions

The model of linear information filtering has the characteristics of simple realization and fast speed, but it is ineffective for the filtering of information elements with complex relationship between the eigenvalues. The improved model based on neural network not only greatly broadens the application range of the model, improves the recognition efficiency, but also has strong self - learning function and adaptive ability, and has some fault tolerance and flexibility. The next step, we will be committed to the extraction of information, semantic analysis and other aspects of the study.

6   Conclusion

With the development of social informatization, especially the continuous development of Internet, the proportion of redundant information in information is increasing. How quickly and effectively to obtain their own expectations of information from the massive information has become an urgent problem to be solved. The proposed model will greatly accelerate the speed of information acquisition and reduce the cost of information acquisition, thus greatly improving the efficiency, so it has a broad application prospects.

Acknowledgements. The work is supported in part by Department of Education of Guangdong Province under Grant 2015KQNCX188.

References

1. Zhu, L., Bai, L.: Web information filtering technology based on mutual information. Appl. Mech. Mater. **687–691**, 2224–2228 (2014)
2. Hellmich, M.: Statistical inference of a software reliability model by linear filtering. J. Stat. Manage. Syst. **19**(2), 163–181 (2016)
3. Li, T., Corchado, J.M., Bajo, J., et al.: Effectiveness of Bayesian filters: an information fusion perspective. Inf. Sci. **329**, 670–689 (2016)
4. Sven, B., Wei, W., Benjamin, L., et al.: Information filtering in resonant neurons. J. Comput. Neurosci. **39**(3), 349 (2015)
5. Liu, J.Q., Luo, M.: Research on internet monitoring system based on multi-layer text information filtering method through artificial neural networks. Adv. Mater. Res. **532–533**, 1036–1040 (2012)
6. Hinton, G., Vinyals, O., Dean, J.: Distilling the knowledge in a neural network. Comput. Sci. **14**(7), 38–39 (2015)
7. Jia, W., Zhao, D., Shen, T., et al.: An optimized classification algorithm by BP neural network based on PLS and HCA. Appl. Intell. **43**(1), 1–16 (2015)
8. Xu, B., Zhang, H., Wang, Z., et al.: Model and algorithm of BP neural network based on expanded multichain quantum optimization. Math. Probl. Eng. **2015**(12), 1–11 (2015)
9. Sundermeyer, M., Oparin, I., Gauvain, J.L., et al.: Comparison of feedforward and recurrent neural network language models, pp. 8430–8434 (2013)
10. Zhang, J., Huang, L., Xu, H., et al.: An incremental BP neural network based spurious message filter for VANET. In: International Conference on Cyber-Enabled Distributed Computing and Knowledge Discovery, pp. 360–367. IEEE (2012)

The Theory of Basic and Applied Research in Information Retrieval Sorting Algorithm

Xinze Li[✉], Jiying Yang, and Qi Liu

Yunnan Open University, Kunming 650223, Yunnan, China
xinzeli2018@126.com

Abstract. As the computer technology is advancing endlessly and the information quantity is increasing exponentially, people have raised higher and higher requirements on retrieval technique, especially with the appearance of network technique and multimedia technology. The software and hardware environment of information retrieval technique is remarkably improved, making information retrieval technique develop from traditional linear retrieval to non-linear retrieval of hypertext support, and the traditional Boolean logic retrieval model no longer dominates in the information retrieval. It is hard to predict the new changes, new technologies and new ideas induced by technological advancement, yet we can grasp the correct direction for future development of information retrieval technique via comprehensive research, comparison, and analysis.

Keywords: Information retrieval · Sorting algorithm · Intelligent

1 Introductions

Information retrieval mainly refers to the representation, storage, organization and access of information, i.e. retrieving the information materials related to user's query requirements from the information database. The purpose of information retrieval is to obtain the information needed, which requires perfect retrieval techniques. The technology realizing information retrieval is just information retrieval technique, which, in fact, refers to a series of related information retrieval algorithms and their software design [1].

As the computer technology is advancing endlessly and the information quantity is increasing exponentially, people have raised higher and higher requirements on retrieval technique, especially with the appearance of network technique and multimedia technology. The software and hardware environment of information retrieval technique is remarkably improved, making information retrieval technique develop from traditional linear retrieval to non-linear retrieval of hypertext support, and the traditional Boolean logic retrieval model no longer dominates in the information retrieval.

It is hard to predict the new changes, new technologies and new ideas induced by technological advancement, yet we can grasp the correct direction for future development of information retrieval technique via comprehensive research, comparison and analysis.

© Springer Nature Singapore Pte Ltd. 2018
K. Li et al. (Eds.): ISICA 2017, CCIS 874, pp. 228–237, 2018.
https://doi.org/10.1007/978-981-13-1651-7_20

2   Characteristics of the Main Information Retrieval

2.1   Intelligent Information Retrieval

The artificial intelligence technologies such as acquisition and representation of knowledge, natural language processing, machine learning and knowledge inference grow from the increasing demand of social intelligence over the time, and the combination of artificial intelligence and information retrieval is a useful attempt to obtain intelligence of information. In the information retrieval system, the integration of artificial intelligence technology enables the traditional information retrieval system to catch on the user's query needs more accurately, get better retrieval performance and realize higher intelligent [2, 3]. In short, the introduction of artificial intelligence technology will make the traditional information retrieval system towards the direction of higher intelligence. At present, the information retrieval system that modern artificial intelligence technology has been introduced into is called intelligent information retrieval system. The goal of intelligent information retrieval is to realize the intelligent retrieval in the aspects of the processing, the information acquiring, indexing, retrieving and sorting of the user's query contents, thus taking the place of people to complete complicated tasks like information collection, filtering, analyzing and processing. The current intelligent information retrieval system can be divided into the following 3 categories according to different research focuses.

- Semantic Retrieval System

The system has raised the information retrieval from the keywords-based level to the knowledge (or concept) based level. Based on the concepts as well as the relationships between them, the information retrieval is mainly studying three problems: how to analyze the semantic of user input; how to translate the user-submitted query into the semantic meaning through semantic comprehension and calculation, thereby truly understanding and accurately describing the user's query needs from the semantics; how to express the knowledge needed for the retrieval system in order to fully embody the relationship between information. In addition, it can obtain the information that the user can use directly through the query and inference of the knowledge base. The ontology-based intelligent information retrieval system is a semantic retrieval system [4, 5].

- Cross-Media Information Retrieval System

The system allows the use of a variety of media information to express user inquiries, with the ability to output a variety of media-type query results. With powerful search function, it is used more widely and better agrees with the way of human intelligence, enriching the computer services as the extension of computer functions. But now, there has been no mature cross-media information retrieval algorithm and technology both at home and abroad, and no good cross-media information retrieval effect has been accessed [6]. Further research is needed in the unified representation of the cross-media information, semantic annotations and content recognition of cross-media data, sorting of results of cross-media information retrieval as well as related feedback, etc.

- Personalized Information Retrieval System

The system can provide personalized searching results for users with different information needs, that is, it can generate different retrieval results for the same query words submitted by different users according to their different needs. The personalized information retrieval system is mainly to study how to continuously learn, adapt to information and track the dynamic changes of users' interests through intelligent agents, how to extract the information of users' interest or higher-level knowledge and laws based on web mining technology, how to enable the server to notify users automatically of the latest updates in the system by push technology, and collect the information that the user may be interested in and submit it to the user through the intelligent agent, thereby providing personalized information retrieval services [7].

2.2 Visual Information Retrieval

Research shows that 70%–80% of information obtained by people is from vision, which is the main path for people to cognize and transform the world. The information visualization aims to visualize the abstract information to direct and expedite search procedure, then the users can see content invisible in the past, thereby improving the discernment of complicated problems, models or systems. The core of information retrieval visualization is to convert the literature information, user questioning, various information retrieval models and the internal semantic relation invisible in information retrieval process with retrieval model into charts and display them in a low-dimensional visualized space, and to provide information retrieval service for users. The information retrieval visualization has following advantages:

First of all, the general information retrieval results are linear, providing information to the user by a list in most cases, so that the user can check the articles one by one according to the provided list information before determining whether to obtain the full text. And visual information retrieval, by transparentizing the retrieval process and results, helps users directly see the information, which can be displayed by an intuitive graph of the internal relationship between web pages and retrieval requirements, to provide users with a full overview of the retrieval contents and enable users to view them all clearly [8].

Secondly, the general information retrieval lacks the effective feedback mechanism, ignoring the user's interactive function in browsing. But visual information retrieval can provide visualized information, which is helpful for users to discover new searching methods and inspire them to explore further. The most important thing is that visual retrieval allows users to dynamically adjust and filter the searching results, helping them to determine their retrieval strategy and increasing the interaction between them and the system so that the users' information processing and retrieval ability can be brought into maximum play.

Thirdly, the traditional retrieval systems judge the matching degree of the documents and retrieval questions according to their separate similarity matching algorithms, and work out the relevance of the documents. The list is ranked by relevance, but there are differences with the user's understanding because the ranking of each search engine is calculated based on the keyword weight, as a result, in some cases the documents that really meet the needs of the user are ignored by the user as they appear at the back of the search list and some important information is missed. However, as visual information retrieval allows graphical display of the retrieval process and results, users can get involved in the control of the retrieval process based on visual display, thus further determining the value of a word to indexing and retrieval, and catching on the semantic relationship between retrieval and documents from multiple perspectives.

Because of unceasing increase of the information retrieval system database in content, diversification, and complication of the information types, it will be more difficult for the common user with different specialized backgrounds to obtain information. The advantages of visual retrieval mentioned in the above analysis can make up for the insufficiency of the existing information retrieval in the network environment and can improve the utilization of the network information resources to a new level [9].

2.3 Grid-Based Information Retrieval

Grid enables people to transparently use other resources such as calculation and storage resources to "use calculation resources like power grid does" [10]. Its major functions are providing an integrative intelligent information processing platform and adopting interfaces to connect the scattered information resources on the internet to eliminate information isolated island and provide wide-area information sharing service such as grid-based information retrieval, etc.

The grid-based information retrieval system is a retrieval system which takes manifold heterogeneous storage systems such as multiple file systems, record systems and database systems as objects to provide users with uniform interface to retrieve the data in multiple heterogeneous storage systems, and present the retrieval result obtained from multiple data sources to the users in a uniform format, then the user can obtain the information needed without need to know the concrete sources of information. In grid-based information retrieval system, the data resources are dynamically registered and canceled, and the system selects appropriate information resources according to users' retrieval requirements, to retrieve related information from it, while users do not know the concrete source of information.

2.4 One-Stop Information Retrieval

The term "one-stop" comes from the business field of European and American countries. For example, "One-stop Shopping" means that businesses are constantly expanding their business to win customers so that consumers can buy all the goods they need in a store. At present, this concept has been expanded to many areas, such as the one-stop service of government departments, etc.

There are similar cases in the field of information retrieval. In today's society, electronic resources have become the main form of people's use of information. The electronic resources are complex in format and widely distributed on different platforms. However, few users know exactly the specific databases where the content they need is stored, so users have to leap from one database to another, leading to great inconvenience in use. In this case, users hope there would be a unified retrieval method of document information provided so that all digital resources in different formats and types can be linked up seamlessly, and one-stop information retrieval approach is born at the right moment [11].

One-stop information retrieval is based on the unified retrieval technology to construct the system platform. It provides the user with a unified retrieval interface and transforms the user's retrieval demand into the retrieval expression of different data sources, while concurrently retrieving multiple distributed heterogeneous databases locally and over the Internet. It then integrates the retrieval results and finally presents the retrieval results to the user in a uniform format. The unified retrieval platform of one-stop information retrieval realizes the unified retrieval of many kinds of information resources and unified display of results. And these retrieval results can act as the starting points of the link and get linked to the information stored in other resource databases related to its content.

One-stop information retrieval meets the requirements of information retrieval development. As for users, they are tired of querying data in many databases and hope to obtain information from different databases through a single interface; for information service manufacturers and providers, purchase of the database alone cannot meet the needs of users and they must provide a unified platform to reduce the user burden.

2.5 Specialized Information Retrieval

The specialized information retrieval will only involve the information of a certain subject and domain, and the information is relatively more concentrated. Besides, usually, the personnel of this discipline participate in the formulation of the information, which can both raise retrieval speed and improve specificity, increase retrieval depth and intensity, thereby improving recall ratio and precision ratio [12].

The specialized information retrieval system has advantages incomparable by the largely integrated retrieval engine in the provision of specialized information. It adopts basic techniques the same with those of integrated engine, which are basically proven techniques. Their development has no technical obstacles and just fits a trend of Internet development: more specialized and finer division of labor. Thus specialized information retrieval system will provide personalized service for users of different domains in a better way, and will be a trend of future information retrieval.

2.6 Integration Information Retrieval

In substantive numeric resources established by various social organizations and individuals, the organization and management form of data are diverse and the data formats stored in different databases or documents are widely divergent. Besides, these databases or documents may be dispersed over different regions, and database products of different database manufacturers provide different retrieval modes or retrieve software, leading to a lot of inconvenience in information retrieval, which highlights a development trend of information retrieval service, i.e. integration [13].

The process of processing retrieval by integrated retrieval system: Firstly structure a text library, which is used to store the information which the users may retrieve; secondly establish indices, which can significantly raise the velocity of information retrieval, then after indices are established, search can begin; lastly filter and sort the results, and return the filtered and sorted result to the users.

2.7 Personalized Information Retrieval

So-called personalized information retrieval refers to the technology that provides personalized retrieval result for users with different information needs, i.e. it can generate different retrieval results for the same query words submitted by different users according to different users' demands [14]. This is a very promising information retrieval technique and also one of the most focused directions in various research on information retrieval system currently. It plays a vital function to improve users' access result and user experience to perfect existing information retrieval system.

Although the existing experiments and research demonstrated that personalized information retrieval can better users' retrieval experience, there is still an important question to be answered, i.e. whether the personalized information retrieval technique must be used. For current research progress, there are two noteworthy points:

- In Terms of User Use Level

The current Web users tend to regard these personalized technologies as not user-friendly. The more complicated the personalized functions are, the more difficult they are in use.

In Terms of Applicability of Personalized Information Retrieval Technique

Not all conditions need personalized technologies. The common user query can be divided into three types: clear query, semi-fuzzy query, and fuzzy query. Wherein, the personalized information retrieval technique can improve the retrieval quality of semi-fuzzy query and fuzzy query. However, the common search engine seems to be more suitable for clear query.

3 Information Retrieval Sorting Algorithm Application Examples

3.1 Information Retrieval Sort Algorithm

Information retrieval sort technology is one of the most core and key technologies of search engine and one of the problems pressing for further study and perfection in the current search engine. A great many users are dissatisfied with the search result, saying the search result sorting is less than ideal, the search result is too disordered, the search result is not proper, and there are too many advertisements. The unreasonable design of existing search engine sort algorithm floods many results in which the users are really interested in numerous query results of the search engine, lowers the efficiency of search engine and wastes users' time, thereby leading to the dissatisfaction of users with the search engine.

Information retrieval performance of many decision factors, such as query expressions quality, query expansion technology, but fundamentally it is deter-mined by the sort algorithm [15]. Current popular information retrieval sort algorithms fall into three categories: sort algorithm based on the content on Web page, sort algorithm based on the Web page link analysis and sort algorithm based on retrieval users. In practical situations, the search engine usually comprehensively employs above algorithms to sort the query results.

Added to this, in sorting and retrieving, some search engines would further refer to inclusion condition of directory system artificially edited, or sort by expenses for keywords paid out of consideration of own commercial profit.

3.2   Information Retrieval Overall Framework of the Sorting Algorithm

To optimize retrieval performance of network information and solve related problems, we will adopt a practical method-personalized information retrieval, which understands and learns the users' use habit and preference by extracting users' query text, and creates a user model based on users' personal information, then uses a user model to optimize retrieval algorithm and sorts the retrieval result before presenting it to the users. Able to improve users' retrieval experience, the personalized information retrieval is a retrieval technique with great long term potential [16]. The system mainly includes three modules, with the general frame as shown in Fig. 1.

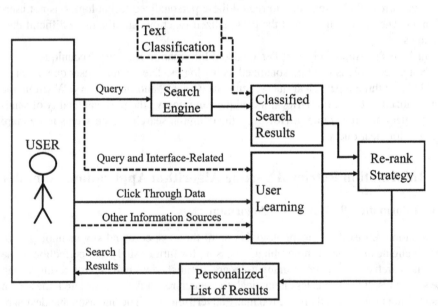

Fig. 1. Overall framework

3.3 Theme Directory-Based User Model

Figure 2 shows the user model based on theme directory. The system automatically records the query history of users on the page and provides corresponding query records for users when they are searching for keywords. After collecting some record data, the page also displays the accumulative clicks by users. The users can click the accumulative clicks or click single page belongs the subject.

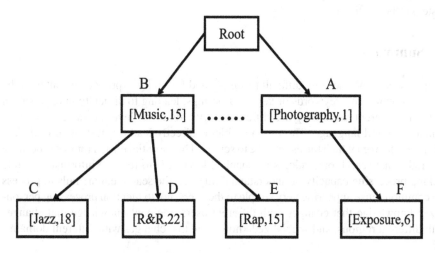

Fig. 2. Theme directory-based user model

3.4 Sort Based on the Contents of the Webpage

The basic idea of sort algorithm based on webpage content is to calculate the index entries for the keywords describing characteristics of subject content on the page and sort the matching webpage. This sorting algorithm is in many types, with the most widely applied being sort algorithm based on word frequency position weighting and sort algorithm based on link analysis. Besides, the sort algorithm based on intellectualization is also developing rapidly.

3.5 Sort Based on the User

Whether query result of the search engine is correlated and whether sorting is reasonable finally rest with the subjective judgment of users. There are large deviations in evaluation on retrieval result due to large differences in users' motive, knowledge, interests, social experience, etc. The ideal retrieval sort algorithm should consider the users' individual factors. The sorting based on retrieval users mainly comprises two types: Sorting based on retrieval user group and sorting based on individual retrieval user (also called personalized retrieval sorting). The users' individuality is expressed in two modes: one is the user actively submits personal information or directly feeds back to the system; the other is the system analyzes the user's retrieval behavior on client side or server side to confirm [17].

It is obvious that the sort algorithm based on user searching can improve the retrieval quality of system, yet this algorithm still has lots of drawbacks in practical situations: the arbitrary user actions unavoidably lower accuracy of analysis on users' behavior; related page feedback which is designed to improve retrieval quality makes operation of users tedious; creating a bulk of user models renders operation and maintenance of system rather difficult.

4   Summary

In many cases, the users are difficult to simply and faithfully express own content to be retrieved with some keywords or keywords strings, leading to difficulty in retrieval. In addition, different users may use different keywords to retrieve the same concept in human's natural language. These two problems directly lead to substantive irrelevant information, from which the users have to screen. The essential reason for this is because the traditional search only adopts mechanical keywords to retrieve information while lacking processing capacity or understandability, i.e. the search engine fails to process the common sense appearing very simple to the users, not to mention processing person-alized knowledge that changes with different users, regional knowledge that changes with different regions, and specialized knowledge that changes with different domains, etc.

In a word, the future information retrieval technique will make breakthroughs comprehensively in the idea, technology, humanization, and intelligentization, etc. to gradually adapt to the thinking mode of the human brain, thereby retrieving information smartly, efficiently, rapidly and flexibly. Lastly, the information can be searched out and rapidly obtained as desired. Of course, these breakthroughs also need related technical support such as computer software and hardware technology, communication tech-nology, AI technology, visualization technology, etc. However, in any case, the future information retrieval technique will be destined to present to people in a brand-new look to promote the ordered organization of disordered information world, so as to develop and utilize the information resources more reasonably.

References

1. Jin, F.: A primary research on information retrieval and information retrieval technology. Shanxi Libr. J. **6**, 22–24+49 (2001)
2. Fan, W., Pathak, P., Zhou, M.: Genetic-based approaches in ranking function discovery and optimization in information retrieval—a framework. Decis. Support Syst. **47**(4), 398–407 (2009)
3. Ganzha, M., Paprzycki, M., Stadnik, J.: Combining information from multiple search engines —preliminary comparison. Inf. Sci. **180**(10), 1908–1923 (2010)
4. Zhou, H., Liu, B., Liu, J.: Research on mechanism of the information retrieval based on ontology label. Procedia Eng. **29**, 4259–4266 (2012)

5. Cao, D.N., Gardiner, K.J., Cios, K.J.: Protein annotation from protein interaction networks and Gene Ontology. J. Biomed. Inform. **44**(5), 824–829 (2011)
6. Zhai, J., Song, Y.: Semantic retrieval based on SPARQL and fuzzy ontology for electronic commerce. J. Comput. **6**(10), 399–402 (2011)
7. Dai, W., You, Y., Wang, W., Sun, Y.: Search engine system based on ontology of technological resources. J. Softw. **6**(9), 1729–1736 (2011)
8. Feng, J.: Current situation and prospect of information retrieval visualization. Doc. Inf. Manag. Sci. Technol. **26**(03), 32–34 (2012)
9. Pan, Q.: Research review on information retrieval visualization. Res. Libr. Sci. (12), 7–9, 14 (2010)
10. Xu, Z., Feng, B., Li, W.: Grid Computing Technology. Publishing House of Electronics Industry, Beijing (2004)
11. Chen, F.: User-oriented one-stop retrieval. Inf. Sci. **28**(12), 1828–1831 (2010)
12. Yao, L., Xie, T.: The status and improvement of specialized information retrieval. Libr. J. Henan **33**(03), 78–79 (2013)
13. He, X.: Design and implementation of integrated retrieval system based on information retrieval. J. Commer. Econ. **14**, 37–38 (2011)
14. Li, S.: A summary of personalized information retrieval technology. Inf. Stud.: Theory Appl. **32**(05), 107–113 (2009)
15. Gao, W., Liang, L., Xia, Y.: An improved algorithm for ranking in information retrieval. J. Yunnan Nationalities Univ. (Sci. Ed.) **19**(1), 52–55 (2010)
16. Yang, J.Y., Zhang, B., Mao, Y.: Study on information retrieval sorting algorithm in network-based manufacturing environment. Appl. Mech. Mater. 484–485, 183–186 (2014)
17. Zhang, X., Yu, J.: Research on model building and personalized search algorithm based on user interest model. Comput. Knowl. Technol. **12**(18), 1–4 (2016)

Summary of Research on Distribution Centers

Zeping Li[(⊠)] and Huwei Liu

School of Information, Beijing Wuzi University, Beijing, China
2953270587@qq.com

Abstract. The development of logistics in developed countries should be earlier than us, and it is more mature in theory and practical application. Domestic logistics development is relatively late, but logistics research has also become a hotspot in recent years. The distribution center of logistics is also becoming a hot spot. The article combines the existing research results and summarizes the selection path, location layout and storage strategy, which can be used as a reference for future research.

Keywords: Picking path · The storage layout · Storage strategy

1 Introduction

With the rapid development of the 21st-century economy, logistics industry as a pillar industry of supporting economic take-off, also plays an important role, the logistics distribution center also gradually for the intensive and scale to make a lot of contribution to the development of logistics. In distribution center, warehousing management plays an extremely important role in the whole system. The literature [1] pointed out that the warehouse as a labor input more intensive areas, picking in the storage work occupies an extremely important position, the cost also accounted for a large proportion, accounting for about 55% of the entire storage operating costs. The literature [2] shows that walking time occupies half of the picking time on the route of picking up the goods. The walking time is a direct part of the cost of picking up, but the cost of consumption is not of additional value.

Picking the job is to choose the goods quickly according to the needs of the customer. The selection strategy mainly involves the study of the selection strategy and path. Chosen strategy and path indeed rule depends on the position layout and storage strategy, this constituted the main part of the distribution center, three parts in this paper summarized the research results of domestic and foreign scholars study theory and reviewed in this paper.

2 Domestic and Overseas Research Status

This paper summarizes the research results of the selection path, location layout and storage strategy from the related literature.

© Springer Nature Singapore Pte Ltd. 2018
K. Li et al. (Eds.): ISICA 2017, CCIS 874, pp. 238–250, 2018.
https://doi.org/10.1007/978-981-13-1651-7_21

2.1 Summary of Domestic and Foreign Literature

At present, the commonly used heuristic path method, as mentioned in the previous section, has S-type method, return type method, midpoint type method, maximum interval method and combination method, and then this article will list the relevant foreign and domestic research results.

(1) Summary of foreign studies

In the literature [3], Christophe concludes that intelligent algorithm can save 47% more than the traditional path through intelligent LKH algorithm compared with traditional model S, traversal type and maximum interval method. At the same time, the algorithm is applied to the problem of picking path optimization, the algorithm can be very effective in the solution of the optimal path, the algorithm also has some shortcomings, when the problem size increases, the algorithm will make the time also increased dramatically. Traditional s-type, return type, midpoint type, maximum interval and hybrid path(see Figs. 1, 2, 3, 4 and 5). In the literature [4], some heuristic algorithms are improved under the constraints of the weight, the vulnerability and the product category of Chabot.

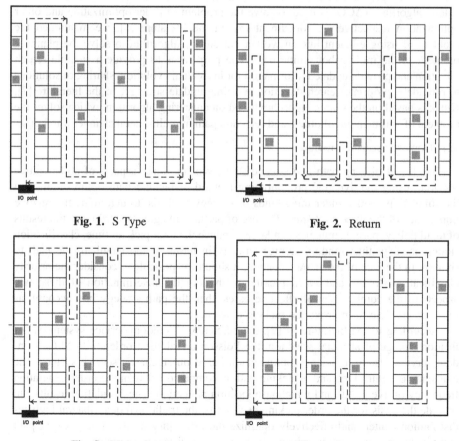

Fig. 1. S Type

Fig. 2. Return

Fig. 3. Mid-point

Fig. 4. Largest gap

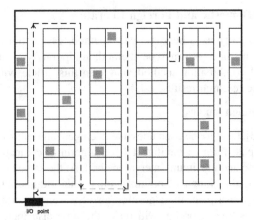

I/O point

Fig. 5. Combined

In recent years, most scholars use artificial intelligence algorithm to solve the classical TSP problem of batch picking. Particle swarm optimization (ACO) and ant colony algorithm (ACO) is used to solve the problem of order optimization and order optimization respectively in document [5]. In the literature [6], we mentioned the problem of using efficient human worker ant colony algorithm to solve the planning problem of the multi-robot online path. The purpose is to provide a suitable search direction and provide updated information for individual evolution, sharing of solutions to provide appropriate search direction and time update strategy. In the literature [7], Kulak is based on the search algorithm based on clustering algorithm, which solves the problem of optimization of orders and selection path in multi-lateral channel warehouse.

(2) Domestic research review

The [8] of picking path and storage strategy of the three aspects of collaborative research, using the dynamic programming method of picking path optimization, and the simulation results under three kinds of strategy analysis, to determine the relative importance of the three strategies. The use of partial strategies, according to the results of in all policy, partial strategies can largely reduce the total picking time, classification storage strategy than random storage strategy in terms of the length of the path to shorten. In the face of the actual situation, it should be possible to consider the priority classification storage and batch picking method in the determination of these two aspects of the combination of effective cases, then the path strategy into account, as much as possible to improve the picking efficiency.

According to the literature of [9], the efficiency of random service system classification storage of artificial selection, the Poisson distribution orders into M/G/1 random service system, efficiency optimization based on stochastic service system. The efficiency of return type and S picking path is compared between the order generation frequency of five goods and warehouse location allocation. The simulation results can provide the basis for the order picking decision of the traditional quadrilateral logistics distribution center, and effectively optimize the order picking the time, picking path and picking mode selection scheme of the logistics distribution center.

Literature [10] Jia and Lan, through analyzing the existing problems of manual order picking a path in the distribution center, find out some deficiencies of manual order picking mode. On this basis, according to the different distribution characteristics of the order to the storage area, the order picking point of the distribution center is designed, and the order picking path is further optimized to improve the picking efficiency. Literature [11] studies the optimization of service efficiency of manual order picking system in logistics distribution centers. The purpose of this study is to select the picking path in the system. The random model of return type and S type path strategy is constructed under random storage strategy, and the effectiveness of the model is verified by simulation. The results show that the number of picking goods is more effective, and the S method is better than the return type. The selection distance, time, and selection of the order in the distribution center can provide a valuable reference for its selection.

Intelligent algorithms mainly include genetic algorithm, ant colony algorithm, neural network algorithm and particle swarm optimization algorithm. Intelligent algorithm is a commonly used algorithm to solve the TSP problem. Its advantage is that it has small-time complexity and space complexity. The disadvantage is that the convergence is slow, and the solution is usually the approximate optimal solution. Intelligent algorithms often iterate more and more closely to the optimal solution, but this also means that the algorithm runs longer.

The literature [12] through the goods picking path optimization model is solved by a genetic algorithm, the simulation of path selection in use of Matlab, by solving that genetic algorithm is better than other algorithms, to achieve the purpose of picking path optimization. In the literature [13], the shortest mathematical model of stacker picking path with container constraints is constructed, and the basic ant colony algorithm and the maximum and minimum algorithm are used to solve the model. The pheromone initialization mechanism of the max-min ant colony algorithm is used to overcome the local extremum when the basic ant colony is stuck in the stagnation state earlier. In the document [14], a flexible and scalable warehouse environment model is established according to the warehouse structure, warehousing and logistics environment characteristics, and the appropriate robot motion rules are formulated, so that the model can be applied to the dynamic environment. The task requirements and environment information are fused to construct a feasible network topology to avoid segmentation task search. Dijkstra algorithm is used to calculate the optimal path on the basis of feasible network topology, but no analysis of running time is given.

The intelligent algorithm is not without shortcomings. There are some limitations. The hybrid intelligent algorithm is used to some extent, which avoids the error caused by using the intelligent algorithm alone. In the literature [15], the author takes the path planning problem of robot picking as TSP problem, combines path planning method - A* algorithm and improved adaptive genetic algorithm to solve, and carries out the experimental simulation. The main representative methods of the literature [16] are intelligent algorithms, such as A* algorithm, artificial bee colony algorithm and so on. A* algorithm, as an intelligent heuristic algorithm, has been successfully applied to the Kiva robot cluster scheduling. The scholar of the literature [17] combines the advantages of genetic algorithm and genetic algorithm particle algorithm for stacker path optimization problems, the simulation results that the convergence speed and the optimization results were significantly improved, reducing the stacker work time. The encoding methods of genetic algorithms include binary coding, sequential encoding and real coding, and so on [18].

The [19] Xu improved adaptive genetic algorithm is applied to autonomous agent dynamic path planning, using one dimensional path encoding, and the use of domain knowledge and local obstacle avoidance technology to generate the initial population, design of crossover and mutation and smoothing operator and proposed crossover probability and mutation probability of the new adjustment method. In the literature [20], the minimization of the selection path of Sun Honghua has established the combination of the order batch and path optimization with the mathematical model. In this paper, the envelope algorithm and genetic algorithm are used to solve the two problems, and the general distance formula between the order items in the two-zone distribution center environment is derived. In the literature [21], the mathematical model is first established for single car picking, and the genetic algorithm and simulated annealing algorithm are used to solve the mathematical model of the problem. Then the mathematical model of multiple vehicles is established, and the hybrid genetic annealing algorithm is used as a whole, and then the genetic algorithm is used to optimize the picking path of each vehicle separately.

Document [22] in order to further improve the efficiency of the warehouse, targeted to warehouse picking work to optimize, improve work efficiency. Using TSP to solve the problem, combined with graph theory to solve Hamilton circuit, the optimal picking path model is established and solved to improve the picking efficiency. The results show that the picking path optimization based on the TSP problem is obviously due to the traditional return type and S type. The literature [23] study in the limited time, the shuttle car driving distance and the influence on product quality optimization problem. The Pareto thinking and the search algorithm and improved, put out strategy of tabu search algorithm, which is more close to the global optimal solution. The optimization method of picking path is relatively broad, and the method is more. According to the literature summary of the optimization of picking the path, there are more researchers abroad and earlier in the optimization of picking path, and the domestic research in recent years has also increased significantly. Most studies have focused on the automatic piece picking path optimization for artificial intelligent algorithm are few, the use of intelligent algorithms or improved, but the mixed two algorithms of picking path optimization is less. There are more constraints in the actual process, and the constraints should be taken into account when building the model. In order to overcome the limitations of the single algorithm, the use of hybrid algorithm will become a trend. The storage layout is less, the storage strategy and the storage layout are combined to identify the chosen path.

2.2   Research on the Layout of the Cargo Space

(1) Summary of foreign studies

Literature [24] the layout design of the warehouse can influence the optimized operation of the future, and has a significant influence on the order picking and the walking path of the picker. The traditional layout is shown in Figs. 6 and 7, with the layout of the direct shelf arrangement, and the traditional layout of the insertion. The literature [25] conducted optimization research on the fish-bone type warehouse, and the optimized structure would reduce the walking path by 10%–15% compared with the traditional structure under the multi-constraint conditions. The layout of the fish-bone layout is shown in Fig. 8. The above literature mainly considers only one P& Amp; D point, and the literature [26] considers the Flying-V and Inverted-V layout of multiple P& Amp; D points, the V layout is shown in Fig. 9. By constructing the path model for the distribution center of suitable size to 1%, the path is not only to save Flying-V 3%-6%, Inverted-V only save path, this is because with the increase of P& Amp; D, operating performance will be reduced, so the layout of P& Amp; D should be as close as possible to the middle position.

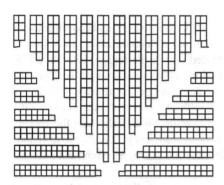

Fig. 6. Traditional layout warehouse

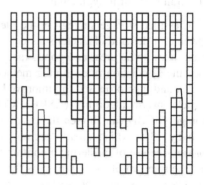

Fig. 7. Add middle span traditional layout warehouse

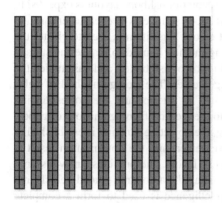

Fig. 8. V layout

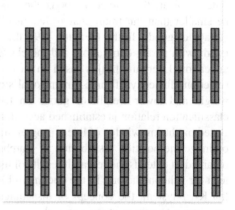

Fig. 9. Fishbone layout

(2) Domestic research review

The [27] of goods to the people of fishbone mode of operation in the layout optimization problem. According to the order frequency and correlation between items by clustering, the mathematical model is established to minimize the chosen distance as the goal. It puts forward the TS-SC algorithm and then generates a number of orders for the experiment on the performance of the algorithm and the effect. The experimental results show that the TC-SC algorithm has a faster convergence rate and good optimization ability. Literature [28] combined with foreign new Fishbone warehouse layout, the optimization design of logistics warehouse internal layout is studied, and the mathematical model is established with the minimum total moving distance as the goal. The results of actual case analysis show that the method can effectively shorten the total picking distance under the premise of ensuring the utilization of warehouse space. In order to ensure the storage of goods, the storage center of fishbone layout is expected to be smaller than the traditional warehouse, but more storage is needed to ensure the storage of goods. In the literature [29], the ant colony algorithm is used to cluster the goods with strong correlation, and the clustering method is used to realize the storage allocation. Literature [30] is based on the distribution of fish bones, and has studied the selection efficiency of three-dimensional storage center. The storage method of light-weight and lower weight of goods on the shelf is set up, and the corresponding classification relation is established according to the record of storage history of goods. Based on the knowledge of linear programming, the objective function is established to optimize the position distribution of warehouse center, and the optimization result is obtained. It extends the research results of fishbone layout, and provides a new research idea for fishbone layout. In the document [31], using the automatic guided vehicle and the Kiva system, the concept of distributed intelligence is used to form a complete warehouse to person storage model.

2.3 Domestic and Foreign Research of Storage Strategy

(1) Summary of foreign studies

Classified storage strategy in the literature [32] is the first goods according to the attribute or demand characteristics can be divided into different classes, each kind of fixed takes up an area, and goods randomly placed within the region. Literature [33] used the association rules mining method to find the correlation strength between the goods, and then proposed the improved classification storage algorithm (MCBH) and particle swarm algorithm (ASBH). Compared with the conventional classification storage, MCBH can shorten the picking path 4%, while MCBH shortens the picking path by 13%, which greatly optimizes the picking efficiency. Literature [34] puts forward the use of clustering algorithm to determine the position assignment strategy, and the standard of measurement is the relationship strength of the frequency that two SKU appears simultaneously in an order. The purpose of clustering is to store the products that are frequently in an order, which can be reduced by 20% to 30%.

(2) Domestic research review

In the study of literature [35] adopted two kinds of clustering analysis and association rules analysis method, according to the order of goods situation, analyze the connection degree of the commodity, combined with outbound quantity, storing, and export to establish an optimization model for the distribution of the storage. Literature [36] based on automated warehouse as the research object, research on the storage location, using the weight coefficient transformation method of the multi-objective genetic algorithm to understand the model, effectively shorten the stacker walking path. In the literature [37], the storage allocation and efficiency estimation model of the automatic storage model for multi-load palletizing machine are put forward. In the literature [38], the authors propose six strategies for location-based allocation based on correlation analysis, which details the algorithms, steps and processes of each allocation strategy, and simulates the model.

3 Conceptual Design

According to the literature at home and abroad, the paper sets up a random model of the selection path S of the fishbone layout of the warehouse. In order to build the probability model of the S type selection path distance, a series of hypotheses are needed for the warehouse operation, which is as follows:

The warehouse has only one I/O point in the middle, that is, the two parts of the warehouse are about central symmetry. Because the storage of each kind of goods is random, only the right half area can be considered. The space between the two shelves is called the picking channel. The space at the back and the front of the warehouse is called the rear aisle of the warehouse and the aisle of the front part. The storage space is calculated according to the length of the shelf, and the height of the shelf is not considered. Picking in the channel can pick up the goods on both sides of the shelf, and the distance between the two sides of the goods can be ignored. The selected items are random and independent of each other. In order, the probability of each item being selected is the same in every category of goods, and in each channel, the length of each item is evenly distributed in the range of its belongings. Do not consider the situation of warehouse congestion.

The definition of symbols used in the model l_1 to represent the channel width. l_2 is the representation of shelf width. b is representing the width of the warehouse. α is the representing the angle of the storehouse oblique channel. α_0 is the representing the diagonal angle of the right half of the warehouse. a is representing half the length of the warehouse.

Because the main channel angle of the fishbone layout is changed, and the range of change is between $0°$ and $90°$.

(1) $\arctan\left(\frac{l_2/2}{a-l_1/2\sin\alpha}\right) < a < \theta$

$$n_1 = \left[\frac{(a-l_1/2\sin\alpha)\tan\alpha - l_2/2}{l_1+l_2}\right], n_1 = \left[\frac{a-l_2/2}{l_1+l_2}\right],$$

(2) $\arctan(\frac{l_2/2}{a-l_1/2\sin\alpha}) < a < \pi/2$

$$n_1 = \left[\frac{r - l_2/2}{l_1 + l_2}\right], n_2 = \left[\frac{(r - l_1/2\cos\alpha)\tan\alpha - l_2/2}{l_1 + l_2}\right],$$

The length of A class item to account for the channel is:

$$m_{aj} = \max(0, \min(a, ca + (l_1 + l_2) \times (j - 0.5) \times \tan(90 + \arctan(\cos(\alpha))$$
$$- (l_1 + l_2) \times (j - 0.5) \times \tan(\alpha)))$$

The length of B class item to account for the channel is:

$$m_{bj} = \max(0, \min(a, cb + (l_1 + l_2) \times (j - 0.5) \times \tan(90 + \arctan(\cos(\alpha))$$
$$- (l_1 + l_2) \times (j - 0.5) \times \tan(\alpha)))$$

The length of C class item to account for the channel is:

$$m_{cj} = a - m_{bj} - m_{aj} - (l_1 + l_2) \times (j - 0.5) \times \tan(\alpha)$$

Among them, for any channel in the right half of the region. In A, B, and C, there are only two kinds of goods to be selected on the shelves. In the case of no picking, there is no need to pick up the passageway, and the picking distance will not be produced. In the case of picking, the channel needs to walk all the distance as long as there is one item to be selected. So just consider the probability of at least one selection in the channel. Assuming T goods in an order, at least one of the goods has a probability of $\bar{p}_{aj}$ in a class a storage area in the j channel. So there is:

$$\bar{p}_{aj} = \left[1 - (1 - p_{aj})^T\right] \quad 1 \le j \le n_1 + n_2$$
$$\bar{p}_{bj} = \left[1 - (1 - p_{aj})^T\right] \quad 1 \le j \le n_1 + n_2$$
$$\bar{p}_{cj} = \left[1 - (1 - p_{aj})^T\right] \quad 1 \le j \le n_1 + n_2$$

The selection distance in the channel J is expected to be:

$$d_j = E(d_{aj}) + E(d_{bj}) + E(d_{cj})$$
$$= (\bar{p}_{aj} * m_{aj}) + (\bar{p}_{bj} * m_{bj}) + (\bar{p}_{cj} * m_{cj})$$

The number of channels in the upper and lower regions of an oblique channel is different. Or the right half of the storage area is divided into 1, 2 and two selected areas. Therefore, the desired picking distance of the 1, 2 regions main channel needs to be separated. The farthest passage of the selection of the goods is expected to be:

$$\bar{j}_{1far} = E(j_{1far}) = \sum_{j_{1far}=1}^{n_1} j_{1far} \times p_{1j_{1far}} / \sum_{j=1}^{n_1} p_{1j}$$

To sum up, the selected walking distance from the farthest passage of the selected goods can be obtained.

$$R_{1far} = (\bar{j}_{1far} - 0.5) * \frac{(l_1 + l_2)}{\cos(\alpha)}$$

In this paper, according to the order frequency of different kinds of goods and the distribution of storage space, five cases are set up. The specific data are shown in Table 1.

Table 1. Order frequency and storage space area ratio of various kinds of goods

Order frequency/ space distribution	A	B	C
1	33.33/33.33	33.33/33.33	33.33/33.33
2	45/30	30/30	25/40
3	60/25	25/30	15/45
4	75/20	20/30	5/50
5	85/15	10/30	5/55

According to the known conditions, the results of the simulation results of the S type selection path random model are obtained in the above five cases, as shown in Table 2.

Table 2. S type selection path random model and simulation result comparison table

	1	2	3	4	5
Model	1743.8	1743.1	1533	1242.9	1168.5
Simulation	1663	1659.1	1559.8	1154.8	1076.7
Absolute error	80.8	84	−26.8	88.1	91.8
Relative error	0.046	0.048	−0.017	0.071	0.079

In this paper, based on the reference literature, the distance of the S type selection path of the fishbone warehouse is modeled and analyzed. The model is simulated and simulated. The simulation results are shown in the simulated quantized data case of different warehouses. We can understand that the storage of the fishbone layout is dominated by the selection path S, and there are different sorting path distances.

4 Conclusion

In fact, distribution center of the lot, this is only for path selection, storage layout and storage strategy research. The optimization method of picking path is relatively broad, and the method is more. According to the literature research on optimization of picking path, in the optimization of the selection path, the study abroad is more and the early start, and the domestic research has increased significantly in recent years. Most studies have focused on the automatic piece picking path optimization for artificial intelligent algorithm are few, the use of intelligent algorithms or improved, but the mixed two algorithms of picking path optimization is less. There are more constraints in the actual process, and the constraints should be taken into account when building the model. In order to overcome the limitations of the single algorithm, the use of hybrid algorithm will become a trend. The storage layout is less, the storage strategy and the storage layout are combined to identify the chosen path.

Acknowledgement. The study is supported by College Students' scientific training program of Information College of Beijing Wuzi University in 2017, and Beijing the Great Wall scholars program (No. CIT&TCD20170317).

References

1. Bartholdi III, J., Hackman, S.T.: Warehouse and distribution science: release 0.94 [DB/OL], 11 January 2011. http://www.covesys.com/docs/appnotes/warehouse_anddistribution_science.pdf. Accessed 28 Sept 2016
2. Bartholdi, J., Hackman, S.T.: Warehouse and distribution science [EB/OL], 24 May 2002 (2003). http://www.tli.gatech.edu/whscience/book/sci.pdf. Accessed Feb 2011
3. Christophe, Th.: Using a TSP heuristic for routing order pickers in warehouses. Eur. J. Oper. Res. **200**(3), 755–763 (2010)
4. Chabot, T., La Hani, R., Coelho, L.C., et al.: Order picking problems under weight, fragility and category constraints. J. Prod. Res. **55**, 6361–6379 (2015)
5. Cheng, C.Y., Chen, Y., Chen, T.L., et al.: Using a hybrid approach based on the particle swarm optimization and ant colony optimization to solve a joint order batching and picker routing problem. Int. J. Prod. Econ. **170**, 805–814 (2015)
6. Liang, J.H., Lee, C.H.: Efficient collision-free path-planning of multiple mobile robot systems using efficient artificial bee colony algorithm. Adv. Eng. Softw. **79**, 47–56 (2015)
7. Kulak, O., Tan, M.E.: Joint order batching and picker routing in single and multiple-cross-aisle warehouses using cluster-based search algorithms. Flex. Serv. Manuf. J. **24**(1), 52–80 (2012)
8. Li, S.Z.: Picking storage strategy and path strategy for collaborative research. Ind. Eng. **14**(2), 37–43 (2011)
9. Zhou, L., et al.: Comparative study on the random model of sorting storage return type and S type picking path. Syst. Sci. Math. **31**(8), 921–931 (2011)
10. Zhou, L., et al.: Study on the efficiency of classified storage manual picking random service system. J. Manag. Sci. 2 (2012)
11. Jia, Z., Lan, F.A.: Research on optimization and decision making of manual order picking in distribution center. Logist. Technol. Appl. (10) (2014)

12. Xia, X.K.: Genetic algorithm overhead warehouse selection path optimization. Manag. Based **09**, 40–42 (2015)

13. Yan, F., Yi, X.: Journal of MMAS algorithm stacker storage optimization path verification center based on. Wuhan University (Engineering Science Edition) (05), 645–647 (2013)

14. Xin, Y.Y., Chen, H., Guo, Y.K., Cheng, C., O Lin, L., Yu, L.: Linear temporal logic theory of warehousing robot path planning of high technology of communication based on **01**, 16–23 (2016)

15. Pan Cheng, H.: Simulation study on picking path planning of warehouse logistics robot. North Central University (2017)

16. Shen, W., Yu, N., Liu, J.T.: Intelligent scheduling and path planning for warehousing logistics robot cluster. J. Intell. Syst. **06**, 659–664 (2014)

17. Liu, K., NI u JC, Shen, Y.J., Li, S.J.: Optimization of stacker operation path based on genetic particle swarm algorithm. J. Shijiazhuang Univ. Railw. (Nat. Sci. Edit.) (02), 67–71 (2016)

18. Chen, Y., Li, Y.Y., peach: The genetic algorithm for solving the traveling salesman problem improved. Control Decis. **29**(8), 1483–1488 (2014)

19. Xu, X., Liang, R.S., Yang, H.Z.: Simulation of agent path planning based on improved genetic algorithm. Comput. Simul. **31**(6), 357–361 (2014)

20. Sun, H.H., Dong, H.H.: Distribution, order batching and building Tun goods road nephew study. Logist. Technol. (7), 179–183 (2014)

21. A Ri GL: A RFID based optimization algorithm for picking route in distribution center. Shenyang University of Technology (2015)

22. Lee, Z.T.F, Li, J.Q.: Electricity, TSP warehouse picking path optimization to research: a case study on a company. Based on Logistics Engineering and Management (6) (2016)

23. Ran, L.H.J.: Optimization of picking the route for shuttle car to a man in dynamic warehouse. Log. Eng. Manag. **06**, 70–73 (2017)

24. Gu, J.X.: Research on warehouse design and performance evaluation: a comprehensive review. Eur. J. Oper. Res. **203**(3), 539–549 (2010)

25. Kevin, R.: A unit-load warehouse with multiple pickups and deposit points and non-traditional aisles. Transp. Res. Part E: Logist. Transp. Rev. **48**(4), 795–806 (2012)

26. Kar, J.: An overview of warehouse optimization. Int. J. Adv. Telecommun. Electrotech. Signals Syst. **3**, 111–117 (2013)

27. Ning, F., He, C.Q., Li, Y.D.: People working under the mode of delivery of fishbone layout space optimization. J. Zhejiang Sci-Tech Univ. (Soc. Sci. Edit.) **04**, 293–298 (2017)

28. Zhou, L., et al.: Fishbone layout design and analysis of storage area in distribution center. Manag. World (5), 184–185 (2014)

29. Shandong, M.: Modeling and optimization of the layer shuttle vehicle storage system. Shandong University (2014)

30. Liu, Y.Q.: Optimization of logistics technology storage location assignment based in Fishbone (12), 66–70 (2014)

31. Zhou, X., Zhang, X.M., Liu, Y.K.: A mobile robot based flexible picking system for distribution centers. Logist. Technol. **07**, 238–240 (2015)

32. Pan, J.C.-H.: Storage assignment problem with travel distance and blocking considerations for a picker-to-part order picking system. Comput. Ind. Eng. **62**, 527–535 (2012)

33. Chiang, D.M.-H.: Data mining based storage assignment heuristics for travel distance reducing. Expert Syst. **31**, 81–90 (2014)

34. Carlos, E.: Warehouse storage location assignment using clustering analysis (2015)

35. Zhao, S.B.: Based on the data mining of the warehouse allocation strategy, the implementation of the strategy. Southeast University (2016)

36. Xia, L., Yu, M.Z.: Dynamic cargo allocation modeling and simulation of automated stereo-bearing library. Logist. Technol. (3), 131–151 (2016)
37. Wang: The storage allocation and efficiency evaluation model of the automatic warehousing system for multi-platform pile machines. Shandong University (2016)
38. Zhu, K.X.: Based on correlation analysis of a mesh products warehouse storage allocation strategy research. Shandong University (2017)

Factorization of Odd Integers as Lattice Search Procedure

Xingbo Wang[1,2]([⊠])

[1] Foshan University, Foshan City 528000, Guangdong Province,
People's Republic of China
xbwang@fosu.edu.cn
[2] State Key Laboratory of Mathematical Engineering and Advanced Computing,
Wuxi, China
http://web.fosu.edu.cn

Abstract. The article puts forward a 3-dimensional searching approach that can factorize odd composite integers. The article first proves that, an odd composite number can be expressed by a trivariate function, then demonstrates that factorization of an odd integer can be turned into a problem of searching a point in a 3-dimensional cube whose points can be searched rapidly via octree search algorithm or other 3-dimensional searching algorithm. Mathematical principles with their proofs are presented in detail, and an algorithm that reaches square of logarithm time-complexity is proposed with numerical examples. The proposed algorithm can be applied both in sequential computation and parallel computation.

Keywords: Cryptography · Integer factorization
Parallel computing · Octree search

1 Introduction

Factorization of large odd integers, especially the large semiprimes or the RSA numbers, has been to topics in cryptography, as overviewed in bibliographies [1–4]. Literatures show that, there has not been an outstanding approach of factoring integers ever since Pollard published his approach of number field sieve (NFS). Since the NFS requires huge memory resource, as was presented in [5], it is hardly applied in ordinary computational environment and new approaches are still under development. This point of view can be verified through the bibliographies from [6–12]. Looking through present literatures, one can see that, classical approaches always focus on analytic relationships among composite numbers and their divisors and attempt to find a factoring method that is either for general purpose or for special purpose; thus the classical methods seldom remind people of some "pure" computer searching procedure such as binary search, quadtree search and so on. It is known for an engineer of computer programmers that, proper design of a "pure" computer searching is sometimes

© Springer Nature Singapore Pte Ltd. 2018
K. Li et al. (Eds.): ISICA 2017, CCIS 874, pp. 251–258, 2018.
https://doi.org/10.1007/978-981-13-1651-7_22

more efficient than a procedure full of pure analytic formulas, as seen also in Skienas book [13]. In February 2017, Wang [14] introduced a new approach that can exactly locate a divisor of a composite odd number in a definite interval so that a search can be performed in the interval. Li [15] and Fu [16] followed the approach and designed their searching algorithms. However, as pointed out in Wang's article [17], the approach might not be of high efficiency when it is applied on factoring a large seimiprime. Following the previous studies, this article proposes a new approach that turns the factorization of odd integers into a problem of searching a point in a 3-dimensional cube whose points can be searched rapidly via octree search procedure or other 3-dimensional searching procedure. This article presents the details of the new approach.

2   Preliminaries

2.1   Symbols and Notations

Throughout this paper, symbol $a|b$ means integer b is divisible by integer a. Symbol (a, b) denotes the *greatest common divisor* (GCD) of integers a and b. Divisors mean the nontrivial ones. Odd integers mean those bigger than 1 unless special mentioned. Symbol $\lfloor x \rfloor$ is to express x's floor function defined by $x - 1 < \lfloor x \rfloor \leq x$ with x being a real number. The time complexity mentioned in this article is by default in searching steps (not in bit-operations), for example, $O(X)$ means a computation requires around X steps of searches. Symbol $A \Leftrightarrow B$ means A is equivalent to B.

2.2   Lemmas

Lemma 1 (see in [18]). For real numbers x and y, the following inequalities and equalities hold.

$$(\textbf{P0})x - 1 < \lfloor x \rfloor \leq x \Leftrightarrow \lfloor x \rfloor \leq x \leq \lfloor x \rfloor + 1$$
$$(\textbf{P1}) \lfloor x \rfloor + \lfloor y \rfloor \leq \lfloor x + y \rfloor \leq \lfloor x \rfloor + \lfloor y \rfloor + 1$$
$$(\textbf{P2}) \lfloor x \rfloor - \lfloor y \rfloor - 1 \leq \lfloor x - y \rfloor \leq \lfloor x \rfloor - \lfloor y \rfloor < \lfloor x \rfloor - \lfloor y \rfloor + 1$$

3   Theorems and Proofs

Proposition 1. *For positive integer k and real number $x > 1$, it holds*

$$\left\lfloor \frac{x+1}{2^k} \right\rfloor = \begin{cases} \left\lfloor \frac{x-1}{2^k} \right\rfloor + 1, k = 1 \\ \left\lfloor \frac{x-1}{2^k} \right\rfloor, k > 1 \end{cases}$$

Proof. By Lemma 1 (**P2**), it holds, it holds

$$\left\lfloor \frac{1}{2^{k-1}} \right\rfloor = \left\lfloor \frac{x+1}{2^k} - \frac{x-1}{2^k} \right\rfloor \leq \left\lfloor \frac{x+1}{2^k} \right\rfloor - \left\lfloor \frac{x-1}{2^k} \right\rfloor < \left\lfloor \frac{1}{2^{k-1}} \right\rfloor + 1$$

Note that, $k = 1$ yields $\lfloor \frac{1}{2^{k-1}} \rfloor = 1$ and $k > 1$ yields $\lfloor \frac{1}{2^{k-1}} \rfloor = 0$; therefore when $k = 1$ it holds

$$\left\lfloor \frac{x-1}{2^k} \right\rfloor + 1 \leq \left\lfloor \frac{x+1}{2^k} \right\rfloor < \left\lfloor \frac{x-1}{2^k} \right\rfloor + 2$$

which is just $\lfloor \frac{x+1}{2^k} \rfloor = \lfloor \frac{x-1}{2^k} \rfloor + 1$ because $\lfloor \frac{x+1}{2^k} \rfloor$ and $\lfloor \frac{x-1}{2^k} \rfloor$ are integers. Similarly, it holds when $k > 1$

$$\left\lfloor \frac{x-1}{2^k} \right\rfloor \leq \left\lfloor \frac{x+1}{2^k} \right\rfloor < \left\lfloor \frac{x-1}{2^k} \right\rfloor + 1$$

which is $\lfloor \frac{x+1}{2^k} \rfloor = \lfloor \frac{x-1}{2^k} \rfloor$. Hence the proposition holds.

□

Theorem 1. *Let p be an odd integer with $p > 1$; then $\frac{p-1}{2}$ and $\frac{p+1}{2}$ cannot simultaneously be even integers.*

Proof. Suppose $\frac{p-1}{2}$ is an even integer, namely, there is a positive k such that $\frac{p-1}{2} = 2k$; then $p = 4k + 1$ which says $\frac{p+1}{2} = 2k + 1$ is an odd integer. In the same way, it knows that $\frac{p+1}{2}$ being an even integer leads to $\frac{p-1}{2}$ being an odd integer.

□

Theorem 2. *An odd integer in the form of $2^k + 1$ with $k > 1$ can always be expressed in the form of $2^\alpha p - 1$ with some positive integer α and some odd integer $p > 1$. An odd integer in the form of $2^k - 1$ with $k > 2$ can always be expressed in the form of $2^\beta q + 1$ with some positive integer β and some odd integer $q > 1$.*

Proof. Take the case $2^k + 1$ as an example and use proof by mathematical induction. Obviously the conclusion holds for $k = 2$ because $2^2 + 1 = 2 \times 3 - 1$. Assume it holds for $k = n$ and let $2^n + 1 = 2^\beta p_n - 1$, where β is a positive integer and p_n is an odd integer bigger than 1; then $2^{n+1} = 1 = 2(2^n) + 1 = 2(2^\beta p_n - 2) + 1 = 2(2^\beta p_n - 1) - 1$. Since $2^\beta p_n - 1$ is odd, the conclusion holds for $k = n + 1$ and thus it hold forever.

□

Theorem 3. *Let p and s be odd integers with $p > 1$ and $p | s$; then $\frac{p-1}{2} | \frac{1}{2}(s - \frac{s}{p})$ provided that $\frac{p-1}{2}$ is odd and it is bigger tan 1, and $\frac{p+1}{2} | \frac{1}{2}(s + \frac{s}{p})$ provided that $\frac{p+1}{2}$ is odd.*

Proof. Without loss of generality, let $s = pt$, where t is an odd number bigger than 1; then it yields $s - t = (p - 1)t$ and $s + t = (p + 1)t$. Since s and t are both odd integers, it knows that $\frac{s-t}{2}$ and $\frac{s+t}{2}$ are positive integers. Since $\frac{s-t}{2} = \frac{p-1}{2} \times t$ and $\frac{s+t}{2} = \frac{p+1}{2} \times t$, substituting t by $\frac{s}{p}$ immediately results in the theorems conclusions.

□

Corollary 1. *Let p and s be odd integers with $p > 1$ and $p < s$; if $p|s$ then $(2^k p - 1)|(2^k s - \frac{s}{p})$ and $(2^k p + 1)|(2^k s + \frac{s}{p})$ for arbitrary positive integer $k > 0$.*

Proof. (Omitted)

$\square$

Corollary 2. *For a given odd integer N, if there exist odd numbers p and s such that $p|s$ and $N = 2^k s - \frac{s}{p}$ or $N = 2^k s + \frac{s}{p}$; then $(2^k p - 1)|N$ or $(2^k p + 1)|N$ respectively.*

Proof. (Omitted)

$\square$

Corollary 3. *Let k be a positive integer, p and s are positive odd integers with $s \geq 3$; if $k > \lfloor \log_2 N \rfloor$ for a given odd integer N with $N > 3$ then $2^k s - \frac{s}{p} > N$ and thus $2^k s + \frac{s}{p} > N$.*

Proof. Let $F_- = 2^k s - \frac{s}{p}$, $F_+ = 2^k s + \frac{s}{p}$. By Lemma 1 (P0), it knows that, $k \geq \lfloor \log_2 N \rfloor + 1$ yields $k > \log_2 N$, and results in $F_- - N = 2^k s - \frac{s}{p} - N > Ns - \frac{s}{p} - N = N(s-1) - \frac{s}{p} > 0$ because $N \geq 3$. That is $F_- > N$ hence $F_+ > N$.

$\square$

4   Algorithm with Analysis and Numerical Experiments

Corollary 2 says that, if $p|s$ and $N = 2^k s - \frac{s}{p}$ (or $N = 2^k s + \frac{s}{p}$) then $(2^k p - 1)|N$ (or $(2^k p + 1)|N$). Note that, $p|s$ yields $s = pq$ for some odd integer $q > 1$; hence for a given odd integer p, letting q vary and $s = pq$, comparing $2^k s - q$ with N can certainly obtain a divisor $2^k p - 1$ of N if p is properly selected. By this means, factorization of N is turned into a trivariate search problem that can be solved by many searching algorithms.

4.1   Guideline of Algorithm Design

According to Corollary 3, $N = 2^k s \pm \frac{s}{p}$ yields $k \leq \lfloor \log_2 N \rfloor$. Let $s = pq$ with odd integer $q > 1$; then $N = 2^k pq \pm q = (2^k p \pm 1)q$. By Proposition 1, this means $2^k p \pm 1 \leq \lfloor \sqrt{N} \rfloor$, namely $p \leq \lfloor \frac{\sqrt{N}+1}{2^k} \rfloor$, yielding $\lfloor \sqrt{N} \rfloor < q = \frac{N}{2^k \pm 1} < N$ or $q \leq \lfloor \sqrt{N} \rfloor$ yielding $p \geq \lfloor \frac{\sqrt{N}+1}{2^k} \rfloor$. Thus p, q, and k change in their range by $[1, \lfloor \frac{\sqrt{N}+1}{2^k} \rfloor] \times [\lfloor \sqrt{N} \rfloor, N] \times [1, \lfloor \log_2 N \rfloor]$, which is a topological cube. Consequently, finding a triple (p, q, k) that matches to $N = 2^k pq + q$ or $N = 2^k pq - q$ is turned to be the problem that searches a lattice N in the cube, as referred to Fig. 1. According to Bergs algorithm introduced in [15], the search is quite efficient.

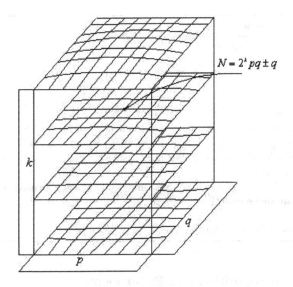

Fig. 1. A cubic searching space

Let
$$F_+(k,p,q) = 2^k pq + q \text{ and } F_-(k,p,q) = 2^k pq - q$$

Then it can see that, for a given k, the two are hyperbolic surfaces of p and q. Meanwhile, two different surfaces of either $F_+(p,q,k)$ or $F_-(p,q,k)$ have no intersection for given p and q. Actually, taking $F_+(p,q,k)$ as an example, it can see that, $F_+(k_1,p,q) - F_+(k_2,p,q) = (2^{k_1} - 2^{k_2})pq > 0$ when $k_1 > k_2$. Hence the maximal values or the minimal values of $F_+(p,q,k)$ or $F_-(p,q,k)$ can be references to determine k's range. Note that $F_+(p,q,k)$ and $F_-(p,q,k)$ are symmetric bilinear functions of p and q. Their maximal values and minimal values are near the corners. Obviously, if $\min(F_+(k,p,q)) \le \max(F_+(k,p,q))$, N is on $F_+(p,q,k)$; if $\min(F_-(k,p,q)) \le \max(F_-(k,p,q))$ N is on $F_-(p,q,k)$.

4.2 Lattice-Searching Algorithm

Based on the analysis above, search algorithms can be designed to find (p,q,k) that matches to $N = 2^k pq + q$ or $N = 2^k pq - q$. The direct and simple one is applying the octree search algorithm on the cube $\Omega = [1, \lfloor \frac{\sqrt{N}+1}{2^k} \rfloor] \times [\lceil \sqrt{N} \rceil, N] \times [1, \lfloor \log_2 N \rfloor]$. However the size of p's varying range being constrained with k restricts the construction of a regular cube. Referring to the approach of stratified tree search that was introduced in [19], here presents an algorithm, which is called *lattice-searching algorithm*, as follows (with C pseudo-languages embedded).

Algorithm 1. Lattice Searching Algorithm

Input: Odd composite number N;

Step 1. Calculate $\sqrt{N}, K_{max} = \lfloor \log_2 N \rfloor$;

Step 2. for $(k = 1; k < K_{max}; k++)$

{

    (QS1).$S_+ = F_+(k, p, q)$ and $S_- = F_-(k, p, q)$ with $p \in [1, \lfloor \frac{\sqrt{N}+1}{2^k} \rfloor], q \in [\lfloor \sqrt{N} \rfloor, N]$

    (QS2). By quadtree subdivision rule, subdivide each of S_+ and S_- into four sub-surfaces, say,$S_{LB}^+, S_{LT}^+, S_{RT}^+, S_{RB}^+$ and $S_{LB}^-, S_{LT}^-, S_{RT}^-, S_{RB}^-$.

    (QS3). Find which sub-surface N lies inside, say, S_{LB}^+; If N is exactly located on it, output(k, p, q) and stop.

    (QS4). Repeat (QS2) and (QS3) on S_{LB}^+ until N is exactly located or not.

    (QS5). If N cannot be exactly located declare failure of k.

}

Step 3. If all k are failure, declare failure of factorization.

4.3   Analysis of Algorithm and Experiments

The lattice-searching algorithm can be applies in either sequential computation or parallel computation. When applied in parallel computation, task of a process can be assigned with k and the quadtree search is performed in the process. This time, the time complexity of the process is $O(0.0625 \log_2^2 N)$. When applied in sequential computation, it time complexity is $o(0.0625 \log_2(\log_2 N) \cdot \log_2^2 N)$. Numerical experiments are done on a personal computer with an Intel Xeon E5450 CPU and 4 GB memory by C++ programming. Conventional data type of unsigned long integer and gmp big number library are tested separately. With conventional unsigned long integer, the maximal factorized integer is 429496729, together with many other integers. Due to limitation of the articles length, here list a few samples taken from source codes of the program.

- //unsigned long N=47871; //OK 591*81
- //unsigned long N=152703; //OK 2679*57
- //unsigned long N=462679; //OK 66097*7
- //unsigned long N=229501; //Ok 12079*19
- //unsigned long N=400789; //Ok 877*457
- //unsigned long N=429496729; //OK 22605091*19
- //unsigned long N=371811; //OK 6523*57

With gmp big number library, a few Mersenne numbers and Fermat numbers are factorized as show in Table 1 and several big numbers are factorized as shown in Table 2.

Table 1. Factorized mersenne numbers and fermat numbers

Big Number N	$N's\ Factorization$
$M_{67}=2^{67}-1$	193707721 × ABigNumber
$M_{71}=2^{71}-1$	228479 × ABigNumber
$M_{83}=2^{83}-1$	167 × ABigNumber
$M_{97}=2^{97}-1$	11447 × ABigNumber
$M_{103}=2^{103}-1$	2550183799 × ABigNumber
$M_{109}=2^{109}-1$	745988807 × ABigNumber
$M_{113}=2^{113}-1$	3391 × ABigNumber
$F_5=2^{32}+1$	641 ×6700417
$F_6=2^{64}+1$	274177 × 67280421310721
$F_9=2^{521}+1$	2424833 × ABigNumber
$F_{10}=2^{1024}+1$	45592577 × ABigNumber
$F_{11}=2^{2048}+1$	319489 × ABigNumber

Table 2. Some factorized big numbers

$N_1 = 1123877887715932507 = 299155897 \times 3756830131$
$N_2 = 1129367102454866881 = 25869889 \times 43655660929$
$N_3 = 29742315699406748437 = 372173423 \times 79915205819$
$N_4 = 35249679931198483 = 59138501 \times 596052983$
$N_5 = 208127655734009353 = 430470917 \times 483488309$
$N_6 = 331432537700013787 = 114098219 \times 2904800273$
$N_7 = 3070282504055021789 = 1436222173 \times 2137748993$
$N_8 = 3757550627260778911 = 16053127 \times 234069700393$
$N_9 = 24928816998094684879 = 347912923 \times 71652460573$
$N_{10} = 10188337563435517819 = 70901851 \times 143696355169$

5   Conclusions

Factorization had long been regarded to be a pure mathematical problem before modern cryptography came into being. However, most researchers of the field have still kept their thinking in the style of pure mathematics. They attempt to derive a good algorithm through a model that is built up by a series of mathematical deductions. Hence one can seldom sees a pure computer searching algorithm. This article demonstrates how a pure computer searching can be applied on the factorization. In fact, the problem can also be regarded as a problem of optimization that finds an objective in an space of solution. Hope more people can concern the topic and achieve expected achievements.

Acknowledgment. The research work is supported by Department of Guangdong Science and Technology[grant number 2015A030401105 and 2015A010104011]; the State Key Laboratory of Mathematical Engineering and Advanced Computing[grant number 2017A01]; Foshan Bureau of Science and Technology [grant number 2016AG100311]; Foshan University[grant number gg040981]. The authors sincerely present thanks to them all. The authors sincerely present thanks to them all.

References

1. Yan, S.Y.: Cryptanalytic Attacks on RSA. Springer, New York (2008). https://doi.org/10.1007/978-0-387-48742-7
2. Surhone, L.M., Tennoe, M.T., Henssonow, S.F.: RSA Factoring Challenge. Springer, USA (2011)
3. Wanambisi, A.W., Aywa, S., Maende, C., et al.: Advances in composite integer factorization. Mater. Struct. **48**(5), 1–12 (2013)
4. Abubakar, A., Jabaka, S., Tijjani, B.I., et al.: Cryptanalytic attacks on Rivest, Shamir, and Adleman (RSA) cryptosystem: issues and challenges. J. Theor. Appl. Inf. Technol. **61**(1), 1–7 (2014)
5. Sonalker, A.A.: Asymmetric Key Distribution. A thesis submitted to the Graduate Faculty of North Carolina State University (2002)
6. Singh, K., Verma, R.K., Chehal, R.: Modified prime number factorization algorithm (MPFA) For RSA public key encryption. Int. J. Soft Comput. Eng. **2**(4), 17–22 (2012)
7. Park, J., Sys, M.: Prime sieve and factorization using multiplication table. J. Math. Res. **4**(3), 7–12 (2012)
8. Wanambisi, A.W., Aywa, S., Maende, C., Muketha, G.M.: Algebraic approach to composite integer factorization. Int. J. Math. Stat. Stud. **1**(1), 39–44 (2013)
9. Sarnaik, S., Gaikwad, D.G.U.: An overview to Integer factorization and RSA in cryptography. Int. J. Adv. Res. Eng. Technol. **2**(9), 21–26 (2014)
10. Wang, X.: Seed and sieve of odd composite numbers with applications in factorization of integers. IOSR J. Math. **12**(5, Ver. 8), 01–07 (2016)
11. Wang, X.: Factorization of large numbers via factorization of small numbers. Glob. J. Pure Appl. Math. **12**(6), 5157–5173 (2016)
12. Kurzweg, U.H.: More on Factoring Semi-primes. http://www2.mae.ufl.edu/uhk/MORE-ON-SEMIPRIMES.pdf
13. Skiena, S.S.: The Algorithm Design Manual. Springer, London (2008). https://doi.org/10.1007/978-1-84800-070-4
14. Wang, X.: Genetic traits of odd numbers with applications in factorization of integers. Glob. J. Pure Appl. Math. **13**(1), 318–333 (2017)
15. Li, J.: Algorithm design and implementation for a mathematical model of factoring integers. IOSR J. Math. **13**(I Ver. 6), 37–41 (2017)
16. Fu, D.: A parallel algorithm for factorization of big odd numbers. IOSR J. Comput. Eng. **19**(2, Ver. 5), 51–54 (2017)
17. Wang, X.: Strategy for algorithm design in factoring RSA numbers. IOSR J. Comput. Eng. **19**(3, Ver. 2), 01–07 (2017)
18. Wang, X.: Brief summary of frequently-used properties of the floor function. IOSR J. Math. **13**(5), 46–48 (2017)
19. Berg, M.D., Kreveld, M.V., Snoeyink, J.: Two-and three-dimensional point location in rectangular subdivisions. Academic Press Inc. (1995)

Research on Key Technology of Distributed Indexing and Retrieval System Based on Lucene

Rongrong Li[✉]

Department of Computer Science, Guangdong University of Science and Technology,
Dongguan, China
408976182@qq.com

Abstract. Taking Chinese as the language object, after analyzing the current Chinese word segmentation algorithm and Lucene relevance ranking algorithm, an improved word segmentation algorithm and an improved relevance ranking algorithm based on Lucene full-text search toolkit were proposed. This paper also uses distributed storage, parallel computing, inverted indexing and retrieval techniques to analyze and design a search engine for digital information in the network to provide users with fast and accurate search service for massive digital information. The experimental analysis compares the speed of word segmentation and word segmentation by comparing various word segmentation algorithms and compares their response time, the number of hits, the accuracy and the recall rate of the keyword search results. The experimental results show that the system greatly improves the information Search speed to ensure the accuracy of search results.

Keywords: Lucene · Distributed · Word segmentation algorithm · Index

1 Introduction

With the popularity of the computer and the rapid development of the network, digital information has exploded, and the digital information on the Web is indeed staggering. On the one hand, these geographically dispersed and heterogeneous documents contain a great deal of valuable information, and users urgently need to find the needed resources from these information; On the other hand, although the processing power of a single computer continues to increase, searching for such a huge amount of data requires only limited processing power of a single computer, and in particular, requires multiple computers for "Team combat". Parallel computing and distributed computing can solve the problem of massive data by using the computing or storage resources of multiple computers or multiple processors [1]. Therefore, it is necessary to introduce parallel processing or distributed processing technology into the information retrieval, resulting in distributed parallel search technology.

As a highly optimized inverted index search engine, Lucene has provided the possibility for search to vertical and industrialize, breaking the high-tech search barrier. However, in practice, the indexing time increases linearly with the number of indexed files, and the retrieval server can't handle the request within a limited time when there are a high number of inbound or indexed data.

© Springer Nature Singapore Pte Ltd. 2018
K. Li et al. (Eds.): ISICA 2017, CCIS 874, pp. 259–270, 2018.
https://doi.org/10.1007/978-981-13-1651-7_23

For the massive digital information, how to search it quickly and accurately has become the research focus. At present, there are not many researches on the search engine for massive web digital information in our country. This paper will focus on this issue and analyze and design a professional search engine to realize massive digital information.

2   Lucene Application and Search Engine

2.1   Lucene Overview

Lucene is a top project of the Apache Software Foundation. It is not a complete full-text search system, is a high-performance. Scalable open-source full-text search toolkit developed in Java. Lucene is divided into five modules: Corpus, Analysis, Index, Storage and Search [2]. Each module can be divided into: mutual agreement part and the concrete realization part. Lucene provided for indexing and query function interface can be easily embedded into various applications, full-text indexing and search capabilities. Projects on Lucene include: Eyebrows (search and archiving in mailing list management system), Jive (web based forum system), Eclipse (full text search section), etc.

2.2   HDFS Overview

Hadoop Distributed File System (HDFS) is a distributed file system. It has a Master in Hadoop system and is mainly responsible for the work of NameNode and coordinating the operation of the job (JobTracker). JobTracker's main responsibility is to start, track and schedule the implementation of the task of each Slave. There are more than one Slave, each Slave usually has the DataNode function and is responsible for running the task after the task (TaskTracker). TaskTracker performs the Map task and the Reduce task in conjunction with the local data based on the application's requirements.

NameNode: manages the namespace of the file system, records the file system tree and all files and indexes in the tree, and also records the data nodes of the block where each file resides.

DataNode: The worker of the file system, which stores and provides the services of positioning blocks and periodically sends the storage list of the blocks to the name node.

Block: HDFS basic storage logic unit, a file may contain multiple blocks, a block can contain multiple files, the size of the file and block size parameters, the default size of the block is 64 MB, if you set Large, it may cause the Map to run slowly; setting small, may result in Map the number of too many.

2.3   Map Reduce Overview

Map Reduce is parallel computing. A Map Reduce job usually divides the input dataset into several independent data blocks. The Map processes the task in parallel and sorts the output of the Map first. The result is then entered into a Reduce job. Normally job input and output are stored in the file system [3].

Map method: applied in parallel to each input data set, each call will generate a (k2, v2) queue, such as: Map(k1, v1) → list(k2, v2).

Reduce method: Gather the data pairs with the same key in the output queue (k2, v2) of the Map side, gather them together, then form the destination data list (k3, v3) when output. For example: Reduce(k2, list(v2)) → list(k3, v3).

2.4 Search Engine Structure

Based on Lucene text information search engine structure showed in Fig. 1, the techniques used include: improved segmentation algorithm, indexing, retrieval and improved relevance ranking algorithm. Text information search engine core modules include: word processing, indexers, and crawlers, the results of relevance ranking, index file, and so on [4].

Word processing: for the text information using an improved segmentation algorithm for word segmentation;

Indexer: index the data after the word segmentation using inverted indexing technology;

Retriever: The search terms entered by the user are retrieved in the index file.

Index Files: Index files are stored on a distributed HDFS file system and are processed using parallel computer Map Reduce to process files on HDFS.

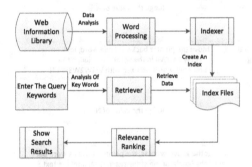

Fig. 1. The structure and process of search engine

3   Search Engine Design

3.1 Word Segmentation Algorithm Improvement

Western languages use spaces and punctuation to separate words, but Asian languages such as Chinese and Japanese can't be separated by spaces. However, Lucene uses Chinese words and double word segmentation methods to achieve the effect that applications require. Therefore, there is a need for an effective segmentation method for Chinese language. At present, Chinese analyzers mainly support StandardAnalyzer, CJKAnalyzer, ChineseAnalyzer, IK_CAnalyzer and ICTCLAS. In order to make the word breaker more accord with the Chinese habit and improve the speed and precision of Chinese word segmentation, based on the statistics and thesaurus, an improved algorithm is proposed for the maximal matching before segmentation algorithm.

Suppose the text S = W1W2W3W4W5 word segmentation, first of all to the text pointer to the beginning of the text S string, according to the distribution of Chinese words, two words is the most word, so the first word of two words, the algorithm set the most The length of a long participle is M words, and the flow of the improved word segmentation algorithm is shown in Fig. 2.

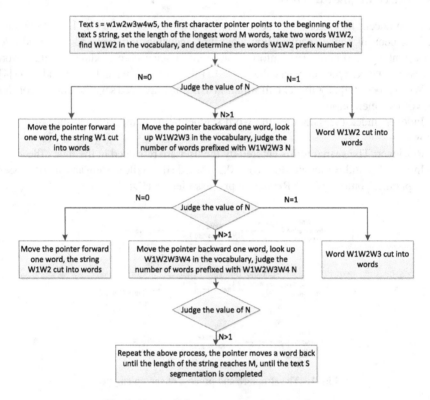

Fig. 2. Improved segmentation algorithm flows

Then the improved word segmentation algorithm can be described as:

(1) The pointer moves backward two words from the beginning of the string, obtaining two words W1W2, finding W1W2 in the vocabulary, and judging the number N of words prefixed with W1W2, and if N is equal to 0, the pointer moves forward One word, and sub-string W1 is divided into words; if N is equal to 1, sub-string W1W2 is divided into words.

(2) If N is greater than 1, the pointer moves backward one word, looks up W1W2W3 in the vocabulary, and judges the number N of words prefixed with W1W2W3. If N is equal to 0, the pointer moves forward one word, and Substring W1W2 is split into words; if N is equal to 1, substring W1W2W3 is split into words.

(3) If N is greater than 1, similar to (2), the pointer moves backward one word, looks up W1W2W3W4 in the vocabulary, and judges the number of words N prefixed with W1W2W3W4. If N is equal to 0, the pointer advances Move a word, and divide the substring W1W2W3 into words; if N is equal to 1, substring W1W2W3W4 is cut into words.

(4) If N is greater than 1, like (2) and (3), the pointer is shifted backward by one word until the length of the resulting substring reaches the maximum length M.

(5) Repeat (1)–(4) until the text S segmentation is completed.

3.2 Index Data Structure

Lucene inverted index data structure shown in Fig. 3, Lucene term is the smallest unit of index, which directly represents a key word and its appearance in the file location and the frequency of occurrence of such [5]; Several items form the field, the domain is an associated tuple, the domain is composed of a domain name and a domain value, the domain name is a string, and the domain value is an item; A number of domain composition documents, the document is the result of extracting all the in a file; Segment a number of documents, the number of segments in memory reaches the specified number (default is 10) will be combined into a segment; Several segments make up sub-indexes, which can be combined as indexes or sub-indexes that can be merged into a new internal element that contains all the merged items. In Lucene section index generation, merge threshold (Merge Factor) affects the number of index files in memory and hard disk. Every time a Document is added, an index is generated and is held in memory. When the number of segment indexes exceeds the merge threshold, an index is merged into a segment index through a merger process.

As shown in Fig. 3 from left to right is a search order, right to left is an index order. This order is to be mentioned below the inverted index.

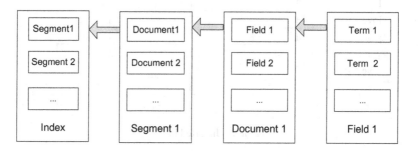

Fig. 3. Inverted index data structure

3.3 Inverted Index

Inverted index is a keyword-oriented indexing mechanism, is currently the most commonly used search engine storage [6]. In this paper, the inverted list to organize the index, the inverted table in the middle of a key word in the document where the occurrence and the number of occurrences and other, this attribute value to determine the record, rather than the records to determine the value of the attribute search called inverted index. The inverted table of the organizational structure is shown in Fig. 4 [8].

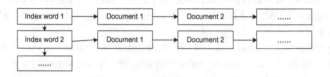

Fig. 4. Inverted index structure table

For example, there are documents 1 and 2:

Documents 1: I love you;
Documents 2: You love me.

Firstly, we extract keywords from document 1 and document 2, and according to the general index, we get the result as shown in Table 1, the results obtained from Lucene inverted index are shown in Table 2. As can be seen from Tables 1 and 2, the number of general index keywords grows linearly with the content of the document, and the inverted keyword only needs to modify the corresponding record when the same keywords appear, saving the storage space of the index file and improving the retrieval efficiency. Lucene in the specific implementation of the index, the Table 2 in the keywords, the location and the number of documents appear in the construction of the index were saved in the dictionary file, location file and the frequency file. Also stored in the dictionary file are pointers to location files and frequency files.

Table 1. General index

Key words	Document number	Location	Frequency
I	1	1	1
love	1	2	1
you	1	3	1
you	2	1	1
love	2	2	1
me	2	3	1

Table 2. Inverted index

Key words	Document number	Location	Frequency
I	1	1	1
love	1, 2	2, 2	1, 1
you	1, 2	3, 1	1, 1
me	1	3	1

3.4   Indexing

Word segmentation text information will be created indexing using inverted index technology. Figure 5 is the establishment of the index module can be divided into the following four steps; Obtain data from the data source (database), parse the resulting Field object, and construct the Document object, IndexWriter index.

First, get the text information from the database, for a data analysis, remove each field and value, according to the corresponding field of the database field Lucene domain to generate Field object;

Then, the generated Field added to the Document to generate Document object;

Then, through the addDocument method of IndexWriter class to create an index, indexing, Lucene data analysis and processing;

Finally, until there are no data to update, the index is created and the indexer is closed.

When large amounts of data are indexed, there is a bottleneck when writing index files to disk and a buffer in Lucene memory addresses this issue [8]. IndexWriter provides three parameters MergeFactor, min-MergeDocs, maxMergeDocs to set the size of the buffer and the frequency of indexing files to disk.

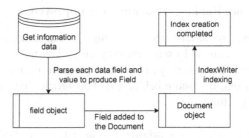

Fig. 5. Creating index module

3.5   Search Module

Lucene provides comprehensive and efficient search capabilities. Figure 6 is based on the index module to achieve the search module can be divided into the following four steps: Parse the query keywords entered by the user, parse and generate the Query object, IndexSearcher searches for the Hits search results, and the relevancy ranking algorithm displays the results.

First of all, Lucene query keywords entered by the user through the query analyzer QueryParser word processing, such as removing extra space;

Next, QueryParser generates a query object after the word segmentation query keywords;

Then, indexSearcher then searches for the Query object and returns the search results in the Hits result set.

Finally, the search result set Hits is displayed to the user through an improved relevance ranking algorithm.

After Lucene completes a search, all the search results will not be displayed, but will have the highest match with the query keyword 100 in the results of the ID on the cache,

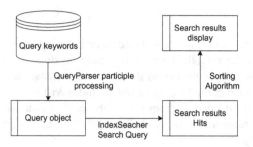

Fig. 6. Searching module

and then returned to the user, if the 100 can't meet user needs, Lucene will search again and usually generate a cache that is twice as big as the last time. When dealing with large amounts of text information, this search technology not only saves space in the result set, but also automatically filters out low-matching results to improve search accuracy and speed.

3.6 Lucene Relevance Ranking Algorithm and Improvement

In information retrieval, we usually get a lot of eligible results. Some of the results may not be what the user wants, but users generally only view the first dozen of results. Therefore, Lucene uses the relevancy ranking algorithm to sort the search results, high score results in the front, low score behind [9]. This article first introduces Lucene's relevance ranking algorithm:

$$\text{Score}(d) = \sum_{t \, in \, q} tf(t \text{ in } d) * idf(t) * boost(t * field \text{ in } d) * lengthNorm(t * field \text{ in } d)$$

The score obtained by this scoring formula is the original score, but the score returned by the Hits object for a document is not necessarily its original score, because if the document with the highest score exceeds 1.0, then all the next scoring will be based on this score for standard calculations. Therefore, all Hits objects score less than or equal to 1.0 [10]. So, Lucene in relevancy ranking algorithm, if a document contains a query more time, the score is higher, if the other query words, the more the other words, and the score will be less. It can be seen that Lucene's relevance ranking algorithm is easily affected by the number of query words, but the appearance of query words in a document is not reflected in the algorithm, nor does it reflect the respective characteristics of different text information. Therefore, the following improved ranking algorithm is proposed:

Score(d) = k1 * LuceneScore + k2 * PageScore + k3 * RepScore + k4 * DataScore, Where k1, k2, k3, k4 are weight coefficients and satisfy the following conditions:

(1) k1 + k2 + k3 + k4 = 1;
(2) 1 > k1 > k4 > k2 > k3 > 0,

Users meet the (1), (2) conditions, you can set the weight coefficient according to their own need. The description of the specific scoring parameters in the formula is shown in Table 3. The improved algorithm introduces PageRank and secondary search algorithm. In order to reflect the characteristics of text information, different text information has its own weight DataScore, so that the user The important text information can be displayed in the front, the non-important information displayed in the back, the final display of search results more in line with the characteristics of text information.

Table 3. Parameter descriptions of improved rating formula

Weights	Description
LuceneScore	Lucene sorting algorithm to calculate the score
PageScore	PageRank calculated score
RepScore	Secondary search RepScore + (hitNum − 1) * increment calculated score
DataScore	Each information data unique weight score

4   Experimental Results and Analysis

Based on the theoretical analysis of this article, 200,000 pieces of data are set up under the software platform of JDK1.7, Tomcat7.0, Oracle11g, Myeclipse10.6, Windows XP and the hardware platform of Intel Core Due CPU i5-2450M (2.5 GHz) and 2 GB memory Index, in order to provide the corresponding search service. From the following four aspects of text search engine experimental analysis:

(1) Comparison of text information search engines and traditional search engines are shown in Table 4. It can be seen from Table 4 that the search engine have a lot of improvement and improvement over the traditional search engines in data sources, word segmentation algorithms, indexes and query analysis.

Table 4. Comparison table of search engines

Technology	Text search engine	Traditional search engine
Incremental and bulk indexing	Support for incremental and bulk index, for large data volume index, interface design optimization batch index and small batch incremental index	Only support batch index, sometimes a little increase in data sources also need to rebuild the index
Data source	No specific data source is defined, is a document structure that can be flexibly adapted to a variety of applications	Only support page format, the lack of other formats of document support, not flexible
Index content	You can filter data source information and even control which fields need to be indexed	Lack of generality, often indexing all information
Index	With inverted index, a dedicated interface for incremental indexing	Adopt a general index, does not support incremental indexing
Query analysis	A unified query analysis interface, rich in content	Lack of common query analysis interface

(2) The improved word segmentation algorithm is compared with several mainstream Chinese word segmentation algorithms mentioned above. For example, the Chinese sentence "Lucene-based search engine design and implementation" in a variety of word segmentation algorithm results are as follows in Table 5. It is not difficult to find that the improved word segmentation algorithm is more in line with the Chinese language habits, more importantly; the improved word segmentation algorithm saves storage space, which also saves the time to create an index.

Table 5. Comparison of segmentation results

Segmentation algorithm	Segmentation results
StandardAnaluzer	[基][于][Lucene][的][搜][索][引][擎][设][计][与][实][现]
CJKAnaluzer	[基于][Lucene][的搜][搜索][索引][引擎][擎设][设计][计与][与实][实现]
IK_CAnaluzer	[基于][Lucene][的][搜索引擎][搜索][索引][引擎][设计][与][实现]
ICTLAS	[基于][Lucene][的][搜索][引擎][设计][与][实现]
Improved word segmenta-tion algorithm	[基于][Lucene][搜索][引擎][设计][实现]

(3) When the text information size is 10 MB and 900 MB, the improved word segmentation algorithm is compared with several mainstream Chinese word segmentation algorithms on the word segmentation speed, as shown in Table 6, One of the first participle is the size of the data is 10 MB segmentation, the second segmentation is the size of the data 900 MB segmentation. The experimental data show that the improved word segmentation algorithm improves the word segmentation speed of 45 times than IK_CAanalyzer, which is similar to the segmentation speed of StandardAnalyzer and ICTCLAS.

Table 6. Comparison of segmentation algorithms speed

Segmentation algorithm	The first time 10 MB/ms	The second time 900 MB/ms
StandardAnaluzer	6755	632907
CJKAnaluzer	5416	412056
IK_CAnaluzer	602439	63189651
ICTLAS	16832	1652843
Improved word segmentation algorithm	12034	102629

(4) In terms of keyword search, experiments show that this search engine does a quick search of text information (200,000 pieces of data) and experimentally analyzes the response time and the number of hits of the search results, etc. As shown in Table 7, Keyword search time all within 1 s to complete.

Table 7. Time and number of keyword retrieval

Key words	Response time	Hit the number
音乐	0.026	652
自由	0.035	763
幸福	0.056	1899
爱	0.105	2551

(5) The above several Chinese word segmentation algorithm in the keyword search, the experiment shows that the search engine does achieve text information (200000 data) search accuracy as shown in Table 8, the experimental analysis of the accuracy of the search results Rate and recall rate and so on.

Table 8. Accuracy and recall rate of retrieval

Segmentation algorithm	Accuracy	Recall rate
StandardAnaluzer	46.40	52.98
CJKAnaluzer	72.9	76.76
IK_CAnaluzer	96.25	97.12
ICTLAS	97.58	97.12
Improved word segmentation algorithm	96.27	98.78

5   Concluding Remarks

Professional search engine with its more accurate, more effective retrieval and more and more by the user's attention, After in-depth study of Lucene, this article first analyzes the overall structure; then analyzes and designs various functional technologies, puts forward corresponding improved algorithms for data indexing and searching; finally designs a feasible and effective solution to mass text information search engine. There are still some areas to be improved in this system, such as the improvement of search accuracy and search efficiency, which are the contents that need to be further studied in the future.

References

1. Erciyes, K.: Parallel and distributed computing. In: Erciyes, K. (ed.) Distributed and Sequential Algorithms for Bioinformatics. Springer, Heidelberg (2015). https://doi.org/10.1007/978-3-319-24966-7_4
2. Ding, G.Q., Lin, M.: Research the key technologies of the Mongolian full-text retrieval based on Lucene. Appl. Mech. Mater. **347–350**, 2185–2190 (2013)
3. Malekimajd, M., Ardagna, D., Ciavotta, M., et al.: Optimal map reduce job capacity allocation in cloud systems. ACM SIGMETRICS Perform. Eval. Rev. **42**(4), 51–61 (2015)
4. Wang, H.W., Wang, W., Meng, Y.: Countering page ranking spam for search engine based on text content and link structure analysis. Syst. Eng. Theory Pract. **35**(2), 445–457 (2015). Xitong Gongcheng Lilun Yu Shijian
5. Gennaro, C.: Large scale deep convolutional neural network features search with Lucene (2016)
6. Stalnaker, D., Zanibbi, R.: Math expression retrieval using an inverted index over symbol pairs. In: Proceedings of SPIE - The International Society for Optical Engineering, vol. 9402, pp. 940207–940207-12 (2015)
7. Procházka, P., Holub, J.: Positional inverted self-index. In: Data Compression Conference, pp. 627–627. IEEE (2016)
8. Wei, D., Hong, M., Song, Y.: Research of the Mongolian synergistic index technology based on Lucene. In: IEEE International Conference on Software Engineering and Service Science, pp. 322–325. IEEE (2015)
9. Gupta, D., Singh, D.: User preference based page ranking algorithm. In: International Conference on Computing, Communication and Automation, pp. 166–171. IEEE (2017)
10. Beebe, N.L., Liu, L.: Ranking algorithms for digital forensic string search hits. Digit. Investig. **11**(S2), S124–S132 (2014)

Intelligent Information Systems – E-commerce Platforms

Research on the Integrated Development Model of e-Commerce Channel and Physical Retail Channel

Sisi Li[✉]

GuangZhou Huashang Vocational College,
Zengcheng, Guangzhou City 511300, Guangdong Province, China
2676501836@qq.com

Abstract. New retail is the trend of future development, currently, whether it is e-commerce channel or physical retail channel, all have some problems, are badly in need of transformation. In the "Internet +", online and offline integration, realize the difference operation; with each other online and offline diversion, increasing flow; With the help of online and offline mutual diversion, the flow is increased continuously; Building a smart logistics distribution system, serving the integration of e-commerce channels and physical retail channels is a common choice for many enterprises, it is the main content of the new retail too. In this regard, this paper analyzed several modes of integrated development on current e-commerce channel and physical retail channel, and put forward the relevant suggestions on the integration development.

Keywords: E-commerce channel · New retail · Physical retail channel
Channel integration

In 2016, Jack Ma pointed out that e-commerce will be disappeared in future and will be replaced it with "new retail", which will integrate online and offline business, and combine the retail development mode with logistics. The integration of offline and online business is bound to require the integration of e-commerce channel and physical retail channel, and integration development.

1 The Current Situation and Shortcomings of Retail Industry Channel Development

1.1 The Physical Retail Industry Faced Difficulties in Operation and Need to Be Transformed Urgently

The physical retail industry in the face of the impact of e-commerce channel, many businesses faced difficulties in operation. On the one hand, many people choose to shop online, their business is shrinking. On the other hand, with the increase of rent and manpower costs, the operating costs of enterprises are rising. As a result, the physical retail industry is in urgent need of transformation.

© Springer Nature Singapore Pte Ltd. 2018
K. Li et al. (Eds.): ISICA 2017, CCIS 874, pp. 273–279, 2018.
https://doi.org/10.1007/978-981-13-1651-7_24

1.1.1 The Costs of Physical Retail Industry Continue to Rise and Revenues Continue to Fall

Began in 2011, sales of large retail enterprises in China have been slowing down for a long time, and negative growth has continued to occur in recent years, the China's retail industry development report released by the ministry of commerce showed that the labor costs in retail top 100 enterprises increased by 4.2%, and rents rose by 8.6% in 2015, which was similar to the increase in 2012, and still in a state of rapid rise. The growth of the two costs was significantly higher than the retail sales growth. New retail stores slow down and profits continue to decline., China's retail sales top 100 reached 4 trillion and 129 billion 260 million yuan in 2015, the growth rate decreased by 3.8% compared with 2014; Among them, 87 retail stores scale of 1 trillion and 736 billion 730 million yuan, accounting for 42.1% of the top 100, the contribution rate is only 7.2%, sales scale growth is also declining. The total retail sales of 2002–2015 years and the increase in the situation as shown in Fig. 1. The retail sales of national hundred key large retail enterprises fell by 0.5% in 2016, it is worth noting that with the trans-formation of physical retail enterprise, in the second half of year 2016, the operating condition improved, and began to a positive growth in July, September, November, December. On the other hand, the costs of physical retail enterprises have been rising. In 2016, the artificial cost of enterprises increased by 4% and grew faster than the same period of sales. All in all, Rising costs and declining sales scale have forced many retailers to find a way out of their predicament.

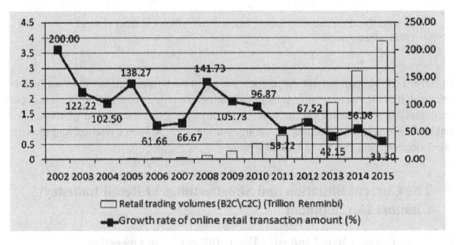

Fig. 1. 2002–2015 total retail sales and growth

1.1.2 The Growth of Physical Retail Industry Shown Polarization

In the past, the homogenization competition development of physical retail industry "thousand shop side", coupled with the impact of online retail channel, consumer demanded for personalized development, made the business faced the difficulties, therefore, many entities began to transform the retail sector. In the process of trans-formation, there must be a different degree of labor pain, which makes the growth of different formats shown polarization. The growth of convenience stores, super-markets and shopping centers is faster, while the growth of stores and specialty stores is slower. 2014–2016 first half years of retail enterprises in China (department stores, supermarkets) closing number statistics as shown in Table 1.

Table 1. The number of business failures

Trade	2014		2015		First half of 2016	
	Retail enterprises	Failures	Retail eEnterprise	Failures	Retail enterprises	Failures
Department stores	Parkson	4	Wanda Dprt Store	48	Parkson	2
	Ito Yokado, Zhongdu Department Store	3	Golden Eagle Marks & Spencer	5	Mopark, NOVO, Laiya Department Store,Friendship Store,Hua Lian, RAINBOW,Hahan Department Store, Xidan Market, Nanjing Yaohan, Century Ginwa, Golden Eagle, Xinhua Department Store	1
	Central,NOVO, Balloon	2	RAINBOW	4		
			Ito Yokado	3		
	Wangfujing, RAINBOW,Simgo Department Store, Mopark,Spring Department Store, Nanning Department Store	1	Far Eastern, NOVO,La Vita	2		
			New Wowld Department Store, Wangfujing,Pang Dong Lai Stroe, Intme Department Store, Ito Yokado, Sunshine Department Store, Winone Department Store, Central, Jiuguang Department Store, Sunlight Department Store	1		
Total	23		83		15	
Supermarket stores	Walmart	16	Lianhua	612	Walmart	10
	Lotte Mart	6	Yun Nan Tian shun Super Market	40	Parknshop Super Market	7
	Ren Ren Le	5	Every day One and one	30	Carrenfour	3
	Vanguard, Carrenfour, Tesco	4	Carrenfour	18	Aeon Mall	2
	New Huadu	3	New Huadu	14	Vanguard,	1
	Yonghui GMS	2	Ren Ren Le	11		
	Metro, Beijing Hualian,	1	Yonghui GMS	8		
			Lotte Mart	5		
			Beijing Hualian	4		
			Metro	2		
	Other	123		1		
Total	178		750		26	

1.2   The Growth of e-Commerce Retail Industry is Slowing Down, and the Experience Feeling of Consumer is not Enough, the Customer Value Needs to Be Further Excavated

1.2.1   The Development Trend of Online Retail Industry

Online retail after more than ten years of development, has become increasingly mature, the growth rate is slow but stable. China's online retail sales in 2016 is RMB5.16 trillion, increased by RMB148 million than RMB3.68 trillion in 2015, increased by 26.2%, growth rate is narrowed by 9.3%. Among them, online retail sales of physical goods are RMB4.19 trillion, accounting for 2.6% of total social consumer goods and increased by 1.8% higher than the previous year.

1.2.2   The Product Quality of Online Retail, Experience Feeling of Consumer Is Insufficient, Service Level Is not High, Need to Be Transformed Urgently

Although online retail has maintained stable development, its competition has been transformed from the past incremental expansion into the stock development. Overall, the product quality of current e-commerce retail enterprise is not high, enterprises in order to avoid the conflict of online and offline channel, often implement the different strategies online and offline, the so-called specifically for e-commerce, the quality of product online is likely to be worse than offline, such as large enterprises: Midea, Pampers and so on. Although this way can effectively avoid channel conflict, the product quality and consumer's experience feeling are falling synchronously, and some enterprises' online services can't be kept up, the consumer's experience feeling is even worse. This leads to the repurchase rate of products is lower, and the advertising costs to acquire new customers is rising, the retailer's profit is constantly squeezed, therefore, the e-commerce retail enterprises have to gradually improve their product quality, consumer shopping experience and the service level of enterprises from the improvement of sales.

2   Analysis on Model of the Integration Development for e-Commerce and Physical Retail Channel

2.1   Take Platform e-Commerce as the Center, Integrate Offline Physical Retail Channel

Taking the platform e-commerce as the center, the integration of offline physical retail channel mode requires the e-commerce enterprises to have a strong strength, and the offline physical shops have abundant resources, which promote each other. Such as the "smart logistics backbone" between Alibaba and Yintai in China, Alibaba used their powerful payment system and large data resources, help Yintai for precision fast transaction, at the same time, numerous online stores of Yintai provided the rich commodity resources for ali to develop new retail business, the cooperation of both effectively enhanced the value of the two channels, gathered a lot of the scenes of the users and offline resources, improved the customer's experience feeling, formed a new retail business system of integration online and offline.

Based on the existing research of Ant-Miner algorithms, we aim to find an algorithm in this paper, which can improve the prediction precision and stability of the algorithm, and simplify the rules. So as to improve the performance of classification min-ing algorithm, and make the improved algorithm have more advantages in massive data classification.

2.2 Proprietary Platform e-Commerce + Logistics Mode

Proprietary platform electricity + logistics model is a proprietary e-commerce plat-form to integrate offline power retail enterprises to carry out the sales in e-commerce platform, it integrates the resources of both sides well, the consumers can buy products on proprietary e-commerce platform, however, distributed by offline physical enterprise, this mode of relying on the offline store, can be the biggest save the cost of warehousing, distribution and logistics e-commerce enterprises. For example, Jing-dong and Wal-mart have teamed up to build the "Jingdong wal-mart franchised supermarket". The cooperation with Jingdong and Wal-mart is a typical strong coop-eration between proprietary platform e-commerce and physical retail channel, Jingdong has large e-commerce users group, and Wal-mart's products are very rich, this is a good solution to the Jingdong supermarket products in the past a single problem, can further excavate the value of the consumers. At the same time, Wal-mart is doing e-commerce at a lower cost and gaining a huge users group base of Jingdong, formed the "new retail" model of online and offline integration.

2.3 Online Self-operation + Offline Experience Store Model

The customer experience degree of e-commerce enterprise is not high, especially for empirical products, such as skin care products and agricultural products, the problem of product quality satisfaction, they don't know whether it is suitable for them only used, but once used, there is no way to return. For this kind of product, build offline expe-rience shops is a very good way, can let the consumer realize online consumption, offline experience, enhance consumer experience, still can use offline shops to grain the online store. For example, Jumei through the opening of offline experience shop, let consumers experience skin care products, also allow consumption to enhance the sense of trust.

2.4 Take Physical Store as the Main Body, Develop the Online Self-operation Business

With the spread of mobile payments, many physical stores are beginning to develop their online businesses. Large businesses even pulled out of proprietary APP, such as Guangzhou Youyi, Wangfujing, Pang and other large stores. Online and offline interaction leads to online purchases, offline consumption or offline delivery of goods or logistics. For some small business opportunities, they will build their own micro stores, taobao stores or into power e-commerce companies. Consumers can also easily achieve the goal of online consumption and offline experience. For example, Dong-guan's Jiarong supermarket has built its own shopping website; consumers can purchase goods on jiarong's website or micro website, and then pick up the goods offline or select the services for the introduction of delivery. This model has greatly improved the convenience of consumers, and can help physical stores to better grasp consumers' habits and carry out personalized marketing.

3   Suggestions on Channel Integration Development

3.1   Online and Offline Business Should Be Integrated with Each Other

The new retail era requires the enterprise to deeply excavate the customer's value, improve the customer experience, and make the online business convenient for customers, and the offline business to improve the customer experience. This requires targeted guidance to consumers and increased turnover. When consumers buy online, they can distribute preferential information online, encourage consumers to experience offline, and effectively promote offline store traffic. If users consume offline, promote online store through QR code, and set O2O experience machine in store, encourage users to buy experiences online. The goal of each other is to foster consumer habits, whether in stores or online. At the same time, the integration of online and offline business also has the benefit for the management of customers and excavation of value, form a perfect user database and then carry out accurate marketing.

3.2   Improve the Logistics System and Improve Users' Shopping Experience and Service Quality

Under the background of new retail, logistics is an integral part, therefore, should build perfect logistics system, large enterprises can use own logistics, small businesses can choose the third party logistics cooperation, to increase the speed of delivery and service level, at the same time, to reduce the logistics cost.

3.3   Properly Handle the Integration of Online and Offline Channels and Avoid Channel Conflicts

Many enterprises have more or less channel conflicts in their online and offline business. For this, enterprises can adopt differentiated management strategies to solve this problem. Product differentiation can be adopted, price differentiation can be implemented, and service differentiation and strategy differentiation also can be adopted. From the point of the current enterprise who carried out the differentiation strategy, part of the enterprise differentiation is a success, such as Fotile offline business products and services all realized the differentiation, is a different type of the online and offline sales, in the late machine cleaning service is also different. While other companies in the product differentiation strategy reduced the consumer's experience, such as the same products of pampers, products quality of e-commerce channel is significantly lower than the traditional channel, the sensitivity and satisfaction of consumers all reduced. Therefore, adopting the differentiation strategy, need to consider the user's experience.

4   Conclusion

New retail is the trend of future development, the integration of e-commerce and physical retail channel is an inevitable trend, therefore, enterprises should choose the right integration mode and choose the appropriate integration strategy.

Acknowledgments. This work was partially supported by Key platform construction leap plan and major project and achievement cultivation plan project, characteristic innovation project of Department of Education of Guangdong Province of China (Grant Nos. 2014GXJK174). We would like to thank the anonymous reviewers for their valuable comments that greatly helped us to improve the contents of this paper.

References

1. Ministry of commerce: China retail industry development report (2016), vol. 7 (2017)
2. Guo, X., Zhang, J.: The main models and countermeasures analysis on online and offline integration development for retail industry in China. J. Beijing Bus. Univ. **5** (2014)
3. Chen, K.: Discussion on the business model of traditional retailers transformed to e-commerce. Commer. Econ. Res. **18** (2016)
4. Jarrett, F.G.: Short term forecasting of Australian wool prices. Aust. Econ. Pap. **4**, 93–102 (1965)
5. Zhou, T., Xiong, J.: Research on the green-roof reconstruction strategy of existing buildings in the Yangtze River basin of China, no. S1, p. 1 (2015). http://kns.cnki.net/kcms/detail/50.1208.TU.20160412.1305.028.html. Accessed 12 Apr 2016
6. Chen, H.: Research on the Development of Shanghai Leisure Sports Industry in the Post-EXPO Era by the 'Integrated 4 Traditional Chinese Diagnoses'. International Research Association of Information and Computer Science. In: Proceedings of 2015 International Conference on Social Science, Education Management and Sports Education(SSEMSE 2015), p. 4. International Research Association of Information and Computer Science (2015)
7. Wang, L.: Relevant research on the development of virtual instrument integrated test system. In: Proceedings of 2016 2nd International Conference on Materials Engineering and Information Technology Applications (MEITA 2016), p. 4 (2016)
8. Xu, M., Yang, B.: An integrated design approach for gated communities: dilemmas and the way forward, no. S1, p. 1 (2015). http://kns.cnki.net/kcms/detail/50.1208.TU.20160412.1305.022.html. Accessed 12 Apr 2016
9. Qin, F., Li, Y., Qi, H., Ju, L.: Advances in compact manufacturing for shape and performance controllability of large-scale components-a review. Chin. J. Mech. Eng. p. 1. http://kns.cnki.net/kcms/detail/11.2737.TH.20161205.1631.032.html. Accessed 05 Dec 2016
10. Bai, G., Hou, Y.-Y., Jiang, M., Gao, J.: Integrated systems biology and chemical biology approach to exploring mechanisms of traditional Chinese medicines. Chin. Herb. Med. **02** (2016). http://kns.cnki.net/kcms/detail/12.11410.R.20160408.1722.004.html. Accessed 08 Apr 2016

Study on Potency of Controlling on Crematogaster Rogenhoferi to Parasaissetia Nigra Nietner

Lihe Zhang, Bin Du, Baoli Qiu$^{(\boxtimes)}$, and Hui Wang

College of Agriculture, South China Agricultural University,
Guangzhou 510642, Guangdong, China
750446620@qq.com

Abstract. To assess the controlling potential of *Crematogaster rogenhoferi* against *Parasaissetia nigra* Nietner 1st instar larva, we determined the functional response and searching efficiency of *C.rogenhoferi* preying1st instar larva of P. nigra and analyzed its predatory behaviors. The results showed that predation functional response of *C.rogenhoferi* conformed to Holling's type II model, and the feeding quantity rose with the increase of density. However, as the 1st instar larva of *P.nigra* increased, the searching efficiency of *C.rogenhoferi* decreased. A mutual interference effect existed among different individuals, and the searching efficiency of *C.rogenhoferi* accosted with a Hassell-Varley model. In the predation, although the feeding and espionage proportion of *C.rogenhoferi* were not high, the proportions of casting and searching behavior were large. Observations of comprehensive feeding behaviors of *C.rogenhoferi* showed *C.rogenhoferi* had certain control function and control potential on the 1st instar larva of *P.nigra*.

Keywords: *Crematogaster rogenhoferi* · *Parasaissetia nigra* Nietner
Functional response · Searching efficiency

1 Introduction

Parasaissetia nigra Nietner is a major insect on rubber trees, which are mainly distributed in South China (Chen 2014). Severe outbreak will cause devastating disaster to the rubber industry, but it still happens in Hainan. With the help of its mutually-beneficial ants, the level of harm is constantly intensified in a seemingly long-term trend.

The use of natural enemies is an effective way to biologically control the spread of *P.nigra* (Zhang et al. 2010). *P.nigra* has many natural enemies, natural parasitic enemies and predators (Zhou et al. 2007). At present, the development and utilization of its natural enemies are mainly concentrated on the natural parasitic natural enemies, and but ignore their natural predatory enemies are rarely studied, Accosting to Wang Qi, *Crematogaster rogenhoferi* are predators of natural enemy of the *P.nigra* on rubber trees (Wang and Zhou 2012). Ants and plants have coevolved over long time (Speight et al. 1999), so ants and their host plants are often species-specific (Tobin 1995).

© Springer Nature Singapore Pte Ltd. 2018
K. Li et al. (Eds.): ISICA 2017, CCIS 874, pp. 280–289, 2018.
https://doi.org/10.1007/978-981-13-1651-7_25

For example, *Crematogaster* is specific for Macaranga plant (Holldobler and Wilson 1990; Fiala et al. 1999), pine trees (Tschinkel 2002) and Acacia (Palmer et al. 2000). In conclusion, *Crematogaster* can build nests in different plant species.

C.rogenhoferi has a strong control effect on the symbiotic dominant species of *P. nigra*, which effectively reduces the population of symbiotic ants, slows the spread of *P.nigra* and reduces the harm level (Zhang et al. 2015). *C.rogenhoferi* also has strong predation ability and plays a role in controlling pests, such as *Dendrolimus punctatus* Walker and coffee stem borers (Huang et al. 1984; Wei and Yu 1998).

At present, there is no report about the control effect of *C.rogenhoferi* on *P.nigra*. Thus, we studied the functional response, searching efficiency and predation behavior of *C.rogenhoferi* to *P.nigra*, and analyzed the control potential of *C.rogenhoferi* on *P. nigra*, aiming to scientifically evaluate the control effect on pests and effectively use natural enemies of *P.nigra*, thus providing a scientific basis.

2 Materials and Methods

2.1 Test Insects

C. rogenhoferi was collected from rubber seedlings in Dafeng Farm, Fushan Town, Chengmai County, Hainan. The ant nests were cut quickly with branches and shears, and put into black plastic bags against agitation, given the very strong attack of *C. rogenhoferi*. Finally, the bags were put together in a square storage box (sealed with 120 mesh head screen) to avoid escape and bite. Then the box was moved to the tropical plant pest control insect room of College of Environmental and Plant Protection, Hainan University, and placed in a plastic square box ($80 \times 40 \times 15$ cm^3). The bottom of the box was laid with the mature coconut bedding, 5 cm of tile and 1 cm of fine sand, and painted in the highest plastic box Vaseline. The box was set to the sink against escape. At room temperature 26 ± 1 °C and humidity 60% $\sim$ 70%, the ants were fed with diluted 20% honey water with *Tenebrio molitor* and artificial ant feeding. When the population was stable, the ants adapted to the tests.

P. nigra Nietner was collected from the more severely damaged rubber plantation in Fushan Town, Chengmai County, Hainan Province. Artificial branches were used to cut off the branches of *P. nigra* adults in the rubber plantation and then brought back to the tropical crop pest control and maintenance insect shelter Hainan University in Environmental and Plant Protection College tropical crop pest control and maintenance of insect shelter. The pumpkin was used as the host to feed and reproduce indoors (temperature 26 ± 1 °C, humidity 70 ± 5%). After the population was established and stabilized, the 1st instar larvae were selected for experiments.

2.2 Experimental Methods

2.2.1 Functional Response of *C. rogenhoferi* on *P. nigra* Nietner

The density of the 1st instar larva of *P.nigra* was set to be 10, 20, 30, 40, 50 and 60 heads, with a total of 6 gradients. Each density treatment was repeated 5 times. The conditions were laboratory temperature at 26 ± 1 °C, relative humidity 70%, and

L:D = 12:12. The 1st instar larva of *P. nigra* were selected in the breeding room because these individuals were too small and able to drive light. Therefore, the instars were first illuminated for 3 to 5 min, and the 1st instar larvae more focused, easy to pick. Following each density treatment described above, the 1st instar larvae were gently picked into 3 × 1.5 cm² glass vials, and the bottle was sealed with a 120-gauge gauze with access to *C. rogenhoferi* that had been starved for 24 h. After that, the predation amount of *C. rogenhoferi* under each density was observed under LED light.

2.2.2   Searching Efficiency of *C. rogenhoferi* on *P. nigra* Nietner

C. rogenhoferi was set with 5 gradient levels of density (2, 4, 6, 8 and 10 heads), and each level was processed 10 times. Other conditions included: test temperature at 26 ± 1 °C, relative humidity 70%, L:D = 12:12, and LED light irradiation 3 ~ 5 min. Under LED light, 50 first instar larvae of *P. nigra* were selected, with 120 mesh gauze bottle mouth for 24 h. Under the LED lamp, the number of pre-nosed 1st instar larvae under the treatment of the density were statistically counted.

2.2.3   Predation Behavior of *C. rogenhoferi*

Fifty 1st instars larvae of *P. nigra* were selected indoor, and placed in 3 × 1.5 cm² small glass tubes. The orifice was covered with 120 mesh gauze, and then access to followed by a hunger treatment for 24 h with *C. rogenhoferi*. Under the LED light, the animals were continuously observed for 4 h, and each prey behavior duration and frequency of occurrence were recosted separately using timers. Experiments were repeated five times. The percentage of feeding duration of *C. rogenhoferi* for different behaviors was calculated as follows: Percentage of duration (BDR) = duration of different behaviors/total observation time (4 h) × 100%.

2.3   Data Processing and Statistics

2.3.1   Functional Response

The functional response of *C. rogenhoferi* preying *P.nigra* 1st instars nymphs functional response equation, using the Holling IItype fitting as follows:

$$Na = \frac{aTtN}{1 + aThN} \tag{1}$$

In the formula, where N is the density of prey, Na is the prey for the corresponding density, a is the instantaneous attack rate, and Tt is the total time available for the predator (24 h in this study), Th are the average of processing times (e.g. the time consumed by the predator to prey on one prey) (Liu et al. 2011).

Sort Eq. (1) into:

$$\frac{1}{Na} = \frac{1}{aTt} * \frac{1}{N} + \frac{Th}{Tt} \tag{2}$$

Set

$$A = \frac{Th}{Tt}, \quad B = \frac{1}{aTt} \tag{3}$$

Transform Eq. (2) into a linear equation:

$$\frac{1}{Na} = B\frac{1}{N} + A \tag{4}$$

By least square method to find A and B, and then find the value of a and Th.

2.3.2 Searching Efficiency

The searching efficiency of *C. rogenhoferi* on the 1stinstar nymphs of P.nigra was fitted by Hassell's (1969) model (1969) as follows:

$$E = QP^{-m} \tag{5}$$

where E is searching efficiency, Q is looking for parameter, P is Predator (parasitic density), and m is the interference parameter.

Through logarithm on both sides, Eq. (5) is transformed into a linear form:

$$LnE = LnQ - m \ln p \tag{6}$$

Let $Y = Lna$, $LnQ = A$, $-m = B$, $LnP = x$. Use linear least squares method to find A, B, and then find the value of Q, m, and then we get the equation.

3 Results

3.1 Control Effect of *C. rogenhoferi* to *P. nigra* Nietner

3.1.1 Functional Response of *C. rogenhoferi* to *P. nigra* Nietner

As showed in Fig. 1, the predation response of *C. rogenhoferi* conformed to the Holling II model. The predation of *C. rogenhoferi* on the 1st instar larva of *P. nigra* increased with the rise of prey density. When prey density reached a certain level, the predation amount of prey decreased and the curve growth was accelerated.

Table 1. Number of preys consumed by *C.rogenhoferi* to *P.nigra* 1st instar larva in different prey densities

Density of C.rogenhoferi to P.nigra 1st instar larva (N)	Measured average predation (Na)	Theoretical predation (Na')
10	3	5.30
20	10.6	10.00
30	16.4	14.71
40	21.2	19.42
50	24.6	24.13
60	26.6	28.84

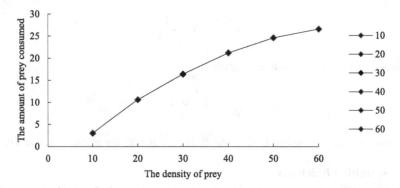

Fig. 1. Function responses of *C.rogenhoferi* to *P.nigra* 1st instar larva.

Accosting to the functional response equation, the predation functional response and related parameters of *C. rogenhoferi* to the 1st instar larva of *P. nigra* were calculated (Table 2).

The correlation coefficient (R = 0.9789**) of *C. rogenhoferi* was significantly correlated with the prey density (Table 2). The chi-square test $x^2 = 1.5698$ $< x^2_{0.05} = 11.07$. The theoretical value agrees well with and is not significantly different form the measured value, indicating the Holling II model can better reflect the change of predation of *C. rogenhoferi* on the 1st instar larva of *P. nigra*.

Table 2. Function response model estimation of *C.rogenhoferi* prey on *P.nigra* 1st instar larva (HollingII).

Functional response linear equation	Correlation coefficient (R)	Functional reaction disk equation	Chi-square (x^2)	Instantaneous attack rate (a)	Processing time (Th)	Up limit predation
$\frac{1}{Na} = \frac{0.4709}{N} + 0.5867$	0.9789**	$Na = \frac{2.1236N}{1+1.2462N}$	1.5698	2.1241	0.5867	1.7

3.1.2 Searching Efficiency of *C. rogenhoferi* to *P. nigra* Nietner

The searching efficiency is a behavioral effect of predator (or parasite) attack on prey (or host) during predation (or parasitism). The prey (or host) density, predator (or parasite) density, spatial distribution, and climate factors are closely related. Therefore, when studying prey-predator or parasite-host systems, researchers must consider factors that improve the searching efficiency.

The relationship of searching efficiency E with prey density and natural enemy density can be described by the following formula:

$$E = \frac{Na}{N * P} \tag{7}$$

where E is searching efficiency, Na is prey the number of prey (or host), N is the density of prey (or host), and P is predator (or parasite) density.

Based on Eq. (7), the searching efficiency of *C. rogenhoferi* on the 1st instar larva of *P. nigra* was calculated (Table 3).

Table 3. Searching efficiency of *C.rogenhoferi* prey to 1st instar larva of *P.nigra*

Density of *C.rogenhoferi*	2	4	6	8	10
Number of predation (head)	23.5	29.0	36.4	36.6	38.8
Searching efficiency	0.2350	0.1450	0.1213	0.0915	0.0776

In a certain space, the number and mutual interference of predators both gradually increase, resulting in a decrease in the searching efficiency of each predator. The relationship between searching efficiency and predators can be fitted using the Hassell-Varley model. The data in Table 2 were used to fit the model:

$$LnE = 0.2446 - 0.01842LnP$$

Results show Q = 1.277 and m = 0.01842. The mathematical model underlying the searching efficiency of *C. rogenhoferi* over 1st instar larva of *P. nigra* was obtained as follows:

$$E = 1.2771P^{-0.01842}$$

Table 4. Theoretical and measured searching efficiency of *C.rogenhoferi*

Density of *C.rogenhoferi*	Measured Searching efficiency	Theoretical Searching efficiency
2	0.235	0.208
4	0.145	0.171
6	0.121	0.134
8	0.092	0.097
10	0.078	0.060

The theoretical and measured values of searching efficiency and predator density in Table 4 are plotted in Fig. 2.

As showed in Fig. 2, the theoretical and measured values are very close. The chi-square test shows $x^2 = 0.0139 < x^2 0.05 = 9.488$, and there is no significant difference between the theoretical and measured values, indicating the Hassell-Varley model can better reflect the searching efficiency of *C. rogenhoferi* on 1st instar larva of *P. nigra*. The searching efficiency of *C. rogenhoferi* decreased with the increase of its density, indicating there is mutual interference between individuals in the population during predation.

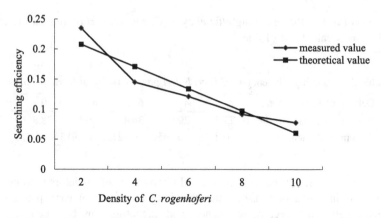

Fig. 2. Relationship between density of *C. rogenhoferi* (P) and searching efficiency (E)

3.2 Observation of Predatory Behavior of *C. rogenhoferi* to *P. nigra* Nietner

The pre-experimental observation about the behavior of *C. rogenhoferi* to 1st instar larva of *P. nigra* was summarized in Table 5. It was found the predation behavior of *C. rogenhoferi* always alternated between resting and locomotion. Its feeding behavior was often accompanied by frequent searching behavior. During the searching, *C. rogenhoferi* usually encountered the first instar nymphs, killing the prey with its upper jaws in very short time. The feeding process is fast and instant. The predation behavior of *C. rogenhoferi* can be transformed (1) from walking to searching; (2) from searching to spying; (3) from spying to feeding; (4) from feeding to grooming; (5) from grooming to walking; (6) from walking to resting.

Table 5. Description of behavioral events of *C.rogenhoferi*

Behavioral events	Watch the contents of the statement
Walking	Predators crawl in glass bottles
Resting	Predator stand motionless
Grooming	Predator rapidly move with fore and hind legs across body surface and antenna
Orienting	Predators crawl in a glass jar without particular direction
Poking	Predators penetrate sting needles into or bite the target, but do not feed
Feeding	Predators bite quickly and eat

In the experimental observation, the numbers of prey behavior and predation behavior of *C. rogenhoferi* were counted. The percentages of duration of different predation behaviors by *C. rogenhoferi* on 1st instar larva of *P. nigra* are shown in Fig. 3.

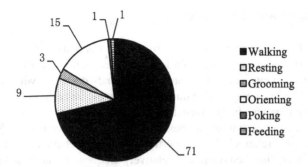

Fig. 3. Ratio of each behavior event to the total duration of predation.

During the whole predation on the first instar nymphs of *P. nigra* in *C. rogenhoferi*, their feeding behavior was similar to the sparrow behavior. After each prey, grooming behavior occurred. *C. rogenhoferi* usually combed the head antennae after searching or preying. This combing behavior kept the black and brown belly ants clean, improving the predation efficiency. Organisms such as maxillary and labial palpi in *C. rogenhoferi* have tasting and smelling receptors that are critical in finding and testing predators in their preying behavior. During the preying behavior of *C. rogenhoferi*, the total walking time is long, but the resting behavior is always accompanied by walking. Rest breaks are more frequent and the duration of each break is shorter. During the *P. nigra* process, the proportion of search duration is relatively high (BDR = 15%), which is related to the smaller and haster search of 1st instar larva of *P. nigra*.

During the experiment, *C. rogenhoferi* was one of the death causes of some nymphs in addition to the death of the 1st instar larva.

Table 6. Frequency of each behavior in predation course of *C.rogenhoferi* to *P.nigra* 1st instar larva

Prey bug state	Frequency of each behavior					
	Walking	Resting	Grooming	Orienting	Poking	Feeding
P. nigra 1st instar larva	59.3 ± 1.8 aA	10.5 ± 1.6 cC	39.5 ± 1.6 bB	33.8 ± 0.8 bB	12.5 ± 0.7 cC	11.3 ± 1.3 cC

The frequencies of prey behavioral events on *C. rogenhoferi* for the 1st instar larva of *P. nigra* vary (Table 6). The frequency of walking is the highest (59.3%), followed by grooming (39.5%) and orienting (33.8%). The frequencies of poking and feeding are similar and low, which accost with the experimental results. Reasons are that *C. rogenhoferi* is an omnivorous insect and *P.nigra* is difficult to search, leading to the decrease in its orienting and feeding frequencies. However, the frequencies of searching and combing all behavioral incidents increase to 39.5% and 33.8% respectively. Thus, *C.rogenhoferi* with omnivorous behavior has certain predatory ability to *P.nigra*.

4 Conclusions and Discussion

The predation functional response of *C. rogenhoferi* accosts with Holling IImodel. When the prey density reaches a certain level, its predation amount begins to decline, and the curve growth rate eases; the searching efficiency decreases with the increase of prey density. The reason may be that during the prey, the individuals of *C. rogenhoferi* have mutual interference. In the predation, the predation amount is affected by the climate and space in addition to the pest density and predator density (Chen 2005). *C. rogenhoferi* has a relatively low proportion of feeding behaviors to *P.nigra*, but its grooming and searching behaviors are relatively frequent throughout the predation, indicating the first-instar nymph density of *C.rogenhoferi* in *P.nigra* is 50 when there is a certain predation potential. The main reason for low feed intake is that the first instar nymphs of *P.nigra* are small and hast to search. Field investigation shows *C. rogenhoferi* often nests *P. nigra* into them or 4 to 4 together with scale insects, usually on the branches of very few first instar nymphs. An average rubbery *P. nigra* spawns 1345 eggs per female (Wu 2008). The first instar nymph has high density and small searching space, which make searching easy. In conclusion, *C. rogenhoferi* has important utilization value and potential as a natural predatory enemy of *P. nigra*. Analysis and research on the predation ability of predatory pests using the predatory function and finding effect become a necessary procedure to test and evaluate the control ability of natural enemies (Liu et al. 2011). Predatory functions of *Coccinella septempunctata*, *Cyrtorrhinus livdipennis* Reuter, *Scolothrips takahashii* Priesner and other natural pest enemies to control the development and utilization of specific pests have been reported (Wu et al. 2012; Huang and Huang 2010; Li et al. 2006). Wei (2009) studied predatory function of *Pseudaulacaspis pentagona* predators of two predominant species-geminus, *Cybocephalus nipponicus* Endrody Yonnga. Yan (2012) studied the functional response of the predatory natural enemy *Chrysopa formosa* Brauer of *Dysmicoccu neobrevipes* Beaidesley. In the future, *C.rogenhoferi* can be released at the peak of *P.nigra* to effectively reduce the population of mutual-beneficial ants, and increase the predation efficiency of *C.rogenhoferi* to the 1st instar larva, thus greatly reducing the damage of *P.nigra*.

References

Chen, P.: Occurrence regularities and control strategies of *Parasaissetia nigra* Nietner in Yunnan. Pract. Tech. Rural Areas **02**, 38–39 (2014)

Zhang, F., Niu, L., Xu, Y., Han, D., Zhang, J., Fu, Y.: Effects of *Metaphyse parasaissetiae* Zhang and Huang on *Parasaissetia nigra* Nietner. Chin. J. Appl. Ecol. **08**, 2166–2170 (2010)

Zhou, M., Li, G.-H., A, H.-C., Li, J.-Z., Duan, B.: Occurrence and control strategies of *Parasaissetia nigra* Nietner. Plant Protection China **03**, 23–24 (2007)

Wang, Q., Zhou, X.: Preliminary report on artificial breeding and release of Cematogaster rogenhoferi. Guangdong Agric. Sci. **10**, 98–100 (2012)

Speight, M.R., Hunter, M.D., Wall, A.D.: Ecology of Cocepts and Application, p. 350. Black Well Science, Oxford (1999)

Tobin, J.E.: Ecology and diversity of tropical forest canopy ants. In: Lowm, M., Nadkarni, N. (eds.) Forest Canopies, pp. 129–147. Academic Press, Sandiego (1995)

Hölldobler, B., Wilson, E.O.: The Ants, p. 732. Springer, Berlin (1990)

Fiala, B., Jakob, A., Maschwitz, U., Linsenmair, K.E.: Diversity, evolutionary specialization and geographic distribution of a mutualization ant-plant complex: Macaranga and Crematogaster South East Asia. Bio. J. Linn. Soc. **66**, 305–331 (1999)

Tschinkel, W.R.: The natural history of the arboreal ant, *Crematogaster ashmeadi*. J. Insect Sci. **2**, 1–15 (2002)

Palmer, T.M., Young, T.P., Stanton, M.L., Wenk, E.: Short-term dynamics of an acacia ant community in Laikipia, Kenya. Oecologia **123**, 425–435 (2000)

Zhang, L.-H., Zhou, X., Chen, T.-L., Zhang, B., Wang, Y.-B.: Category and competition of two kinds of ants on rubber tree. Chin. J. Ecol. **12**, 3424–3429 (2015)

Huang, Y., Wei, J., Xi, F., et al.: Preliminary investigation of *Cematogaster rogenhoferi*. Guangxi For. Sci. Technol. **5**(1), 19–22 (1984)

Wei, J.-N., Yu, X.-W.: Diversity and control evaluation of natural enemies in *coffee stem* borers in Simao region. Biodiversity **4**, 8–12 (1998)

Chen, Y.: *Neochrysoclaris* okazakii Kamijo on the function of *Liriomyza sativae* Blanchast research. J. Fujian Agric. Univ. **20**(2), 80–83 (2005)

Wu, Z., Li, G., Hou, J., et al.: Rubber helmet *Parasaissetia nigra* Nietner scale biological characteristics. J. Yunnan Agric. Univ. **23**(5), 701–704 (2008)

Liu, S., Wang, S., Liu, B.-M., Zhou, C.-Q., Zhang, F.: Prediction and Predation Behavior of *Bemisia tabaci* (Gennadius) Larvae in *Chrysopa pallens* (Rambur). Sci. Agric. Sin. **06**, 1136–1145 (2011)

Wu, D., Wang, S.S., Hu, G., Du, J., He, C.: Predatory functional response of *Coccinella septempunctata to Acyrthosiphon pisum*. Prolls Lawn **01**, 12–17 (2012)

Huang, L.-M., Huang, S.: Functional and numerical responses of *Cyrtorrhinus livdipennis* Reute to preying on eggs of *Nilaparvata lugens* (Stal). Acta Ecol. Sin. **15**, 4187–4195 (2010)

Li, D., Tian, J., Shen, Z.: Functional responses of *Tetranychus viennenses* to *Scolothrips takahashii* Priesner. Acta Ecol. Sin. **05**, 1414–1421 (2006)

Wei, Z.: Study on the predation efficiency of predatory species of *Pseudaulacaspis pentagona* (Targioni Tozzetti) and its dominant species. Xinjiang Agricultural University (2009)

Yan, Z.: Biological characteristics of *Chrysopa formosa* Brauer and its predation efficiency on *Dysmicoccu neobrevipes* Beaidesley. Hainan University (2012)

Research on the Management
and Optimization of Warehouse Location
in e-Commerce Enterprises

Huwei Liu$^{(\boxtimes)}$ and Zeping Li

School of Information, Beijing Wuzi University, Beijing, China
liuhuwei@outlook.com

Abstract. In the e-commerce environment, the quantity of order in e-commerce enterprises is increasing, the number of single order is small, and there are many kinds of them, which puts forward higher requirements for picking operation. The distribution of the cargo location in the e-commerce warehouse is the key factor affecting the selection efficiency, which directly affects the response speed of the customer order. The most effective way to improve the efficiency of storage operation is to optimize the storage location of goods.

Keywords: Cargo location management · Storage strategy
Cargo location optimization

In recent years, e-commerce enterprises have been developing in a wider and larger way, and the logistics center has a wide variety of goods. Meanwhile, the volume of enterprise orders has exploded. Problems in e-commerce enterprises are becoming more and more prominent, especially the problem of unreasonable allocation of location. Unreasonable distribution of location will not only extend the time to pick up the goods but also increase the error rate of picking the goods, make order response slow, cause the goods cannot be timely delivery, leading to the result that customer satisfaction decreases and directly affect the operation of the enterprise. Therefore, it is an urgent task for e-commerce enterprises to solve the problem of irrational distribution in logistics center, improve the efficiency of selection and enhance the service capability of logistics center. The key to improve the profit and reduce the cost is to optimize the storage location in the warehouse.

1 The Meaning and Steps of Cargo Management

1.1 The Meaning of Location Management

The allocation of storage, in simple terms, is one of the reasonable planning and work of the warehouse. Things like how to place goods and where to place in the warehouse all belong to the allocation of storage. The literal meaning of storage distribution is to make reasonable arrangements of storage for goods that enter the storage center so that goods can be safely stored, easy to choose and to meet the requirements of efficient

© Springer Nature Singapore Pte Ltd. 2018
K. Li et al. (Eds.): ISICA 2017, CCIS 874, pp. 290–302, 2018.
https://doi.org/10.1007/978-981-13-1651-7_26

operation. The methods of distribution of storage can affect the smooth operation of the warehouse in general.

E-commerce enterprise location management refers to the storage of goods within the warehouse for the planning, distribution, use, adjustment and so on. Each store stored in the warehouse has their own characteristics in its origin, physical and chemical properties, batch number, whereabouts, shelf life and other aspects, each warehouse needs to determine a reasonable position for each good, and it should not only meet the needs of custody, more to facilitate the operation and management of warehouse. According to the way of use of cargo location, it is divided into free and fixed position.

Free Position. Free position is also called "random location". Each location can be stored in any kind of materials (except the bad). As long as a certain cargo space is free, storage of goods can be deposited.

Fixed Position. The fixed position is also known as "fixed material position". That is, a certain cargo space can only store the goods of a certain specification, but not other goods. Bulk goods are stacking in place of fixed position. In order to facilitate the management of the location of goods, numbering method can be used, e-commerce warehouses generally use the computer for cargo management. It can assign the cargo space according to the conditional assignment and can carry out various kinds of queries, and understand the situation of the position of the goods at any time.

1.2   The Steps of the Cargo Management

See Fig. 1.

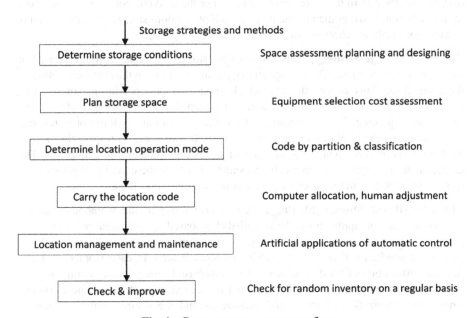

Fig. 1. Cargo storage management flow

1.3   Cargo Storage Strategy Commonly Used

In a competitive marketplace where services and products are clearly homogeneous, e-commerce enterprises who want to get a better development must be better and faster meet the requirements of end users, and make quick response to consumer demand, which brings higher requirements in management level and operation efficiency of enterprises. Scientific and reasonable storage strategy can not only save a lot of warehouse operation time, but also reduce directly in the process of equipment and personnel movement distance, and be able to more effective use of existing space location.

At present, in the e-commerce enterprise warehouse, the commonly used cargo allocation strategy is divided into six categories in related literature: Random storage, Closest open location storage, Dedicated storage, Class-based storage, and Full turn-over storage. The storage strategies that are commonly used in enterprises are Random storage, Dedicated storage, Shared storage, and Class-based storage [1–3]. Among:

Random Storage Strategy. This strategy refers to the position of the goods in the warehouse randomly assigned and the randomly assigned application is more. The advantage of its distribution is the efficiency of goods allocation is high, the disadvantage is that for higher frequent outbound goods may be relatively far from the warehouse entrance. Therefore, the random storage strategy can ensure the efficient utilization of storage space, but cannot guarantee the minimization of the picking cost. On the other hand, random storage of goods has no laws at all, which brings great inconvenience to the picking personnel. For warehouses that use random storage strategies, it may lead to the result that the higher frequent goods being removed from the entrances and exits, or where the entrances and exits are placed with large quantities of cargo and the rear of the warehouse is empty. For the convenience of operation, most of the early warehousing management system adopts random storage strategy, which is suitable for small warehouse space and less variety of goods.

Dedicated Storage Strategy. This is one of the storage strategies commonly used in the traditional warehouse. The storage strategy is mainly to keep the storage position of the goods fixed. This storage strategy, which requires storage picking personnel to be familiar with the types of goods and storage location, has the advantages of the minimum cost of picking. The disadvantage of it is that cannot make full use of warehouse space, and the flexibility of storage space is small. It also needs to re-plan the storage location of the goods when the production of goods types changes in enterprises. The dedicated storage strategy is applicable to situations where the quantity of goods need to store is large and the number of goods in the warehouse is large.

The Class-Based Storage Strategy. This is also more in the warehouse storage location assignment application. The so-called class-based storage strategy is to sort the goods, and then to designate fixed storage area according to the results of the classification of goods. Each item corresponds to a storage area. In each storage area, the position distribution of goods is random. The class-based storage strategy combines the advantages of dedicated storage strategy and random storage strategy, and also considers the optimization of space and picking to find the optimal balance point of optimization between space and picking [4, 5]. At present, there are three main types of classification storage standards for goods: the turnover rate of unit time, the maximum inventory and the order volume index. Each classification criteria will be applied to the corresponding warehouse layout. Among them, the unit time turnover mainly takes into account the outbound rate of the goods in the warehouse over a period of time, and the goods will be classified according to the level of turnover; the maximum inventory classification is in accordance with the size of storage space that occupied by the goods in the warehouse to sort the goods (Table 1).

Table 1. Storage strategy

Storage strategy	Meaning	Advantages	Disadvantages
Dedicated storage strategy	The goods are stored in a fixed location, which cannot be replaced optionally, and the goods must meet the size and weight requirements of the location	It reduces the distance and time of loading and unloading goods, saves the cost, keeps the goods in a fixed storage area, which is easy to keep the location of goods. It can help locate the locations of goods quickly, and the efficiency of picking and checking is high	The number of storage location depends on the kinds of goods and the maximum amount of inventory, the use efficiency of the warehouse storage space is low
The class-based storage strategy	Classify the goods need to be put in storage according to the kinds, quantity, and attributes of them, and each kind of goods is stored in a relatively fixed storage space	Flexible management, high loading and unloading efficiency	The requirements of the number of a storage location in the warehouse are relatively high, and the utilization rate of the overall storage space is low
Random storage strategy	When the goods are in the receiving period, according to the frequency of the goods in the past, the WMS will allocate the corresponding storage space randomly	The utilization rate of warehouse space is high	It is not convenient to manage and find goods in the inventory
Shared storage strategy	For goods that need to be warehoused, the cargo space can be randomly selected, and the cargo space can be shared between different types of goods	Effectively save storage space and delivery time	It is inconvenient to manage the goods

2 Position Optimization Management

The main purpose of the position optimization is to improve the efficiency of warehousing operation, shorten the distance and time of manual and equipment movement, and save manpower and material resources. The customer order must be the center of the distribution optimization of the goods, and the important factors such as the demand correlation of commodity and the frequency of the output must be considered. In addition, commodity weight, volume, shelf stability, capacity and other general factors should also be considered. The effect of the position optimization can be evaluated from the selection efficiency of the goods, the amount of the inventory, the distance and the time of movement.

2.1 Principle of Space Allocation for Storage

Domestic and foreign scholars have studied the allocation of goods from different angles. According to the layout of the warehouse, a mathematical optimization model for the allocation of goods has been set up by them [6, 7]. In the actual storage environment, the eight principles of the allocation of storage should also be considered in addition to the several commonly used storage strategies. They are as follows:

The Principle of Goods Turnover Rate. Goods for the distribution of storage can be classified according to the turnover rate of them, and complete the storage allocation of goods in accordance with the level of cargo turnover rate;

The Principle of Cargo Identity. The same kinds of goods must be stored in the same storage area in the warehouse and cannot be stored randomly in the warehouse.

Cargo Correlation Principle. Several interrelated goods need to be concentrated in the same storage area. Such as screws and nuts, two interrelated goods appear in the order, need centralized distribution to the same storage area;

Cargo Compatibility Principle. For chemicals, two goods should be stored separately if they can cause chemical reactions. The goods with low compatibility should be kept away from each other;

Cargo Complementary Principle. The so-called cargo complementarity refers to the two goods that can be replaced between each other. When one of the goods is lacking, another kind of goods can be used to replace it. Highly complementary goods should be stored in the adjacent space position;

Cargo Weight Principle. The heavier goods should be stored at the bottom of the cargo space;

Cargo Size Principle. When the size of the goods that need to be stored is not the same, the goods can be stored in different storage areas according to the size of them;

The Principle of Cargo Characteristics. For the special goods, storage areas can be divided into special areas in order to store the special goods.

2.2   The Basic Steps of Position Optimization

The basic steps of position optimization are as follows (Fig. 2):

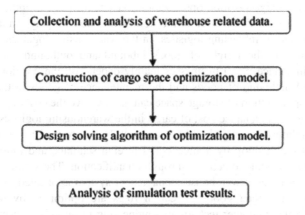

Fig. 2. Basic steps of position optimization

Collection and Analysis of Warehouse Related Data. Summarize the basic information and data of the basic situation, the location parameter and the basic characteristics of the goods, the turnover rate and so on, which are the important reference for the scientific and reasonable cargo allocation.

Construction of Cargo Space Optimization Model. Establish a reasonable mathematical model after the mathematical abstraction according to the type and size of the shelf and the size, weight, quantity, and others objects of the goods in e-commerce enterprise warehouse. The model is constructed to meet the constraints and efficiency of the two factors.

Design Solving Algorithm of Optimization Model. Design and solve algorithm, which is suitable for the actual needs, according to the constructed mathematical model, the optimal situation of the cargo needs to be optimized. The key to this step is how to determine the optimal algorithm, which also is the core of the entire distribution of the allocation process. Only select the efficient algorithm, a satisfactory solution can be ensured.

Analysis of Simulation Test Results. Use the MATLAB software to carry out computer simulation experiment of the mathematical model, verify the mathematical model and algorithm according to the simulation results, and analyze the effects of position after optimization comparatively, which is one of the most intuitionistic and effective methods.

2.3   Analysis of Storage Allocation Optimization Management

The optimal management of cargo location is different from the storage strategy. The storage strategy mainly chooses some kind of way to carry on the cargo allocation to the goods, and the cargo position optimization management is the optimization of the space on the basis of the distribution of the storage for goods. It plays an important role in supply chain and warehousing logistics. In the warehousing logistics, warehousing costs mainly include the freight of goods inbound and outbound and the cost of warehouse maintenance. Such businesses mainly depend on the position of goods as the inbound and outbound of goods and the maintenance of goods in the warehouse. Therefore, the optimization of storage space can greatly save the cost of goods inbound and outbound and maintenance cost of cargo in the warehousing logistics, thus reduce the cost of warehouse logistics highly as a result There are many theories about the optimization of the position by domestic and foreign scholars, and most of them are theoretical and impossible to achieve through quantification. The slotting optimization management, that modern warehouse management system applicated, needs quantifiable parameters, and then goods allocation rules are optimized by the numerical analysis through the computer technology engineering to optimize the parameters.

Chan and Chan [8] proposed according to the human engineering to deal with storage rules such as the goods position and the complementarity of goods, some of the rules can be quantified, but no quantitative slotting optimization management concept; In Yu and De Koster [9], the article puts forward the principle of the gold region and the balanced workload, but there is no specific quantitative method and quantitative results; Gagliardi et al. [10] proposed cargo location optimization management, but there is no precise quantification of cargo location optimization management. Through the in-depth study and analysis of the storage environment, the optimization of cargo location in this paper mainly focuses on the following aspects (Table 2):

Table 2. ABC classified storage strategy.

Class	Turnover rate	Storage strategy
A	High	Put on the storage near the entrance of the warehouse
B	Middle	Place in the middle area
C	Low	Keep in the locations that far from the entrance of warehouse

ABC Classified Storage Strategy. The goods are divided into three categories: A, B, and C:

Compatibility of Goods. According to the compatibility of goods stored goods, less compatible goods cannot be stored together.

Frequency of Goods. Goods, whose frequency of outbound and warehousing is high, should be stored in different roadways as much as possible, so as to avoid reducing the picking efficiency caused by the blockage of roadways.

Complementarity of Goods. Highly complementary goods should be assigned to adjacent cargo locations to facilitate temporary replacement when the goods are in short supply.

The management of optimization of the cargo location is to allocate and use the cargo space reasonably. It is to consider how to allocate the cargo space to ensure the stability of the goods shelf, and to ensure the efficiency of the warehouse's entry and outbound. The distribution of cargo location generally follows that "gently under heavy" and goods that often used should be placed close to the entrance and exit of the warehouse. Different products also have different rules for the allocation of position [11, 12]. The goods that have expiration date should be changed the freight of the goods frequently, and the goods with short time of the shelf life should be shifted to the cargo position near the entrance of cargo offset, and that with long shelf life should be shifted to the cargo position away from the entrance [13, 14]. To optimize the cargo position on the shelves regularly can improve the goods delivery efficiency and save the cost of production.

3   Study on the Optimization of Location Assignment

3.1   Location Assignment Optimization Problem

Set a row of shelves with a p column q layer, and the nearest column of the distance from the roadway is listed as column 1, and the lowest level is the 1st layer, and the position of the item in the i column j layer is denoted as $(i,j)(i = 1, \cdots, p; j = 1, \cdots, q)$. According to the gravity coordinate formula of the object, we know that the center of gravity of the object can be divided into several small individuals to find the center of gravity, and we can use this method to find the center of gravity of the shelf:

$$X_c = \frac{\sum_{i=1}^{p} \sum_{j=1}^{q} G_{ij} * i}{\sum_{i=1}^{p} \sum_{j=1}^{q} G_{ij}} \qquad G_{ij} \le G_{max}$$

$$Y_c = \frac{\sum_{i=1}^{p} \sum_{j=1}^{q} G_{ij} * j}{\sum_{i=1}^{p} \sum_{j=1}^{q} G_{ij}} \qquad G_{ij} \le G_{max}$$

G_{max} is the maximum weight each shelf shelves can bear, G_{ij} is cargo weight on any one cargo, X_c and Y_c are the overall coordinates of the shelf.

To meet the needs of shelf stability, the objective function is to minimize the center of gravity of the Y-axis of the shelf and make the X-axis coordinates of the shelf at the center of the shelf.

For goods of the following weight:

$$\begin{bmatrix} 16 & 17 & 18 & 19 & 15 \\ 6 & 7 & 8 & 9 & 9 \\ 1 & 2 & 3 & 4 & 5 \\ 8 & 12 & 13 & 10 & 6 \end{bmatrix}$$

In order to improve the storage efficiency of the warehouse, the frequently used goods with high frequency of storage and delivery are usually placed near the starting point. Suppose a parcel number is R (R = 1, 2, ⋯, q), there are P kinds of goods, one of the goods COI value is I_i (COI is the cube index, COI that Heskett given is calculated as $I_i = C_i/f_i$, where C_i is the inventory capacity required for the total amount of storage for a given item and f_i is the frequency of a given item's storage and exit frequency). The smaller the inventory capacity required for the total quantity of goods stored and the higher the frequency of entry and exit, the smaller the COI value. According to this principle, the small COI value goods are stored on the shelves near the entry and exit, and the following target conditions are obtained:

$$minCOI = \sum_{i=1}^{m} I_i = I$$

In order to reduce the access time of goods, according to the principle of goods storage, there should be less time for storage and delivery for goods with higher frequency of storage and transportation, and for goods with relatively low frequency of storage, they can have relatively long time for entry and exit. In this way, the entire storage and retrieval operations can be guaranteed to have the shortest total time. Therefore, the objective function is the requirement for the efficiency of access. The frequently accessed items should be put on the locations that can be quickly taken, and the following formula is established.

$$minT = \sum_{i=1}^{p} \sum_{j=1}^{q} t_{ij} = T$$

t_{ij} is the time spent transporting items on the jth floor and the ith column to the alleyway.

$$t_{ij} = \max \left(\frac{L * i}{V_x}, \frac{H * (j - 1)}{V_y} \right) (s)$$

In order to facilitate the processing, the starting and braking time of the stacker is not considered.

V_x and V_y are the horizontal and vertical speed of the stacker (m/s), l and h are the length and height of the cargo space (m).

Because the above two objective conditions are aimed at improving the access efficiency of goods and reducing the access time of goods, they can be merged into one objective function:

$$minE = I * T$$

Obviously, the location-allocation should consider both the shelf stability and access efficiency, which is a combination of multi-objective optimization problem. For multi-objective optimization problems, the objectives are conflicting in many cases.

Generally, there is no unique global optimal solution. Instead, there exists a set of optimal solutions. The elements in the optimal solution set are not comparable to all objectives. The solution to achieving the optimal performance of a target may well mean that the performance of other objectives is poor. There is not much practical significance in pursuit of one of the objective optimization. It is of great practical significance to seek a satisfactory solution (which may not be optimal for a certain target) of all the objective functions. After comparing various algorithms, we choose the clonal selection algorithm to solve the multi-objective optimization problem.

The concept of Pareto optimal solution is widely used in multi-objective optimization problems. For $\min f(X) = [f_1(X), \cdots, f_n(X)]$, set its domain as a, $X^* \in a$, if there is no $X \in a$ makes $f_i(X) \leq f_i(X^*)(i = 1, \cdots, n)$, says X^* is Pareto optimal solutions of the problem (valid solution or non-inferior solution).

3.2 Simulation Verification

Because the Pareto multi-objective optimization problem does not have a unique solution, it has a set of solutions, which can satisfy various constraints and obtain multiple optimal solutions.

The following matrix stores the cargo matrix for the flat shelf, where each element represents the weight of the goods stored in each place.

$$\begin{bmatrix} 16 & 17 & 18 & 19 & 15 \\ 6 & 7 & 8 & 9 & 9 \\ 1 & 2 & 3 & 4 & 5 \\ 8 & 12 & 13 & 10 & 6 \end{bmatrix}$$

The center of gravity for the initial placement of the goods is as follows: $X_c = 3.06, Y_c = 2.85$.

The center of gravity optimized by the clonal selection algorithm is: $X_c = 3.01, Y_c = 2.00$.

$$\begin{bmatrix} 5 & 2 & 3 & 4 & 1 \\ 8 & 8 & 7 & 6 & 6 \\ 19 & 9 & 10 & 13 & 12 \\ 19 & 16 & 17 & 18 & 15 \end{bmatrix}$$

The matrix below shows the frequency of goods in and out of each initial position.

$$\begin{bmatrix} 0.112 & 0.165 & 0.185 & 0.269 & 0.328 \\ 0.835 & 0.456 & 0.265 & 0.498 & 0.921 \\ 0.756 & 0.743 & 0.621 & 0.321 & 0.429 \\ 0.557 & 0.287 & 0.198 & 0.676 & 0.697 \end{bmatrix}$$

It is assumed that the type of cargo loaded at each cargo space varies. The COI value of the goods is as follows:

$$\begin{bmatrix} 8.929 & 6.061 & 5.405 & 3.717 & 3.049 \\ 1.198 & 2.193 & 3.774 & 2.008 & 1.086 \\ 1.323 & 1.346 & 1.610 & 3.115 & 2.331 \\ 1.795 & 3.484 & 5.051 & 1.479 & 1.435 \end{bmatrix}$$

Optimal results of cargo optimization based on COI value by clonal selection algorithm are as follows:

$$\begin{bmatrix} 8.929 & 6.061 & 5.405 & 3.717 & 3.115 \\ 3.774 & 2.193 & 1.795 & 2.008 & 1.346 \\ 3.484 & 3.049 & 1.610 & 1.479 & 1.323 \\ 5.051 & 2.331 & 1.435 & 1.198 & 1.086 \end{bmatrix}$$

It is assumed that the unit cargo space has a length of 1 m and a width of 1 m respectively, and ignores the acceleration movement after the start of the stacker and the deceleration movement before the stop, and the horizontal uniform speed and the vertical speed of the stacker are each 1 m/s. The starting and ending points are in the lower right corner. The matrix below is the time it takes for the stacker to reach each location. The unit is second.

$$\begin{bmatrix} 5 & 4 & 4 & 4 & 4 \\ 5 & 4 & 3 & 3 & 3 \\ 5 & 4 & 3 & 2 & 2 \\ 5 & 4 & 3 & 2 & 1 \end{bmatrix}$$

Since the optimal solution of the second objective function is the product of the COI value of the goods and the time of goods arrival and departure, the optimal result of the second objective function can be obtained according to the above solution. The following matrix shows:

$$\begin{bmatrix} 44.645 & 24.244 & 12.46 & 14.868 & 12.46 \\ 18.87 & 8.772 & 5.385 & 6.024 & 4.038 \\ 17.42 & 12.196 & 4.83 & 2.985 & 2.646 \\ 25.255 & 9.324 & 4.305 & 2.396 & 1.086 \end{bmatrix}$$

It is concluded from the optimal results of the above two objective functions that the shelf is the most stable when the center of gravity of the longitudinal coordinates of the cargoes is the smallest and the center of gravity of the abscissa is close to the middle. But the way to store the goods at this time cannot be stored in the way that the small COI value goods are placed closer to the entrance of the warehouse (the optimal result of the second objective functions). Therefore, the Pareto multi-objective optimization problem can only get the set of the optimal solution, and it is difficult to obtain an optimal solution that satisfies all the objective functions.

4   Conclusion and Prospect

For a long time, the general logistics warehousing center has accumulated a lot of experience and lessons in the storage and distribution. Some research shows that nearly 60% of the modern warehouse operations are mainly concentrated on some non-primary operations, such as loading and unloading, handling and workers walking. Electrical business logistics is not exceptional also, that is to say, the efficiency of warehousing center and picking operations needs to be improved urgently. Therefore, it plays an import role to make reasonable and effective storage and distribution in the entire storage center's operations.

Location assignment of warehouse management system occupies an important position in the entire warehouse operations. The results of storage allocation not only affect the operational efficiency of warehouse and the cost of the entire warehouse but also affects the control and management of products in the warehouse. Reasonable location allocation and storage strategy not only facilitate the warehouse operators to manage inventory but also improve the efficiency of warehouse operations and save enterprises' costs.

In the actual work, e-commerce enterprises can choose the appropriate cargo management method according to the characteristics of their own business. At present, most e-commerce enterprises have realized the importance of location meticulous management to the overall operation and began to use storage strategy to improve cargo location assignment in accordance with the principle of storage. But in the practice, most e-commerce enterprises make adjustments to the location according to the classification of goods or on the basis of outbound frequency. They do not combine with the characteristics of e-commerce enterprises such as a large number of orders, a few goods on single order and high frequency of picking, and not make more refined optimization to the layout of cargo position with order demand as the center. Fine cargo management will be helpful for enterprises to optimize the location, which can shorten the picking time, improve the efficiency of selection, and thus improve customer satisfaction.

Acknowledgement. The study is supported by College Students' scientific training program of Information College of Beijing Wuzi University in 2017, and Beijing the Great Wall scholars program (No.CIT&TCD20170317).

References

1. Tian, G., Liu, C., Lin, J., et al.: Research and development on optimization and scheduling problems in an automated warehouse. J. Shandong Univ. Technol. **31**(1), 12–17 (2001)
2. Yin, G., He, F., Sheng, D.: Research on the storage alignment optimization model for automated stereoscopic warehouse. J. Fujian Univ. Technol. **4**(3), 347–350 (2006)
3. Frazele, E.A., Sharp, G.P.: Correlated assignment strategy can improve any order-picking operation. Ind. Eng. **21**(4), 33–37 (1989)
4. Petersen II, C.G.: An evaluation of order picking routing policies. Int. J. Oper. Prod. Manag. **17**(11), 1098–1111 (1997)

5. Van den Berg, J.P., Zijm, W.H.M.: Models for warehouse management: classification and examples. Int. J. Prod. Econ. **59**(1), 519–528 (1999)
6. De Koster, R., Le-Duc, T., Roodbergen, K.J.: Design and control of warehouse order picking: a literature review. Eur. J. Oper. Res. **182**(2), 481–501 (2007)
7. Hausman, W.H., Schwarz, L.B., Graves, S.C.: Optimal storage assignment in automatic warehousing systems. Manag. Sci. **22**(6), 629–638 (1976)
8. Chan, F.T.S., Chan, H.K.: Improving the productivity of order picking of a manual-pick and multi-level rack distribution warehouse through the implementation of class-based storage. Expert Syst. Appl. **38**(3), 2686–2700 (2011)
9. Yu, Y., De Koster, M.B.M.: Designing an optimal turnover-based storage rack for a 3D compact automated storage and retrieval system. Int. J. Prod. Res. **47**(6), 1551–1571 (2009)
10. Gagliardi, J.P., Renaud, J., Ruiz, A.: On storage assignment policies for unit-load automated storage and retrieval systems. Int. J. Prod. Res. **50**(3), 879–892 (2012)
11. Yu, Y., De Koster, M.B.M.: On the suboptimality of full turnover-based storage. Int. J. Prod. Res. **51**(6), 1635–1647 (2013)
12. Pohl, L.M., Meller, R.D., Gue, K.R.: Turnover-based storage in non-traditional unit-load warehouse designs. IIE Trans. **43**(10), 703–720 (2011)
13. Accorsi, R., Manzini, R., Bortolini, M.: A hierarchical procedure for storage allocation and assignment within an order-picking system. A case study. Int. J. Logist. Res. Appl. **15**(6), 351–364 (2012)
14. Ang, M., Lim, Y.F., Sim, M.: Robust storage assignment in unit-load warehouses. Manag. Sci. **58**(11), 2114–2130 (2012)

A New SOC Estimation Algorithm

Weihua Zhong, Fahui Gu, and Wenxiang Wang[✉]

College of Information Engineering, Jiangxi College of Applied Technology,
Ganzhou 341000, Jiangxi, China
gufahui@139.com, 406873165@qq.com

Abstract. The DE algorithm has strong global search ability and robustness, but also has the shortcoming of slow convergence speed and local search ability is insufficient, and TLBO algorithm has the advantage of strong local search ability and faster convergence speed, but will be fall into the local optimum when dealing with complex problems. In this paper, the DE algorithm and TLBO algorithm are combined to construct a two-population co-evolutionary algorithm based on the DE and TLBO algorithm (DPCEDT). By theory analysis, the proposed DPCEDT algorithm can be used to improve the SOC estimation algorithm of power battery which is an extremely complex problem.

Keywords: DE · TLBO · DPCEDT · SOC · SOH

1 Introduction

The DE algorithm has simple structure, easy to realize, low time complexity, and because of the advantages of superior performance it show the strong robustness and heuristic search in solving problems, but due to the local search ability is not strong, late algorithm to the slow convergence speed, easy to fall into local optimum; TLBO algorithm have a few parameters, structure is simple, which make the algorithm is easy to implement, has a strong ability to solve, and the solving speed and precision is high, but it's easy to fall into local optimum for high-dimensional multimodal, multi-objective optimization problem.

Many scholars try to mix the two algorithms, such as Jiang et al. combine DE algorithm and TLBO algorithm on short-term optimization problems for fire, water, and electricity, and established a hDE - TLBO algorithm [1], which can effectively deal with the total fuel costs, emissions and other constraint conflict; Wang and others proposes a TLBO - DE algorithm for chaotic time series prediction problem [2], by using DE algorithm's strong searching ability, and updating the current optimal individual location information, forcing TLBO out of stagnation, so as to speed up the convergence rate, and improve the overall performance of the algorithm; Zhu et al., in order to overcome the premature and slow convergence, parameter setting complex, combining TLBO and DE, free search model is set up to replace traditional roulette wheel selection method, constructed a TLDE algorithm [3], etc.

This article also try to build DPCEDT algorithm considering from a single group algorithm insufficiency, the mechanism of cooperative co-evolution, the size settings of

© Springer Nature Singapore Pte Ltd. 2018
K. Li et al. (Eds.): ISICA 2017, CCIS 874, pp. 303–311, 2018.
https://doi.org/10.1007/978-981-13-1651-7_27

double population, application, various perspectives such as hybrid intelligent optimization algorithm to, at the same time, the algorithm is applied to estimation of state of charge (SOC) of power battery [4], thus further to verify the validity and superiority of the algorithm.

2   The Basic Principle of Differential Evolution Algorithm and Teaching-Learning-Based Optimization Algorithm

2.1   Differential Evolution Algorithm

DE algorithm is a kind of swarm intelligence optimization algorithm, the main idea of the algorithm is based on the random optimization, the concrete process is produced by individual variation based on difference vector, and then tested hybrid get individual, finally selected operation to choose better individuals into the next generation, generation optimization on the finish.

The steps involved in DE algorithm are summarized as follows:

1. *Encoding and initializing the population*

DE algorithm adopts real number coding, and the initial population is randomly generated. Formula (1) represents each individual:

$$X_i = \{x_i(j), x_i(j), \dots, x_i(j)\} \tag{1}$$

Among them, $x_i(j) \in [l_j, u_j]; i = 1, 2, \dots, NP; j = 1, 2, \dots, D; x_i(j) = randreal[l_j, u_j]$, that is, $x_i(j)$ is a random real number that follows uniform distribution on $[l_j, u_j]$.

2. *Operation of differential mutation*

The core of DE algorithm is the differential mutation operation, which is also the key to determine the performance of the algorithm. In order to improve the algorithm performance, therefore, researchers put forward different difference mutation types of DE algorithm, in order to better distinguish between these algorithms, often using DE/a/b/c uniform related strategy to represent algorithm, in which a for how to choose the base vectors, can be random (rand), also can be to choose the best (best); b how many vectors to do the difference operation; c means hybridization, generally divided into binomial and exponential hybridization.

3. *Hybrid operation*

DE algorithm hybrid operations are done through hybrid operator, test the individual after hybridization, in standard DE algorithm, the individual test is made by mutated and the parent individual discrete hybridization, in order to avoid the same individual test and the parent, be sure to test at least one dimensional vector is from mutations in individual, specific according to the formula (2) hybrid:

$$U_i(j) = \begin{cases} V_i(j), & if(randreal_j[0, 1] < CR \ \ or \ \ j = j_{rand}) \\ X_i(j), & otherwise \end{cases} \quad (2)$$

Among them, V_i is the individual obtained after the mutation, $j = 1, \dots, D$, j_{rand} is a random integer between $[1, D]$.

4. Select operation

DE algorithm through after the pilot individual variation, hybrids will test the individual wheel on one-on-one way and the parent individual comparison, better retain selected to the next generation, specific selection method according to the formula (3):

$$X_i = \begin{cases} U_i, & if(f(U_i) \leq f(X_i)) \\ X_i, & otherwise \end{cases} \quad (3)$$

In which, $f(X_i)$ is the adaptive value of the individual.

2.2 Teaching-Learning-Based Optimization Algorithm

TLBO algorithm mainly "teaching" and "learning" the two stage, specific for the first phase is the teacher's "teaching", teachers teaching for students, a good teacher can provide students overall achievement; the second phase is the student's "learning", where students learn from each other by randomly selecting another student in the class to improve their academic performance.

In the TLBO algorithm, the following key steps are included:

1. Initialization

Using $X^j = (x_1^j, x_2^j, \dots, x_d^j)(j = 1, 2, \dots, NP)$ to represent each student, the decision variables are generated by the formula (4):

$$X_i^j = X_i^L + rand(0, 1) \times (X_i^U - X_i^L) \quad (4)$$

Where $j = 1, 2, \dots, NP, i = 1, 2, \dots, d$.

2. "Teaching" stage

At this stage, selecting the best individual in the population as a teacher, the teacher tries his best to get the students to approach their own level and improve the average grade of the subjects. Each trainee is based on the difference between the average teacher's $X_{teacher}$ and the average student *mean*. The specific learning process is carried out in accordance with the formula (5) and (6):

$$X_{new}^i = X_{old}^i + difference \quad (5)$$

$$differece = r_i \times (X_{teacher} - TF_i \times mean) \quad (6)$$

Among them, X_{old}^i and X_{new}^i respectively represent the achievements of student i before and after the study. $mean = \dfrac{1}{NP} \sum\limits_{i=1}^{NP} X^i$ is the average achievement of all students, $TF_i = round[1 + rand(0, 1)]$ is the learning factor and $r_i = rand(0, 1)$ is the learning step. After the "teaching" activity is completed, it is necessary to compare the results of each student before and after the study, and to select the best results.

3. *"Learning" stage*

Students through the class members are randomly selected as the study object, might as well assume that selected object is: $X^j(j = 1, 2, \ldots, NP, j \neq i)$, chooser is: $X^i = (x_1^i, x_2^i, \ldots, x_d^i)(i = 1, 2, \ldots, NP)$, then by analyzing their learning process is X^i and X^j to purposefully learning gap, the specific study way as shown in Eq. (7):

$$X_{new}^i = \begin{cases} X_{old}^i + r_i \times (X^i - X^j) \ f(X^j) < f(X^i) \\ X_{old}^i + r_i \times (X^j - X^i) \ f(X^i) < f(X^j) \end{cases} \qquad (7)$$

Among them, $r_i = rand(0, 1)$ is the study step of student i. The "learning" activity is a circular process. It is necessary to compare the results of each student before and after the study until the academic performance is improved.

3   The Dual-Population Co-evolution Algorithm Based on DE and TLBO Algorithm

3.1   The Thought of DPCEDT Algorithm and It's Description

It is well known that the DE algorithm has strong global search ability and robustness in solving engineering problems, but often prone to slow convergence speed in the process of search, and weak local search ability. And TLBO algorithm has the advantage of strong local search ability, can under the smaller computational resources at a rapid speed of convergence to the local extremum, but in the face of high-dimensional multi-modal, multi-objective and complex optimization problems, TLBO algorithm is easy to fall into local optimum.

Their respective advantages and disadvantages for DE and TLBO algorithm, adopts the mechanism to foster strengths and circumvent weaknesses, will have a clever union DE algorithm and TLBO algorithm and build a co-evolutionary algorithm based on DE and TLBO (Dual-population Co-evolution algorithm based on DE and TLBO, DPCEDT). DPCEDT algorithm makes full use of the global search ability, simple and easy to implement and the advantages of strong robustness of DE algorithm and the local search ability, simple and easy to realize and the advantages of fast convergence of TLBO algorithm, the algorithm describes as follows:

Step 1: *Initialization*

For specific optimization problems, 2N individuals were randomly generated and the related parameters of DPCEDT algorithm were set, such as number of iterations, cross-factor, mutation probability, selection mode, etc.

Step 2: *Divide two subpopulations of size N*

Considering the diversity of the population, the average 2N individuals were divided into two identical subpopulations, and the selection of individuals was randomly selected.

Step 3: *Two subpopulations evolve independently*

The DE subpopulation evolves independently by mutation, crossover, selection and other steps to produce the next generation population and optimal individual DB^t. TLBO subpopulation evolved independently through the process of teaching process and learning process, which produced the next generation population and optimal individual TB^t. By calculating the adaptive value of DB^t and TB^t, the optimal individual in the whole population is recorded as GB^t.

Step 4: *Individual information fusion after the evolution of two subpopulations*

In the evolution of each generation of DE sub-population, the optimal individual in the two subpopulations is first used to generate a composite individual CI^t according to formula (8):

$$CI^t = DB^t \times W + TB^t \times (1 - W)$$
$$W = \frac{f(DB^t)}{f(DB^t) + f(TB^t)} \tag{8}$$

From the formula (8), the compound individual CI^t is made up of two sub population in the respective optimal individuals, and the weight of linear combination was decided by their respective optimal individual fitness, so the two subpopulations integrate and share their excellent searching. After the composite individual is generated, the DE subpopulation will evolve by formula (9):

$$X_i^{t+1} = X_{r_1}^t + F \times (CI^t - X_{r_2}^t) + F \times (X_{r_3}^t - X_{r_4}^t) \tag{9}$$

After the operation of the variation operation, the DE subpopulation can generate test individuals through the traditional crossover operation, and select the new generation sub-population by selecting the operation. There are two parameters F and CR in the DE subpopulation, and the values of these two parameters affect its search performance on a certain range. In order to reduce the influence of these two parameters on the search performance of the DE subpopulation, we adopted the *jDE* parameter adjustment strategy proposed by Brest et al. to adapt the values of parameters F and CR.

In TLBO child population of each generation of evolution in the process of operation, first calculate the average value of all individuals in TLBO subpopulation to generate a average individual $X_{TSubMean}$, then the global optimal individual of the entire population is set to the teacher. In the process of teaching, TLBO subpopulations generate new individuals by formula (10):

$$X_i^{t+1} = X_i^t + r_1 \times (GB^t - TF \times X_{TSubMean}^t) \tag{10}$$

The GB^t is the optimal individual of DE subpopulation and TLBO subpopulation, $DE/rand/1/\exp$ is learning step, $r_1 = rand(0, 1)$, TF is learning factor, $TF = round[1 + rand(0, 1)]$, formula (10) using the entire population of the best individual GB^t as the search direction, by sharing good search information to speed up the convergence rate of the whole population. After the "teaching" search process, the TLBO subpopulation performs the "learning" search process of the basic TLBO algorithm for each individual.

Step5: *Save the optimal individuals of the two subpopulations*

When the DPCEDT algorithm has completed the search operation of two sub-populations of DE and TLBO, the global optimal individual in the whole population is saved to GB^t.

Step 6: *Judgment stop condition*

To determine whether to meet the stop condition, namely whether to reach the precision requirement, to achieve the stop, or to perform the maximum number of iterations.

4   The Application of the DPCEDT Algorithm in SOC Estimation of Power Battery

4.1   Overview of SOC Estimation of Power Battery

Power battery is the power supply, generally refers to the battery used in electric bus, electric car, electric bicycle and other vehicles. The power battery charge state SOC is used to describe the percentage of the battery's remaining power as its rated power, and the definition is as the formula (11):

$$SOC = \frac{Q_l}{Q_o} \tag{11}$$

With Q_l for battery remaining power, the unit ampere hour (Ah), Q_o for battery rated capacity, unit Ann (Ah), SOC unit%, the state when the battery full charge the SOC = 100%, when it used up SOC = 0. In actual use, the value of SOC is between 20% and 80%, because both overcharge and discharge can shorten battery life.

Because the SOC estimation not only involves the schedule estimation and remaining power estimates, still involves the safety management of the battery and charge and discharge management, obviously, the SOC estimation has obvious nonlinear characteristics, therefore, SOC estimation has been a difficult problem in the field of power battery research.

Hu et al. researched the charging and discharging characteristics of NI-MH batteries under the condition of different temperature conditions [6]. On this basisa four dimensional Map graph model of NI-MH batteries which set the current, voltage, temperature of the battery as 3d coordinate was extension set up, and the estimation accuracy up to 3%. However, their research also lacks consideration for battery aging. The research shows that power batteries' state of health (SOH) has great influence on SOC estimation, and the maximum error can be up to 25% in the battery scrap stage.

4.2 The Application of DPCEDT Algorithm in SOC Estimation of Power Battery

The previous SOC estimation method based on the 4-dimensional Map can be expressed as the function of the SOC relation between voltage, current and temperature [7]

$$SOC = f(V, I, T) \tag{12}$$

In practice, when the battery management system collects the voltage, current and temperature of the battery, it can find the SOC of the battery. After considering the battery aging factor, the formula (12) can correct the aging battery SOC estimation:

$$SOC' = \frac{1}{SOH} * SOC = \frac{1}{SOH} * f(V, I, T) \tag{13}$$

1. SOH estimation method

The research of Shi showed that the SOH decreased with monotonically decreasing and approximate linear relationship with the impedance [8], and the mathematical model was as formula (14)

$$SOH = k_1 * \frac{r}{r_0} + k_2 \tag{14}$$

In the top form, k_1 and k_2 are constants, r_0 is the new battery internal resistance, r is the internal resistance of the battery after aging. A first-order model of power cell, as shown in Fig. 1:

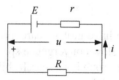

Fig. 1. The first-order model of power battery

Using E to describe the electromotive force of the battery, use u to describe the voltage of the external work, describe the current of the work with i, and describe the load resistor with r, then the power cell first-order model can be formulated (15) description:

$$E = u + i * r \tag{15}$$

From formula (14) and (15) we get:

$$SOH = \frac{k_1(E - u)}{i * r_0} + k_2 \tag{16}$$

In formula (16), E, k_1, k_2 and r_0 are constants, and the value of u and i is collected in real time through the battery management system. Formula (16) with u, i two observed variables, SOH as estimated variables, according to the engineering experience voltage u easier to obtain high measurement precision, due to the hall sensor is vulnerable to outside interference, lead to current i difficult to get good accuracy of measurement, direct calculate SOH by using formula (16) will have a larger error. For the estimation of state variables, DPCEDT algorithm has a fast convergence speed and avoids the trap of local optimal solution. The accuracy of SOH estimation by using DPCEDT algorithm can be reflected by the root mean square error:

$$fitness(SOH) = \sqrt{\frac{1}{n} \sum_{k=1}^{n} (i_k - \hat{i}_k)^2} \tag{17}$$

In the above formula, i_k is the real value and $\hat{i}_k$ is the calculated value, and the combination formula (16) can be seen that:

$$\hat{i}_k = \frac{k_1(E - u_k)}{r_0 * (SOH - k_2)} \tag{18}$$

Therefore, the new SOC estimation algorithm can be expressed as: firstly, the initial SOC can be calculated by using the 4-dimensional map, and then work out SOH by using DPCEDT algorithm, and finally SOC was corrected by SOH by using formula (12).

5   Conclusion

For the reason that DE algorithm have global searching ability and strong robustness but also with the disadvantage of slow convergence speed and insufficient local search ability, and TLBO algorithm have strong local search ability and faster convergence rate but will fall into the local optimal when dealing with complex problems, this article proposed a dual-population co-evolutionary algorithm based on DE and TLBO (DPCEDT) by combing DE algorithm and TLBO algorithm skillfully. By theory analysis, the proposed DPCEDT algorithm can be used to improve the SOC estimation algorithm of power battery which is an extremely complex problem.

Acknowledgement. This work was jointly supported by Natural Science Foundation of China (61773296), the Education Department of Jiangxi Province of China Science and Technology research projects with the Grant No. GJJ151433, GJJ161687, GJJ161688 and GJJ161691.

References

1. Jiang, X., Zhou, J.: Hybrid DE-TLBO algorithm for solving short term hydro-thermal optimal scheduling with incommensurable objectives. In: Control Conference, pp. 2474–2479. IEEE (2013)
2. Wang, L., Zou, F., Hei, X., et al.: A hybridization of teaching–learning-based optimization and differential evolution for chaotic time series prediction. Neural Comput. Appl. **25**(6), 1407–1422 (2014)
3. Zhu, C., Yan, Y., Haierhan, et al.: Teaching-learning-based differential evolution algorithm for optimization problems. In: Eighth International Conference on Internet Computing for Science and Engineering, pp. 139–142 (2015)
4. Feder, D.O., Hlavac, M.J.: Analysis and interpretation of conductance measurements used to assess the state-of-health of valve regulated lead acid batteries. In: Proceedings of the 16th International Telecommunications Energy Conference (1994)
5. Zhang, J., Sanderson, A.C.: JADE: adaptive differential evolution with optional external archive. IEEE Trans. Evol. Comput. **13**(5), 945–958 (2009)
6. Hu, Z., Wang, W., Lin, Y., et al.: SOC estimation method for NI-MH battery based on 4-dimensional map diagram. J. Mot. Control **16**(2), 83–89 (2012)
7. Wang, W.: Research on power battery management system and its SOC estimation method. Master's thesis, Central South University (2013)
8. Xu, X., Wang, L., Shi, H.: Study on battery aging life based on electrochemical impedance spectroscopy. Res. Des. Power Supply Technol. **39**(12), 2579–2583 (2015)

Analysis on Current Situation of E-Commerce Platform for the Development from C2M Model to C2B Model

Bo Yang[✉]

Huashang College, Guangdong University of Finance and Economics, Zengcheng
Guangzhou 511300, China
375442583@qq.com

Abstract. The development from C2M model to C2B model is the main trend in the development of current e-commerce platform, C2M e-commerce model eliminated the intermediate circulation of commodities, to make consumers and manufacturers realized the docking and communication directly, manufacturers can make production according to the personalized needs of consumers. This model has been obtained a very good development in the e-commerce platform, such as clothing industry, automobile industry, necessary malls and so on, also exposed some problems, need other industries to continue to improve, to ensure the stability and maturity of the development for C2M e-commerce model.

Keywords: E-commerce platform · C2B · C2M · Current situation

C2M model is the fourth type of e-commerce model which born under the background of Internet +, is a new type of e-commerce model which the development based on new technologies, such as cloud computing, big data and information technologies, etc., is also an e-commerce model which is the most complete and thorough customization development for C2B e-commerce model. With the deepening of the development for supply-side reform in manufacturing industry in China, C2M mode has become one of the many choice directions for the development of the manufacturing industry, and given birth to a variety of e-commerce platform, to provide a more personalized service for consumers [1].

1 The Concept of C2B E-Commerce Model and C2M E-Commerce Model

1.1 C2B E-Commerce Model

C2B e-commerce model that is Customer-to-Business is the business model for consumers to connect with enterprises, through the mass aggregation of consumers to form a large number of purchasing groups, thus transform the mainstream business model. This e-commerce model has made consumers' personalized needs and low prices become important characteristics [2].

K. Li et al. (Eds.): ISICA 2017, CCIS 874, pp. 312–321, 2018.
https://doi.org/10.1007/978-981-13-1651-7_28

1.2 C2M E-Commerce Model

C2M e-commerce model that is Customer-to-Manufactory, is a new e-commerce model based on the background of industrial 4.0, eliminated the intermediate circulation of the entire supply chain for suppliers-manufacturers-distributors-wholes alers-retailers-consumers, realized the direct docking of manufacturers and customers, made the information transmission efficiency greatly improved, through the model of first sale and post production, the manufacturers can complete the production of the goods in accordance with the customers' personalized needs and orders, and distribute to the consumers. C2M model needs to meet several key problems: the production process can carry out single production, the entire supply chain has a flexible adaptive system, under the premise of the whole production is in order, can ensure the automatic adjustment of any faults or defects on the whole production chain and the whole production system in the production process, can optimize energy consumption and improve efficiency of energy using and management in single production or small batch production, to ensure low energy consumption and high efficiency in the production process, can be able to make effective forecast before the completion of production and failure occurs, and carry out predictive maintenance.

E-commerce platform of C2M presents two forms, one is the third-party e-commerce platform of C2M, and the other is the enterprise e-commerce platform of C2M. C2M e-commerce platform makes full use of the intelligent platform ecosystem and the advantages of the Internet, to achieve client (C) and manufactory (M) real-time interaction and interoperability [3] (Fig. 1).

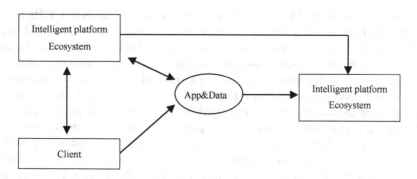

Fig. 1. E-commerce platform of C2M

The businesses to the e-commerce platform of C2M, directly from the manufacturer to consumers, including dealers and cut the middle part of the brand operation, inventory, warehousing, taxes, profit, only need to consider the manufacturer's profit (Fig. 2).

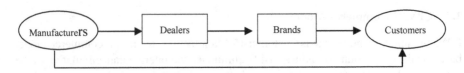

Fig. 2. Businesses to the E-commerce of C2M

1.3   The Origin of C2M E-Commerce Model

After the steam engine, power, IT technology, Internet technology and advanced manu-facturing technology combined, the development of the Industry 4.0 under the guidance of Internet, It will be another industrial revolution to realize the efficiency of large-scale manufacturing and the integration of individual workshops and the automation, intelli-gence, networking and customization of modern industry. The eighteen reports put forward the integration of industrialization and informatization and government depart-ments are also introducing relevant policies and measures to promote the Internet upgrade transformation of industry. Compared with C2B, C2M's product personaliza-tion and end-to-end sales are more thorough and thorough. Therefore, C2M e-commerce model should gradually be in the manufacturing industry, clothing industry, telecom-munications industry, automobile industry and so on.

2   The Inevitability of Development from C2B to C2M

C2M model is the inevitable result of the development of C2B model, C2B model contains three types of group-buying, personalized customization and C2B model based on SNS, among of them, the latter two types are very similar to C2M in nature. Group-buying is based on the third-party e-commerce platform for many consumers aggrega-tion, to generate more purchase orders, Taobao bargain, Meituan and other e-commerce platform are a group-buying platform in primary stage of this type. Some consumers have higher requests on quality and characteristics, will aggregate on the SNS platform or the third-party platform, put forward more personalized requirements, thereby make the enterprise gradually pay attention to this type of consumer demand, and according to the demand of this kind of consumer groups to design and produce new products, such as Jingdong mall scheduled group representatives, that is more than a certain number and then re-production, this kind belongs to the advanced stage of group-buying. Personalized customization in the early stage of development is mainly based on customize ticket, hotel, personalized clothing and gifts, etc., respond to the bid quotation of the hotel on line, the design of IDX independent and fashion shoes, simulation of full room personalized design on Shangpinzhaipei and so on, which is successful represen-tative of this kind of business model. The model integrates C2B and SNS is the consumer to use the SNS communication platform to launch group-buying products to conform to their own needs, to make the companies can launch group-buying which meet the requirements of the consumers, Mogujie, Pingduoduo are the representative of this kind of mode. The latter two types of C2B have already started to have C2M property, and Shangpinzhaipei have already had all the characteristics of C2M and development is mature. From the point of development trend, C2B e-commerce platform will gradually increase the integration with SNS platform, make the consumers more directly involved in the commodity production experience, to enable businesses to understand the specific needs of consumers before commodity production, to make the production is more reasonable, it also makes C2M mode become the necessity of development.

3 C2M E-Commerce Platform Developed Mature in China

3.1 The Revolution of C2M E-Commerce Model in the Clothing Industry

3.1.1 A Brief Introduction to the General Situation of the Development of C2M Pattern in Clothing Industry

Clothing industry is one of the earliest industries which used C2M e-commerce model, in 2003, Shandong Red Collar Group changed the main direction of enterprises into personalized customization, through using a large number of information technologies, such as data collecting, software development, and so on, developed C2M e-commerce platform - RCMTM system platform, based on the clothing CAD system, set up database of all kinds of formal versions which covered a variety of shapes, involved five types including suits, trousers, shirts, vest, and coat, contained hundreds of millions of a plate types. The longest delivery from receiving orders to dispatching the products is 7 days, in this mode, Red Collar Group successfully get rid of the predicament for overall orders declining in clothing manufacturing industry, annual profit growth above 150%, annual revenues amounted to more than RMB2 billion [4]. C2M model was reduced by 10% of the production cost, improved the production efficiency of 50 times, under the support of intelligent technology, 3D printing mode, informatization, big data and cloud computing technology, etc., to make a lot of disperse data of the needs for the consumers into effective production data successfully, to create a good example of "Internet + industry". Based on the rapid development of C2M model in clothing industry, on June 29, 2017, Hongmeishuma as the first digital textile of C2M platform, and put into service for home textile personalized custom designs, using 3D printing and digital printing technology, to provide personalized custom services in a wide range of designs for the consumers, not only reduced the cost, but also improved the efficiency, at the same time, it had the advantage of zero pollution [5].

3.1.2 Analysis of the Advantages of C2M Business Model in Clothing Industry

Fashion designers must predict fashion trends and customers' preferences in the process of designing clothing at first (Fig. 3), the whole process of designer clothing also needs to include the clothing designed by the consumer to be bought and accepted in order to reflect the value of the clothing. So it can be seen that the prediction for the clothing design and even the whole clothing enterprise is crucial.

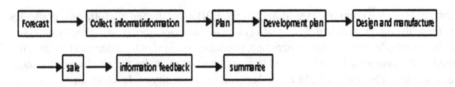

Fig. 3. Fashion design process

Normally, the designers of a clothing company predict the trend and customers' preferences based on the data in the past few years and the dressing styles of some leading figures in the fashion industry. But in the current situation that each individual seeks individualization in clothing, such a prediction method has its drawbacks. We can imagine designers and fashion leaders and many consumers in the market as independent points. The connection between them corresponds to the connection between dots and points through some social tool or method (Fig. 4). As a result of the restriction of conditions, designers can only obtain information of fashion leaders and very few customers through some channels.

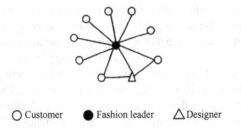

○ Customer   ● Fashion leader   △ Designer

Fig. 4. The contact of the designer in the information network under normal circumstance

Normally, the designers of a clothing company predict the trend and customers' preferences based on the data in the past few years and the dressing styles of some leading figures in the fashion industry. But in the current situation that each individual seeks individualization in clothing, such a prediction method has its drawbacks. We can imagine designers and fashion leaders and many consumers in the market as independent points. The connection between them corresponds to the connection between dots and points through some social tool or method (Fig. 4). As a result of the restriction of conditions, designers can only obtain information of fashion leaders and very few customers through some channels.

When using the C2M business model of the fashion designer in predicting, they can get rid of the shackles of historical data, that every one of the potential consumer preferences, as an important part of the garment industry C2M business model is to use 3D body scanning technology. Through 3D human scanning technology, clothing companies can get a large amount of data from each consumer and potential consumer's body shape. When designers get a large number of individual physical appearance indicators, they can be personalized in the fashion design. Because the clothing design on body size indicators throughout the details is not a precise absolute value, but one can error range, so you can put the individual into different bust to the same standard range, when the individual production data enough can be applied to large scale, design process this is actually equivalent to a fixed version of the original will be integrated in the numerical index is changed into a standard with smaller, can satisfy the single index range of more individuals. Another important link of C2M business mode of clothing industry is to set up a display shop. Clothing companies choose different styles of clothing to get consumers' preferences information through display shop. In this way, the clothing company can connect with each individual consumer directly, not only with a small number of consumers and fashion leaders (Fig. 5).

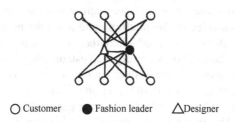

○ Customer   ● Fashion leader   △Designer

Fig. 5. The contact of the designer in the information network under C2M business model

Comparing the positions of designers in Figs. 4 and 5, we can see that the clothing business of C2M business mode in Fig. 4 can directly connect their designers with more individuals in the network. According to the structure of Tong theory, the traditional model of the number of structure Tong garment enterprises of the designer occupied by Fig. 4 using network structure of copper C2M business model designer clothing enterprises accounted for to the network is shown in Fig. 5. This means that clothing companies using the C2M business model can gain more information benefits. Not only that, because the clothing business of C2M business mode is connected with all the individual and fashion leaders, designers, as the "bridge" as called in the structural hole theory, can also get control benefits.

3.2 The Current Situation of Development for C2M E-Commerce Model in Automobile Industry

In November 2015, Geely automobile and the necessary shopping mall have launched the C2M customized e-commerce model in the automotive industry [6]. In April 2016, C2M e-commerce platform of Changan mall under the independent operation of Changan automobile introduced a whole personalized custom services, took its CS15 new models as the only order platform of individualized choice models, and gradually integrated online orders, factory production and offline 4S shop service, to ensure that each link all can provide high quality service for consumers, to improve the competitiveness of Changan automobile. In the first half of 2017, SAIC also introduced personalized customization services, and consumers can participate in the design of SUV.

3.3 The Emergence and Stability of New E-Commerce Platforms – Necessary Mall

On July 2, 2015, the world's first C2M e-commerce platform - necessary mall online, based on this platform, consumers and manufacturers can directly contact, the factory can directly get the personalized needs of consumers, and according to this demand to produce the corresponding products for the consumer, there is no need to go through the complicated circulation of goods, consumers through this platform will be able to enjoy the specificity and high quality goods. Platform can provide the 3D view model to consumers, the consumers through the platform to choose the specific requirements of goods, for example, choosing a pair of glasses, can choose the specific design and color of picture frame, lens and leg on the platform, and remit the prepaid payment, manufacturers according to the order for production, then dispatch products to consumers by Courier service [7]. In the early of necessary mall establishing, it covered the manufacturers of Samsonite luggage, woman's shoes and man's shoes of Burberry and Prada, Under Arm our sports products, Cartierde serve and Essilor glasses, in November 2015, cooperated with Geely automobile to develop the personalized customization in the automotive industry. The purpose of necessary mall is to let consumers enjoy luxury with cabbage price, according to the requirements of rigid regulations and flexible production chain for price and quality, to develop a flexible manufacturing e-commerce platform which can cover more than 1000 categories.

4   Problems Exposed in the Development from C2B to C2M

4.1 The Contradiction Between the Dilemma of Experience Shopping and the Authenticity of High Quality Low Price

C2M model is a direct docking between consumers and the manufacturers based on e-commerce platform, manufacturers' production is conducted around the individual needs of consumers, directly dispatch to consumers' hands after finished production, commodities uniqueness makes the performance of use, quality problems and problems for conformity of quality and price all cannot be corroborated, consumer cannot enjoy the advantage of experiential shopping, also can't understand whether the goods meet the requirements of the high quality low price before use. For example, two kinds of garment, through the customization of C2M, in the absence of proof of brand labels, how the consumer to know whether the garment has the same raw material or process as the original brand? [8] With the development of C2M model, the contradiction between this kind of luxury and cabbage price will gradually broke out, manufacturers will face the problems to prove themselves, and the third party platform also will face the problems to prove the manufacturers, this is also the difficulty of choosing a manufacturer in the necessary mall.

4.2   The Contradiction Between Freedom and Operability of Personalized Requirements

C2M is based on personalized custom, consumers can enjoy a larger degree of freedom in this model, according to their own needs and imagination to put forward the requirements to manufacturers, this degrees of freedom is not only the requirements of consumers, also need to satisfy the requirements of database provided by business platform, before this kind of service is online, manufacturers need to select a database that the feasibility and operability is stronger, which is validated by a number of technical for consumers to choose, therefore, consumers' freedom is also limited to this range, beyond this range, manufacturers can't guarantee the feasibility and performance of technical, this kind of freedom whether can be able to meet the personalized expectations of consumers is an important problem that manufacturers need to consider when launched the service, is also an important factor whether can make C2M service really meet the personalized needs of consumers [9]. Especially in car industry, intelligent manufacturing industry, etc., need to determine its rationality after integrating all parts and testing, this need to be proven by actual implementation, when the database of manufacturers is collecting such data, also need to prove the feasibility of combination through technology research and experiment, therefore, the choice is more freedom in appearance, take Changan mall personalized customization as an example, its optional program primarily includes gearbox, models, appearance, skylights and glass, fundamentally speaking, is mainly to meet appearance [10]. When manufacturing industry responds to supply-side reform through the C2M model, whether the technical requirements of each industry have the technical requirements of the customization requirements is the primary problem that must be solved.

4.3   Problems for the Integration of Industrial Chain in C2M Mode

The product control of C2M is the consumer, and the manufacturer needs to coordinate the whole supply chain according to the demands of the consumers, the integration difficulty has great problems in industry, agriculture and service industry. From the point of manufacturing industrial, for a complete industrial chain, when it is reduced or missed in a link, must be perfected and supplied from other links, this can cause the problems of other link cost increasing and coordination degree increasing, a manufacturing company who wants to really develop C2M model, must make the optimal integration for the whole supply chain, only in this way can ensure really mature for this model, and bring enough profit for the enterprise, to satisfy the cost-benefit principle for the business enterprise development. Manufacturers need to make production more lean and cost management more perfect, so as to complete the production of variety, small batch and customization, using intelligent storage, logistics technology to ensure the cost reduction of delivery and efficiency enhancement [11]. The after-sales service of customization commodity is also the problem that manufacturer must consider, provide after-sales service for small batch and unique products, is a great test for manufacturer, whether cost expenditure, after-sale personnel protection and workload all are faced with great challenges [12].

5   Conclusion

With the advent of the era of Industrial 4.0 and Internet +, make C2M e-commerce mode into more industries, there are also a number of enterprises in the appliance, household, pharmaceutical, education and other industries are accelerating the research and application of C2M model, Haier, Dima bath, Handan iron and steel group, and other enterprises have made good results of technology and research and development in the development of C2M mode, YiPuZan network platform through the data mining and analysis technology, is committed to providing consumer demand data for more business enterprises, make it into production data effectively [13]. With the development of C2M e-commerce platform, enterprises should continue to absorb experience in their development, further perfect the enterprises' organization structure, management mode and technical ability, to provide better technical storage for the stable operation of C2M mode.

Acknowledgment. This work was partially supported by Key platform construction leap plan and major project and achievement cultivation plan project, characteristic innovation project of Department of Education of Guangdong Province of China (Grant No.2014GXJK174), Higher education innovation strong school project of Huashang College, Guangdong university of finance and economics (Grant No.HS2014CXQX10). We would like to thank the anonymous reviewers for their valuable comments that greatly helped us to improve the contents of this paper.

References

1. Li, M.: Who works in the Institute of industry and planning for Institute of Information and Communication in China. From C2B to C2M. Posts and Telecommunications, 2016-07-19008
2. Newspaper reporter Chong Ren. Is the manufacturer ready from C2B to C2M?. Liberation daily, 2015-07-23015
3. Xu, J.: Opportunities and challenges for the transformation of e-commerce platform from C2B to C2M. Enterp. Reform Manag. **04**, 58 (2017)
4. Xiao, W.: Hongmeishuma builds C2M platform of home textile clothing. China Fashion daily, 2017-07-07 007
5. Duan, J., Pan, H.: Research on the application of e-commerce platform for C2M model. Mod. Comput. (Prof. Ed.) **18**, 44–48 (2017)
6. Chen, Q.: Research on the strategy of C2M e-commerce model which facilitates the transformation and upgrading for traditional manufacturing enterprises. Econ. Trade Pract. **08**, 177 (2017)
7. Ruo, S.: The necessary mall C2M: achieve the price of luxury goods into the price of cabbage. Commer. Cult. Mon. **28**, 42–45 (2016)
8. Jiang, L.: Research on the model of "personalized customization" of C2M in red collar group. Econ. Trade Pract. **1X**, 340 (2016)
9. Jin, S.: Research on the model of "mass customization" of C2M for clothing industry in China. Art Educ. **7**, 116 (2015)
10. Qi, D.: Research on the current situation and development strategy of e-commerce C2B model in China. Fujian Qual. Manag. **9**, 23 (2015)

11. Qian, W., Ji, S.: New trend for the development of e-commerce: research on C2B model. Bus. Age **33**, 63–64 (2013)
12. Mou, Y., Zhang, W.: Research on the development of C2B e-commerce model. Mod. Econ. Inf. **10**, 195 (2015)
13. Bee fan world: From C2M to see how personalized intelligent manufacturing landing. Enterp. Res. **5,** 14–16 (2016)

On the Artistic Characteristics of Computer Aided Design in Fashion Design

Ping Wang[(✉)]

Department of Art, Guangdong University of Science and Technology, Dongguan, China
28993716@qq.com

Abstract. Today, fashion design has basically matured. Based on the interaction and internal connection between computer clothing design and fashion design, this paper also put forward a series of suggestions for the computer clothing design in the apparel industry played a reference role. In addition, this article also studied how to develop fully automated garment manufacturing, clothing design provides a certain way of development. Clothing CAD/CAM has three meanings, namely, intelligent design, graphic design and virtual simulation. Major issues in apparel design technology include design database construction, 3D modeling, mesh generation, fabric drape modeling and crash testing. In the process of researching and analyzing garment manufacturing automation system, the author also analyzes these technical parameters emphatically. All in all, when the garment is fully automated, the era of mass customization of apparel is imminent.

Keywords: Artistic characteristics · Computer aided design · Fashion design CAD

1 Introduction

With the rapid development of society, the computer has become an indispensable tool in the creative design of art. The use of computers can be graphic design, interior design, architectural design, costume design, industrial design, animation design. He Yuhan in the "computer-aided design approach in apparel design and application of" one article, clearly pointed out that clothing as an art, but also a technology [1]. Therefore, the actual process of fashion design, computer-aided design tools should be used as soon as possible to the production. Sun Jianghong in the "Computer Aided Design - AutoCAD 2009 Practical Guide," a book for readers to introduce the AutoCAD 2009 Chinese version of the basic application technology. The book step by step, from the perspective of mechanical mapping to explain the relationship between the software and engineering drawings, drawing management, plane view and three-dimensional view operation, plane drawing and labeling, three-dimensional object drawing.

Computer-aided design refers to the use of compBaek, Hyun et al. [12] examined the utilization value of Art Print shown as the convergence phenomenon with art, the main flow of the 21 century fashion design and analyzed the characteristics and utilization method of Art Print that can be used as Subgraphic Element as well as Visual Identity differentiation strategy of fashion brand.

© Springer Nature Singapore Pte Ltd. 2018
K. Li et al. (Eds.): ISICA 2017, CCIS 874, pp. 322–328, 2018.
https://doi.org/10.1007/978-981-13-1651-7_29

2 Computer-Aided Design Overview

Computer aided design is mainly used in two aspects. One is for the engineering field, the other for the art and design field. Computer-aided ad production is an application in the field of art design, which refers to the use of multimedia computer technology and software products to create or create all or part of the advertising works [2]. Compared with the traditional advertising production, there are high efficiency, strong performance, simple image processing and so on. There are five steps while using the computer to create an ad: selection, selection of drawing software, analysis of pictures, image processing, print output.

Their graphic devices to help designers to design work, and it refers to as CAD. In engineering and product design, the computer can help designers to undertake computing, information storage and mapping work. In the design of the computer, it is usually used to calculate a large number of different programs, analysis and comparison to determine the optimal program and a variety of design information. Whether digital and text or graphics, it can be stored in the computer's memory. The designer can usually start with the sketch design, the sketch into the work of the heavy work can be handed over to the computer to complete. By the computer automatically generated design results, you can quickly make a graphical display, so that designers Timely to make the design to determine and modify. The use of the computer can be edited and edited, zoom, pan and rotate the graphics data processing work [3]. CAD can reduce the designer's work, shorten the design cycle and improve the design quality. The following graphic is a computer-aided design (Fig. 1).

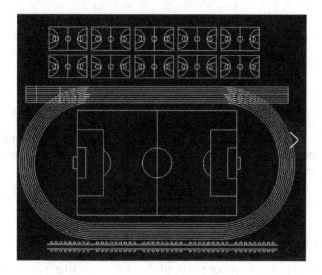

Fig. 1. Computer-aided design draft

3   Computer Aided Design Components

It is usually based on an interactive computer system with graphical functions. The main devices are: computer host, graphic display terminal, tablet, plotter, scanner, printer, tape drive, and all kinds of software. The following Fig. 2 shows the components of the computer-aided design.

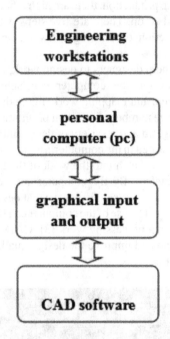

Fig. 2. Component of a computer-aided design

3.1   Engineering Workstations

Generally refers to a single-user interactive computer system with super minicomputer functionality and 3D graphics processing capabilities. It has a strong computing power with the standard graphics software, a high-resolution display terminal. You can work together in the sharing of resources on the LAN, and has formed the most popular cad system [4].

3.2   Personal Computer (Pc)

System is cheap and easy to operate and flexible use. 80 years later, pc machine performance is constantly renovated. There are the rapid development of hardware and software, coupled with graphics cards, high-resolution graphics display applications, and pc machine network technology development. The pc system has become a large number of cad and the emergence of a rise trend.

3.3   Graphical Input and Output Devices

In addition to the computer host and general peripherals, the computer-aided design mainly uses graphical input and output devices. Interactive graphics system is particularly important for cad. The general function of the graphical input device is to send the coordinates of the points on the plane to the computer. Common input devices are keyboard, light pen, touch screen, joystick, trackball, mouse, tablet and digitizer. Graphical output device is divided into soft copy and hard copy two categories. Soft copy device refers to a variety of graphic display devices, human-computer interaction is essential; hard copy device commonly used as a graphical display of ancillary equipment, it copies the image on the screen in order to save. There are three commonly used graphical displays: directed beam display, memory tube display, and raster scan display. In order to make the image clear, the electron beam must constantly redraw the graphics [5], it is also called refresh display, it is easy to erase and modify the graphics, suitable for interactive graphics means. The storage tube shows that the saved image needs to be refreshed, so it can display a lot of data and the price is low. Raster scanning system can provide color images, image information can be stored in the so-called frame buffer memory, the image resolution is higher.

3.4   CAD Software

In addition to the computer's own software such as the operating system, compiler, cad mainly use interactive graphics display software, cad application software and data management software class 3 software. Interactive graphics display software for graphics display of the window, editing, viewing, graphics transformation, modification, and the corresponding human-computer interaction. Cad application software to provide geometric modeling, feature calculation, drawing and other functions to complete the various fields for the specialized design. The four elements of building application software are: algorithm, data structure, user interface and data management. The data management software is used to store, retrieve and process large amounts of data, including text and graphics information. To this end, the need to build engineering database system. It has the following characteristics compared with the general database system: the data type is more diverse, the design process of complex physical relations, library values and data structures often change, the designer's operation is mainly a real-time interactive processing.

4   The Artistic Expression of Computer Aided Design in Costume Design

4.1   The Use of Reverse Thinking in Costume Design

The trend of clothing with the progress of human civilization continue to change the transformation. From the 1950s Dior's elegant "new style" style to the 60's hippie style; from the 50's Dior's "new style" style to the 60's hippie style; From the mid-90s minimalism to the beginning of the new century, the rise of luxury after the rise of romanticism, clothing in the world stage can be described as a wave by wave, blissful. This popular change in clothing depends largely on the designer's reverse design thinking, reverse thinking to promote the continuous development of the world's clothing and progress for the overall concept of clothing opened a new space. So, what is the reverse thinking? The so-called reverse thinking, that is, the thinking of the opposition, reverse, reverse, reverse the point of change, standing on the opposite of habitual thinking, breaking the original way of thinking, the state of things And the characteristics of the push to the opposite or limit, to find things in the new point of view. In the costume design process, rational and correct use of reverse design thinking can make the design more creative and personality, the impact of people's usual visual experience, and thus creatively solve the design problems.

4.2   Computer-Aided Design of Warp Knitting Seamless Clothing

Computer aided design (CAD) refers to the design process of computer and graphic equipment by designers, and computer aided design is the core technology of changing traditional design pattern. The application of CAD system effectively improves productivity, reduces development costs, improves product quality and speeds up product updates. At the same time, CAD has the advantages of visualization, real-time adjustment of new product design, improved accuracy and reduced error; Through the use of CAD systems, the existing design data reusability has also been greatly improved, speed up the design process [6], so that the design work more convenient. Computer-aided design is the extension of computer graphics industry in the textile and garment industry, the traditional way to design a clothing from the early preparation to complete the need to prepare a lot of physical material, spend half a day or even a few days to complete the final design, and the application of CAD The system only takes a few hours or even ten minutes to complete, greatly shorten the clothing design cycle, while design patterns and patterns can be reused and the development of a series of products, computer-aided design of the development of clothing has brought to the garment industry Great influence, the future will play a greater role.

4.3   The Impact of Design Automation on Fashion Design

The computer's own cad is designed to automate or semi-automate the design and development of the computer itself. The research includes automation of functional design and automation of assembly design, involving computer hardware description language, system level simulation, automatic logic synthesis, logic simulation, microprogramming automation, automatic logic partitioning, automatic placement and routing, and corresponding interactive graphics systems and engineering databases system. The integrated circuit cad is sometimes also included in the scope of computer design automation [7].

4.4 Computer-Aided Design to Improve the Level of Clothing Design Cut

4.4.1 Clothing Lines

Clothing outline, skirts, ferry, seams, and ribs, zippers, lace and so give the concept of lines. Lines can be divided into two types of straight lines and curves. Straight line gives a strong, concise, solemn sense, apply to men's, sportswear, to show vigorous strong. Curve gives a successful, lively, gentle feeling, apply to women's and children's clothing, to show gentle or naive and lively. Clothing design should be based on the characteristics of the line, the appropriate choice to a main, the rest as a supplement, so as to form the ideal style.

4.4.2 Clothing Color

Costume design focuses on the feeling of color, feelings and symbols, heat and heat, and color. There are four kinds of color perception: First, cold and warm feeling. Blue, blue, easy to make people think of the sky and the sea, resulting in cold feeling, known as cool, suitable for summer; red, orange, yellow three people easy to make people think of the sun and fire, resulting in warm, In the winter clothing; green and purple easy to associate the green plants, flowers, etc., known as neutral color, commonly used in spring and autumn equipment. Second is the sense of severity. Lightness of high color people have a sense of light, low light color makes people feel heavy. So the general download is lower than the color of the coat, to maintain a stable, solemn [8]. Three is forward and backward and expansion and contraction. Brightness, high purity and warm color has a sense of forward and expansion, the strongest white; brightness, low purity and cool colors have a sense of retreat and shrinkage, the strongest black. Four hard and hard. Bright and low color, high or low purity of the color has a rigid, serious sense, should be used for men, the elderly installed, the use of high-quality, Winter clothes or coat.

4.4.3 Clothing Materials

Constitute the material material of clothing. According to the purpose of the design, the need and conditions of choice of materials. That is, according to the material or the type of material, according to the existing material design appropriate clothing or according to the requirements of clothing to choose the appropriate material. The overall outline and appearance of the garment shape shaping garment in three - dimensional space. Clothing modeling is the core of fashion design [9]. Clothing design principles, the principle elements throughout the entire process of clothing modeling. Therefore, the clothing modeling involves the body size, volume, visual error, clothing composition, and clothing design and other issues.

5   Conclusion

With the continuous development of computer processing technology, the scope of its application is also expanding, the visual arts had a great impact, including art design [10]. At present, the art of various colleges and universities have set up computer-aided design courses, and the proportion of this course is very large [11]. Computer-aided design in the modern art design industry has gradually replaced the traditional hand-drawn tools for designers to provide a broader space for artistic expression, as well as a huge potential to achieve creativity. At the same time, computer-aided design is increasingly showing great superiority [12]. At present, computer-aided design is widely used in the field of art design, it has been with the traditional hand-painted has formed a very significant difference. Therefore, we must keep up with the pace of development of the times, as soon as possible to adapt to the new trend.

Acknowledgments. The dissertation is part of the Guangdong institute of science and technology college scientific research project 《Based on the analysis and research of the transformation and upgrading of the dongguan garment industry to the flexible production mode》 (item number:GKY-2017KYYB-27) phased research fruits.

References

1. A Study on the Practical Teaching Mode of Art Design Specialty - A Case Study of Provincial Demonstration Practice Teaching Center of Arts and Design in Changsha University, Yihai (2015)
2. Zhang, H.: Thinking in the teaching of artistic design practice. Jiannan Literature (classical teaching), February 2015
3. Liao, J.: Have the characteristics of vitality - the design and construction of professional characteristics of our hospital review and thinking. In: National Textile Education Society of Textile Education Exchange Conference (2014)
4. College of Education, Zhejiang University, Hangzhou 310027, China: Research on Teaching Reform and Teaching Reform in Art Design Education in Colleges and Universities (2014)
5. He, Z.: China Higher Vocational Art Design Education Research. Nanjing Normal University (2015)
6. Wang, G.: On the cultivation of creative ability of art and design students. Fujian Normal University (2014)
7. Qu, Y.: School of Economics and Management, Suzhou University, Hangzhou 310027, China: Discussion on the advantages and disadvantages of computer aided design to art design. J. Suzhou Univ. (Eng. Sci. Ed.), June 2004
8. Deng, Z.-M., Feng, C., Xiao, J., Cui, W.-G.: Computer aided design system for textile. Appl. Technol. Market, June 2015
9. Huang, Z.: Ever-changing beauty – Computer-Aided Design Charm. Ornaments, January 2016
10. Chen, Y.,Qian, J.: Modern Computer Aided Design Research. Today's Print, April 2015
11. TCAD Xplore, Site, E.T.: IEEE Transactions on Computer-Aided Design of Integrated Circuits and Systems. Micro IEEE, 7(1), 50 (2010)
12. Baek, H., Bae, J., Jeong, S.: A study on the utilization characteristics of art print for strengthening visual identity of fashion brand. J. Korean Soc. Des. Cult. 23 (2017)

Artificial Intelligence and Robotics – Query Optimization

Rock-Paper-Scissors Game Based on Two-Domain DNA Strand Displacement

Wendan Xie[1], Changjun Zhou[1], Xianwen Fang[2], Zhixiang Yin[2], and Qiang Zhang[3]([✉])

[1] Key Laboratory of Advanced Design and Intelligent Computing (Dalian University), Ministry of Education, Dalian, China
[2] School of Mathematics and Big Data, Anhui University of Science and Technology, Huainan, China
[3] School of Computer Science and Technology, Dalian University of Technology, Dalian, China
zhangq30@gmail.com

Abstract. Based on two-domain DNA strand displacement, a computing model is proposed. The model is used as a "referee" for two players in a well-known Rock-Paper-Scissors game, which can be utilized as an example of the study of game theory and artificial intelligence (AI). A molecular model based on Two-domain strand displacement is applied to emulate the process of the game. The output of the circuit shows the final result of the underlying game, that is, each player's win, lose and draw. The two players hold a win of one inning and two win of three innings which are simulated by employing Visual DSD software. The simulation results show that the molecular model is correct and feasible. The establishment of the computing model is hoped to provide some new insights for the AI in the field of nanotechnology.

Keywords: Two-domain · DNA strand displacement · DNA nanotechnology
Artificial intelligence · Rock-Paper-Scissors

1 Introduction

In the field of molecular biology, DNA nanotechnology is a common method for studying biological computing using DNA as engineering material. With the development of nanotechnology, its applications are becoming more and more widespread, such as DNA self-assembly, DNA origami, and medical treatment. DNA nanotechnology studies the designs and constructions of artificial nucleic acid structures which have specific functions [1] and is also used to build controlled DNA nanostructures. In 1956, the AI was first proposed in the Dartmouth Conference [2] which is one of the important fields in scientific research, and has been paid more and more attention in the computer field. The combination of nanotechnology and AI is in line with the development trend of "emerging technology".

The DNA has three characteristics: the nanometer scale, the complementary pairing of the base and the super parallelism of hybridization between strands, which makes it the best material for self-assembly research. DNA self-assembly technology has attracted widespread attention in the field of DNA computing. Stefan [3] reported a

© Springer Nature Singapore Pte Ltd. 2018
K. Li et al. (Eds.): ISICA 2017, CCIS 874, pp. 331–340, 2018.
https://doi.org/10.1007/978-981-13-1651-7_30

method for folding a custom template strand by binding individual staple sequences to multiple locations on the template. The combination of molecular self-assembly based on DNA origami and photoetch pattern brings about hierarchical nanometer systems, in which a single molecule is located at precise positions of multiple length scales. Scheible [4] demonstrated a DNA origami microarray that is compatible with monomolecular fluorescence and super-resolution microscopy requirements. In [5] the self-assembled nanostructures with complex surfaces are designed and constructed in 3D space using origami technique. Zhang [6] used a series of DNAzyme-based logic gates to control the DNA patch self-assembly onto the specified DNA origami frame. "YES", "OR", "AND" and other logical systems and logical switches are recognition of tiles based on DNAzyme-mediated implementation and DNA origami framework.

DNA strand displacement technology has been widely used in the field of biological computing and has achieved many results, such as solving the NP-complete problem [7–9] logical gate [10]. Zhang [11] constructed the molecular computing model of logic "AND" and "OR" based on DNA strand displacement for the first time. In 2015, Rogers [12] prepared programming colloidal phase transitions with DNA strand displacement. Sawlekar [13] implemented nonlinear feedback controllers via DNA strand displacement reactions. Zhu [14] realized a novel aptamer-based sensing platform using three-way DNA junction-driven strand displacement, and on the basis of which, a label free fluorescent detection platform based on aptamer was fabricated. Lakin [15] achieved supervised learning in adaptive DNA strand displacement networks. There are many kinds of single strand, such as Two-domain strand [16] and Three-domain strand [17]. Cardelli [16] proposed a Two-domain strand displacement and constructed some logical gates on the basis of the Two-domain strand, such as the Transducer gate, the Join gate and the Fork gate. The Two-domain include the identification and the toehold. The long strand must pass toehold to identify double strand when it combines with double strand to replacement for short chain. Otherwise a series of strand displacement reactions will not occur.

The purpose of this paper is to devise a DNA computing model based on Two-domain strand displacement, and to realize the competition process of Rock-Paper-Scissors. This DNA computing model plays the role of "referee". The "converging technologies" has enormous advantages. The "referee" will announce the winner of the game in accordance with the rules of the game after watching the whole course of the two player matches. This work can be seen as beginning to explore the direction of combining nanotechnology with AI agent design. Specifically, this article takes Rock-Paper-Scissors as an example, which is a well-known game and can achieve intelligent "referee" by designing a powerful circuit. The rest of this paper is organized as follows. The second part introduces the Rock-Paper-Scissors, rules of a win of one inning and two win of three innings and the intelligent "referee" circuit based on the two-domain DNA strand displacement. The third part takes use of Visual DSD software to carry on the simulation experiments. And the fourth part summarizes the related work.

2   Rock-Paper-Scissors

This section describes the technical details of our "referee" devise that will determine the winner after watching the entire race of the Rock-Paper-Scissors game. Before introducing the circuit, let's take a look at the game and give some basic symbols for later use.

The rules of Rock-Paper-Scissors game are that "Rock" is better than "scissors", "Scissors" is better than "Paper", and "Paper" is better than "Rock". The game is executed by two players Y and Z who make gestures at the same time. The "referee" will decide the winner according to the rules of the game. For example, when Y and Z respectively make gestures of "Rock" and "Scissors", the player Y is the winner. From the Rock-Paper-Scissors, it can be seen that each player has three possibilities to win. Here, "Rock", "Scissors" and "Paper" are separately denoted with "a", "b" and "c".

For convenience, three games are designed for each player, which are denoted by Yi and Zj, where $i \in \{a, b, c\}$, $j \in \{a, b, c\}$. In addition, some triumphant groups are set to Wij, $Wij = (yi, zj)$, $i \in \{a, b, c\}$, $j \in \{a, b, c\}$. The victory group of player Y is $WY = (Wab, Wbc, Wca)$, and the winning group of player Z is $WZ = (Wba, Wcb, Wac)$. That is to say, $WY = (ya, zb; yb, zc; yc, za)$ and $WZ = (yb, za; yc, zb; ya, zc)$, where WY stands for that player Y is the winner, and WZ represents that player Z is the winner.

2.1   Circuit Design of a Win of One Inning

The main question to be solved by an intelligent "referee" is how to correctly distinguish which player wins the game in a win of one inning. In order to achieve this goal, our idea is to devise a gate, trigger, for two players. The function of the trigger is to identify the input signal and to determine which player is the winner, and then sends a signal of victory, which is represented by WY (or WZ). From the game, it can be observed that there is a situation of the draw, that is, $KD = (ya, za; yb, zb; yc, zc)$, in addition to each player's triumphant group $W = 3$. As long as Y and Z do the same gesture, the game results are the draw. So seven triggers are devised for the player, which are common parts shared by two player. A trigger needs to be designed for each victory group, where different triggers release different winning signals. The function of the players' gate is abstracted as shown in Fig. 1 DNA implementation of the "referee" circuit of Rock-Paper-Scissors is done by a series of DNA strand displacement reactions.

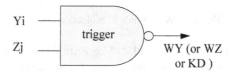

Fig. 1.   The gate of a win of one inning.

In Fig. 1, the trigger receives the input signals Yi and Zj, and makes a judgment to output a signal WY (or WZ or KD).

A DNA specie is devised for every game piece ((yi, zj), i∈{a, b, c}, j ∈ {a, b, c}) of each player, where "t" is on behalf of the toehold as shown in Fig. 2(a). The strands for triggers of Y, Z and the draw are shown in Fig. 2(b). Figure 2(c) indicates the victory signals WY, WZ, and the draw signal KD, whose molecular signal strands are < kyi >, < kzj >, and < kd > respectively.

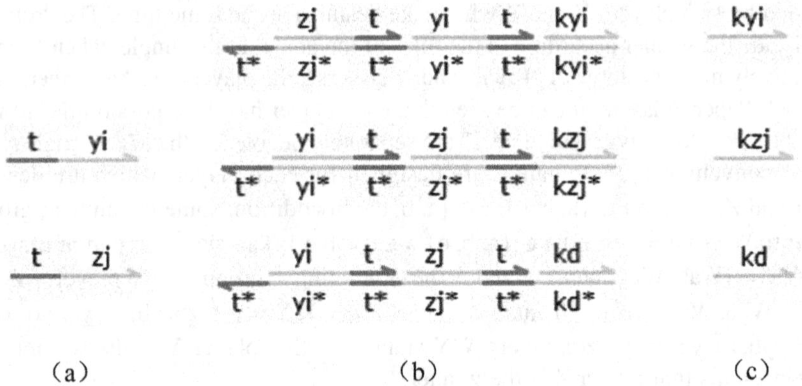

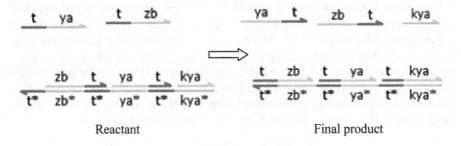

(a)   (b)   (c)

Fig. 2. DNA strands of molecular circuit.

The overall reactions process of molecular circuit is as follows. Figure 3 is YaZb-trigger. The example of YaZb-trigger illustrates the response of the trigger gate. Since other triggers are similar to YaZb-trigger, the reactions of other YZ triggers are not described.

Reactant   Final product

Fig. 3. DNA strands of molecular circuit.

In Fig. 3, strand < t^ ya > and strand < t^ zb > are input signal. The YaZb-trigger receives the input signals, releases a winning output signal strand < kya > and uses WY to represent the result strand. Other useless DNA strands are used as garbage products.

From the Fig. 4, it can be seen that strand < t^ ya > and strand < t^ za > are input signals. The KD-trigger receives the input signals and releases a draw output signal strand < kd > , and uses KD to represent the result strand. Other useless DNA strands are used as garbage products.

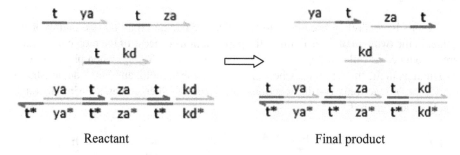

Reactant   Final product

Fig. 4. DNA strand displacement reactions process of the KD-trigger.

2.2   Circuit Design of the Two Win of Three Innings

The rule of the two win of three innings is that the final winner has two win in the three innings. The main problem to be solved by "referee" is that how to automatically detect players to achieve two games and determine which player to win the final victory. To realize this goal, a detector gate is designed for each player on the basis of Fig. 1 circuit. The function of the detector is to identify the Y (or Z) player's winning signal for two times and release an output signal to formally announce the ultimate winner, who is determined by testing the concentration of the final output strand of two players. The circuit of two win of three innings is shown in Fig. 5. The match times of two players is indicated with "m" in the three inning, $m \in \{1, 2, 3\}$. This article uses game piece (ymi, zmj) to represent the process of the two win of three innings. For example, Y1b delegate that the "Scissors" gesture made by player Y in the first inning. Z2c stands for the "Paper" gesture, which is done by player Z in the second inning.

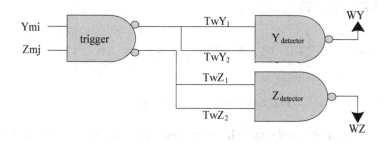

Fig. 5.   Gates of the two win of three innings.

The trigger receives input signal Ymi and Zmj, and makes a judgment in Fig. 5. If Y (or Z) achieve win in both the first and second inning, the trigger will continuously output two temporary winning signals TwY1 and TwY2 (or TwZ1 and TwZ2). The detector will output the final victory signal WY (or WZ) when it receives the two temporary victory signals from the trigger. And then the detector can end the game and do not need the third inning. If the winner of the first inning and the second inning is respectively Y and Z, "referee" will determine the result of the game through the third inning.

Results have already been described, which are decided only by one inning of the game. In the two win of three innings, the part reactions process of the molecular circuit are introduced as follows. Figure 6 shows the reactions of $Y_{detector}$ when player Y wins two innings in the three innings where "h" denotes the toehold, and Fig. 7 demonstrates the $Z_{detector}$'s reactions. As the reactions processes of other games are similar, other reactions are ignored.

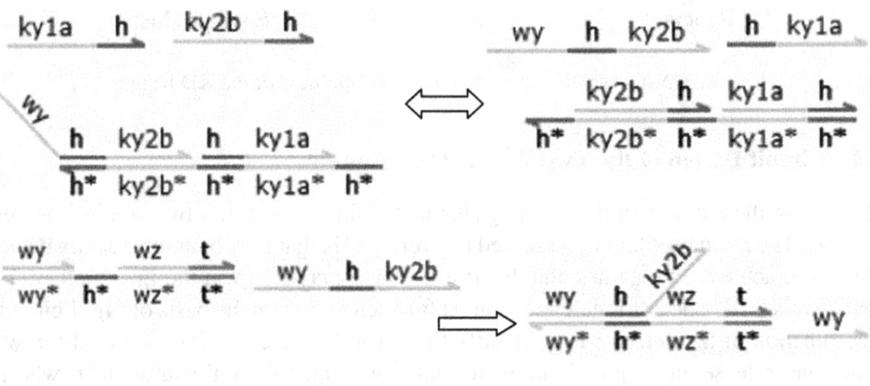

Fig. 6. The reactions process of $Y_{detector}$.

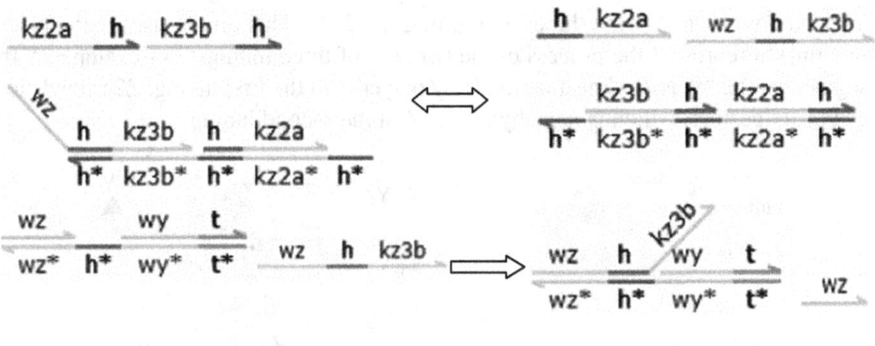

Fig. 7. The reactions process of $Z_{detector}$.

Strand < ky1a h^ > and < ky2b h^ > represent that Y achieves victory in both the first and second innings in Fig. 6. Only when Ydetector receives two temporary triumphant signals TwY1 and TwY2, it will release a final winning signal strand < wy >. The result strand is also represented by WY.

In Fig. 7, strand < kz2a h^ > and < kz3b h^ > stand for Z win in both the second and third innings. Zdetector will release a final triumphant signal strand < wz >, only when it accepts two temporary victory signals TwZ1 and TwZ2. The result strand is also said by WZ.

3   Simulation Experiment Based on Visual DSD Softwareout

This section describes the simulations of the "referee" circuit of Rock-Paper-Scissors. Results include two situations, which are a win of one inning and two win of three innings. All possible game scenarios are tested, and the feasibility and accuracy of the designed circuits are verified by using Visual DSD software. The following are the simulations of several typical cases. Other cases of the game are omitted due to their similarity. Let's first introduce the results of a win of one inning, as shown in Figs. 8 and 9.

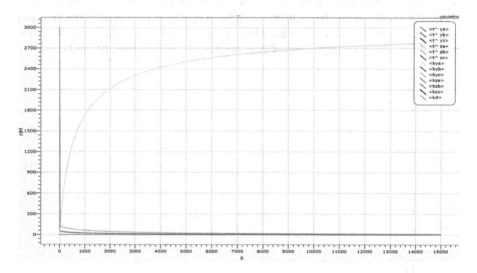

Fig. 8. Simulation results of the player Y win.

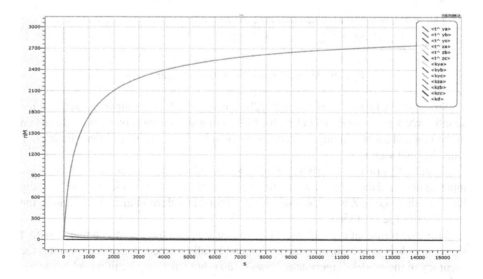

Fig. 9. Simulation results of the draw.

In the Fig. 8 strand $< t^\wedge ya >$ and $< t^\wedge zb >$ are input signals, and the strand $< kya >$ is output signal. The strand $< kya >$ stands for that the Y win the game due to make the "Rock" gesture, which is indicated by the ascendant blue curve.

As long as two players make the same gesture, the game results will be a draw in Fig. 9. The simulation demonstrates that since the Y and Z do simultaneous gestures, result is a draw.

From the simulation of Figs. 8 and 9, it can be seen that the reactions products are fully reacted and the production rate of the product is very high. The experimental results indicate the correctness and feasibility of the molecular circuit model of a win of one inning.

In order to validate that the circuit model of two win of three innings has correctness and feasibility, the circuit model is simulated by using software Visual DSD. The following is the simulation results of the two win of three innings, as shown in Figs. 10 and 11.

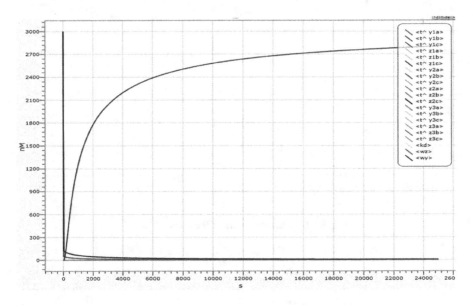

Fig. 10. Simulation results of the the player Y win.

Strand $< t^\wedge y1a >$, $< t^\wedge z1b >$, $< t^\wedge y2b >$, and $< t^\wedge z2c >$ are input signals, and the strand $< wy >$ is output signal in Fig. 10. This picture shows that player Y wins the match in the first and second inning. According to the rule of the two win of three innings, Y wins the final victory.

Strand $< t^\wedge y1a >$, $< t^\wedge z1b >$, $< t^\wedge y2b >$, $< t^\wedge z2a >$, $< t^\wedge y3c >$ and $< t^\wedge z3b >$ are input signals, and the strand $< wz >$ are output signals as shown in Fig. 11. The drawing indicates that player Z wins the game in the second and third inning. According to the rule, Z gains the final victory.

From the results in Figs. 10 and 11, it can be seen that the generation rate of output signal strands is very high, and the input strand is fully reacted, which demonstrates the accuracy of the model. Experimental results also show that the circuit model has feasibility.

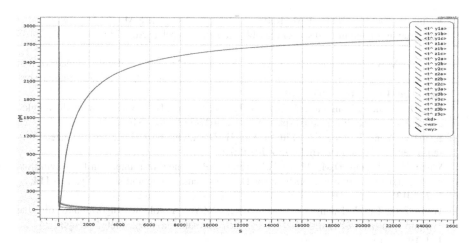

Fig. 11. Simulation results of the the player Z win.

The "referee" circuits based on the Two-domain strand displacement reactions are accomplished, which are simulated by using Visual DSD software. These results illustrate that the molecular models are feasible. It is also shown that the whole strand displacement reactions are completely autonomous process without any external force.

4   Conclusions

Since the great potential of two advanced fields of DNA technology and artificial intelligence, this paper presents a method of building intelligent agent based on Two-domain strand displacement. The game of Rock-Paper-Scissors is emulated through Visual DSD software. Experimental results illustrate that circuits are correct and the "referee" circuits devised in this paper are feasible. The intelligent "referee" of Rock-Paper-Scissors based on the Two-domain is a starting point towards the correlation between the two areas, DNA technology and artificial intelligence, which proposes an interesting research direction. The work of this article can be used as an example to make a useful exploration for the further research of "converging technologies" with enormous advantages.

Acknowledgment. This work is supported by the National Natural Science Foundation of China (Nos. 61425002, 61751203, 61772100, 61702070, 61672121, 61572093), Program for Changjiang Scholars and Innovative Research Team in University (No.IRT15R07), the Program for Liaoning Innovative Research Team in University (No.LT2015002), and the Program for Liaoning Key Lab of Intelligent Information Processing and Network Technology in University.

References

1. Seeman, N.C.: Nucleic acid junctions and lattices. J. Theor. Biol. **99**(2), 237–247 (1982)
2. Mccarthy, J., Minsky, M.L., Rochester, N., Shannon, C.E.: A proposal for the Dartmouth Summer research project on arterial intelligence. J. Mol. Biol. **278**(1), 279–289 (2006)

3. Stefan, N., Katy, B., Nafisi, P.M., Kathy, T., John, G., Douglas, S.M.: Folding complex dna nanostructures from limited sets of reusable sequences. Nucleic Acids Res. **44**(11), 102–108 (2016)

4. Scheible, M.B., Pardatscher, G., Kuzyk, A., Simmel, F.C.: Single molecule characterization of DNA binding and strand displacement reactions on lithographic DNA origami microarrays. Nano Lett. **14**(3), 1627–1633 (2015)

5. Han, D.G., Pal, S., Nangreave, J., Deng, Z.T., Liu, Y., Yan, H.: DNA origami with complex curvatures in three-dimensional space. Science **6027**(332), 342–346 (2011)

6. Zhang, C., Yang, J., Jiang, S.X., Liu, Y., Yan, H.: DNAzyme-based logic gate-mediated DNA self-assembly. Nano Lett. **16**(1), 736–741 (2016)

7. Wang, Z.C., Huang, D.M., Meng, H.J., Tang, C.P.: A new fast algorithm for solving the minimum spanning tree problem based on DNA molecules computation. Bio Systems **114**(1), 1–7 (2013)

8. Wang, Z.C., Tan, J., Huang, D.M., Ren, Y.C., Ji, Z.W.: A biological algorithm to solve the assignment problem based on DNA molecules computation. Appl. Math. Comput. **244**(2), 183–190 (2014)

9. Adleman, L.M.: Molecular computation of solutions to combinatorial problems. Science **266**(5187), 1021–1024 (1994)

10. Chen, Y.Q., et al.: A DNA logic gate based on strand displacement reaction and rolling circle amplification, responding to multiple low-abundance DNA fragment input signals, and its application in detecting miRNAs. Chem. Commun. **51**(32), 6980–6983 (2015)

11. Zhang, C., Ma, L.N., Dong, Y.F., Yang, J., Xu, J.: Molecular logic computing model based on DNA self-assembly strand branch migration. Sci. Bull. **58**(1), 32–38 (2013)

12. Rogers, W.B., Manoharan, V.N.: DNA nanotechnology. Programming colloidal phase transitions with DNA strand displacement. Science **347**(6222), 639–642 (2015)

13. Sawlekar, R., Montefusco, F., Kulkarni, V.V., Bates, D.G.: Implementing nonlinear feedback controllers using DNA strand displacement reactions. IEEE Trans. Nanobiosci. **15**(5), 443–454 (2016)

14. Zhu, J.B., Zhang, L.B., Zhou, Z.X., Dong, S.J., Wang, E.K.: Aptamer-based sensing platform using three-way DNA junction-driven strand displacement and its application in DNA logic circuit. Anal. Chem. **86**(1), 312–316 (2014)

15. Lakin, M.R., Stefanovic, D.: Supervised learning in adaptive DNA strand displacement networks. ACS Synth. Biol. **5**(8), 885–897 (2016)

16. Cardelli, L.: Two-domain DNA strand displacement. Math. Struct. Comput. Sci. **26**(2), 247–271 (2010)

17. Cardelli, L.: Strand algebras for DNA computing. Nat. Comput. **10**, 407–428 (2009)

A Business Resource Scheduled Algorithm of TD-LTE Trunking System Based on QoS

Qiutong Li, Yuechen Yang[✉], and Baocai Zhong

Chengdu Neusoft University, Dujiangyan, Chengdu, Sichuan, China
{liqiutong,yangyuechen,zhongbaocai}@nsu.edu.cn

Abstract. On the precondition of analyzing the algorithm of GBR and PDB, a business resource scheduled algorithm of TD-LTE trunking system based on QoS is presented in this paper. The simulation suggests that this algorithm can serve mixed styles of business better and meet different QoS requirements, especially to the throughput of high rate GBR business. It can make the system gain a higher throughput as well as decreasing the delay of the business.

Keywords: TD-LTE · QoS · Resource scheduled

1 Background

1.1 A Subsection Sample

LTE uses channel sharing and resource scheduled mechanism to promote the utilization of resource between different users. Downlink channel scheduled algorithm is the core function of the MAC layer. The key function of the down scheduled is to allocate the resource of the physical download sharing channel-PDSCH to send system information and users' data. This module affects the system function a lot.

The task of scheduled are making relevant choices of the physical resource, resource allocation strategy, and necessary resource management. Resource allocation should take user' requirements of QoS and the cache state as well. Wireless channel quality and area interference into consideration as well as the differentiation and fairness for RB resources allocation between users. In the TD-LTE trunking system, the broadband makes the system transmission rate increase greatly, and the user's business is diversified. These businesses are also guaranteed by different QoS. The existing resource scheduled algorithm fails to take these hybrid businesses into account. In this paper, an improved scheduled strategy for QoS is proposed based on the existing algorithm.

A Project Supported Scientific Research Fund of Sichuan Provincial Education Department under Grant No. 15ZB0359.

K. Li et al. (Eds.): ISICA 2017, CCIS 874, pp. 341–353, 2018.
https://doi.org/10.1007/978-981-13-1651-7_31

2   Algorithm Based on GBR and PDB

At present, many scholars have proposed scheduling algorithms for QoS to meet the time delay and speed guarantee requirements of the user's business.

The formula of its weight factor algorithm is shown as (2-1):

$$P_i = \frac{R_c(t)}{\bar{R}_i(t)} \cdot \max\left(1, \frac{GBR_i}{\bar{R}_i(t)}\right) \tag{2-1}$$

Pi represents the user's priority, the maximum rate that it can be obtained on the resource block c is $R_c(t)$, $\bar{R}_i(t)$ is the average transmission rate of user i [1]. GBR_i represents the lowest limit of transmission rates that the business can tolerate. The average rate of the user scheduling can be expressed as:

$$\bar{R}_i(t) = \left(1 - \frac{\lambda_i[t]}{t_c}\right) \cdot \bar{R}_i(t-1) + \frac{\lambda_i[t]}{t_c} \cdot R(i) \tag{2-2}$$

$R(i)$ is used to represent the rate at the time slot of t. When the user is unable to get scheduling at time t, $\lambda_i = 0$ or else $\lambda_i = 1$. So it's only going to be updated when $\lambda_i = 1$, $NS = 30TTI$ [2]. But the comprehensive business has a large demand for transmission rate, which results that this algorithm has a very large defect. For example, when the user has a great different requirements for the rate of GBR load, it's going to make it possible that user of the high GBR business is allocated insufficient resource block while the user of the low GBR business gets too much. For this defect, GBRwt algorithm also gives a calculation of priority weighting factors for scheduling users as shown in (2-3):

$$P_i = \frac{R_c(t)}{\bar{R}_i(t)} \cdot \frac{GBR_i}{\bar{R}_i(t)} \tag{2-3}$$

According to the (2-3), the average transmission rate will increase when lower GBR business user gets too much resource block, at the same time it makes the priority of the resource block which is obtained decline. In this way, it is fair to users with different GBR business requirements to be scheduled. But this algorithm is not a perfect solution.

When $GBR = 0$, the weight factor of user become to 0. So this algorithm cannot support the requirement of non-GBR business [3, 4].

Also, we need to take time delay into account. Padovani proposed an M-LWDF (Modified Largest Weighted Delay First) algorithm based on a classical algorithm PDB which can support different real-time stream business. That's to say that assigning different time slots to transfer for different user businesses. According to the reference [6], it can be deduced that:

$$P_i = -\frac{\log(\delta_i)}{T_i} \cdot D_i \cdot \frac{R_c(t)}{\bar{R}_i(t)} \tag{2-4}$$

T_i is the maximum time delay which can be tolerated by a user. The smaller the number is, the higher for the priority the user can get for scheduling. δ_i is a threshold value represents the parameter of QoS. In situations that with same businesses of different users, the bigger δ_i is, the lower priority will be get. The second term D_i is used to indicate the queuing delay of the business. The smaller D_i is, the user will get the lower priority of scheduling. The third item $\bar{R}_i(t)$ is the average transmission rate of the user, $R_c(t)$ is used to represent the maximum transmission rate that can be obtained. Both of these two terms take quality and fairness of the channel into account.

M-LWDF algorithm cannot meet the requirements of different GBR business resource scheduling as well.

3   Improved QoS - Based Enhanced Scheduling Algorithm

3.1   Improved Algorithm Design

Reference [7] considers several aspects of GBR scheduling. The priority of GBR described in this paper is based on channel quality and delay scheduling. Try to track the fast fading of user channels as much as possible to schedule the user at the peak while ensuring that the delay is not greater than the PDB. The GBR connection is based on the channel quality and the delay scheduling. The GBR connection is assigned according to the following four factors to obtain the scheduling priority of the connection.

(1)   Channel quality fluctuations: that is, considering the instantaneous channel quality and average channel quality. For GBR connection, taking into account the fast fading characteristics of the channel, the priority is lower when the user's channel quality is in deep fading. Therefore, it should be prevented to schedule users when the channel quality is in a deep decline. That is, the $R_c(t)$ is the maximum rate user can obtain and $\bar{R}_i(t)$ is the average transmission rate of the user, both of which take into account the fairness of scheduling and channel quality.

(2)   Delay: The closer the priority of the first packet in the connection buffer is at the Packet Delay Budget (PDB), the higher the priority is. In formula (2-4), it is the maximum delay that a user's service can tolerate. The smaller the value is, the higher priority of available scheduling is. It is a threshold value, which indicates the QoS parameters. In the case of different users with the same business, the greater δ_i equals to, the lower the priority of user will get. The queuing delay employed to D_i which represents the services. Smaller Di is the lower scheduling priority the user will get. As a result, $-log(\delta_i) \cdot D_i/T_i$ can be used to represent the bit error rate and delay of resource scheduling.

(3)   GBR business compensated priority. While scheduling resources, when the connection transmission rate obtained reaches the transmission rate required by the GBR service, the priority for this connection should be reduced. However, if the current transmission rate does not meet the transmission rate required by the GBR service, priority should be increased. When $GBR_i/R_i(t) > 1$, it indicates that the current transmission rate does not meet the requirements which need to increase the

priority; when $GBR_i/R_i(t) < 1$, it indicates that the current transmission rate has exceeded the GBR requirements, the priority of the bearer should be reduced.

(4) Relative priority: Since the GBR service requires a very large span for the transmission rate, for example, the voice service requires a relatively low transmission rate and the video service requires a large transmission rate. In order to balance the priority among the GBR service of QCI levels, it can be dynamically adjusted. This paper adds a factor, according to QCI level requirement to adjust the priority between GBR businesses.

In this way, based on formula (2-4), the priority of the user's GBR service can be expressed as:

$$P_i = -\frac{\log(\delta_i)}{T_i} \cdot D_i \cdot \frac{R_c(t)}{\overline{R_i}(t)} \cdot \frac{GBR_i}{\overline{R_i}(t)} \cdot k_{pf} \tag{3-1}$$

For non-GBR connection priority calculation, there is no GBR compensation priority because it does not guarantee the rate. In this way, the priority of the entire non-GBR can be expressed as:

$$P_i = -\frac{\log(\delta_i)}{T_i} \cdot D_i \cdot \frac{R_c(t)}{\overline{R_i}(t)} \tag{3-2}$$

3.2   Improved Downlink Resource Algorithm Allocation Steps

After the downlink data arrives, the downlink dynamic scheduling allocates the time-frequency resources for the data on the candidate connection according to the connection priority from high to low.

The scheduling algorithm includes four steps: judging user connection validity, calculating connection priority, constructing candidate user set and allocating resources. Each TTI will repeat these four steps.

(1) Determine the validity of the connection

The purpose of connection validity judgment is to check which users' connection in the following TTI may send data. In identifying the candidate connection set from all users to filter out a valid user and connection. When selecting a user's connection, exclude the connection that has no data in the buffer.

(2) Calculate the connection priority

For GBR business, priority weighting factors for GBR connection and non-GBR connection are calculated according to (3-1) and (3-2) respectively.

(3) Construct candidate user set

Select the valid connection according to priority from a high to low order. Assume $R = \{rb_c, 1 < c < m\}$ represents the available resource size and $U = \{u_i, 1 < i < n\}$ represents a valid set of users; $P = \{P(i), 1 < i < n\}$ represents a set of priorities.

Selecting $P(i)$ as the largest element in the P set, algorithm estimates the number of RBs rb_c to be used in the TTI for the connection according to the CQI of the user to which the connection belongs and the data amount of the connection;

Assuming that the numbers of RBs in the total resource TTI are sufficient at this point, the connection is put into the candidate connection set and the following sets or parameters are updated:

$$P = P - P(i) \tag{3-3}$$

$$R = R - rb_i \tag{3-4}$$

$$U = U - u_i \tag{3-5}$$

Go to step (1) until $R = \emptyset$ or $U = \emptyset$.

(4) Allocate resources

The resource allocation process determines the number of RBs used by the user and the TBS. The way of downlink resource allocation is limited by the type of resource allocation. Resource allocation for a user is based on Resource Block Group (RBG) or Physical Resource Block (PRB). Select the highest priority users of the connection set to allocate resources until the available RB resources are allocated. Figure 1 demonstrates the process of the improved downstream scheduling algorithm.

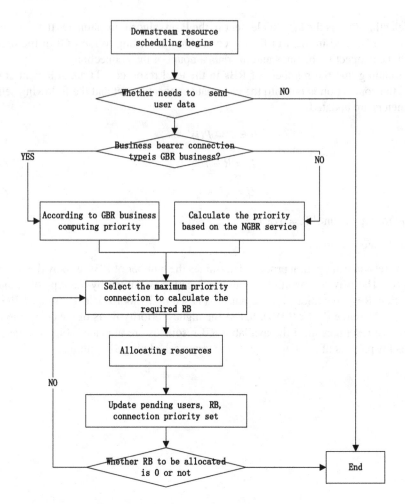

Fig. 1. Improved downstream scheduling algorithm flow

4 Algorithm Simulation Analysis

4.1 System Simulation Processes

This section does a simulation analysis on several basic LTE downlink resource scheduling algorithm and the enhancing scheduling strategy proposed in this paper, and demonstrate a comprehensive comparison of the advantages and disadvantages of various algorithms. Also in this section, it examines whether the algorithm proposed in this paper can solve the user resource scheduling problem of the trunking system.

In the process of implementing packet scheduling algorithm simulation, this paper will be divided into four sub-module parts to achieve the simulation process, the functions of each module is are as follows [8]:

(1) Input sub-module: Simulation packets generated and the arrival of these packets arrived obeying the packet rules. Each user's channel quality will change randomly to simulate the actual change of channel quality.

(2) The scheduling algorithm sub-module calculates the priority of the data packet generated by the input module according to the priority algorithm of the downlink resource scheduling and then sorts and forms the data packets according to the calculated priority.

(3) Statistics submodule: According to the resources allocated by the scheduling algorithm submodule, gathering the system throughput by the algorithm counted, and the statistical results are stored in the corresponding statistics Table

(4) Display Sub-module: based on the statistical data table obtained by the statistical module, the simulation result of the schedule is drawn.

Figure 2 shows TD-LTE system downlink scheduling algorithm in Matlab simulates of an overall flow chart.

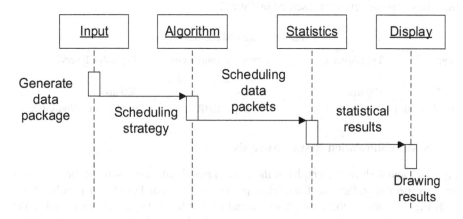

Fig. 2. Flowchart of the overall simulation

4.2 System Simulation Parameters

The simulation platform used in this paper is based on the MATLAB platform of an open source project and is built under the Matlab 2010a environment on the basis of the needs of LTE system-level simulation. In the simulation, the number of users in the cell is set to 10–60 and the total duration of the simulation is 20000 TTIs. In order to simplify the problem, the influence of the error rate of the packet error rate on the throughput will be temporarily ignored, and the values will be unified as 1. Packets delay do not conform to the requirement will be discarded directly before sending. This paper simulates the system throughput of MAXwt, M-LWDF, and the improved resource scheduling algorithm proposed in this paper. Other system-level simulation parameters are given in Table 1.

Table 1. Simulation parameter configuration

Simulation parameter	Parameter value
Send bandwidth	20 MHz
Subcarrier spacing	15 kHz
Frame configuration	2 Downstream: 1 Special: 2 Up
Number of RBs	100follows
TTI length	1 ms
Users in Cell	10–60
The antennas number of eNB	2 × 2

In this simulation, in order to verify scheduling of GBR and Non-GBR business, three types of services are considered. The ratio of three types of services is 1: 1: 1. To simplify the problem, one user only selects one of the business, non-GBR business arrival rate obeys uniform distribution (0–512 kbps), and updates every 10 TTI. The three business models are presented in Table 2.

Table 2. Simulated business types

Type	TypeA/voice	TypeB/discontinuous video	TypeC/TCP service
PDB	100 ms	300 ms	300 ms
BR (Bit rate)	64 kbps (GBR)	512 kbps (GBR)	0–512 kbps (Non-GBR)

4.3 System Simulation Results Analysis

Figures 3 and 4 show the graphs of the throughput simulation results of the service A and B respectively. First, set the relative priority of GBRs of TypeA and TypeB service to 1. Figure 3 shows the throughput simulation results of TypeA. As observed in the figure, since the TypeA service does not require high transmission rate, all of the three algorithms can basically guarantee the transmission rate of this type of service. With the increase of the number of users, the system throughput is also obvious improved. However, when the number of users increases to a certain amount, the resources tend to be inadequate. The EAG algorithm raises the priority of the GBR service which has high requirements on the transmission rate on the precondition of guaranteeing the delay. The rate of TypeB is far greater than TypeA while the relative priority is the same for GBR. TypeB preempt TypeA's resources. Therefore, the throughput of TypeA with EAG is slightly less than the other two algorithms when there are more users.

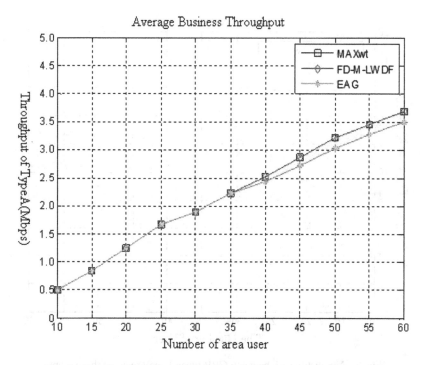

Fig. 3. Not distinguish between GBR types TypeA average throughput

Figure 4 is a graph of throughput simulation results for high-rate GBR services. When there are a lot of users, EAG algorithm compensates for high-rate GBR services and thus has higher throughput. The throughput of M-LWDF algorithm is quite different from EAG for typeB service. We can see that the M-LWDF algorithm only consider the delay factor and needs to be further improved in supporting the service requiring high transmission rate. In addition, MAXwt allocates more resources to TypeA service that require lower transmission rates. The disadvantage of not supporting high-rate GBR services is also apparent in the throughput curve of TypeB service.

Although the result of EAG in Fig. 3 shows that EAG can basically meet the demand for voice service TypeA, it is slightly lower than the other two algorithms. This is because the low-speed voice service TypeA is preempted by the high-speed service TypeB. When the number of users increases, if the voice service needs to be guaranteed first according to user's requirements, we can modify the GBR relative priority factor of the service TypeA k_{pf} to 2 according to the QCI level and the relative priority factor of TypeB is k_{pf} still set to 1. Therefore, the relative priority of TypeA GBR can offset the preemption caused by the high rate of TypeB to ensure the transmission of voice services. Advanced that the voice service requires lower rate, it will bring little impact on the throughput of TypeB services. Figures 5, and 6 are the simulation results.

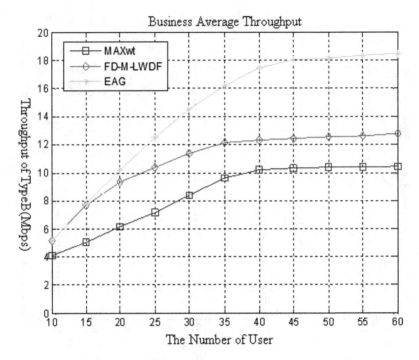

Fig. 4. Not distinguish between GBR types TypeB average throughput

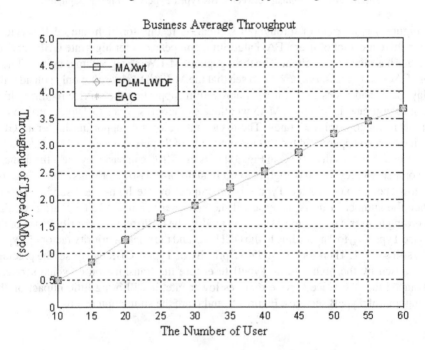

Fig. 5. Distinguished GBR types TypeA average throughput

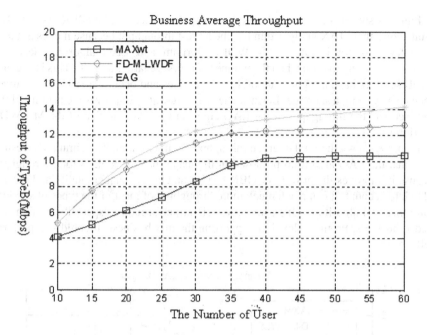

Fig. 6. Distinguished GBR types TypeB average throughput

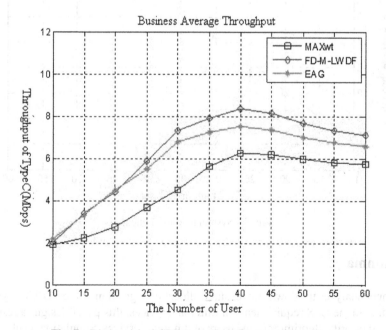

Fig. 7. Distinguished GBR types TypeC average throughput

Figure 7 shows the simulation result of TypeC service of Non-GBR service. The result shows that MAXwt algorithm has the lowest throughput because it does not take the delay into consideration. M-LWDF algorithm and EAG algorithm lead the throughput starting to drop when the number of users are larger than 40. This is because TypeB services have a higher rate of TypeC services and occupy non-GBR resources when resources are a shortage. Because the EAG algorithm improves the priority of the GBR service, the throughput of the EAG algorithm is less than that of the M-LWDF algorithm.

Figure 8 shows the total average throughput of the system. The simulation results show that the EAG algorithm proposed in this paper has the highest total throughput because it improves the priority of GBR service and takes channel quality and service delay into account. It can reach much more throughput of most of the types of services than other two algorithms. It shows an excellent performance especially in high rate needed business. From the level of supporting the mix business, it could still perform well.

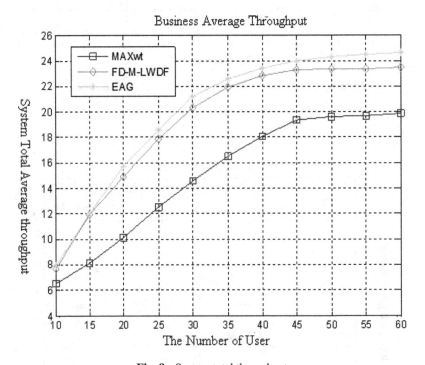

Fig. 8. System total throughput

5   Summary

This paper analyzes the downlink resource allocation algorithm of the TD-LTE system. According to the QoS requirements of different services, this paper designs a resource scheduled priority algorithm for the users of different services. Simulation results show that the algorithm proposed in this paper is better for mixed-type services than other algorithms. It guarantees different QoS business requirements as well as service queuing delay in each service. It can further access to higher system throughput and also reduce the delay of each user's services.

References

1. Kenington, P.B.: Linearized transmitters: an enabling technology for software defined radio. IEEE Commun. Mag. (10), 56–61 (2003)
2. ITU-R M. Requirements, evaluation criteria and submission templates. The development of IMT-Advanced, no. 11, pp. 144–147 (2008)
3. 3GPP TS 36.304 V8.6.0. Evolved Universal Terrestrial Radio Access (E-UTRA) User Equipment (UE) procedures in idle moden scbsc; Protocol specification
4. 3GPP TS 36.331. Evolved Universal Terrestrial Radio Access Network; Radio Resource Control (RRC); Protocol specification
5. Evolved Universal Terrestrial Radio Access (E-UTRA); Radio Resource Control (RRC) protocol specification. 3GPP TS36.222 (2007)
6. Padovani, R., Pankaj, R.: Data throughput of CDMA-HDR a high efficiency-high data rate personal communication wireless system. In: VTC, vol. 3, 1845–1858 (2000)
7. Calabrese, F.D.: Scheduling and link adaptation for uplink SC-FDMA systems (2009)
8. Asheralieva, A., Mahata, K.: A two-step resource allocation procedure for LTE-based cognitive radio network. Comput. Netw. (2013)
9. Li, Z.H., Wang, H., Pan, Z.W., Liu, N., You, X.H.: QoS and channel state aware load balancing in 3GPP LTE multi-cell networks. Sci. China Inf. Sci. (4) (2013)
10. Hsu, L.-H., Chao, H.-L., Liu, C.-L.: Window-based frequency-domain packet scheduling with QoS support in LTE uplink. In: IEEE 24th International Symposium on Personal, Indoor, and Mobile Radio Communication: MAC and Cross-Layer Design Track (2013)
11. The Vienna LTE simulators - Enabling reproducibility in wireless communications research. EURASIP J. Adv. Sig. Process. (1) (2011)

Assumption Queries Processing of Probabilistic Relational Databases

Caicai Zhang[1], Zongmin Cui[2(✉)], and Hairong Yu[3]

[1] Zhejiang University of Water Resources and Electric Power, Hangzhou, China
[2] School of Information Science and Technology, Jiujiang University, Jiujiang, China
cuizm01@gmail.com
[3] Qufu Normal University, Qufu, China

Abstract. Many prevail applications, such as data cleaning, sensor networks, tracking moving objects, emerge an increasing demand for managing uncertain data. Probabilistic relational databases support uncertain data management. Informally, a probabilistic database is a probability distribution over a set of deterministic databases (namely, possible worlds). Assumption queries in probabilistic relational databases have natural and important applications. To avoid unnecessary updates of probabilistic relational databases in existing general methods of assumption queries processing, an optimization method by computing conditional probability is proposed to handle assumption queries. The effectiveness of the optimization strategies for assumption queries is demonstrated in the experiment.

Keywords: Probabilistic relational databases · Representation
Possible worlds · Assumption queries · Probability

1 Introduction

Many modern applications, such as data cleaning [1], sensor networks [2], need to process uncertain data that are retrieved from diverse and autonomous sources [3]. Informally, a probabilistic database is a probability distribution over a set of deterministic databases (namely, possible worlds) [4].

It is common in applications for probabilistic databases that users would have *priori knowledge* (defined as C) about the database through other sources [5]. *Assumption queries* are useful in scenarios that when different users with their different priori knowledge or assumptions query over a probabilistic database, and prefer the query answer taking their assumption into account without affecting the state of the database. Koch and Olteanu [6] proposed two methods for assumption queries. The first method is to generate a temporary database conditioned on the assumption of the query, then evaluate query over the temporary database. Two types of assumption queries will benefit by this method, one is multiple assumption queries with the same assumption clauses, the other is the assumption clause affects a large number of data in the database, such as

© Springer Nature Singapore Pte Ltd. 2018
K. Li et al. (Eds.): ISICA 2017, CCIS 874, pp. 354–364, 2018.
https://doi.org/10.1007/978-981-13-1651-7_32

functional dependency constraints. While the data affected by the assumption is small or the assumption clauses vary frequently, the cost of generating temporary conditioned probabilistic database is too much. The second method is to combine the assumption into the query and compute the result probability directly. However, the second method is not able to compute the correct answer for all the assumption clauses, such as the existence of specific tuples [7]. In fact, the second method does not work correctly over correlated probabilistic databases. Therefore, a more general method for assumption queries is proposed in this article, where the assumption is represented by lineage.

The main contributions of this paper are as follows:

(1) A new evaluation approach is proposed for processing *query with assumptions*. The conditional probability of result of the conventional query under the given priori knowledge is calculated. The result of *query with assumptions* is obtained based on the result of conventional query and the conditional probabilities.
(2) The correctness of the proposed approach for processing query with assumptions is proved by showing that it obtains an equivalent result of a conventional query over the conditioned probabilistic database.
(3) An extensive experimental study is conducted to evaluate our approach for assumption queries processing in different configurations, showing its efficiency and scalability.

The remainder of this article is organized as follows. Section 2 describes related works. Section 3 gives necessary background on probabilistic databases. Section 4 gives the proposed assumption query processing method. Section 5 is our experimental study. Finally, Sect. 6 concludes the article with final remarks.

2   Related Work

In this section, we review related work in the areas of probabilistic database and conditioning.

There are significant amounts of work on representation formalism for uncertain data. In general, these work can be divided into two categories [8] based on different correlation assumption. The simple correlation model [9] associates existence probabilities with individual tuples and assumes that the tuples are mutually independent or exclusive; While the richer representation formalism allows complex correlations between tuples [10]. Many application domains naturally generate correlated data. Furthermore, dependencies among tuples arise naturally during query processing, even when the base data tuples are assumed to be independent [11]. There are several formalisms that can express complex correlations between tuples, lineage-based [12], possible worlds decompositions [13], U-relations [13], the variable-based representation [14].

The addition of knowledge to a probabilistic database instance is called conditioning which removes possible worlds that do not satisfy the given constraint, and thus possibly reduces uncertainty.

Koch and Olteanu [6] studied the conditioning problem for probabilistic relational databases over possible worlds decompositions. Tang et al. [15,16] proposed a framework called *D_based* for conditioning over the general uncertain data model. They designed optimized conditioning algorithms in P-TIME for mutually exclusive constraints and referential constraints. For arbitrary constraints, *D_based* encodes the set of satisfactory assignments of variables in the database (satisfactory possible worlds). These mutually exclusive satisfactory possible worlds are mapped to the formulae of a set of new Boolean variables. Each tuple updates its formula to a disjunction of mappings of satisfactory possible worlds where this tuple exists. Zhu et al. [17] proposed an approach called *C_based* that decreases the time complexity of the conditioning algorithm for an arbitrary constraint. Furthermore, the *C_based* approach results in concise conditioned probabilistic databases and efficient probabilistic query processing.

3   Preliminary

3.1   Probabilistic Databases

Definition 1. *A probabilistic database in the extensional representation* [18] *is a discrete probability space* $PDB = (W, P)$, *where* $W = \{w_1, \ldots, w_n\}$ *is a set of traditional databases, called possible worlds, and* $P : W \to [0,1]$ *is a function mapping from possible worlds to probability values, such that* $\sum_{j=1}^{n} P(w_j) = 1$.

The extensional representation of probabilistic databases can express any finite set of deterministic databases, however, it is not compact to enumerate all possible worlds and hence there is a need to have a concise representation formalism.

Definition 2. *A probabilistic database* $\mathbb{D}$ *in the intensional representation is a quadruple* $\langle \Re, E, P, f \rangle$, *where* $\Re$ *is a traditional relational database, E is a set of Boolean independent variables* $\{e_1, \ldots, e_m\}$, *P specifies the probability value of each variable being true. f associates each tuple t with a variable or a Boolean expression composing of variables, namely f(t), whose truth value determines the presence of t in the actual world and whose probability is defined by probabilities of composed variables* [19].

A possible world w_i is a traditional database instance such that the expression associated with the tuple in the possible world is true, the expression associated with the tuple not in the possible world is false. A joint value $V_j(E)(j \in [1, 2^m])$ of all variables in E determines a possible world w_i of the probabilistic database. $P(w_i) = \sum_{w_i \sim V_j(E)} P(V_j(E))$ and $\sum P(w_i) = 1$, where $w_i \sim V_j(E)$ represents the joint value $V_j(E)$ is a joint value leading to the possible world w_i.

3.2   Lineage

Lineage captures the dependencies that are needed for correct confidence computation, without restricting which query plans can be used. The confidences of the result tuples of a query are computed based on the probability of the lineage. A lineage of tuple t intuitively captures "where t came from" [20]. A lineage is represented in [20] as a mapping L that associates with each tuple identifier a Boolean expression whose symbols are other tuple identifiers in a database.

Initially, the lineage of each base tuple is its unique tuple identifier. When new data is generated by queries over probabilistic database, its lineage is a Boolean formula of DNF form over the lineages of tuples which lead to the generation of new data. Suppose a relation R is the result of a query over other relations $R_1, \cdots, R_m$. Then for each $t \in R$, $L(t)$ is a formula of DNF form, where each conjunctive clause involves tuple identifiers from $R_1, \cdots, R_m$, and the co-existence of all tuples appeared in one conjunctive clause lead to the existence of t.

A lineage represents only those possible worlds of a probabilistic database that are consistent with respecting to a lineage. For a given possible world w_i of a probabilistic database PDB, the value of $L(t)$ in PDB is true if t appears in w_i and false if it does not. The confidence of a result tuple t is the probabilistic evaluation of the Boolean formula $L(t)$, using the confidence values for the tuples identified in $L(t)$. Joins produce conjunctive lineage, duplicate eliminating projection produces disjunctive lineage.

4   Assumption Query Optimization

Definition 3. *Uncertain Query with Assumptions (AQ(Q, C)). Let $W_{\{D,C\}} = \{w_1, \cdots, w_m\}$ is the set of possible worlds of $\mathbb{D}$ that satisfy C, $P_{\{D,C\}}$ is the probability distribution of $W_{\{D,C\}}$, where for each $w_i \in W_{\{D,C\}}$, $P_{\{D,C\}}(w_i) = P_{\mathbb{D}}(w_i) / \sum_{w_j \in W_{\{D,C\}}} P_{\mathbb{D}}(w_j)$. A query with assumptions $AQ(Q, C)$, returns a set of tuple RT_{AQ}, $RT_{AQ} = \{t | w_i \in W_{\{D,C\}}, t \in Q(w_i)\}$, with the probability $P(t \in AQ) = \sum_{t \in Q(w_i)} P_{\{D,C\}}(w_i)$.*

Theorem 1. *Suppose that Q is a query over the probabilistic relational database $\mathbb{D}$, RT_Q represents the set of result tuples, and C is an assumption condition. If Q_C is a Boolean query with the assumption in the WHERE clause, and $L_{Q_C}(true)$ is the lineage of the result tuples. Then RT_{AQ}, which is the result of query Q with assumption C, satisfies the following rules.*
   (1) $RT_{AQ} \in RT_Q$ // (2) $\forall t \in RT_Q$, if $P(L(t)|L_{Q_C}(true)) \neq 0$, then $t \in RT_{AQ}$, and $P(t \in RT_{AQ}) = P(L(t)|L_{Q_C}(true))$
   where $L(t)$ is the lineage of tuple t.

Proof. Suppose that $\mathbb{D}'$ is the conditioned probabilistic relational database of $\mathbb{D}$, $PDB_{\mathbb{D}}(W_{\mathbb{D}}, P_{\mathbb{D}})$ is the equivalent possible world space of $\mathbb{D}$, and $PDB_{\mathbb{D}'}(W_{\mathbb{D}'}, P_{\mathbb{D}'})$ is the equivalent possible world space of $\mathbb{D}'$. Based on the definition of conditioned probabilistic relational databases [17], we have
   $W_{\mathbb{D}'} = \{w_i \mid w_i \in W_{\mathbb{D}}, C \sim w_i\}, P_{\mathbb{D}'}(w_i) = P_{\mathbb{D}}(w_i)/P(C)$
   where $P(C) = P(L_{Q_C}(true))$, $C \sim w_i$ represents that w_i satisfies C.
   Since $AQ_{PDB}(Q, C)$ is equal to executing the conventional query Q over $PDB'\{W', P'\}$, we have

$$RT_{AQ} = \{t \mid w_i \in W_{\mathbb{D}'}, t \in Q(w_i)\}$$
$$= \{t \mid w_i \in W_{\mathbb{D}}, C \sim w_i, t \in Q(w_i)\}$$
$$P(t \in RT_{AQ}) = \sum_{w_i \in W_{\mathbb{D}'}, t \in Q(w_i)} P_{\mathbb{D}'}(w_i)$$
and $RT_Q = \{t \mid w_i \in W, t \in Q(w_i)\}$
$\forall t \in RT_{AQ}$, we have $t \in RT_Q$, therefore
$$RT_{AQ} \subseteq RT_Q$$
$\forall t \in RT_Q$, given a w_i, if $L(t)$ is true, then $t \in Q(w_i)$, therefore, we have,
$$P(t \in RT_Q) = P(L(t)) = \sum_{L(t) \sim w_i, w_i \in W_{\mathbb{D}}} P_{\mathbb{D}}(w_i)$$
$\forall t \in RT_{AQ}$
$$P(t \in RT_{AQ}) = \sum_{L(t) \sim w_i, w_i \in W_{\mathbb{D}'}} P_{\mathbb{D}'}(w_i)$$
$$P(L(t) \wedge L_{Q_C}(true)) = \sum_{L(t) \sim w_i, C \sim w_i, w_i \in W_{\mathbb{D}}} P_{\mathbb{D}}(w_i)$$
$$P(L(t) \mid L_{Q_C}(true)) = \sum_{L(t) \sim w_i, C \sim w_i, w_i \in W_{\mathbb{D}}} P_{\mathbb{D}}(w_i) / P(L_{Q_C}(true))$$
$$= \sum_{L(t) \sim w_i, w_i \in W_{\mathbb{D}'}} P_{\mathbb{D}'}(w_i)$$

Therefore, we have the rule (2) in the theorem.
$\forall t \in RT_{AQ}, P(t \in RT_{AQ}) = P(L(t) \mid L_{Q_C}(true))$

By Theorem 1, a *query with assumptions* $AQ_{PDB}(Q, C)$ can be processed by the following steps:

1. The conventional query Q is evaluated without probability computation over PDB, a set of result tuple RT_Q and their lineage L are obtained.
2. For each tuple in RT_Q, its probability of being result of $AQ_{PDB}(Q, C)$ is computed.
3. Only the tuples with probability greater than 0 will be returned.

5   Experiments

In this section, we describe an experimental study on the query optimization strategies for assumption queries. This section is organized as follows. Section 5.1 describes the experimental environment setup. Section 5.2 study the performance of optimization strategies for assumption queries.

5.1   Experimental Environment Setup

A probabilistic relational database $\mathbb{D} = \langle D, E, P, f \rangle$ is actually stored as follows. D is stored as a traditional relational database (We use Oracle 10g), with an additional column f in each probabilistic relation; E and P are stored in another relation V_P. We implement three user-defined function: OR, AND. As described in query lineage, function OR generates an expression by linking each f with $\vee$, function AND generates an expression by linking each f with $\wedge$. For a query, the set of result tuples and the corresponding lineage expression are obtained by executing the query over the database management system, then the probabilities of answering tuples are obtained by computing the probabilities of lineage expression being true.

All of the algorithms are implemented in Java, and all of our experiments are conducted on a Pentium 2.5 GHz PC with 3G memory, on Windows XP.

5.2   Comparison of MV and Conditional for Assumption Query

In our proposed assumption query method (refer as Conditional), the lineage of the general query and the lineage of query where the conditioning constraints are true are computed individually, then the conditional probability of these lineage expressions are computed as the probabilities of answering tuples of the assumption query, which is the general query conditioned on the conditioning constraints. Koch and Olteanu [6] proposed two approaches for evaluating assumption queries. The first method (refer as MV) is to generate a new probabilistic relational database by updating the original probabilistic relational database on the conditioning constraints, and evaluate the general query over the new probabilistic relational database. The second method is to combine the conditioning constraints in the WHERE clause of the general query. Since the second method limits the types of the conditioning constraints, and can not apply on correlated probabilistic relational databases, it is not a general assumption query evaluation method. Therefore, in this experiment, we compare the Conditional method with the MV method only, and show their performance on different settings.

For the MV method, in the step of updating probabilistic relational databases, we apply the current most efficient conditioning method introduced in [17].

Correlated Parameters. The two methods have the same time cost on obtaining the set of answering tuples, because the conditioning updates the correlations among tuples only, not any attribute values. They have different time cost of computing probabilities of answering tuples. The time complexity of the MV method includes the time cost of updating probabilistic relational databases and evaluating the probability computation of lineage of answering tuples. The time complexity of database updating is $O(2^{|V_{L_{Q_C}(true)}|} + |V_{L_{Q_C}(true)}| \cdot |D|)$, and the time complexity of probability computation is $\sum_{t \in RT_Q} O(2^{|V_{L'(t)}|})$, where L' represents the lineage of query over the conditioned probabilistic relational database. The time cost of the Conditional method is $\sum_{t \in RT_Q} O(2^{max\{|V_{L(t)}|, |V_{L_{Q_C}(true)}|\}})$. It depends on the specific variables appeared in the conditioning constraints and lineage of the general query.

Let AsgnSet be the set of assignments of variables appeared in the conditioning constraints $V_{L_{Q_C}(true)}$ which can make the constraints true. If $|AsgnSet| > |V_{L_{Q_C}(true)}|$, then the variables in the conditioned probabilistic relational database is more than the original probabilistic relational database, and $|V_{L'(t)}| \geq |V_{L(t)}|$; If $|AsgnSet| < |V_{L_{Q_C}(true)}|$, then the variables in the conditioned probabilistic relational database is less than the original probabilistic relational database, and $|V_{L'(t)}| \leq |V_{L(t)}|$.

Data Sets. The data sets also include three tables $R(A, B)$, $S(B, C)$ and $T(C)$, where the f value of each tuple is an expression composed randomly by variables in $\{e_1, \ldots, e_{14}\}$.

For the assumption queries in the experiments, the query part Q is a traditional query over relational databases, and the number of answering tuples $|RT_Q|$ ranges from 2 to 14. The number of variables in the lineage of answering tuples $|V_{L(t)}|$ is fixed to be 14. There are two types of $L_{Q_C}(true)$, one is composed of variables in $\{e_1, \ldots, e_{12}\}$, the other is composed of variables in $\{e_1, \ldots, e_{15}\}$. In both types, the number of the satisfactory assignments of involved variables ranges from 8 to 15, and the number of variables in $|V_{L'(t)}|$ ranges from 11 to 18, where $L'(t)$ is the lineage of query over the conditioned probabilistic relational database.

Results and Analysis Effect of $|V_{L'(t)}|$. Figure 1 shows the time cost of Conditional and MV approaches when $|V_{L(t)}| = 14$, $|RT_Q| = 10$ and $|V_{L'(t)}|$ ranges from 11 to 18. Figure 1(a) shows the time cost of Conditional and MV approaches when $V_{L_{Q_C}(true)} = 12$, and Fig. 1(b) shows the time cost of Conditional and MV approaches when $V_{L_{Q_C}(true)} = 15$. **(1) Result.** The time cost of Conditional barely changes as $|V_{L'(t)}|$ does. The time cost of MV increases fast as $|V_{L'(t)}|$ increases. When $|V_{L'(t)}| > 14$, the time cost of MV approach nearly increases exponentially. As shown in Fig. 1(a), when $|V_{L'(t)}| < 12$, the time cost of MV is less than Conditional; When $|V_{L'(t)}| > 12$, the time cost of MV is greater than that of Conditional. As shown in Fig. 1(b), when $|V_{L'(t)}| < 14$, the time cost of MV is less than that of Conditional; When $|V_{L'(t)}| > 14$, the time cost of MV is greater than that of Conditional. **(2) Analysis.** When $V_{L_{Q_C}(true)} < |V_{L(t)}|$, the time cost of Conditional is affected by the number of variables in the lineage of query only, while when $V_{L_{Q_C}(true)} < |V_{L(t)}|$, the time cost of Conditional is affected by the number of variables in the conditioning constraint only. Therefore, the time cost of Conditional has nothing to do with $|V_{L'(t)}|$. The time cost of MV includes the cost during updating the probabilistic relational database based on the constraint, and the cost of query processing over the updated probabilistic relational database. The time complexity for probability computation of each answering tuple is $O(2^{|V_{L'(t)}|})$, then the time cost of Conditional increases fast as $|V_{L'(t)}|$ increases. When $|V_{L'(t)}| \geq max\{|V_{L(t)}|, |V_{L_{Q_C}(true)}|\}$, the time cost on probability computation of MV will surely be equal to or greater than that of Conditonal. Furthermore, MV needs the extra cost of updating the probabilistic relational database. Therefore, the time cost of MV is greater than Conditional, and their difference increases fast as $|V_{L'(t)}|$ increases. When $|V_{L'(t)}| < max\{|V_{L(t)}|, |V_{L_{Q_C}(true)}|\}$, the time cost on probability computation of answering tuples of MV is less than that of Conditional, the time cost of MV may be less that of Conditional, although MV needs extra cost of updating probabilistic relational databases.

Effect of Varying $|RT_Q|$. Figure 2 shows the time cost of Conditional and MV when $|V_{L(t)}| > |V_{L_{Q_C}(true)}| = 14$, $|V_{L'(t)}| = 12$, and $|RT_Q|$ ranges from 2 to 14. **(1) Results.** The time cost of Conditional increases linearly as $|RT_Q|$ increases, while the time cost of MV increases slowly as $|RT_Q|$ increases. When $|RT_Q| < 12$, the time cost of MV is larger than that of Conditional, and their difference decreases slowly as $|RT_Q|$ increases. When $|RT_Q| \geq 12$, the time cost of MV is

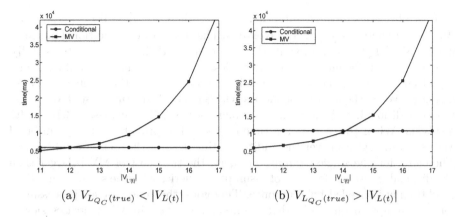

(a) $V_{L_{Q_C}(true)} < |V_{L(t)}|$ (b) $V_{L_{Q_C}(true)} > |V_{L(t)}|$

Fig. 1. The time cost of MV and Conditional when varying $|V_{L'(t)}|$.

less than that of Conditional, and their difference increases as $|RT_Q|$ increases.
(2) Results analysis. Since $|V_{L'(t)}| < max\{|V_{L(t)}|, |V_{L_{Q_C}}(true)|\}$, the time cost
of probability computation of MV is less than that of Conditional. However, when
$|RT_Q|$ is small, the time cost saved in probability computation is not enough to
cover the additional time cost in the step of conditioning in MV. Therefore, when
$|RT_Q| < 12$, the time cost of MV is larger than that of Conditional. As $|RT_Q|$
increases, the time cost saved in probability computation in MV increases. When
$|RT_Q| \geq 12$, the time cost saved in probability computation is enough to cover
the additional time cost in conditioning, then the time cost of MV is less than
that of Conditional.

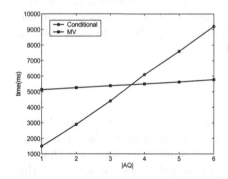

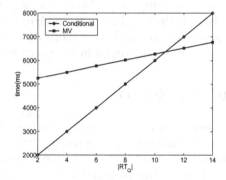

Fig. 2. The time cost of Conditional and MV when varying $|RT_Q|$.

Fig. 3. The time cost of Conditional and MV when varying the number of assumption queries.

Effect of Varying the Number of Assumption Queries. Figure 3 shows the
time cost of Conditional and MV when fixing the assumption and varying the

number of queries in the assumption queries, where $|V_{L(t)}| > |V_{L_{Q_C}(true)}| = 14$, $|V_{L'(t)}| = 12$ and $|RT_Q| = 1$. (1) **Results.** The time cost of Conditional increases linearly as $|RT_Q|$ increases, while the time cost of MV increases slowly as $|RT_Q|$ increases. When the number of queries is less than 4, the time cost of MV is larger than that of Conditional, and their difference decreases as the number of queries increases. When the number of queries is larger than 4, the time cost of MV is less than Conditional, and their difference increases as $|RT_Q|$ increases. (2)**Results analysis.** When $|RT_Q| = 1$, the time cost saved in probability computation of MV is not enough to cover the additional time cost in the step of conditioning. Therefore, for one single assumption query, the time cost of MV is larger than Conditional. However, for a set of assumption queries with the same assumption, conditioning is needed for only once. Therefore, the time cost for processing a set of assumption queries with the same assumption in MV increases slowly, since it mainly depend on the time cost of probability computation of lineage expressions of result tuples. However, the time cost of processing each assumption query in Conditional method barely changes, then the time cost of processing a set of assumption queries increases linearly as the number of queries increases. When the number of assumption queries is larger than 4, the time cost saved in probability computation of updated lineage expressions in MV is enough to cover the additional time cost of probabilistic databases updating conditioned on the assumption, then the time cost of the probability computation in MV is less than that of Conditional.

The suitable applications of MV and Conditional are concluded as follows based on the experimental results. MV is suitable for processing assumption queries when (1) the updated probabilistic database conditioned on the assumption includes less variables than the original probabilistic database; and (2) there are many tuples in the query result, or there are a set of assumption queries with the same assumption. Conditional is suitable for processing assumption queries when (1) the variables in the updated probabilistic database conditioned on the assumption is equal to or more than in the original probabilistic database; or (2) there are few tuples in the query result.

6   Conclusion

A new evaluation approach for processing assumption queries is proposed. The conditional probability of result of the conventional query under the given priori knowledge is calculated. The naive approach for assumption queries is that for each query with assumptions, a posteriori probabilistic database satisfying the given assumption is generated, then deleted after the query is finished. It is impractical to generate new database for each assumption. Our approach obtains result of query with assumptions without generating new databases.

There are several problems that emerge from this work and remain open. The assumption in this work limits to that which can reduce the number of possible worlds of the probabilistic database, so the assumption might be extended to that which introduces new possible worlds to the database or replace possible worlds of the database in the future work[1].

References

1. Ayat, N., Akbarinia, R., Afsarmanesh, H., Valduriez, P.: Entity resolution for probabilistic data. Inf. Sci. **277**, 492–511 (2014)
2. Škrbić, S., Racković, M., Takači, A.: Prioritized fuzzy logic based information processing in relational databases. Knowl.-Based Syst. **38**, 62–73 (2013)
3. Yang, F.P., Hao, M.I.: Effective image retrieval using texture elements and color fuzzy correlogram. Information **8**(1), 27 (2017)
4. Sen, P., Deshpande, A., Getoor, L.: PrDB: managing and exploiting rich correlations in probabilistic databases. VLDB J. **18**(5), 1065–1090 (2009)
5. Miklau, G., Suciu, D.: A formal analysis of information disclosure in data exchange. J. Comput. Syst. Sci. **73**(3), 507–534 (2007)
6. Koch, C., Olteanu, D.: Conditioning probabilistic databases. Proc. VLDB Endow. **1**(1), 313–325 (2008)
7. Yue, K., Wu, H., Liu, W., Zhu, Y.: Representing and processing lineages over uncertain data based on the bayesian network. Appl. Soft Comput. **37**, 345–362 (2015)
8. Dalvi, N., Ré, C., Suciu, D.: Probabilistic databases: diamonds in the dirt. Commun. ACM **52**(7), 86–94 (2009)
9. Cormode, G., Srivastava, D., Shen, E., Yu, T.: Aggregate query answering on possibilistic data with cardinality constraints. In: The 29th IEEE International Conference on Data Engineering (ICDE), pp. 258–269. IEEE Computer Society, Arlington (2012)
10. Fink, R., Olteanu, D., Rath, S.: Providing support for full relational algebra in probabilistic databases. In: The 27th IEEE International Conference on Data Engineering (ICDE), pp. 315–326. IEEE Computer Society, Hannover (2011)
11. Sen, P., Deshpande, A.: Representing and querying correlated tuples in probabilistic databases. In: The 23rd IEEE International Conference on Data Engineering (ICDE), pp. 596–605. IEEE Computer Society, Istanbul (2007)
12. Aggarwal, C.C.: Trio a system for data uncertainty and lineage. Manag. Min. Uncertain Data **2006**, 1151–1154 (2006)
13. Dan, O., Koch, C., Antova, L.: World-set decompositions: expressiveness and efficient algorithms. Theoret. Comput. Sci. **403**(2), 265–284 (2008)
14. Fink, R., Olteanu, D.: Dichotomies for queries with negation in probabilistic databases. ACM Trans. Database Syst. **41**(1), 4–47 (2016)
15. Tang, R., Cheng, R., Wu, H., Bressan, S.: A framework for conditioning uncertain relational data. In: Liddle, S.W., Schewe, K.-D., Tjoa, A.M., Zhou, X. (eds.) DEXA 2012. LNCS, vol. 7447, pp. 71–87. Springer, Heidelberg (2012). https://doi.org/10. 1007/978-3-642-32597-7_7

---

[1] This research was supported by the Science Project of Department of Water Resources of Zhejiang Province [grant number RC1746]; the National Natural Science Foundation of China [grant number 61762055]; the Jiangxi Provincial Natural Science Foundation of China [grant number 20161BAB202036]; and the Jiangxi Provincial Social Science "13th Five-Year" (2016) Planning Project of China [grant number 16JY19].

16. Tang, R., Shao, D., Ba, M.L., Wu, H.: Conditioning probabilistic relational data with referential constraints. In: Han, W.-S., Lee, M.L., Muliantara, A., Sanjaya, N.A., Thalheim, B., Zhou, S. (eds.) DASFAA 2014. LNCS, vol. 8505, pp. 413–427. Springer, Heidelberg (2014). https://doi.org/10.1007/978-3-662-43984-5_32

17. Zhu, H., Zhang, C., Cao, Z., Tang, R.: On efficient conditioning of probabilistic relational databases. Knowl.-Based Syst. **92**, 112–126 (2016)

18. Soliman, M.A., Ilyas, I.F., Chang, K.C.C.: Probabilistic top-k and ranking-aggregate queries. ACM Trans. Database Syst. (TODS) **33**(3), 13–19 (2008)

19. Fuhr, N., Rölleke, T.: A probabilistic relational algebra for the integration of information retrieval and database systems. ACM Trans. Inf. Syst. (TOIS) **15**(1), 32–66 (1997)

20. Sarma, A.D., Theobald, M., Widom, J.: Exploiting lineage for confidence computation in uncertain and probabilistic databases. In: The 24th IEEE International Conference on Data Engineering, pp. 1023–1032. IEEE Computer Society, Cancun (2008)

Design and Implementation of Self-balancing Robot Based on STM32

Ling Peng$^{(\boxtimes)}$ and Chunhui Zhou

Guangdong University of Science and Technology, Dongguan 523083, China
4829294@QQ.com

Abstract. The two-wheeled self-balancing robot has a simple structure, low cost and high flexibility, which is very suitable for indoor space. In this paper, we designed a self-balancing robot, by using the STM32 microprocessor as the main controller, and the attitude sensor MPU6050 is used to collect the obliquity and angular velocity. However, the gyroscope and the accelerometer make noise interference and drift error, the Kalman filter algorithm, therefore, is used to fuse the obliquity and angular velocity, in order to obtain the optimal obliquity. The PID control algorithm will combine the optimal obliquity and the real-time speed obtained by the high-precision encoder of the coaxial motor, to output the stable and reliable PWM signal, which can be sent to the motor drive chip. The motor drive chip can drive the operation of the two motors, to obtain the more ideal operation control effect. The results showed that the self-balancing robot could achieve stable self-balancing control.

Keywords: The STM32 microcontroller · The attitude sensor
The Kalman filter algorithm · The PID control algorithm

1 Introduction

With the continuous expansion of mobile robot applications, hardware and software research to deepen, the robot is facing the environment and the task that are more and more complex, and demanding more and more. The robot often encounters an amount of relatively narrow space and the need for a large corner of the workplace, such as a large amount of indoor space. In order to implement the task flexibly and efficiently in a more complex environment, the two-wheeled self-balancing robot [1] is born for such mission. In addition, due to the progress of society and science and technology, the advancement of urbanization makes the development of manned balance car very hot, which is based on the development of self-balancing robot. With its advantages of being flexible and lightweight, low cost, fast and smooth mobile, this new technology products' research and development will be fully utilized in many areas to improve people's work efficiency and quality of life.

Therefore, the study of the far-reaching technology as well as the development of intelligent technology are expected to achieve in the future stable and reliable carrier-based automation platform. Of course, due to technical and economic constraints, the main body of this paper is to explore its more stable control methods and

K. Li et al. (Eds.): ISICA 2017, CCIS 874, pp. 365–375, 2018.
https://doi.org/10.1007/978-981-13-1651-7_33

algorithms. The two-wheeled self-balancing robot designed in this paper combines the idea of autonomous movement, which overcomes the limitations of robot in many applications, and obtains the ideal operation control effect and achieves self-balancing.

2　Overall Design

This paper uses the MPU6050 sensor to collect the obliquity and angular velocity of the robot, using the serial port to transmit to the STM32 main controller. The STM32 will use the Kalman filter algorithm to fuse the obliquity and angular velocity, in order to obtain the optimal obliquity. The two motor coaxial high-precision encoder will obtain the real-time speed of the two motors, calculating the average of the two motor real-time speed as the real-time speed of the robot. The real-time speed of the robot and the optimal estimate of the obliquity will together constitute a closed-loop control system. The closed-loop control system uses the PID control algorithm to output the stable and reliable PWM signal, which can send to the motor drive chip. And then the motor drive chip will control the two motor operation, in order to achieve the robot's self-balancing control. The power supply module is responsible for the power supply of each module control circuit of the system. The overall framework of the system is as shown in Fig. 1.

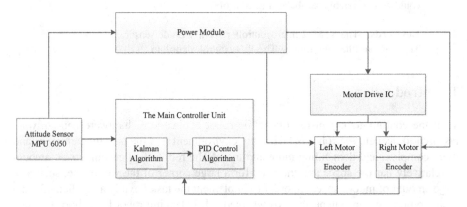

Fig. 1. The overall diagram of the system

3　Hardware Design

3.1　Master Chip STM32F103C8T6

The system uses the STM32F103C8T6 as the master chip, based on the ARM Cortex-M3 core, with 72 MHz maximum operating frequency, 64 K bytes of Flash program memory and 20 K bytes of RAM as well as 48 fast I/O ports. The STM32 supports a variety of communication interfaces, such as I2C, USART, SPI, CAN, etc. It can collect the MPU6050 attitude data through the I2C or USART to the STM32 main controller. The internal also contains two 16-channel AD converter and four 16-bit timers, through the algorithm, which will output a stable and reliable PWM signal to the motor drive module to control Motor stable operation [2].

3.2   Attitude Sensor MPU6050

The system uses the sensor MPU6050 to collect the robot's obliquity and angular velocity [2]. The MPU6050 is a six-axis motion handling sensor that integrates a three-axis gyroscope, a three-axis accelerometer, and an expandable digital motion DMP processor to detect the robot's attitude state accurately. The accelerometer is mainly used to detect the obliquity, and the gyroscope is mainly used to detect angular velocity. The three axes are the X-axis, Y-axis, Z-axis voltage values, and then through each axis corresponding to the 16-bit AD converter, the analog signal into digital output, and finally through the serial port to the main controller STM32. The angular velocity of the MPU6050 sensor is ±250, ±500, ±1000 and ±2000°/sec (dps), which can accurately track fast and slow motion, the user-programmable accelerator full-range sensing range is ±2 g, ±4 g, ±8 g and ± 16 g [3].

3.3   Motor Drive Chip

The STM32 main controller will process the collected data, and output a stable and reliable PWM signal to the motor driver chip. Theoretically, each channel of the module outputs up to 1 A continuous drive current, starting peak current up to 2A/3A (continuous pulse/single pulse). There are four modes controlling the motor: forward/reverse/brake/stop. The TB6612FNG chip can change the PWM duty cycle to achieve the two DC motor speed control. At the same time, the STM32 main controller with the motor coaxial high-precision encoder can obtain the speed of feedback. The modified speed can be obtained again by calculating, which lets the motor drive chip to control the motor.

3.4   Power Module Design

The STM32 main controller and the attitude sensor MPU6050 need 5 volts power supply, and the motor driver chip TB6612FNG and the DC motor need 12 V power supply. The system power supplies from a group of 3S battery pack, which can constitute a 12 V DC power supply. In order to meet the different hardware demand, the system selects the LM2596S_DC-DC buck chip, the chip interface leads to 3 pairs of 12 V pin and 5 pairs of 5 V pin to get a different voltage. The LM2596S_DC-DC reduces the voltage in the range of 3.2–46 volts to a specified voltage range in the range of 1.25–35 volts. The converted voltage value is changed by twisting the sliding rheostat. The input voltage range: DC 3.2 volts to 46 volts. The output voltage range: DC 1.25 V to 35 V, the voltage continuously adjustable, the maximum output current of 3A.

4 Software Design

4.1 Attitude Acquisition Algorithm

When the robot is in a stationary state, the accelerometer of the MPU6050 can be used to obtain a more stable obliquity angle. But when the robot is in a dynamic state, it is susceptible to gravity, or other equipment acceleration, or external vibration, mechanical noise, it can not get the exact obliquity angle. Coordinating with the gyroscope data of the MPU6050, it can make the tilt angle smoother because the gyroscope has better dynamic performance providing an instantaneous dynamic change of angle. However, the gyroscope is prone to be bias or drift. After the integral calculation, the bias is getting bigger and bigger, and it is not suitable for long-time dynamic measurement [4]. Based on the above reasons, this system uses the Kalman filter algorithm to accelerate the output value of accelerometer and gyroscope and obtain the optimal obliquity.

The Kalman filter is an effective recursive filter, through a string of measurement data containing noise, predicting the internal state of linear dynamic system. The Kalman filter calculates the weighted average of the predicted value and the measured value by predicting a certain value and evaluating the uncertainty of the predicted value, resulting in an estimate of the true value of the measured data and its associated calculated value. The maximum weight gives the least value of uncertainty. The Kalman filter output estimates are more likely to approach the true value than the original measurement data, since the weighted average has better estimates of uncertainty, as compared to the predicted value or compared to the measured value. The Kalman filter only needs to estimate the estimated value $X_{k|k-1}$ of the current state by the state $X_{k-1|k-1}$ estimated by the previous time and the current observation Z_k. It can obtain the optimal estimation of the current time finally [5].

Therefore, based on the optimal obliquity estimation of the $K - 1$ time and the obliquity of the accelerometer, the recursive algorithm is continued according to the covariance of the two until the optimal obliquity estimation of the K time is calculated. The filter block diagram is shown in Fig. 2.

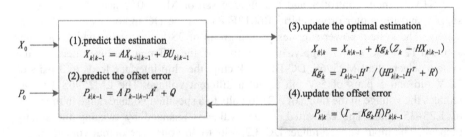

Fig. 2. The flow diagram of the attitude acquisition algorithm

(1) Using the optimal estimation of the $K - 1$ time, that can predict the estimation of the K time:

$$X_{k|k-1} = AX_{k-1|k-1} + BU_{k|k-1} \tag{1}$$

$$X_{k-1|k-1} = \begin{vmatrix} Angle \\ Q\_bias \end{vmatrix} \tag{2}$$

$$A = \begin{vmatrix} 1 & -dt \\ 0 & 1 \end{vmatrix} \tag{3}$$

$$B = \begin{vmatrix} dt \\ 0 \end{vmatrix} \tag{4}$$

In the formula(1), $X_{k-1|k-1}$ and $X_{k-1|k-1}$ are the estimation of the K and K − 1 time respectively, as a two-dimensional column vector, such as the formula(2), A is two-dimensional matrix as the formula(3), B is a two-dimensional column vector as the formula(4), $U_{k|k-1}$ is the state control quantity of the K time. The angular velocity Gyro measured by the gyroscope at the K − 1 time, is used as the control quantity of the predicted obliquity. In the formula(2), Angle is the obliquity estimation of the K time, Q_biasis the gyroscope bias of the K time.

(2) based on the actual offset error at k − 1 time,it can predict the offset error of the K time:

$$P_{k|k-1} = A P_{k-1|k-1} A^T + Q \tag{5}$$

$$Q = \begin{vmatrix} cov(Angle, Angle) & cov(Q\_bias, Angle) \\ cov(Angle, Q\_bias) & cov(Q\_bias, Q\_bias) \end{vmatrix} \tag{6}$$

$$Q = \begin{vmatrix} D(Angle) & 0 \\ 0 & D(Q\_bias) \end{vmatrix} \tag{7}$$

In the formula(5), $P_{k|k-1}$is the offset error covariance matrix of prediction at the K time,$P_{k|-1k-1}$ is the offset error covariance matrix of verity at the K − 1 time. A can refer to formula(1), A^T is the transposed matrix of A. Q is the covariance matrix of the vector in the formula(2), that is the formula(6). Because the bias noise and the angle noise are independent of each other, so cov(Angle, Q_bias) = 0 and cov(Q_bias, Angle) = 0. Owing cov(X, X) = D(X), the formula(6) can be simplified to the formula (7), D(Angle) is the process noise of the covariance at the obliquity state, D(Q_bias) is the process noise of the covariance at the gyroscope bias state. The different values will have different effects on the filtering effect as the Kalman filter is running continuously.

(3) The estimation value$X_{k|k-1}$ of K time is always combined with the measured value Z_k of the accelerometer to obtain the optimal estimate value $X_{k|k}$ of K time.

$$X_{k|k} = X_{k|k-1} + Kg_k(Z_k - HX_{k|k-1}) \tag{8}$$

$$Kg_k = P_{k|k-1}H^T/(HP_{k|k-1}H^T + R) \tag{9}$$

$$H = |1 \quad 0| \tag{10}$$

In the formula(8), $X_{k|k}$ is the optimal estimate value of K time, $X_{k|k-1}$ is the estimation value, according to the formula(1). Z_k is the angle measurement of the accelerometer. Kg_k is a Kalman gain. In the formula(9), $P_{k|k-1}$ is the offset error covariance matrix of prediction at the K time, according to the formula(5). H is a single dimension vector, that is the formula(10). H^T is the transposed matrix of H. R is the angle measurement noise value, that is, measurement bias.

(4) Update the actual offset error of K time.

$$P_{k|k} = (I - Kg_kH)P_{k|k-1} \tag{11}$$

$$I = \begin{vmatrix} 1 \\ 1 \end{vmatrix} \tag{12}$$

In the formula(11), $P_{k|k}$ is the offset error covariance matrix of verity at the K time. I is a two-dimensional column vector, such as the formula(12). Kg_k, H and $P_{k|k-1}$ see formula(9).

According to the initial obliquity X_0, the initial value Z_0 measured by the accelerometer, and the initial value of the offset error P_0, and then recursively use the above formulas until the obliquity estimation of the robot is obtained.

4.2   PID Control Algorithm

In the robot's self-balancing control, the PID control algorithm is used: P is the ratio, I is the integral, D is the differential, the three control algorithms have their own role. The ratio P reflects the current error of the system, the bigger the coefficient, the faster the adjustment is, the lower the error will be. The integral I is to reflect the cumulative error of the system, that can be used to eliminate the steady-state error. The differential D reflects the error of change ratio, that can be said to be predictable [6].

The two motor coaxial high-precision encoder will produce pulses, the pulse will be converted to obtain the real-time speed of the two motors, calculating the average of the two motor real-time speed as the real-time speed of the robot. The real-time speed of the robot and the optimal estimate of the obliquity that is obtained by the Kalman algorithm will together constitute a closed-loop control system. The closed-loop control system will output the stable and reliable PWM signal to send to the motor drive chip, and then to control the two motor operation, achieving the robot's self-balancing control, as shown in Fig. 3.

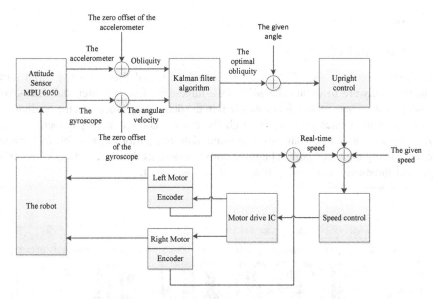

Fig. 3. The flow diagram of the PID control algorithm

The robot uses the PD control algorithm in the aspect of upright control. According to the optimal estimation of obliquity and the given obliquity, the deviation is calculated by the ratio and differential to obtain the balance control PWM:

$$Balance = K_p \times Angle + K_d \times Angle_{set} \tag{13}$$

In the formula(13), K_p is the p coefficient of the PD control algorithm, and K_d is the D coefficient of the PD control algorithm.

The robot uses the P control algorithm in the aspect of speed control. According to the real-time speed of the robot and a given speed to compare, the deviation is calculated by the ratio, in order to obtain the speed control PWM:

$$Velocity = K_{sp} \times Speed \tag{14}$$

In the formula(14), K_{sp} is the P coefficient of the P control algorithm.

The robot to achieve self-balancing control, the need for vertical and speed aspects of the PWM signal data fusion, and then output the PWM signal to DC motor drive chip by the STM32, as the formula(15).

$$PWM = K_p \times Angle + K_d \times Angle_{set} + K_{sp} \times Speed \tag{15}$$

5   Testing

5.1   Obtain MPU6050 Data

The STM32 connects the PC through the serial port, and then uses the serial port debugging software to obtain the MPU6050 three-axis accelerometer and three-axis gyroscope data. There were a lot of problem in the debugging process at the beginning, such as no output, messy code. Through the circuit, configuration parameters were modified time after time, including baud rate, data bit, parity bit, stop bit and frame rate, and finally the serial port debugging software outputs the correct data. Three-axis angle and three-axis acceleration data, as shown as Fig. 4.

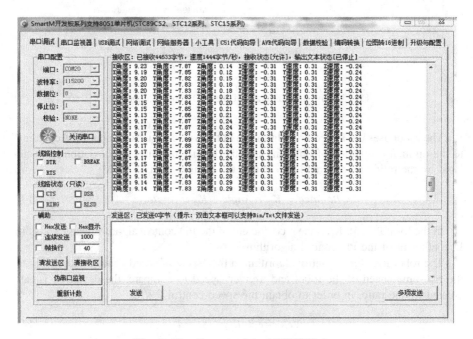

Fig. 4. The diagram of the MPU6050 data

5.2   Kalman Filter Algorithm Fusion Attitude Data

In order to verify the feasibility and effectiveness of the Kalman filter algorithm integrating the angle and angular velocity. The data fusion algorithm is transplanted into the STM32 main controller for debugging, and the parameters of the formula are set to the initial value, as shown in Table 1.

Table 1. Setting the Kalman filter parameter.

Parameter	X_0	dt	A		B	P_0	
Value	$\begin{vmatrix} 0 \\ 0 \end{vmatrix}$	0.002	$\begin{vmatrix} 1 & -0.002 \\ 0 & 1 \end{vmatrix}$		$\begin{vmatrix} 0.002 \\ 0 \end{vmatrix}$	$\begin{vmatrix} 1 & 0 \\ 0 & 1 \end{vmatrix}$	
Parameter	H	Q	R		I		
Value	$\begin{vmatrix} 1 & 0 \end{vmatrix}$	$\begin{vmatrix} 0.001 & 0 \\ 0 & 0.003 \end{vmatrix}$	0.5		$\begin{vmatrix} 1 \\ 1 \end{vmatrix}$		

The experimental results as shown in the Fig. 5, there is a large noise disturbance at the obliquity observed by the accelerometer, when not using the Kalman filter algorithm, and the output of the obliquity is $\pm 8°$. By using the Kalman filter algorithm, the output of the obliquity is $\pm 4°$, which can filter out the noise interference signal effectively.

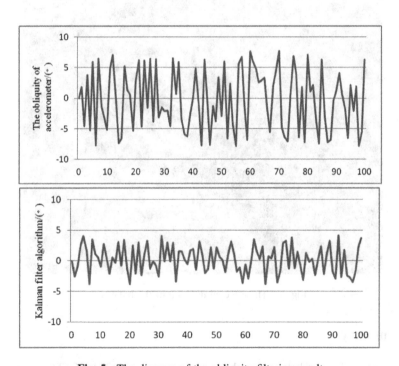

Fig. 5. The diagram of the obliquity filtering result

5.3 Setting PID Control Parameter

Through a number of experiments, the best parameters of the PID control algorithm can be obtained to achieve the robot self-balance. In the aspect of upright control, according to the formula(13), firstly the K_p is increased slowly, and the robot will stand up slowly. With the K_p value increasing, the robot will oscillate, so the K_p value needs to be reduced slightly. In addition, the K_d value is increased slowly, with the K_d value increasing, the robot will produce high-frequency jitter, so the K_p value needs to be reduced slightly. Finally set $K_p = 6.68$, $K_d = 0.67$, which can ensure that the robot is in equilibrium.

In the aspect of speed control, according to the formula(14), the K_{sp} is increased slowly, and the robot will move slowly. With the K_{sp} value increasing, the robot will lose balance and lean forward, so the Ksp value needs to be reduced slightly. Finally setting $K_{sp} = 6.81$, which can ensure that the robot move forward stably.

After the robot is energized, on an ordinary terrazzo floor, the robot needs 15 s to achieve self-balance, and the obliquity in equilibrium state is within $\pm 10°$. In order to maintain the dynamic balance of the robot itself, when the robot tilts forward, it needs to move forward to reduce the obliquity. Similarly, when the robot dumped, it needs to move backwards to reduce the obliquity. The effect of the robot is shown as the Fig. 6.

Fig. 6. The effected diagram of the balance

6 Summary

In this paper, the self-balancing robot chooses the STM32F103C8T6 as the main controller. The angle and angular velocity are collected by MPU6050. The self-balancing control of the robot is realized by the Kalman filter algorithm and the PID control algorithm. The experimental results show that the Kalman filter algorithm can effectively filter out the interference and deviation of the MPU6050 in the acquisition process, and obtain the optimal estimation value of the obliquity. The PID control algorithm integrates the optimal estimation of the obliquity with the robot real-time speed to achieve the robot self-balance.

References

1. Yuan, J.: The research of the two-wheel self-balancing robot on modeling, control and experimental research. Xi'an University of Electronic Science and Technology (2014)
2. Wang, S., Xiong, W.: The design of the two-wheel self-balancing car system based on STM32. Research and Exploration in Laboratory, no. 5 (2016)
3. Jun, W., Zhimin, L., Fangguo, W.: The design and implementation of the two-wheeled balancing mobile robots based on STM32. Electron. World **7**, 145–147 (2016)
4. CSDN. http://blog.csdn.net/zhuanghe_xing/article/details/7935251
5. Zhang, J., Lu, B., Wu, D.: Research on angular acceleration estimation method based on Kalman filter. Ind. Control Comput. (11) (2015)
6. Yaxin, H., Ankun, G., Lou, C., Bing, L.: The PID control of two-wheel self-balancing car. Inf. School Yangtze Univ. **9**, 3–4 (2013)

The Design and Implementation of a Route Skyline Query System Based on Weighted Voronoi Diagrams

Jiping Zheng[1,2(✉)], Yiwei Ding[1], Shunqing Jiang[1], and Zhongling He[1]

[1] College of Computer Science and Technology, Nanjing University of Aeronautics and Astronautics, Nanjing, China
{jzh,jiangshunqing}@nuaa.edu.cn, dingyiwei999@foxmail.com,
hebertclock@outlook.com
[2] Collaborative Innovation Center of Novel Software Technology and Industrialization, Nanjing, China

Abstract. Many applications of road networks are emerging nowadays and skyline routes in road networks are important for users. In this paper, we design and implement a route skyline query system based on weighted Voronoi diagrams. After introducing related techniques such as route skyline, weighted Voronoi diagram and Dijkstra's algorithm, we provide the design of our route skyline query system. The implementation of our system mainly focuses on user interfaces and integration of the weighted Voronoi diagram constructing algorithms as well as skyline route query processing algorithms. In addition, our system is able to display executing strategy of Dijkstra's algorithm and parallelize network weighted Voronoi diagram algorithm by using thread mechanism.

Keywords: Road network · Skyline route · Weighted Voronoi diagram

1 Introduction

Inspired by the advances in positioning technology, route search in road networks is one of the most commonly used location based services. It is not only important in transportation planning but also is being integrated into part of our daily life for widely used online location based services e.g. Google/Baidu Maps [1,2], Bing Maps [3], MapQuest [4] etc. With the help of GPS navigators, a user can conveniently find a desired route by specifying a source address and a destination address. However, route search is usually limited to a single criterion. In many travel scenarios, multiple criteria such as time, distance, toll charge need to be considered. Consider the scenario that users want to travel several places sequentially, e.g. a user wants to go to a supermarket, and s/he must find a parking lot first. That means, s/he needs to find a route passing through a parking lot and then a supermarket. As shown in Fig. 1, suppose that the user is at location q, and the gray areas R_1 and R_2 are close to supermarkets. Black

© Springer Nature Singapore Pte Ltd. 2018
K. Li et al. (Eds.): ISICA 2017, CCIS 874, pp. 376–389, 2018.
https://doi.org/10.1007/978-981-13-1651-7_34

points in the figure represent parking lots. To arrive at the destination as fast as possible, the user will choose p_3, p_4, p_5 or p_6 as the target parking lot. Actually, there are some criteria to evaluate a route for this, such as the prices of the parking lot and the supermarket, or the distances from the user to the two places. For given preferences and sequenced categories of places to go, it is necessary to find a route to meet the requirements.

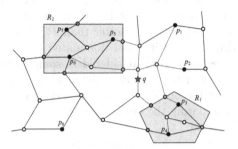

Fig. 1. Route query in a road network

In this case, it is unrealistic or meaningless to evaluate the quality of places to go with Euclidean distance. Instead, considering routes from the source point to the destination is a more reasonable choice. Factors including the costs of the routes, the distances, the traffic conditions and so on need to be considered. As it turns out, the problem becomes intractable to some extent.

In order to solve this problem, we design and implement a system that can obtain a set of optimal sequenced routes (OSRs) that are relatively superior, and they will not dominate each other in the factors mentioned above. To get OSRs on an attribute, we use the network weighted Voronoi diagram (NWVD) which has been discussed in our previous research [9,10] to organize the data points. Then we build Voronoi diagrams according to the point sets in reversed sequence and get routes through finding the nearest Voronoi cell from the previous point in the target route that has been found. Furthermore, we process route skyline queries on constructed NWVDs and design and implement our system based on all above techniques.

The rest of this paper is organized as follows. Section 2 discusses related techniques. Route skyline, weighted Voronoi diagram and Dijkstra algorithm are detailed for our system. The design of our system is provided in Sect. 3. Section 4 provides the description and implementation of our system. In addition, NVD and skyline route related algorithms as well as user interfaces and parallelization of NVD algorithm are proposed. We conclude this paper in Sect. 5.

2   Related Techniques

2.1   Route Skyline

Tian et al. [17] gave the definition of skyline route. In a road network with all points of the same category, s, t are source point and destination point respectively. The n-dimensional attributes of each route in a road network are represented by an n-dimensional vector $< c_1, c_2, \ldots, c_n >$. When the attribute value of any dimension of the route P is not less than the attribute value of route P', and P has at least one dimension better than P', then P dominates P'. A route is called a skyline route if it is not dominated by any other route. However, multiple attributes should be considered in some route skyline problems in road networks. These attributes include distance, time spent on the way, gas emission, traffic lights and so on. Each route has various values on their different attributes. Through comparing the dominance of attribute values, a number of skyline routes will be found.

Kriegel et al. [13] proposed the route skyline queries in planning multi-preference route. They proposed a route skyline query algorithm based on multiple optimization criteria (Basic Route skyline Computation, BRSC) and its improved algorithm (Advanced Route skyline Computation, ARSC). Since the number of candidate routes is huge, it is unrealistic to compute all possible routes between the starting point and the destination point. Therefore, the algorithm defines the lower bound estimate as the global clipping strategy and explores a local clipping strategy based on the Pareto-optimal sub-route to prune away sub-routes that cannot be extended into skyline routes. The algorithm considers a road network as a multi-attribute graph, each multiple criteria edge of which corresponds to a vector. In order to compute route skyline with high efficiency, the algorithm estimates the lower bound of each optimal criteria. If the estimate of a sub-route has been dominated by another one, the sub-route cannot be extended into the route skyline and will be pruned. Aljubayrin et al. [6] focused on the field of search for routes. They proposed the route skyline query of multi-category points of interest (POI). For example, there is a starting point s and a destination point d and a number of POIs with three categories (gas station, restaurant and supermarket). The value of each POI is defined based on the price and quality of the items to be purchased. When a user wants to go through each category of POIs and d from s, there may be a lot of possible routes to choose. Supposing route t_1 with short distance and high cost, route t_2 with long distance and low cost and route t_3 with relatively short distance and relatively low cost, we can see t_1, t_2 and t_3 all belong to route skyline since they do not dominate with each other. Users can choose the most satisfied route from them. They also proposed an off-line framework to estimate road network distance between a POI and a predefined geographical area containing querying points. The result of the proposed method is better than that of estimating based on Euclidean distance. They proposed two algorithms, namely WPOIs and CWPOIs. WPOIs algorithm re-defines weights of POIs based on the distance and cost between each POI and the querying point. It gets candidate skyline points of each category through changing the weight of two attributes. CWPOIs algorithm not only considers distances between each POI and the querying point, but also polymerizes some points close to each other with appropriate data structure (such as QuadTree or R-tree). It is more accurate to get route skyline when a number of POIs are gathered in a position.

We adopt above route skyline query techniques to build our system. The difference between our route skyline algorithm and the algorithms in [6,13] is that our route skyline algorithm is based on the network weighted Voronoi diagrams instead of the whole route space.

2.2 Weighted Voronoi Diagram

In mathematics, a Voronoi diagram is a partitioning of a plane into regions based on distance to points in a specific subset of the plane. That set of points is specified beforehand, and for each point there is a corresponding region consisting of all points closer to that point than to any other. The Voronoi diagram has a wide range of applications in route planning [8,14,15,18,19]. A road network can be represented by a two-dimensional undirected weighted graph $G(N, P, E)$, where N is a set of nodes in the road network, E is a set of edges and P is a set of POIs. Figure 2 shows a typical road network which can be found in the literature [11,12,14,16]. Its N is $\{n_1, n_2, n_3, \cdots, n_{13}\}$, E is $\{n_1n_2, n_1n_3, \cdots, p_3n_{13}\}$ and its P is $\{p_1, p_2, p_3\}$. When constructing an NVD from a road network, we divide network Voronoi polygons (NVP) based on the position of the target POI. Each POI p_i has its corresponding Voronoi polygon $NVP(p_i)$. To any point d' in the network, if d' belongs to the area of $NVP(p_i)$, the network distance between d' and p_i is less than that between d' and other target POI. The points in $\{b_1, b_2, \cdots, b_7\}$ in Fig. 2 are the boundary points of NVPs.

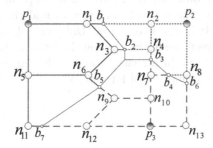

Fig. 2. Voronoi diagram of a road network

A weighted Voronoi diagram (WVD) in n dimensions is a special case of a Voronoi diagram. The Voronoi cells in a WVD are defined in terms of a distance function. Usually, the distance function is a function of the generator points' weights. Network weighted Voronoi diagram (NWVD) is a special form of weighted Voronoi diagram. In a road network, similarly, given a set of points P, we can construct an NVD. We divide the network so that each $p \in P$, there is a network Voronoi cell $NVC(p)$, where p is the generator of the cell. Differing from Euclidean space, a network Voronoi cell consists of a number of sections in the network, where the nodes in the sections meet the condition that the distance between one of them and the generator of NVC is the shortest. When points are weighted, the distance measurement between a node in the section and a point is the real distance in a road network between them plus the weight of the point. Figure 3 displays network weighted Voronoi cells of p_1, p_2 and p_3. The detailed construction procedure of network weighted Voronoi diagrams is described in Sect. 4.2.

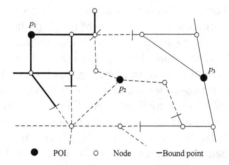

Fig. 3. Network weighted Voronoi diagram

2.3   Dijkstra's Algorithm

The optimal sequenced routes are the routes consist of points with different types instead of real routes in the network. In order to find the real ones, Dijkstra's algorithm is the best. It is well-known that Dijkstra's algorithm is used to find the shortest path between points in a graph. Dijkstra's algorithm is optimized for getting a shortest route in a sparse graph with a given source node and a given target node. Please refer to [5] for details.

3   System Design

In this section, we describe several use cases involved in the system. We outline the architecture used before presenting its implementation.

3.1   Use Cases

Preferences Input. Preferences here include a user's location, the sequence of places that the users want to go, and one or more preferred attributes. As a route-planning system, it must know where the user is. Usually the location should be a coordinate selected in the map or obtained through GPS. Since the requirement of getting a series of places is with different categories, it is necessary to select categories from a given list sequentially, so that target routes can be presented. Besides, considering various scenes, there are several attributes to choose from such as distance, convenience and quality of the place. Choosing an attribute means increasing the weight of it when planning routes.

Viewing the Result Routes. After getting the result routes, the system should present them to the user. Unlike Google Maps, which can provide a route from a starting point to a destination point, our system provides routes from a starting point to more than one destination point. Any place corresponding to the category in the sequence is also considered as a destination, and it may not be the last one. Therefore, it is necessary to mark the destinations to remind users. Moreover, since our system may present more than one route according to users' preferences in some cases, we need to list optimized routes for users to choose.

View the Map. As a map-like system, such as Google Maps, Baidu Maps and MapQuest so on, it is convenient for users to move the map not only by scrollbars but also by dragging the map directly. In addition, users should be able to zoom in and out the map by clicking "+/−" buttons or scrolling the mouse wheel, as we can see in Google/Baidu Maps.

3.2 Architecture

According to the technologies mentioned above, the part of the system for processing input data can be divided into five modules, namely, inputting module, NWVD building module, route skyline querying module, the module of finding routes in network with Dijkstra's algorithm and routes presenting module. In order to implement our system efficiently, our system is designed using a Model View Controller (MVC) architecture as shown in Fig. 4.

The model part is a data provider interacting with the data. The view includes preference inputting and route displaying models. According to the techniques mentioned above, the controller includes three modules, which are NWVD building module, route skyline querying processing module and the module of finding routes in network with Dijkstra's algorithm, which is the core of the system. The interfaces among the three modules has predefined formats for input and output. In addition, they will be executed sequentially. Hence, the architecture of pipes and filters ought to be applied in the part of the controller. Filter are used to convert a formatted input to a formatted output, and pipes are used to connect filters in order to ensure the communication of the three modules.

Actually, when the system is deployed as a sever instead of as a desktop software, it should interact with web applications, and the data flows between the view and the controller as well as between the model and the controller could be differentiated. For example, we will get user inputs from web browsers and need to show the routes on Google Map, or we may fetch map data from different data providers. Therefore, the interfaces between MVC are the same, regardless of where the data format is a Vector, JSON (JavaScript Object Notation) or XML. In the actual implementation, we will implement the common interfaces into the most suitable ones. This will increase the extensibility of our system.

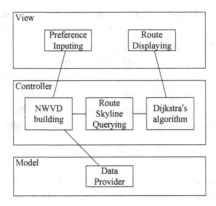

Fig. 4. Architecture based on MVC

4   System Description and Implementation

4.1   GUI Designs

When designing the style of graphic user interfaces (GUI) for the system, we follow the principle of Material Design proposed by Google as much as possible since it looks comfortable. Figure 5 shows the GUI of the navigation bar in our system. It will pop up when the menu button on top right clicked. The options in it are all functions of the system. Definitely the navigation bar is able to be extended, and we can add more options to it for new functions of the system in the future. Actually we added a "View NWVD" in the developing mode to see how NWVDs work and whether they work correctly. Figure 6 shows the GUI of inputting preference described in Sect. 3.1.

Figure 7 shows the GUI of the main window of our system. The map component is on the left side, which presents the map and the result routes described in Sect. 3.1. On the right side are the routes for users choosing to show on the map. More details about route demonstration are in Sect. 4.5.

Our system is designed to solve the problem of finding route skyline, hence it can be integrated into any map application including Google Maps and Baidu Maps. Actually, as an individual system, for the purpose of demonstrating it, it is necessary to design a set of user interfaces in order to interact with users easily. Our system is established based on Qt. Qt is a famous software development

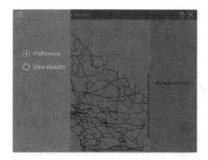

Fig. 5. GUI - navigation bar   **Fig. 6.** GUI - preference

framework with not only common interfaces but also flexible GUI design systems. We use QML (Qt Meta Language or Qt Modeling Language), a language looking like a combination of Javascript and Cascading Style Sheets (CSS), to construct GUI according to the style of Material Design. Another reason of using Qt is its excellent class libraries. It is convenient to develop our system based on Qt's libraries since it has encapsulated a large amount of useful APIs. Moreover, because it is written by C++ and optimized, programs based on Qt execute efficiently. Thus, our system can be migrated to server leaving GUI without any modification.

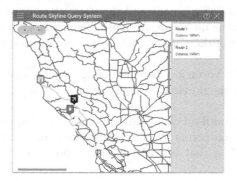

Fig. 7. GUI - main window

4.2   Build NWVDs

After the user inputs preferences, we get the location and a serial of categories of points defined as the sequence $C = (c_1, c_2, \ldots, c_m)$. We select an area by the location and the sets of points in the area with corresponding categories. Each category corresponds to a set of points.

The steps of constructing NWVD in Multi-Preference Sequenced Route Skyline (MPSRS) are as follows:

1. Divide the information into two parts. One of them is the information of points with different categories. Another part is the information of the road network, which can be constructed as a multi-attribute adjacency list (MAL). The MAL can be represented by $L(V, E)$, where V is a set of nodes in the network and $E \subset V \times V$ is a set of edges. W is a set of attribute vectors with d-dimensional positive weights on E. Each edge e has an attribute vector w.
2. Select all points with the last category in the sequence and add them to the MAL. Since the points are the destinations of traversal process, It is unnecessary to construct a WVD. Considering the points as nodes in the network, traverse the MAL to update the nodes and edges of it.
3. Select all points with the previous category in the sequence and add them to the MAL. Update the MAL. According to the characteristic of WVD, suppose a point p in the set of points P has the weight $w(p) \in R^d$, define the measurement of distance $d(x, p)$ as $D(x, p) + w(p)$, where $D(x, p)$ is the distance in the road network. For each node v in V, the algorithm records a tuple with three elements (v, p_v, d_v), where p_v is the nearest point to v and d_v is the network distance $D(v, p_v)$ between v and p_v respectively. If the nearest point of two nodes u and v in an edge $[u, v]$ is not the same one, i.e. $p_u \neq p_v$, p_u and p_v are the boundary points of two Voronoi cells on the edge $[u, v]$. The algorithm maintains a min heap h, in which are the entries of $< v, p >$. An entry has a weight $\Delta = D(v, p) + w(p)$, where $v \in V$ and $p \in P$. The nearest point to each node and the shortest distance between them are updated through the min heap. All nodes belonging to the same point set are added into the set of the Voronoi nodes of the point. Please refer to Algorithm 1 for details.
4. After constructing Voronoi cells of all points with the category, find the nearest point p'_j with the previous category to each point p_i. Set $D(p_i, p'_j)$ as the weight of p_i.
5. Repeat steps 3–4 until all categories of points are added into the MAL. Now an NWVD based on points of an attribute has been constructed completely.
6. Repeat steps 1–4 so that NWVDs based on points of all attributes will be constructed.

For a weighted point set, we use Algorithm 1 to construct NWVDs. To the NWVDs of an attribute, we can find a point of the Voronoi cell where the starting point is in. Iterate the process so that we will find an optimal sequenced route of the attribute. In Algorithm 2 we define $sfx(C, i) = (c_{i+1}, \ldots, c_m)$ is the suffix sequence of C with the size $m - i$.

We build NWVDs in reverse order of the sequence of categories. In order to find the required routes, the algorithm finds every point except the start one through traversing the NWVDs in order. Suppose a point in a route was found, the algorithm would find the nearest point to it in the previous NWVD as the next point in the route. Obviously, the first point is the location given. In consequence, we get a route from a serial of NWVDs.

Algorithm 1. Construct NWVC

Input: a set of point P in the network, $p \in P$
Output: NWVC of p

1   *initialize $L(V, E)$ and add points with a category $p_i \in P$ into L;*
2   **for** *each $v \in V$* **do**
3   **if** $v \in P$ **then**
4         $p_v = v$;
5         $d_v = 0$;
6         $w(v) = D(v, v')$;
7         $enheap(h, v)$;
8   **else**
9         $d_v = \infty$;
10        $p_v = -1$;
11 **while** *h is not empty* **do**
12     $v = $deheap$(h)$;
13   mark v;
14   **for** *each $e(v, w)$ and w is not marked* **do**
15        $\Delta = d_v + length(e) + w(v)$;
16   **if** $d_w = \infty$ **then**
17           $d_w = \Delta - w(v)$;
18           $p_w = p_v$;
19           $enheap(h, w)$;
20   **else if** $d_w < \infty$ *and* $\Delta < d_w$ **then**
21           $d_w = \Delta - w(v)$;
22           $p_w = p_v$;
23   update d_w in h;
24 return all $v \in V$ and $p_v = p$;

Algorithm 2. Preference Optimal Sequenced Route

Input: querying point q, sequence M
Output: an optimal sequenced route of the attribute l

1   $VC(p) = $ the Voronoi cell containing q in NWVD;
2   $p = $ the generator of $VC(p)$;
3   **if** $|M| = 1$ **then**
4   return p;
5   **else**
6   return $(p, O_l(p, sfx(M, 1)))$;

Figure 8 shows an NWVD of a category with multiple attributes in the map. The red circles represent POIs of the category, radii of which are the comprehensive weights of the attributes. The blue lines represent weighted Voronoi cells. Using Algorithm 2 we get an optimal route - the green line in the figure.

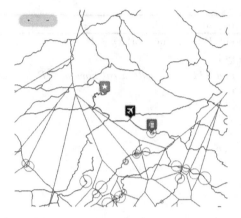

Fig. 8. Network weighted Voronoi diagram (Color figure online)

4.3 Route Skyline

For an MAL with d-dimensional route attribute, route r in it can be indicated as a point in d-dimensional route attribute space (RAS). Given an MAL, route $r = (s, v_1, v_2, \ldots, v_k)$ is a route starting from s and going through node v_1 to node v_k. $r' = (s, v_1, v_2, \ldots, v_i)$ is a sub-route of r. $r'.attr[\ ]$ is the attribute vector of r'. $r'.lb[\ ] = r'.attr[\ ] + (D_1(v_i, v_k), \ldots, D_d(v_i, v_k))^T$, where $D_l(v_i, v_k)$ is the lower bound of the network distance between v_i and v_k. The details of finding route skyline can be found in Algorithm 3.

Algorithm 3. Preference Optimal Route $O_l(q, M)$

Input: intialize candidate queue Q_{cand}, set of route skyline S_{RS} and set of sub-route skyline S_{SRS}.

Output: result set of route skyline

1 **while** Q_{sand} *is not empty* **do**
2     $r = Q_{rand}.top()$;
3   **if** *the category of r.lastNode is the last one in the sequence* **then**
4   **if** *r is not dominated by any route in* S_{RS} **then**
5             $S_{RS}.insert(r)$;
6   remove all route dominate by r in S_{RS};

7   **else**
8   compute attribute vector $r.lb[]$;
9   **if** $r.lb[]$ *is not dominated by any sub-route in* S_{SRS} **then**
10            $vecr = r.expand()$;
11   insert sub-routes in $vecr$ into Q_{cand};

12 **return** S_{RS};

4.4   Parallelization

Since the categories are sequenced, NWVDs of the same attribute are not independent but correlated. However, NWVDs of different attributes are irrelevant. Therefore, it is suitable to construct them with multi-thread or distributed computing. We define a class named *SkylineThread* inherited from *QThread*. *QThread* is a class provided by Qt to manage threads. Thus, we take full advantage of CPUs to parallelize the algorithms so that improving the speed of processing.

4.5   Route Demonstration

After finding the routes needed, what users want is to choose a route to go. However, the routes are made up of points, each of which is represented as a place. That means, when drawn on the map, the routes will not fit the road network. It is like two places are connected directly with a line. This is not what we expect. To solve this problem, we use Dijkstra's algorithm in the system. Because the location and the coordinate of each point in the routes are known, we are able to select the information of the area, including nodes and edges of the roads. Then it becomes a single source shortest path problem in a graph to find a path between one point and the next one. In our system, Dijkstra's algorithm is used to convert the routes to the paths fitting the road network. Qt provides some interfaces in order to draw lines and polylines on a canvas, which can draw a path consisting of a set of nodes.

To remind users of the destinations, icons standing for different categories are attached to the location in the map, for example Fig. 9 gives some helpful hints for users. Exactly the routes presented will go through these icons.

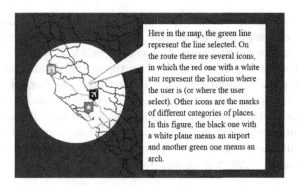

Fig. 9. Routes and icons in the Map (Color figure online)

Generally the icons are mapped with the categories one by one. Instead of writing hard code (such as *Enum* type or macro *define*), we create a table in the database to store the path of the icons and categories as well as the strings of them in the implementation. This will ensure the correction of the mapping and avoid the occurrence of missing one of them. It is also convenient to modify the system in case that we need to add or remove categories. The only difference is that the starting point is shown as a red icon with white star in it.

Most of the time users will get more than one route after querying. Our system makes the routes that are sorted by the preference chosen on the right of the map (as shown in Fig. 7), so that users are able to select their favorite route when they may not like the top one.

To achieve a good experience between users and our system, the map of our system is refreshed while some specific signals are emitted, such as zoom out button clicked, mouse wheel scrolled and mouse dragged/dropped. The system will call the corresponding method for each signal to zoom out the map or move it. Consequently, our system can get the same user experience as those common map applications (e.g. Google Maps, Baidu Maps).

When given a large amount of data, our system will spend a lot of time to execute algorithms if the running machines are not powerful enough. In that case, GUI of our system will be blocked by expensive calculation and will response slowly, which means users cannot do anything with it except waiting. A common solution for this problem is to set GUI thread as the main thread and open an extra thread to execute these time-consuming algorithms. We encapsulate all algorithms into several methods and integrate them into the pipe mentioned in Sect. 3.2. The pipe will be activated when receiving the start instruction. In the GUI thread, we add a waiting animation into the system. It is a *gif* image with a circle turning around so that users will know that the system is running to get the best solution for them.

5   Conclusion

In this paper, we combine the network weighted Voronoi diagram and the route skyline query to design and implement a route skyline query system in road networks. When given preferences, the system builds NWVDs and works out the route skyline which can be displayed on the map and different categories of places, such as school, hospital or supermarket etc., are marked with different icons. Our system is valuable to practical map applications for our techniques can be integrated into the existing map applications easily. In our future work, we focus on route skyline queries while query points are moving, historical query results can be utilized to avoid heavy computation. Our research about continuous skyline queries [7, 20] will be integrated to our system.

Acknowledgements. This work is partially supported by the National Natural Science Foundation of China under grant No. U1733112, 61702260, the Natural Science Foundation of Jiangsu Province of China under grant No. BK20140826, the Fundamental Research Funds for the Central Universities under grant No. NS2015095.

References

1. http://maps.google.com/
2. http://maps.baidu.com/
3. http://maps.bing.com/
4. https://www.mapquest.com/
5. https://en.wikipedia.org/wiki/Dijkstra's_algorithm
6. Aljubayrin, S., He, Z., Zhang, R.: Skyline trips of multiple POIs categories. In: Renz, M., Shahabi, C., Zhou, X., Cheema, M.A. (eds.) DASFAA 2015. LNCS, vol. 9050, pp. 189–206. Springer, Cham (2015). https://doi.org/10.1007/978-3-319-18123-3_12
7. Chen, J., Zheng, J., Jiang, S., Qiu, X.: Distance-based continuous skylines on geo-textual data. In: Morishima, A., Chang, L., Fu, T.Z.J., Liu, K., Yang, X., Zhu, J., Zhang, R., Zhang, W., Zhang, Z. (eds.) APWeb 2016. LNCS, vol. 9865, pp. 228–240. Springer, Cham (2016). https://doi.org/10.1007/978-3-319-45835-9_20
8. Fang, Z., Tu, W., Li, Q., Shaw, S., Chen, S., Chen, B.Y.: A voronoi neighborhood-based search heuristic for distance/capacity constrained very large vehicle routing problems. Int. J. Geogr. Inf. Sci. **27**, 741–764 (2013)
9. Jiang, S., Zheng, J., Chen, J., Yu, W.: Efficient computation of continuous range skyline queries in road networks. In: Huang, D.-S., Han, K., Hussain, A. (eds.) ICIC 2016. LNCS (LNAI), vol. 9773, pp. 520–532. Springer, Cham (2016). https://doi.org/10.1007/978-3-319-42297-8_48
10. Jiang, S., Zheng, J., Chen, J., Yu, W.: K-th order skyline queries in bicriteria networks. In: Li, F., Shim, K., Zheng, K., Liu, G. (eds.) APWeb 2016. LNCS, vol. 9932, pp. 488–491. Springer, Cham (2016). https://doi.org/10.1007/978-3-319-45817-5_52
11. Kolahdouzan, M., Shahabi, C.: Voronoi-based K nearest neighbor search for spatial network databases. In: International Conference on Very Large Data Bases, VLDB, pp. 840–851 (2004)
12. Kolahdouzan, M.R., Shahabi, C.: Alternative solutions for continuous K nearest neighbor queries in spatial network databases. Geoinformatica **9**(4), 321–341 (2005)
13. Kriegel, H.-P., Renz, M., Schubert, M.: Route skyline queries: a multi-preference path planning approach. In: IEEE International Conference on Data Engineering, ICDE, pp. 261–272 (2010)
14. Okabe, A., Boots, B., Sugihara, K.: Spatial Tessellations: Concepts and Applications of Voronoi Diagrams. Wiley, Hoboken (1992)
15. Safar, M., Ebrahimi, D., Taniar, D.: Voronoi-based reverse nearest neighbor query processing on spatial networks. Multimed. Syst. **15**(5), 295–308 (2009)
16. Sharifzadeh, M., Shahabi, C.: Processing optimal sequenced route queries using voronoi diagrams. Geoinformatica **12**, 411–433 (2008)
17. Tian, Y., Lee, K.C., Lee, W.-C.: Finding skyline paths in road networks. In: The ACM SIGSPATIAL International Conference on Advances in Geographic Information Systems, GIS, pp. 444–447 (2009)
18. Wallgrün, J.O.: Autonomous construction of hierarchical voronoi-based route graph representations. In: Freksa, C., Knauff, M., Krieg-Brückner, B., Nebel, B., Barkowsky, T. (eds.) Spatial Cognition 2004. LNCS (LNAI), vol. 3343, pp. 413–433. Springer, Heidelberg (2005). https://doi.org/10.1007/978-3-540-32255-9_23
19. Zhao, G., et al.: Voronoi-based continuous K nearest neighbor search in mobile navigation. IEEE Trans. Industr. Electron. **58**(6), 2247–2257 (2011)
20. Zheng, J., Chen, J., Wang, H.: Efficient geometric pruning strategies for continuous skyline queries. ISPRS Int. J. Geo-Inf. **6**(3), 91 (2017)

Improved RFID Anti-collision Algorithm Based on Quad-Tree

Hui Guan[✉], Zhaobin Liu, and Yan Zhang

Suzhou Vocational University, Suzhou, China
16938298@qq.com

Abstract. With the wide application of Radio Frequency Identification (RFID) technology in many fields, anti-collision algorithm to solve the problem of multi-tag identification becomes more and more important. The current RFID anti-collision algorithm is mainly divided into two categories: ALOHA based algorithm and tree based algorithm. The traditional tree based anti-collision algorithm has a long time and low efficiency. Based on this, this paper proposed an improved RFID anti-collision algorithm based on quad-tree. It can eliminate idle timeslots of the identification process by grouping and re-encoding the original ID code of electronic tag. The mathematical analysis and simulation results show that the identification performance of the proposed algorithm is greatly improved compared with other traditional tree based algorithms.

Keywords: Radio frequency identification (RFID) · Anti-collision · Quad-tree
Tag identification

1 Introduction

RFID is the acronym for radio frequency identification technology. It uses radio frequency signals through space coupling (alternating magnetic or electromagnetic fields) to realize the contactless transmission of information between the reader and the electronic tag, and achieve the purpose of automatic identification by the information transmission. As a major technology in the perception layer of Internet of Things, RFID is now widely used in logistics, warehousing, transportation and many other areas. However, within the identification range of a reader if there are multiple tags exist, the reader will detect conflicts when these tags simultaneously transmit information to it. This is called "collision" and can cause tags to be identified failure. To solve this problem, various anti-collision algorithms have been proposed.

The current RFID anti-collision algorithm is mainly divided into ALOHA based algorithm and tree based algorithm. ALOHA based algorithm based on the idea of using time division multiple access. Tag randomly selects a timeslot to send message, and if a collision occurs it randomly delayed for some time and then re-send. Because of the uncertainty of the time to read tags, this algorithm is prone to "starved" situation that tag is not read by reader for a long time. Tree based algorithm uses the idea of polling. It is in accordance with the laws of the binary combination, uses tree traversal algorithm to search for all possibilities until you identify the correct data. This algorithm's time to

© Springer Nature Singapore Pte Ltd. 2018
K. Li et al. (Eds.): ISICA 2017, CCIS 874, pp. 390–399, 2018.
https://doi.org/10.1007/978-981-13-1651-7_35

read the tag is determined, and can effectively solve the electronic tag's "starved". But because it has to traverse every possibility, it will cause delay longer to read tags. As the number of tags increases, the efficiency of the algorithm will reduce significantly.

On the basis of tree based anti-collision algorithm, an improved RFID anti-collision algorithm based on quad-tree is presented. By improving the structure of quad-tree, it reduces tag's identification time and increases the identification efficiency.

2 Tree Based Anti-Collision Algorithm

Tree is an important non-linear data structures. It is mainly composed of nodes and branches. The node is a data element in the tree. Each node has a parent node and child nodes besides the root and leaf nodes. Branch is a branch pointing to its child nodes. Tree based anti-collision algorithm groups tag's ID according to a certain length. When the group length is 1, it is a binary tree algorithm. When the group length is 2, it is a quad-tree algorithm. When the group length is 3, it is a oc-tree algorithm. And so on. Among them, the binary tree algorithm is the most widely used in RFID system. Its basic principle is as follows: When a reader requests tags to transmit their ID, it sends a prefix P_k of k bits together. Then, each tag within the response range of the reader is confirmed whether it is the same as the beginning part of its own ID. If it is the same, it responds its own ID to the reader. At this point, the three cases can occur. They are identification(only one tag beginning with P_k), collision(two or more tags beginning with P_k), and idle(no tag beginning with P_k). If a collision occurs, firstly add "0" to the rear of the original prefix P_k, to form the new prefix P_{k+1} (k + 1 bits). The reader then sends new prefix to tags, look for the left sub-tree. If a collision occurs again, repeat the above procedure until the successful identification of a tag. Then, add "1" to the rear of the original prefix P_k, to form the new prefix P_{k+1} (k + 1 bits). The reader sends new prefix to tags, and identifies by looking for the right sub-tree. Suppose there are three tags, its ID are: 0100, 0111, 1101. Figure 1 shows the identification process of these tags if using a binary tree anti-collision algorithm.

As can be seen from the above identification process, in the tree based anti-collision algorithm the entire tree nodes can be divided into four kinds. They are initial node, collision node, identified node, and idle node. The initial node has one and only one. The number of identified nodes is equal to the number of tags. The number of this two kinds of nodes is determined. Therefore, to improve identification efficiency, the key is to find ways to reduce the number of collision nodes and idle nodes.

In tree based RFID anti-collision algorithm, multi-tree is also applied in many cases besides the applications of binary tree, such as quad-tree. Tag's group length in quad-tree algorithm is 2. There are four kinds of encoding combinations: 00, 01, 10, 11. As mentioned earlier the same three tags: 0100, 0111, 1101, Fig. 2 shows the identification process of these tags using quad-tree algorithm.

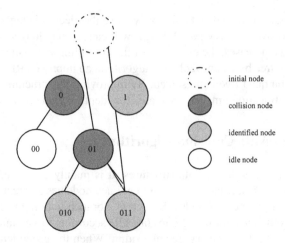

Fig. 1. Binary tree anti-collision algorithm's identification process

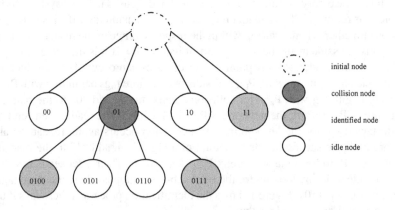

Fig. 2. Quad-tree anti-collision algorithm's identification process

By contrasting the identification process of quad-tree algorithm and binary tree algorithm can be found: fewer collisions nodes and more idle nodes in the quad-tree algorithm. If by improving the quad-tree structure, reduce or even remove the idle nodes, it will greatly improve the identification efficiency.

3   Improved Anti-collision Algorithm Based on Quad-Tree

Improved anti-collision algorithm based on quad-tree mainly through the introduction of grouping re-encoding mechanism, improves spanning tree structure, removes idle timeslots, and shortens the identification time, thereby improves the identification efficiency. The whole algorithm consists of two parts: grouping re-encoding and tag identification.

3.1 Grouping Re-Encoding

The algorithm firstly groups the original tag ID code. The original ID code from highest bit to lowest bit is divided into pairs. If the last group less than two bits, then fill '0'. Thus, there are four kinds of combinations in each group: 00, 01, 10, 11, corresponding to the four branches of the quad-tree. Then re-encode the ID code of each group, replace it with new code. Re-encoding rules is shown in Table 1.

Table 1. Re-encoding rules.

Two bits ID code of each group	The corresponding decimal number d	Substitute code(d-th bit set to '1',low bits all set to '0')
00	0	1
01	1	10
10	2	100
11	3	1000

As can be seen, the maximum length of the new code is 4 bits, the minimum length is 1 bit. So the average length is 2.5 bits, slight increase over the two bits ID code of each group. The new code has a remarkable feature: the highest bit is 1, the low bits are all 0. Reason for using this coding rule is to make the reader have a better discrimination in the identification process. Using the position of collision can directly determine what substitute code is involved in the collision.

The above operation of grouping re-encoding is realized by the tag, thus requiring tags to have the ability to store and generate the substitute code.

3.2 Tag Identification

In the identification process, the electronic tag can be in three kinds of state:

Active state: When the reader transmits initialization command to the tags in the response range, all tags will enter into the active state. In addition, when the tag code and the request code sending from reader are consistent, the tag will also be in active state.

Quiet state: When the tag code and the request code sending from reader are different, the tag will temporarily exit communication connection, go to the quiet state, wait for the next to be activated.

Sleep state: When a tag has been identified, it enters into the sleep state. In the following communication process it does no longer respond to anything until the end of whole identification process.

Transformation relationship between these three kinds of state is shown in Fig. 3.

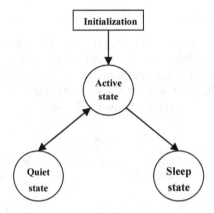

Fig. 3. Transformation relationship between tag's states

Specific identification process is described as follows:

1. Reader sends initialization commands to all tags entering response range, and let n = 0.
2. Tags in active state respond. If the reader detects no collision occurs, the identification is successful, and flow goes to 4. If a collision occurs, so that the tag make the group's serial number add 1(let n = n + 1), and reader sends substitute code of the n-th group.
3. Reader receives substitute code sent from tags. By identifying the position and quantity of collisions, reader determines the specific value of the tag's substitute code, and pushes it and n onto the stack s.
4. The reader checks whether the stack s is empty. If not empty, flow goes to 5, otherwise goes to 6.
5. The code and n pop up from the top of the stack s. Reader requests tags by sending the code on the top of the stack. Flow goes to 2.
6. All tags identification is completed.

In identification process, we specify identification sequence of four substitute codes is: 1, 10, 100, 1000.

Figure 4 shows the entire identification process.

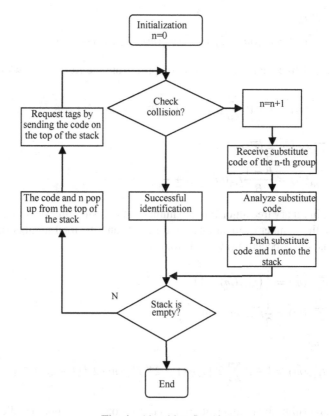

Fig. 4. Algorithm flowchart

4 Performance Analysis and Simulation

The time spent by reader on the identification of all electronic tags in the response range generally uses the number of timeslots as a unit. It is one of the important criteria to measure the performance of anti-collision algorithm. The less the number of timeslots, the more excellent performance of the algorithm. So, next we deduce the number of the total timeslots of improved RFID anti-collision algorithm based on quad-tree by using method of mathematical analysis, compare and simulate it with the binary tree algorithm and quad-tree algorithm.

4.1 Mathematical Analysis

Set m as the number of tags to be identified, $\overline{t_{TA}}(m)$ be the mathematical expectation of the total number of timeslots, $\overline{c_{TA}}(m)$ be the mathematical expectation of the number of collision timeslots, $\overline{z_{TA}}(m)$ be the mathematical expectation of the number of idle timeslots, $\overline{s_{TA}}(m)$ be the mathematical expectation of the number of identification timeslots. In the anti-collision algorithm based on B-tree,

$$\overline{s_{TA}}(m) = m$$

$$\overline{t_{TA}}(m) = \overline{c_{TA}}(m) + \overline{z_{TA}}(m) + \overline{s_{TA}}(m) = \overline{c_{TA}}(m) + \overline{z_{TA}}(m) + m$$

Let L be the B-tree's number of layers (depth), from the Ref. [7] shows that:

$$\overline{t_{TA}}(m) = 1 + B \sum_{L=0}^{\infty} B^L[1 - (1 - B^{-L})^m - mB^{-L}(1 - B^{-L})^{m-1}]$$

$$\overline{c_{TA}}(m) = \frac{1}{B}\left(\overline{t_{TA}}(m) - 1\right)$$

$$\overline{z_{TA}}(m) = \frac{B-1}{B}\overline{t_{TA}}(m) - m + \frac{1}{B}$$

Therefore, in binary tree algorithm the number of collision timeslots, the number of idle timeslots, the number of identification timeslots and the total number of timeslots (both mathematical expectation) were as follows:

$$\overline{c_{BTA}}(m) = \frac{1}{2}\left(\overline{t_{BTA}}(m) - 1\right)$$

$$\overline{z_{BTA}}(m) = \frac{1}{2}\overline{t_{BTA}}(m) - m + \frac{1}{2}$$

$$\overline{s_{BTA}}(m) = m$$

$$\overline{t_{BTA}}(m) = 1 + 2\sum_{L=0}^{\infty} 2^L[1 - (1 - 2^{-L})^m - m \cdot 2^{-L}(1 - 2^{-L})^{m-1}]$$

in quad-tree algorithm the number of collision timeslots, the number of idle timeslots, the number of identification timeslots and the total number of timeslots (both mathematical expectation) were as follows:

$$\overline{c_{QTA}}(m) = \frac{1}{4}\left(\overline{t_{QTA}}(m) - 1\right)$$

$$\overline{z_{QTA}}(m) = \frac{3}{4}\overline{t_{QTA}}(m) - m + \frac{1}{4}$$

$$\overline{s_{QTA}}(m) = m$$

$$\overline{t_{QTA}}(m) = 1 + 4\sum_{L=0}^{\infty} 4^L[1 - (1 - 4^{-L})^m - m \cdot 4^{-L}(1 - 4^{-L})^{m-1}]$$

In improved anti-collision algorithm based on quad-tree, the original ID code of tags were grouped(two bits each group) and re-encoded, and were replaced with new code. The distribution of "1" in new code of each group is the only, thereby achieve purpose of reducing the number of idle timeslots to 0 in the identification process of tags. Its search tree structure is optimized into a non-idle node quad-tree. Compared with traditional quad-tree algorithm, there are:

$$\overline{z_{IQTA}}(m) = 0$$

$$\overline{c_{IQTA}}(m) = \overline{c_{QTA}}(m) = \frac{1}{4}\left(\overline{t_{QTA}}(m) - 1\right)$$

$$\overline{s_{IQTA}}(m) = m$$

$$\overline{t_{IQTA}}(m) = \overline{c_{IQTA}}(m) + \overline{z_{IQTA}}(m) + \overline{s_{IQTA}}(m)$$

$$= \overline{c_{IQTA}}(m) + \overline{z_{IQTA}}(m) + m$$

$$= \frac{1}{4}\left(\overline{t_{QTA}}(m) - 1\right) + m$$

$$= \frac{1}{4}\left\{1 + 4\sum_{L=0}^{\infty} 4^L[1 - (1 - 4^{-L})^m - m \cdot 4^{-L}(1 - 4^{-L})^{m-1}]\right\} + m - \frac{1}{4}$$

$$= \sum_{L=0}^{\infty} 4^L[1 - (1 - 4^{-L})^m - m \cdot 4^{-L}(1 - 4^{-L})^{m-1}] + m$$

4.2   Algorithm Simulation

Next, we use MATLAB to simulate the above mathematical analysis.

Figure 5 shows comparison of collision timeslots between the binary tree algorithm, the traditional quad-tree algorithm and improved quad-tree algorithm. As can be seen, in the same number of tags, the improved quad-tree algorithm's number of collision timeslots is equal to the traditional quad-tree algorithm's, and they are much less than the binary algorithm's.

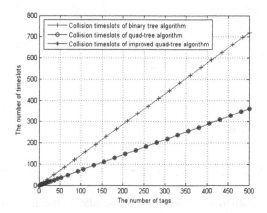

Fig. 5. Simulation comparison chart of tree algorithms' collision timeslots

As can be seen in Fig. 6, the traditional quad-tree algorithm's number of idle timeslots is more than the binary tree algorithm's. But improved quad-tree algorithm through the introduction of grouping and re-encoding mechanism, improves the structure of quad-tree, thus completely eliminate the idle timeslots.

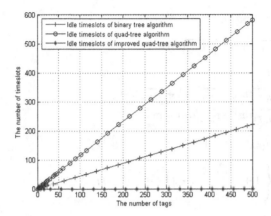

Fig. 6. Simulation comparison chart of tree algorithms' idle timeslots

Figure 7 is the simulation comparison chart of three algorithms' total number of timeslots. Through the simulation results can be seen, improved quad-tree algorithm's total number of timeslots is less than the binary tree algorithm's and the traditional quad-tree algorithm's, about 60% of the binary tree algorithm's and the traditional quad-tree algorithm's. Moreover, the more the number of tags, the wider the gap.

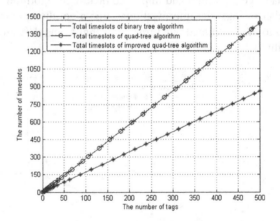

Fig. 7. Simulation comparison chart of tree algorithms' total timeslots

From the above mathematical analysis and simulation it can illustrate that tag identification performance of improved RFID anti-collision algorithm based on quad-tree is significantly better than the other two algorithms, it has a higher identification efficiency.

5   Conclusion

By the means of grouping and re-encoding the tag's original ID code, improved RFID anti-collision algorithm based on quad-tree improves the structure of quad-tree, eliminates idle timeslots, and greatly improves the identification efficiency of tags. Mathematical analysis and algorithm simulation further proves that the identification time of this algorithm is significantly less than the traditional binary tree and quad-tree algorithm in the case of same number of tags. With the increase in the number of tags, this advantage becomes more apparent. Therefore, this algorithm has a certain practical significance in solving the RFID anti-collision problem under dense tag environment.

Acknowledgment. This research was supported by the National Science Foundation of China (61672372 and 61472211). Special thanks to Professor Zhaobin Liu, head of the project team, for my concern and assistance, and for the good advice and strong technical support given by the project team colleagues.

References

1. Klair, K.D., Chin, K.W., Raad, R.: A survey and tutorial of RFID anti-collision protocols. IEEE Trans. Wirel. Commun 105–117 (2010)
2. Finkenzeller, K.: RFID Handbook: Fundamentals and Applications in Contactless Smart Cards and Identification. Wiley, New York (2003)
3. Myung, J., Lee, W., Srivastava, J.: Adaptive binary splitting for efficient RFID tag anti-collision. IEEE Commun. Lett. **10**, 144–146 (2006)
4. Myung, J., Lee, W.: Adaptive binary splitting: a RFID tag collision arbitration protocol for tag identification. Mob. Netw. Appl. **11**, 711–722 (2006)
5. Shin, J.-D., Yeo, S.-S., Kim, T.-H., Kim, S.K.: Hybrid tag anti-collision algorithms in RFID systems. In: Shi, Y., van Albada, G.D., Dongarra, J., Sloot, P.M.A. (eds.) ICCS 2007. LNCS, vol. 4490, pp. 693–700. Springer, Heidelberg (2007). https://doi.org/10.1007/978-3-540-72590-9_100
6. Yeh, K.-H., Lo, N.W.: An efficient tree-based tag identification protocol for RFID systems. In: 22nd International Conference on Advanced Information Networking and Applications, WAINA, pp. 966–970 (2008)
7. Hush, D.R., Wood, C.: Analysis of tree algorithms for RFID arbitration. In: Proceedings of IEEE Symposium on Information Theory, Cambridge, MA, USA, pp. 107–116 (1998)
8. Landaluce, H., Perallos, A., Zuazola, I.J.G.: A fast RFID identification protocol with low tag complexity. IEEE Commun. Lett. **17**, 1704–1706 (2013). Publication of the IEEE Communications Society
9. Li, Z., He, C., Li, J., Huang, X.: RFID reader anti-collision algorithm using adaptive hierarchical artificial immune system. Expert Syst. Appl. **41**, 2126–2133 (2014)

Artificial Intelligence and Robotics – Intelligent Engineering

Discussion on the Important Role of Computer-Aided Intelligent Manufacturing in the Transition of Garment Industry to Softening Production

Ping Wang(✉)

Department of Art, Guangdong University of Science and Technology,
Dongguan, China
28993716@qq.com

Abstract. Nowadays, with the quick development of the whole society, the development of computer technology is also rapid. At the same time, computer aided technology plays an more and more important role of CAM (Computer aided manufacturing) in human production and life in the process of technology, which uses the computer aided manufacturing to complete the whole process from the production activities to help improve the efficiency.

Keywords: Computer aided design · Fashion design garment industry
CAM in human production

1 Introduction

In recent years, all walks of life in our country have enjoyed rapid development. Whether manufacturing or agricultural products, our products are sold at home and abroad. But, only the clothing industry in our country is relatively backward, to train technical personnel. It has a wealth of experience and work independently for long periods of time. On the one hand, in the clothing industry training cycle is the workshop scheduling process, such a limited number of personnel, on the other hand. The process design is not normative and inconsistent. By repeating work process, to bring a lot of waste, especially China's garment industry has been the trend of small batch and multi varieties, short cycle, high quality change direction, so the traditional process design has to meet the requirements, not market of enterprises so the research on computer aided process planning, to improve the garment industry to meet the market demand, strengthen the standard of garment design, improve product quality and production efficiency is very important [1].

Therefore, China's clothing market becomes more and more broad, in order to enhance the comprehensive strength, occupy the market share, all the clothing enterprises have a major breakthrough in the form of production and production equipment. The clothing CAD is an inevitable trend, its efficiency and accuracy will be widely used in practical applications.

K. Li et al. (Eds.): ISICA 2017, CCIS 874, pp. 403–412, 2018.
https://doi.org/10.1007/978-981-13-1651-7_36

The clothing industry is different from other manufacturing industries. It has many kinds of fabrics and different performances. The varieties, styles and uses of the clothing products are also various. It is a product with complicated shape and different structure. Different types of garment factories have their own characteristics in the organization and management of production, and the production processes of different clothing products are not exactly the same. However, from the direction of view, clothing production has common laws, the production process is roughly the same, so the flow of garment manufacturing, layout and design of the guide process of machining process, directly affect the quality of the production smoothly and guarantee products, is a key link in garment processing.

For example, in 2012, the book "Intelligent Computing and Application in Intelligent Manufacturing System of Computer Clothes" published by China Textile Press introduced in detail the principle and application range of intelligent manufacturing of computer-aided clothing. The authors of this book are Wang Dongyun, Ouyang Ling and Wang Yonglin, all three of whom are pioneers in the computer-aided clothing manufacturing industry. The author of this book uses the intelligent computing technology to study the optimal scheduling and optimizing nesting in the computer intelligent garment manufacturing system, and elaborates on the intelligent computing method and its application in the computer intelligent manufacturing system.

2   The Basic Theory of Computer Aided Manufacturing Technology

Computer aided manufacturing technology is mainly divided into two parts, one is computer aided design, namely CAD technology and the other is computer aided manufacturing, namely CAM technology. The so-called CAM technology use computers instead of people to complete the work related to manufacturing and manufacturing system [1]. Computer aided manufacturing (CAM) design is the core. CAD computer integrated manufacturing system (CIMS in the parts after the CAPP process scheduling model), production process, and ultimately generate and simulate the machining track in CAM, NC code generation, so as to control the NC machining. It can be said that the strength of direct CAM system function Then determines the success or failure of the entire design process, the effectiveness of CAD is ultimately reflected by CAM [2]. The principle and structure of computer-aided manufacturing are more complex, the links and principles involved are also very much. For example, the following Fig. 1, is the computer-aided intelligent manufacturing in the operation of a principle diagram.

Therefore, under the situation of manufacturing industry of that the current product update cycle is short, the requirements quality is high, parts of shape are complex, the data processing will become the most important means of processing. What is more, the function of the CAM system is also very important. The general CAM understand the contents for the use of the computer aided design of CNC machining instructions. So the CAM system generally includes parts of geometric modeling, machining path definition, process of machining simulation, processing code (NC code) and other functions [3].

Fig. 1. Computer-aided intelligent manufacturing in the operation of a principle diagram

The so-called fashion CAD technology refers to computer aided garment design, the process of clothing using computer hardware and software, new products and clothing in accordance with the basic requirements of fashion design, to carry out the corresponding design, technical input and output. The clothing CAD technology is an effective combination of the traditional and modern fashion design the computer is the product from design to manufacturing process information integration and information automation. Intelligent CAD system is based on intelligent human-computer interface technology, pattern recognition, cognitive expression of design thinking mechanism, clothing printing expert, knowledge base, 3D human body measurement, human body modeling and so on. There are a lot of research contents in every link. The system supports the designer's traditional way of working and uses multiple design methods to do the job. The intelligent systems make the professional level of the user's request is not high, as long as the general description of design ideas, the system can produce 2D model and three-dimensional style effect through intelligent function, modified by intelligent interaction, to complete the final design. And then using the network technology to connect the system with the consumer terminal, we can realize the online design of clothing consumers, customize their own styles, and realize the rapid response of the garment industry.

The following table is the main component of the digital control management system (Table 1).

Table 1. Digital control management system

Digital control management system		
Core functions	Measurement data acquisition	SA
	Moving target planning	Collision detection analysis, the formation of industrial computer motion data
	Information inquiry and measurement data export	Product number tooling, preassembled video process file measurement data
	Original information entry	Product, digital tooling, measuring theoretical data, preassembled video process files

3 Computer Aided Design Accelerates the Process of Garment Design

The clothing industry has long used the traditional manual design, layout and layout, which wastes more labor, leads low efficiency and long production cycle. This has greatly affected the competitiveness of China's garment industry in the international market. As the requirements of the market for garment fit, diverse, beautiful and novel grow, it makes the clothing industry transform to the direction of the apparel industry more small batch to develop. Clothing CAD research belongs to the computer industry in the garment industry is more mature areas, including: printing, layout, nesting system, part of the CAD company launched the style design system and three-dimensional fitting system. The former provides various drawing tools, image processing and fabric technology, clothing designer clothing effect diagram drawing good effect diagram by a review, not satisfied with the style he will be too satisfied to modify or use a version of the division system re creation. The printing system provides a variety of auxiliary operations to improve the model production speed and save the cost of the enterprise.

The industrial-developed countries in the world are speeding up the application of high-tech industrial development–the CAD technology of clothing. In the recent years, during the process of the application of the clothing CAD, not a little clothing enterprises promote the producing efficiency, strengthen the quality of production, which achieves some effects. Below can explain.

The use of clothing CAD technology can help a fashion designer dispense from the hard daily work. Setting out a series of pieces often need several weeks and several people, but the use of this technology can quickly provide a complete set of specifications or non-standard pieces, in this way, the layout quality uniforms, the clothing patches are of good quality. As a result, clothing manufacturers can meet the needs of the market in time, enhance the competition ability of the product [4].

The computer can lift the artificial heavy labor and conduct multiple rearrangements, to reach a higher rate of materials. In this way, materials can be saved. The application of computer layout is conducive to the management, and layout information can be passed to the CNC cutting machine directly in the machine cut.

The model of the human body can be stored in the computer, which can change the size and design clothing styles adjusted all body types, and print every newest clothing type quickly to be modified in the screen. The use of clothing CAD system graphics and color change function can complete color clothing effect diagram on the screen in order to improve the efficiency of garment design.

In summary, it is useful for clothing manufacturers to use CAD technology of clothing, which can help to improve product quality, constantly update product varieties, enhance market competition ability and make clothing manufacturers achieve greater efficiency. The CAD technology of clothing has been the strong tool for clothing manufactures to improve the market competitiveness.

4 Computers Provide Large Amounts of Data for Intelligent Manufacturing

As early as 1970s', in order to make garment production adapt to the changeable market, some developed countries and regions began to study the production organization and management system with high efficiency, short cycle and good product quality. They have introduced the development of rapid response to the same goal of production systems, injection units, production systems, etc., these systems are called flexible production systems. Flexible production means that you can change the process flow and its combination system more freely. Now, in fact, it means to shorten the time of putting sew from textile to clothing, and then send the ready-made clothes to retailers as much as possible.

40 years has passed, nowadays people improve that system. They use the clothing CAD technology which can replace the papers and pencil in the electronic computer, designing the receipt in a relatively short period of time out of fashion sketch, or according to the requirements of users to design the corresponding style map, can modify at any time in the design process and accelerate the process of costume design.

The color of the matching can be displayed on the color screen of the CAD system, and the color can be modified at any time until it meets satisfactory. The special color matching device can be used to save the output of the better colors.

Using this technology can help conduct the clothing size grading. The body, height and weight of each person are different [5], and these features can be combined to many types again, if the designers decide to design the same type personally for many different body types, which is a heavy work.

Moreover, the changes of clothing is accelerating day by day, the popular cycle is increasingly short, which makes a lot of time and energy spent on the clothing design. When designing different sizes with a style of clothing design, the creative play of designers will also be affected. However, clothing CAD technology can be very convenient in the design of a moderate size clothing, according to a few differences of different size help to design garments of all sizes.

In addition, this technology can provide interactive nesting, tablet was presented above in fluorescence run, beneath the simulation of cloth, the operator can the material is defined as group nesting, nesting pieces can easily rotate, flip, reaching the appropriate position.

Finally, it can also help to adjust the model of garment. The garment can be sleeved by adjustment function into short, the collar will be changed into round sleeve and so on. These measures are really very convenient to the garments manufactures.

5 Computers Aid Garment Industry Manufacturing, Providing Intelligent Manufacturing Processes, Saving Manpower and Improving Production Efficiency

Intelligent clothing CAD is to make the CAD system to a certain extent with designers in general intelligence and thinking, rather than just a replay process, which will lead to the design automation depth.

First of all, the traditional machine equipments are mostly high energy consuming equipment, by compressing and elimination of small high energy consuming equipment, purchasing the energy-saving new intelligent equipment and new materials or equipment upgrades to reduce energy consumption and achieve energy-saving effect, such as cotton group introduced the new technology for producing new material consumption a source of energy in the process is greatly reduced, and the processing of new materials in cotton without boiling, bleaching, dyeing, not only saves the processing [6], but also reduces the consumption of water, electricity and steam, which can save energy and reduce costs.

Secondly, waste materials produced in the clothing manufacturing process are recycled and recycled so that waste resources can be renewed. According to Scientific Outlook on Development and the requirements of sustainable development, improve the ability of independent innovation, energy conservation, improve resource utilization, increase environmental protection efforts and the responsibility of the enterprise, is the theme in the future development of textile and garment.

Finally, by reducing the cost of management and control of operational risk, the enterprise should change the idea, to develop a more comprehensive strategy, change the mentality, facing the rising exchange rate, interest rate and improve the deposit reserve rate and the pressure of micro management of operational risk, control risk seeking ways to avoid risks, reduce the cost and improve the enterprise profit space.

With the help of the system of CAD, CAM, CAPP, the computers aid garment industry manufacturing really simplify the production process, improve the efficiency of the related work, add the interests finally. However, because of the clothing is a flexible body set function, comfort and aesthetics in one, has the particularity and difficulty of the application of intelligent technology in garment industry than in machinery manufacturing, construction and other fields. First of all, the clothing style design system is analyzed. Fashion design in human intelligence thinking belongs to the creation of the image thinking, the existing AI technology to the human logical thinking can be expressed as symbolic reasoning, relatively more easily by computer simulation, but the creative thinking is an extremely complex process, with unpredictable and not repeatable, human beings have to take the research for this kind of thinking, the computer simulation is more difficult. Therefore, there are still many things to do to realize the intelligent clothing design system with computers.

6 Computer Aided Intelligent Manufacturing Accelerates the Transformation and Upgrading of Garment Industry to Flexible Production

Computer aided garment industry production includes traditional modules such as computer aided design, computer aided process planning and computer aided manufacturing. With the development and application of the network technology, intelligent technology, an information flow as the core, from the design and manufacture, production management to the new mode of production and marketing of digitization and integration is becoming the development direction of the apparel industry. Industrial upgrading is an important direction of the current Chinese social and economic development, it is a necessary way for the garment industry facing rising labor costs, enterprises can only rely on existing resources and intelligent equipment, through continuous integration to realize industry upgrading. This chapter will focus on the key aspects about clothing flexibility, such as the CAD, CAPP, CAM stage. This effectively cope with rising labor costs in the following way:

First, the intelligence of clothing CAD system. In the process of industrial upgrading, through intelligent equipment, intelligent production to improve production efficiency instead of manual efficiency, monitor the whole production from textile fabrics and garment processing process, intelligent production can not only improve the production level and production efficiency of textile and clothing industry, and the indirect alleviate the shortage of hard labor the dilemma. Generally speaking, the existing CAD system mainly provides auxiliary drawing and computing functions, and stereoscopic visualization and intelligence are becoming the development trend of CAD system. Generally speaking, the existing CAD system mainly provides auxiliary drawing and computing functions, and stereoscopic visualization and intelligence are becoming the development trend of CAD system. At present, the research results of visualization are more. For intelligent research, we have achieved some local intelligent functions. However, there are still many problems to be solved in order to achieve fully intelligent CAD system.

Secondly, the intelligence of the CAM system. From the late 60s to the early 70s of last century, clothing machinery suppliers began to introduce computer-aided cutting and sewing technology. The technology greatly improves the cutting accuracy and sewing efficiency, but there are still a lot of artificial assistant operations, which affects the efficiency of the whole system. In the process of industrial upgrading, technology and system, improving the efficiency of production is no longer rely on the extensive human put into production, but through the effective personnel training system strictly, make the limited staff to achieve the production efficiency with the corresponding systems and smart devices, which can reduce labor costs rise as a whole.

Thirdly, the intelligence of CAPP system. Clothing CAPP is a bridge connecting CAD and CAM. The direct meaning of clothing CAPP is to use computer to assist in the arrangement and design of garment manufacturing process documents". In fact, in this way, the computer automatic search, inference and decision making of information and knowledge, technological processes and content generation can guide the actual production, computer production information feedback, can help the division process evaluation of production and quality, and to provide a rapid and accurate management of data, in order to control production the line, to achieve "more, high, short and small targets.

According to the processing requirements of different design styles, the equipment and personnel are balanced and arranged reasonably, so that the design information of clothing is transformed into the processing information of clothing. The initial research in computer aided garment industry is not very clear from the CAPP module, but with the development of garment industry integrated manufacturing, agile manufacturing requirements, an important part of the process design as preparation for manufacturing technology, it is necessary to the development of informatization and intellectualization. In the current situation, the world economy is in the stage of slow recovery under the influence of the financial crisis, the growth of domestic economic slowed down, the trend shift from rapid growth to low growth trend, China garment industry is facing competitive pressures from both developed and developing countries, facing the rising cost and labor shortage [7].

7 Conclusion

With the accession of China to the WTO, the digital technology becomes more and more popular and the trend of the global economic comes to integration, informatization, digitization, which affects all aspects of human life. At the same time, the clothing has also undergone tremendous changes. The impact of information technology on the traditional manufacturing industry production and development process, mainly through the auxiliary technology of computer automation, industrial design and production of high efficiency and flexibility, through the implementation of intelligent machines and products of elector-mechanical integration. The development of computer technology is also rapid. At the same time, computer aided technology plays an more and more important role of CAM (Computer aided manufacturing) in human production and life in the process of technology, which uses the computer aided manufacturing to complete the whole process from the production activities.

As an important livelihood industry in our country, the clothing industry takes up a large amount of marketing share. However, in recent years, with the rising labor costs, which has resulted in the increased cost of production factors, along with the gradual increase of trade barriers, China's garment manufacturing cost advantage is no longer obvious. In the future, with the problem of aging population becoming more and more serious and the population structure unbalanced, the labor cost will continue to rise. At the same time, developed countries in Europe and America "re industrialized" to seize the high-end market share of manufacturing, developing countries, with low-cost advantage to seize the low-end market share, China is facing "front and rear pincer attack" double challenges. The clothing industry has a strong advantage of traditional elements, but with the development of science and technology, a variety of forms of consumer diversity, traditional elements of advantage gradually into a disadvantage of intelligent manufacturing technology innovation as a new situation, with high added value, not easy to be imitated the remarkable characteristics, like a stream into the textile and garment industry, which improves the production efficiency again, the promotion of innovation competitiveness. The change from traditional manufacturing to intelligent manufacturing transformation will actually promote the flexible transformation and upgrading of garment industry [8].

According to above, the design ability of a clothing manufacture is an important provenience to balance the ability of a enterprise. Fashion design in human intelligence thinking belongs to the creation of the image thinking, the existing AI technology to the human logical thinking can be expressed as symbolic reasoning, relatively more easily by computer simulation, but the creative thinking is an extremely complex process, with unpredictable and not repeatable, human beings have to take the research for this kind of thinking, the computer simulation is more difficult. Therefore, there are still many things to do to realize the intelligent clothing design system with computers.

The rapid development of information technology has fundamentally changed the way to collect, process and utilize information, and also put forward new requirements for decision-making and response speed, thus leading to great changes in the form of organization. In this change, the computer system can strengthen the direct communication between the executive layer, and greatly reduce the role of middle management, thus reducing the management level and information distortion, and weakening the scale of the institution. On the other hand, management will also be transformed, and fully realize the decentralization. As a result, the enterprise can be optimized in a larger scope. The industrial production of garment gradually develop from the labor-dispensed types to knowledge and technology-types. The application of computer technology has totally changed the process of production of a clothing enterprise. If the clothing CAD system wants to be intelligent, there are still many problems to be solved. Among which, the popular application of the clothing CAD is one of the core parts, the computer aided manufactures of intelligence do play an imperative role in the transformation and upgrading of garment.

Acknowledgments. The dissertation is part of the Guangdong institute of science and technology college scientific research project 《Based on the analysis and research of the transformation and upgrading of the dongguan garment industry to the flexible production mode》 (item number: GKY-2017KYYB-27) phased research fruits.

References

1. Woods, G.P.: Computer aided pattern generation for the garment industry. Queen's University of Belfast (1989)
2. Hardaker, C.H.M., Fozzard, G.J.W.: Computer-aided designers? A study of garment designers' attitudes towards computer-aided design. Int. J. Cloth. Sci. Technol. 7(4), 41–53 (1995)
3. Liu, Y.J., Zhang, D.L., Yuen, M.F.: A survey on CAD methods in 3D garment design. Comput. Ind. 61(6), 576–593 (2010)
4. Cheng, W., Cheng, Z.L.: Applications of CAD in the modern garment industry. Appl. Mech. Mater. 152–154, 1505–1508 (2012)
5. Jayaraman, S.: Computer-aided design and manufacturing: a textile-apparel perspective. In: Acar, M. (ed.) Mechatronic Design in Textile Engineering. Springer, Dordrecht (1995). https://doi.org/10.1007/978-94-011-0225-4_17

6. Fang, J.J., Tien, C.H.: Customized garment creation with computer-aided design technology. In: Handbook of Research in Mass Customization and Personalization (In 2 volumes), pp. 833–851 (2009)
7. Felix, A.: The effect of computer aided design technology (CAD) on employee job satisfaction among users of Computervision Personal Designer Software application. Universidad De Chile (2009)
8. Barfield, W., Shieldst, R., Cooper, S.: A survey of computer-aided design: implications for creativity, productivity, decision making, and job satisfaction. Int. J. Hum. Factors Manuf. 3 (2), 153–167 (1993)

A Study of Miniaturized Wide-Band Antenna Design

Rui Zhang[1], Jianqing Sun[2(✉)], Yongzhi Sun[2], Bin Lan[1], and Sanyou Zeng[1(✉)]

[1] School of Mechanical Engineering and Electronic Information,
China University of Geosciences, Wuhan 430074, China
zrkeen@foxmail.com, 874432599@qq.com, sanyouzeng@gmail.com
[2] No 8511 Research Institute of CASIC, Nanjing 210007, China
sunjianqing126@126.com, nanshen01@126.com

Abstract. In this paper, an antenna with a miniature structure and wide-band is presented. We designed a two-arm conical spiral antenna according to the structure features of the Archimedes spiral and the conical helical antenna, and proposed an exponential asymptote balun to match the impedance. Unlike traditional antenna designs which optimize antenna and matching module separately, we adopted the differential evolution (DE) algorithm to optimize both the antenna and balun simultaneously. In addition, the peak radiation direction of the antenna was added as a constraint when evolving the antenna, which is usually ignored in normal evolutionary antenna designs. Simulation results indicate that the evolved antenna can basically fulfills the requirements. And the evolved antenna with the additional constraint has smaller deviation angle between the peak radiation direction and the antenna's axis than that without the constraint.

Keywords: Wide-band antenna · Antenna design · Spiral antenna
Differential evolution algorithm · Peak radiation direction

1 Introduction

Wide-band antenna is a kind of antenna with very wide working band, the ratio between its working band and center frequency is usually more than 20%. It has some characteristics such as wide communication range, large message capacity, strong anti-interference and great security performance. So wide-band can work better in complex and volatile communication environment. With the development of communication technology, the original antenna could not meet the increasing demand on communication system. To meet the communication requirements, antenna design technology has also been improved. In 1939, P.S.Carter put forward conical bipolar antenna and modified biconical antenna based on Oliver's biconical antenna model [1]. In 1957, Rumsey proposed the theory of non-variable frequency antenna, which opened a new era about wide-band antenna study. Since then, the designers applied the frequency independent theory and similarity principle to antenna design, and had put forward a

© Springer Nature Singapore Pte Ltd. 2018
K. Li et al. (Eds.): ISICA 2017, CCIS 874, pp. 413–426, 2018.
https://doi.org/10.1007/978-981-13-1651-7_37

series frequency independence antennas such as equiangular spiral antenna with structural similarity and log-periodic antenna with equal ratio transformation [2]. Unlike the normal antenna's structure, frequency independence antenna has self-similarity on structure like biconical antenna and horn antenna which have wider band. However, they did not have been applied widely as the size of those antennas are too large.

Hereafter, the development of wide-band antenna came to miniaturize. In recent years, with the development of antenna theory and the applied environment of wide-band antenna having been more complex, the requirements about wide-band antenna is becoming more strict at same time. We need a antenna with wider band, smaller geometric construction, greater gain and some other new characteristics. In order to meet the practical application requirements, the study on wide-band antenna becomes more diversified. For instance, printed slot antenna widen the band by slotting [3], band elimination wide-band antenna can work in different frequency band and avoid disturbance to some narrow-band25. In addition, some antennas increase their gains by adding Electromagnetic Band-Gap (EBG) around the antenna [4].

Spiral antenna is little bigger than planar antenna on volume, but spiral has higher gain than the latter. With the miniaturized technology applied on spiral antenna, spiral antenna has been becoming smaller and smaller, which make it possible to apply spiral antenna in miniaturized communication system [5]. Apart from high gain, spiral antenna also have wide band and circular polarization. Thus, there are increasing studies on spiral antenna in recent years [6]. The antenna performed in this paper is a kind of spiral antenna. Conditional antenna design requires the designers have rich experience on antenna design and solid theoretical foundation of antenna, it is usually difficult to design a perfect new antenna [7]. Fortunately, evolutionary algorithm applied into antenna design made it much easier. Evolutionary algorithm is derived from natural selection and evolution theory. It is a adaptive global optimization searching algorithm which can solve large complex optimization problems. Different from traditional antenna design, evolutionary antenna design use computer automated design to substitute for artificial design, which can reduce the design complexity and shorten the design time. So far, evolutionary antenna design has been proved valid in some major antenna research projects. In the mid 1990s, evolutionary algorithms began to be applied to antenna design and optimization [8–10]. There are some typical cases about evolution algorithm being used to overcome some difficulties on antenna design: Haupt designed the antenna array [9,11], and Anselmi designed the phased array [12], Lohn designed the LADEE satellite antenna with many constraints [13], Ricardo designed antenna by multiobjective evolutionary algorithm [14], Zhang, Zeng and Liu [15,16] designed the patch antenna by evolution algorithm of model learning, Jiao and Zeng [17] presented a dynamic multi-objective evolutionary algorithm to solve an antenna array problem, the dynamic and multi-objective methods are used to balance the exploration and exploitation. Koziel designed antenna by an expensive multi-objective evolutionary algorithm [18]. In June 2013, NASA launched a satellite of IRIS carrying an antenna designed by evolutionary algorithm, and in September, the NASA LADE lunar probe carried an improved evolutionary antenna [19]. In recent years, evolutionary algorithms such as differential evolution (DE) and particle swarm optimization (PSO) have been applied in antenna design and had a good performance. DE is used widely for its' great convergence efficiency. The antenna introduced in this paper is evolved by DE.

Juring evolving antenna, we have used the evolutionary antenna auto-design platform developed by our laboratory. There is a interface about electromagnetic simulation soft on the platform. After constructing a raw spiral antenna model by Ansoft HFSS and generating some optimized parameters about the antenna by DE, we can call the interface to evaluate the antenna. In this way, we convert antenna design into a constrained optimization problem (COP) on mathematics.

After a brief introduction in Sect. 1, this paper is organized as follows: Sect. 2 gives the information about spiral antenna and how to build a model of it. In Sect. 3, we give formation of antenna design as COP and the result of simulation. Finally, Sect. 5 summarizes the conclusion of this paper.

2 Constructing Antenna Model

Considering that conical spiral antenna have good performance on high gains and wide band. Our antenna is based on a conical spiral antenna combined with the characteristics of the Archimedean Spiral Antenna. The global shape of the antenna is a cone, the major radiator is made of double helixes covering on the conical surface from top to bottom, and one of the helixes is rotated 180° by the other one around the z-axis. At the same time, the antenna uses the exponential asymptote balun as impedance matching and non-equilibrium to equilibrium conversion device, the top two spirals in respectively connect with the positive and negative end of exponential asymptote balun, and it is fed by coaxial at the bottom of the balun. Different from ordinary equispaced Archimedean Spiral Antenna, the helix in our design is linked by several short helixes with individual rise angle, and we set the rise angles as optimization variables to optimize the performance. The overall size of the model is 30 mm high and the floor diameter is 60 mm. It can be said an antenna with miniaturization.

2.1 Build of Conical Helix

The equation of the Archimedean Spiral Antenna on rectangular coordinate system is:

$$\begin{cases} X = (\alpha + \beta\theta)cos(\theta) \\ Y = (\alpha + \beta\theta)sin(\theta). \end{cases} \tag{1}$$

Combined with the structure of cone, the equation of the Antenna as fellows when it is expanded a three-dimensional conical spiral antenna:

$$\begin{cases} X = (\alpha + \Delta r \times \frac{\theta}{2\pi})cos(\theta) \\ Y = (\alpha + \Delta r \times \frac{\theta}{2\pi})sin(\theta) \\ Z = \gamma + \Delta h \times \frac{\theta}{2\pi} \end{cases} \tag{2}$$

where α is initial radius of the helix, Δr is radius growth rate, γ is initial altitude, and Δh is the height of the helix for one lap. Considering the practical situation, the spiral in our model is a band with certain width. Hence, the practical spiral signature equation is:

$$\begin{cases} X = (y_{s(n)} + delta\_r_{(n)}(\theta/(2\pi)) + temp\_r * rate)cos(\theta) \\ Y = (y_{s(n)} + delta\_r_{(n)}(\theta/(2\pi)) + temp\_r * rate)sin(\theta) \\ Z = z_{s(n)} - delta\_h_{(n)}(\theta/(2\pi)) - temp\_h * rate \end{cases} \quad (3)$$

where θ and $rate$ are the independent variables, θ is the angle of helix rounds which ranges from 0 to 2π, and $rate$ is the width rate of the whole band width ranging from 0 to 1. $y_{s(n)}$ and $z_{s(n)}$ are the original coordinate of each helix, n is the turn number of helixes (n = 0, 1, 2, ...N. and the whole spiral is made of N small helixes, one helix rounds a circle, N is set 10 in our design.). $delta\_r_{(n)}$ and $delta\_h_{(n)}$ respectively are the variation of spiral radius and height with rounding one circle. $temp\_r$ and $temp\_h$ are parameters related to spiral band width, which are determined by the corn shape. The relationship between them is illustrated in Fig. 1. Each single of helix has their own radius growth rate with individual structure equation. The start points of connected two helixes follow the next mathematics equation:

$$\begin{cases} y_{s(n+1)} = y_{s(n)} + delta\_r_{(n)} \\ z_{s(n+1)} = z_{s(n)} + delta\_h_{(n)}. \end{cases} \quad (4)$$

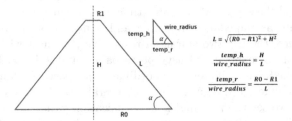

Fig. 1. Relationship of $temp\_r$, $temp\_h$ and spiral band width (wire-band).

According to the Fig. 1, we can conclude that the values of $delta\_r_{(n)}$ and $delta\_h_{(n)}$ obey the follow equation. Where $delta\_h$ is a average height change of spiral rounds one turn, and $ratio\_h_n$ is the practical height change ratio to the average for each helixes. In this paper, $ratio\_h_{(n)}$ is a optimizable variable when evolving the antenna.

$$\begin{cases} delta\_h = H/N \\ delta\_r_{(n)} = \frac{R0-R1}{H} * delta\_h_{(n)} \end{cases} \qquad delta\_h_{(n)} = delta\_h * ratio\_h_n \quad . \quad (5)$$

We use Eqs. (3) and (4) build a spiral on Ansoft HFSS, and rotate it 180 degrees around z-axis to get another one. In this way, we can get a two-arm spiral. Figure 2 shows the single spiral model.

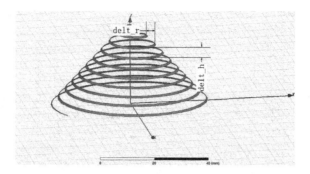

Fig. 2. Spiral band model.

2.2 Build of Balun Model

In our design, we use the coaxial-cable which is an unbalanced RF device as feeding connection. However, two-arm spiral antenna is a balanced structure. Therefore, there requires a conversion between the antenna and coaxial interface, and balun is a kind of imbalance-balance converter, which can be used as intermediate components connecting spiral antenna and feeder coaxial interface. At the same time, balun can also work as an impedance matching device, which enables the antenna to match the 50 ohm input impedance of the coaxial feeding mode to improve the antenna performance. In this paper, the double micro-strip exponential gradient balun is used as the switching element, balun and the antenna are optimized as a whole to achieve the optimal performance. The balun's overall model is shown as Fig. 3. The two-sided balun consists of a medium media and two metal pieces, which is connected to the antenna on the top, the coaxial cable attached below. The medium media adopts Telfon material with permittivity 2.1 and dielectric loss 0.001. Telfon is the commonly used high frequency circuit board material which can maintain stable permittivity and electric loss in high frequency range. The parameters of medium media structure are shown in Table 1. The boundaries of metal plates covering on the medium media are exponential lines, which values are determined by length of up and bottom. The curve equation is:

$$
\begin{cases}
X = \frac{1}{2}balun\_th \\
Y = \frac{1}{2}balun\_a * exp(\frac{balun\_b}{2balun\_a}(sin\frac{t}{balun\_h} - 0.5)\pi + 1)) \\
Z = t
\end{cases}
\tag{6}
$$

where $balun\_a$ and $balun\_b$ are respectively the up and bottom lines length of the metal plates. In order to match the input impedance of coaxial cable, the lower margin of the positive metal strip is $balun\_a2 = 3.1$ mm, and the bottom side of the negative metal strip is $balun\_a3 = 18$ mm. The upper edge length of metal sheet on both side is consistent, which determines the input impedance of balun upper interface matching. It is set $balun\_b2$ and can be optimized while antenna evolving so as to match impedance better and reduce

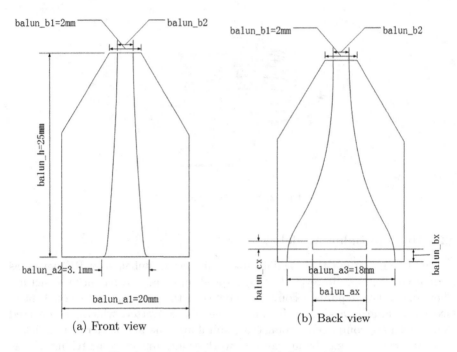

Fig. 3. Front and back views of balun model.

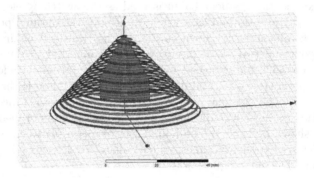

Fig. 4. Final antenna model.

VSWR. In order to improve the wide band performance of balun, a groove is set on the back sheet metal, which located by *balun_ax* and *balun_bx*, and the size determined by *balun_ax* and *balun_cx*. *balun_ax*, *balun_bx* and *balun_cx* are optimizable variables since the location and size of the groove both can affect the performance of balun.

The final model of the antenna are shown as Fig. 4.

Table 1. Balun structural parameters.

Parameters	Values
Thickness (balun_th)	1 mm
Length of bottom side (balun_a1)	20 mm
Length of up side (balun_b1)	2 mm
Height of balun (balun_h)	25 mm

3   Formulating Antenna Design as COP

To evolve antenna, antenna design must be transformed into a COP in mathematics fist.

3.1   Antenna Requirement

Our purpose is to design a wide-band antenna. The requirements are shown in Table 2.

Table 2. Antenna requirements.

Parameters	Requirements
Frequency	Range from 2.7 GHz to 6.3 GHz
VSWR	≤ 2
Axial Ratio (AR)	≤ 3 dB
Gain	≥ 3 dB; $Gain_{max} \geq 6$ dB
Pattern range	$0° \leq \phi \leq 360°; 35° \leq \theta \leq 35°$
Size	60 mm × 60 mm × 30 mm

3.2   Solution Vector and Solution Space

As described in Sect. 2, $ratio\_h_n$, $balun\_ax$, $balun\_bx$, and $balun\_cx$ are set as optimizable variables. As a result, the vector to present the structure is $(ratio\_h_{(0)}, ratio\_h_{(1)}, \ldots, ratio\_h_{(10)}, balun\_ax, balun\_bx, balun\_cx, balun\_b2)$, where the ranges of these variables are $0.5 \leq ratio\_h(n) \leq 1.5$, $3\,\text{mm} \leq balun\_ax \leq 7\,\text{mm}$, $2\,\text{mm} \leq balun\_bx \leq 4\,\text{mm}$, $0.3\,\text{mm} \leq balun\_cx \leq 2\,\text{mm}$, $0.3\,\text{mm} \leq balun\_b2 \leq 1.7\,\text{mm}$. All those 14 variables make up of the solution vector $x = (ratio\_h_{(0)}, ratio\_h_{(1)}, \ldots ratio\_h_{(10)}, balun\_ax, balun\_bx, balun\_cx, balun\_b2)$. And the solution space $\mathbf{X}$ just like Eq. (7):

$$\begin{aligned}
\mathbf{X} &= \{x | l \leq x \leq u\} \\
x &= (ratio\_h_{(0)}, ratio\_h_{(1)}, \ldots, ratio\_h_{(10)}, balun\_ax, \\
&\quad balun\_bx, balun\_cx, balun\_b2) \\
l &= (0.5, 0.5, \ldots, 0.5, 3.0, 2.0, 0.3, 0.3) \\
u &= (1.5, 1.5, \ldots, 1.5, 7.0, 4.0, 2.0, 1.7)
\end{aligned} \tag{7}$$

here l and u are the boundaries of solution space.

3.3 Objective and Constraints

As showed in Table 2, on antenna design problem, the objective is normally gain of antenna, the higher the better for an antenna gain. And there are some constraints about antenna performance such as VSWR and Axial Ration(AR). Thus, an antenna design problem can be formed as a COP like Eq. (11):

$$
\begin{aligned}
min\ f(\boldsymbol{x}) = &-Gain_{(\varphi,\theta,frq)} \\
st:\ &gGain_{(\varphi,\theta,frq)}(\boldsymbol{x}) = 3 - Gain_{(\varphi,\theta,frq)} \leq 0 \\
&gVSWR_{frq}(\boldsymbol{x}) = VSWR_{frq} - 2 \leq 0 \\
&gAR_{frq}(\boldsymbol{x}) = AR_{frq} - 3 \leq 0 \\
&gGain\_max\_angle(\varphi,\theta,frq)(\boldsymbol{x}) = \|Gain\_max\_angle(\varphi,\theta,frq)\| - 0 \leq 0
\end{aligned} \tag{8}
$$

where
$\boldsymbol{x}$ is the solution vector;
φ is the azimuth angle, and the value of φ is between $0°$ and $360°$ with gradient of $5°$;
θ is the elevation angle, and the value of θ is between $-35°$ and $35°$ with gradient of $5°$;
$freq$ is the frequency, and the value of $freq$ is between $2.5\,\text{GHz}$ and $6.5\,\text{GHz}$ with gradient of $0.2\,\text{GHz}$;

According to requirements, there are some constraints about Gain, Axial Ratio and VSWR. Thus, we set them as the constraints of the COP as shown in Eq. (11). In this way, we build a mathematical model of antenna design, and we can evolve it by DE [20] next. For the minimization COP, the solution with smaller objective value and smaller violation value(s) are preferred. Therefore, two corresponding objectives are taken into consideration: one is the original function objective, and the other is the violation objective. We usually consider a normalized violation objective, the normalized violation objective is:

$$
\psi(\overrightarrow{x}) = \frac{\sum\limits_{i=1}^{m} \dfrac{G_i(\overrightarrow{x})}{\max\limits_{\overrightarrow{x}\in P(0)}\{G_i(\overrightarrow{x})\}}}{m} \tag{9}
$$

P is for population, and $P(0)$ is for initial population. m is number of constraints, m is 4 in this problem. $G_i(\overrightarrow{x})$ is defined as:

$$
G_i(\overrightarrow{x}) = \max\{g_i(\overrightarrow{x}), 0\}, i = 1, 2, \ldots, m. \tag{10}
$$

$$
g_i(\overrightarrow{x}) = \begin{cases}
3 - Gain_{(\varphi,\theta,frq)}, & i = 1 \\
VSWR_{frq} - 2, & i = 2 \\
AR_{frq} - 3, & i = 3 \\
\|Gain\_max\_angle(\varphi,\theta,frq)\| - 0, & i = 4
\end{cases} \tag{11}
$$

Here, $g_i(\overrightarrow{x})$ represent the constraints functions shown in Eq. (11). A solution $\overrightarrow{x} \in \mathbf{X}$ is said to be **feasible** if the violation objective $\psi(\overrightarrow{x}) = 0$; otherwise, $\overrightarrow{x}$ is said **infeasible**. As a precondition, satisfying the constraints may be more important than obtaining a small objective value. Given two different solutions, we use the following rules to judge which one is better:

1. The solution with smaller violation objective is always preferred.
2. If two solutions have the same violation objective, the one with smaller original objective is preferred.

4   Solving Antenna Design by DE

4.1   Setting DE Parameters

The parameters for DE algorithm in solving spiral antenna are listed as follows:

1. Evolutionary generations $T = 400$.
2. Population size $NP = 50$.
3. Crossover rate $CR = 0.9$.
4. Scaling factor $F = 0.5$.

4.2   Result and Discussion

We first evolved the antenna without the constraint about the angle deviation between peak radiation direction and central axis of the antenna, and the evolved antenna can basically fulfill the design demands. However, what defective is that there is a deviation angle between peak radiation direction and the central axis of the antenna, the ideal deviation angle should be 0. To deal with this problem, we set a new constraint about the angle deviation apart from normal constraints, the new constraint is defined as follow:

$$st :gGain\_max\_angle(\varphi, \theta, frq)(\boldsymbol{x}) = \|Gain\_max\_angle(\varphi, \theta, frq)\| - 0 \leq 0$$

Here $Gain\_max\_angle(\varphi, \theta, frq)$ is a value of θ, which denotes the peak radiation direction. We need the smallest value of θ, and 0 is best. With the new constraint, we get a evolved antenna with better performance on peak radiation direction while basically fulfilling the design demands simultaneously. The parameters of the antenna after optimizing are shown as Table 3.

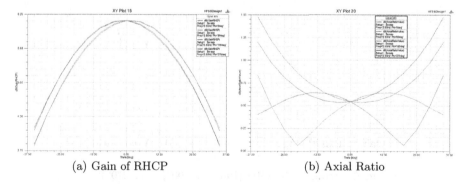

(a) Gain of RHCP   (b) Axial Ratio

Fig. 5. Frequency 2.5 GHz (Phi $= 0°$, $90°$, $180°$, $270°$).

Table 3. Optimized parameters of antenna.

Variables	Evolve without deviation angle constrain	Evolve with deviation angle constrain
ratio_h0	1.038	1.099
ratio_h1	1.157	0.898
ratio_h2	1.131	0.958
ratio_h3	1.104	0.622
ratio_h4	1.054	1.318
ratio_h5	1.101	1.492
ratio_h6	1.181	0.890
ratio_h7	0.896	1.457
ratio_h8	1.025	1.151
ratio_h9	0.860	0.843
balun_ax	4.393 mm	4.738 mm
balun_bx	2.167 mm	3.896 mm
balun_cx	0.549 mm	1.218 mm
balun_b2	1.191 mm	0.335 mm

The right-handed-circular polarization (RHCP) gains and axial ratio of the antenna evolved with new constraint simulated by Ansoft HFSS are shown as Figs. 5, 6, 7, 8 and 9. The five figures present five main working frequencies results respectively. In each figures, figure (a) shows RHCP gains result, and figure (b) shows axial ratio result. Here, results in 4 typical directions (Phi = 0°, 90°, 180°, 270°, represented by four different color lines respectively) are presented in the graphs.

According to the simulation results, we can find that the evolved antenna can almost meet the requirements. The maximum RHCP gains are over 6dB, and the minimum RHCP gains can be 3 dB in most working area. The Fig. 10 shows the direction angle between peak radiation direction and central axis. In the graph, blue line denotes the direction angle result of antenna evolved with the

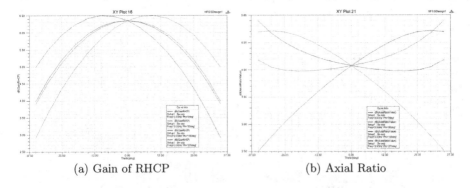

(a) Gain of RHCP   (b) Axial Ratio

Fig. 6. Frequency 3.5 GHz (Phi = 0°, 90°, 180°, 270°).

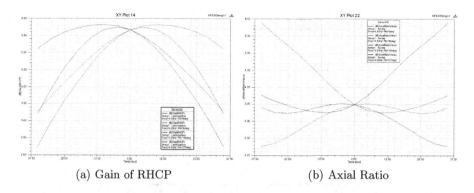

(a) Gain of RHCP   (b) Axial Ratio

Fig. 7. Frequency 4.5 GHz (Phi = 0°, 90°, 180°, 270°).

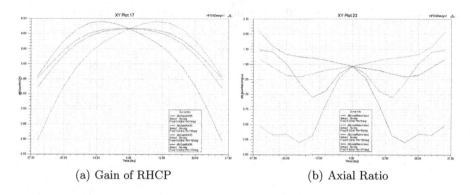

(a) Gain of RHCP   (b) Axial Ratio

Fig. 8. Frequency 5.5 GHz (Phi = 0°, 90°, 180°, 270°).

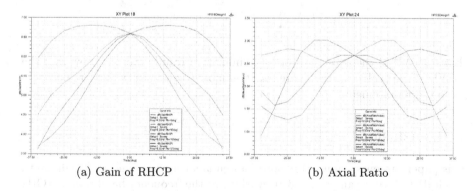

(a) Gain of RHCP   (b) Axial Ratio

Fig. 9. Frequency 6.5 GHz (Phi = 0°, 90°, 180°, 270°).

constraint about angle deviation, and red line denotes result of antenna evolved without the constraint. From the picture, we can see that the antenna evolved with the angle deviation constraint has better performance on peak radiation direction comparing to the antenna evolved without that constraint. The Fig. 11

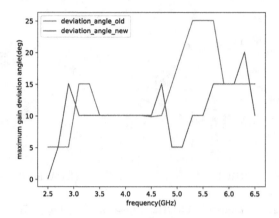

Fig. 10. The deviation of maximum gain direction angle. (Color figure online)

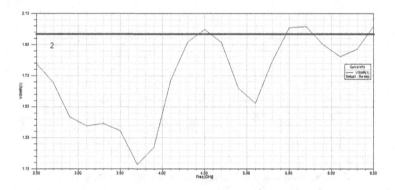

Fig. 11. VSWR of the evolved antenna.

shows VSWR of the antenna, from the simulation results, we can know that the VSWR of the evolved antenna are less 2 on most frequencies, which are basically up to the requirements.

5   Conclusion

This paper presents a miniature spiral antenna with wide-band and high gains, which RHCP gains can be over 3 dB working at the frequency band from 2.5 GHz to 6.5 GHz in most working space.

1. The antenna design is developed from the Archimedean Spiral Antenna. It is composed of two part: two-arm spirals which are the radiation of antenna and balun for feed and impedance matching. The overall shape of antenna is a 30 mm high corn which basal diameter is 60 mm.

2. The shape of spiral and the size of balun are parameterized as the solution vector which are to be optimized, and the requirements of the spiral are coded as the objective and constraints. A constrained optimization problem about antenna design is then formulated.
3. The COP is solved by the differential evolution (DE). Some optimized antennas are found, which meet the requirements in the simulations.

In this paper, the difference from the traditional antenna design is that we optimized the antenna and matching module as a whole; and the difference from the typical evolutionary antenna is that we added a new constrain about the angle deviation.

The simulation results indicated that the antenna can basically fulfill the demands and performance better on the peak radiation direction of antenna.

Acknowledgment. The authors are very grateful to the anonymous reviewers for their constructive comments to this paper. This work is supported by the National Science Foundation of China under Grant 61673355, 61271140 and 61203306.

References

1. Schantz, H.G.: A brief history of uwb antennas. IEEE Aerosp. Electron. Syst. Mag. **19**(4), 22–26 (2011)
2. Rumsey, V.H.: 2-basic features of frequency-independent antennas. In: Frequency Independent Antennas, pp. 13–21 (1966)
3. Fereidoony, F., Chamaani, S., Seyed, A.M.: Uwb monopole antenna with stable radiation pattern and low transient distortion. IEEE Antennas Wirel. Propag. Lett. **10**(4), 302–305 (2011)
4. Shuai, C.Y., Wang, G.M.: A simple ultra-wideband magneto-electric dipole antenna with high gain. Frequenz **72**, 27–32 (2017)
5. Qing, X.M., Chen, Z.N., Chia, M.Y.W.: Characterization of ultrawideband antennas using transfer functions. Radio Sci. **41**(1), 1–10 (2006)
6. Rahman, N., Afsar, M.N.: A novel modified archimedean polygonal spiral antenna. IEEE Trans. Antennas Propag. **61**(1), 54–61 (2013)
7. Eubanks, T.W., Chang, K.: A compact parallel-plane perpendicular-current feed for a modified equiangular spiral antenna. IEEE Trans. Antennas Propag. **58**(7), 2193–2202 (2010)
8. Rahmatsamii, Y., Michielssen, E.: Electromagnetic optimization by genetic algorithms. Microwave J. **42**(11), 232–232 (1999)
9. Haupt, R.L.: Thinned arrays using genetic algorithms. IEEE Trans. Antennas Propag. **42**(7), 993–999 (1994)
10. Linden, D.S., Altshuler, E.F.: Automating wire antenna design using genetic algorithms. Microwave J. **39**(3), 7 (1996)
11. Wen, Y.Q., Wang, B.Z., Ding, X.: A wide-angle scanning and low sidelobe level microstrip phased array based on genetic algorithm optimization. IEEE Trans. Antennas Propag. **64**(2), 805–810 (2016)
12. Anselmi, N., Rocca, P., Salucci, M., Massa, A.: Irregular phased array tiling by means of analytic schemata-driven optimization. IEEE Trans. Antennas Propag. **65**(9), 4495–4510 (2017)

13. Lohn, J.D., Linden, D.S., Blevins, B., Greenling, T., Allard, M.R.: Automated synthesis of a lunar satellite antenna system. IEEE Trans. Antennas Propag. **63**(4), 1436–1444 (2015)
14. Ramos, R.M., Saldanha, R.R., Takahashi, R.H.C., Moreira, F.J.S.: The real-biased multiobjective genetic algorithm and its application to the design of wire antennas. IEEE Trans. Magn. **39**(3), 1329–1332 (2003)
15. Zhang, J., Zeng, S., Jiang, Y., Li, X.: A Gaussian process based method for antenna design optimization. In: Li, K., Li, J., Liu, Y., Castiglione, A. (eds.) ISICA 2015. CCIS, vol. 575, pp. 230–240. Springer, Singapore (2016). https://doi.org/10.1007/978-981-10-0356-1_23
16. Liu, B., Aliakbarian, H., Ma, Z.K., Vandenbosch, G.A.E., Gielen, G., Excell, P.: An efficient method for antenna design optimization based on evolutionary computation and machine learning techniques. IEEE Trans. Antennas Propag. **62**(1), 7–18 (2013)
17. Jiao, R.W., Zeng, S.Y., Alkasassbeh, J.S., Li, C.H.: Dynamic multi-objective evolutionary algorithms for single-objective optimization. Appl. Soft Comput. J. **61**, 793–805 (2017)
18. Bekasiewicz, A., Koziel, S., Leifsson, L.: Sequential domain patching for computationally feasible multi-objective optimization of expensive electromagnetic simulation models. Procedia Comput. Sci. **80**, 1093–1102 (2016)
19. Lohn, J.D., Hornby, G.S., Linden, D.S.: Evolution, re-evolution, and prototype of an X-band antenna for NASA's space technology 5 mission. In: Moreno, J.M., Madrenas, J., Cosp, J. (eds.) ICES 2005. LNCS, vol. 3637, pp. 205–214. Springer, Heidelberg (2005). https://doi.org/10.1007/11549703_20
20. Price, K.V., Storn, R., Lampinen, J.A.: Differential Evolution: A Practical Approach to Global Optimization. Natural Computing Series. Springer, Heidelberg (2014). https://doi.org/10.1007/3-540-31306-0

Yagi-Uda Antenna Design Using Differential Evolution

Hai Zhang[1,2], Hui Wang[1,2(✉)], and Cong Wang[1]

[1] School of Information Engineering, Nanchang Institute of Technology,
Nanchang 330099, China
huiwang@whu.edu.cn
[2] Jiangxi Province Key Laboratory of Water Information Cooperative
Sensing and Intelligent Processing, Nanchang Institute of Technology,
Nanchang 330099, China

Abstract. Differential evolution (DE) is an efficient optimization technique, which has been applied to solve various engineering optimization problems. In this paper, DE is used to optimize the element spacing and lengths of Yagi-Uda antennas. An internal system with interactive simulation is developed based on C++ and CST Microwave Studio. To verify the performance our approach, the Yagi-Uda antenna for 60 GHz communications is designed in the experiments. Simulation results show the effectiveness of our approach.

Keywords: Yagi-Uda antenna · Differential evolution · Optimization

1 Introduction

In recent years, many scholars have tried to analyze and optimize the electromagnetic structure with the new algorithm [1–6]. Evolutionary algorithms are suitable optimization techniques for solving such problems, through the natural law "survival of the fittest, superior bad discard". Storn and Price [7] put forward differential evolution (DE) algorithm, which was based on the vector differences in the population. It solved the optimization problem by the cooperation and competition. DE is simple to implement and performs well. Natural selection and genetic variation are used to search the solution globally, in order to find better solutions and optimize complex problems.

The antenna design, such as 60 GHz communication antenna has the characteristics of discreteness, multi-constraint, computational complexity and uncertainty. With the increasing of the frequency , the size of antenna becomes smaller, and the number of units becomes larger. In the design and synthesis of antennas, the goal is to find a radiating structure that meets a set of performance criteria that usually include gain, input impedance,maximum sidelobe level, beamwidth and physical size. There are a lot of design parameters that affect the antenna performance. It is often difficult to provide good initial values for the optimization and searching. A considerable amount of attention has been focused on

© Springer Nature Singapore Pte Ltd. 2018
K. Li et al. (Eds.): ISICA 2017, CCIS 874, pp. 427–438, 2018.
https://doi.org/10.1007/978-981-13-1651-7_38

optimizing the Yagi-Uda antenna since its typical structure and gain performance. Yagi-Uda antenna is an ideal optimized object for differential evolution algorithm because of its large for number of units as well as its design parameters.

By using a design framework based on DE algorithms, this paper presents a method to optimize the element spacing and lengths of Yagi-Uda antennas. The main program of DE algorithm are compiled using MS Visual C++ 2010, then the parameters-length and spacing were transformed to the electromagnetic model in the CST. During the optimization, CST performs the task of electromagnetic simulation and evaluating each of the antenna designs generated by the DE.

2   Differential Evolution

With the development of society, the optimization problems we are confronted become more complex. To solve these problems, stronger optimization algorithms are required. In the past several years, some bio-inspired optimization algorithms/techniques have been proposed, such DE, particle swarm optimization (PSO) [8,9], artificial bee colony (ABC) [10–14], cuckoo search (CS) [15,16], firefly algorithm (FA) [17–23], and bat algorithm (BA) [24]. Among these algorithms, DE is a competitive to other algorithms.

DE is an efficient population-based optimization algorithms, which has shown good optimization performance on many engineering problems [25–31]. The standard DE has three important operators: mutation, crossover, and selection. Initially, all individuals in the population used the mutation operation to generate mutant vectors. Then, the mutant vectors and their corresponding parent individuals are recombined by the crossover operation, and new trial vectors are generated. Finally, the greedy selection operation is used to choose a better one between a new trial vector and its parent individual as the current individual.

Assume that there are N individuals in the population, and N is the population size. Let $X_i(t) = (x_{i,1}(t), x_{i,2}(t), i, x_{i,D}(t))$ be the ith individual at generation t in the population, and D is the dimensional size. A brief introduction of those three operators is described as follows.

Mutation: For each individual $X_i(t)$, a new mutant vector $V_i(t)$ is generated by

$$V_i(t) = X_{i1}(t) + F(X_{i2}(t)) - X_{i3}(t) \tag{1}$$

where $X_{i1}(t), X_{i2}(t), X_{i3}(t)$, and $X_i(t)$ are mutually different individuals selected from the current population. The parameter F is known as a scale factor. In many cases, F is set to 0.5. There are several different mutation schemes, such as DE/rand/1, DE/rand/2, DE/best/1, DE/best/2, and DE/current-to-best/1, and Eq. 1 is called DE/rand/1.

Crossover: A mutant vector $V_i(t)$ and its corresponding parent $X_i(t)$ are recombined, and their offspring $U_i(t) = (u_{i,1(t)}, u_{i,2(t)}, i, u_{i,D(t)})$ is generated as follows.

$$u_{i,j}(t) = \begin{cases} v_{i,j}(t), & \text{if } rand_j \leq CR \vee j = l \\ x_{i,j}(t), & \text{otherwise} \end{cases} \tag{2}$$

Algorithm 1. Differential Evolution (DE)

---

1 Randomly initialize the population;
2 **While** $t \leq MaxGen$ **do**
3   **for** $i = 1$ to N **do**
4   Generate a mutant V_i according to Eq. 1;
5   Generate U_i according to Eq. 2;
6   Calculate the fitness value of U_i;
7   Conduct the selection according to Eq. 3;
8   **end**
9      $t = t + 1$;
10 **End**

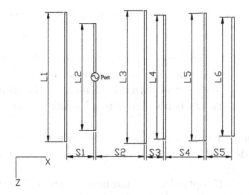

Fig. 1. The structure of antenna

where $randj$ is a random value in the range $[0, 1]$, l is a random integer between 1 and D, and CR is a crossover rate. In many references, CR is set to 0.9.

Selection: A selection method is used to choose a better one between $U_i(t)$ and $X_i(t)$ as the new $X_i(t)$. This process can be implemented by the following equation (this paper only considers minimization problems).

$$X_i(t+1) = \begin{cases} U_i(t), \text{ if } f(U_i(t)) \leq f(X_i(t)) \\ X_i(t), \text{ otherwise} \end{cases} \tag{3}$$

The main steps of the standard DE are presented in Algorithm 1, where $MaxGen$ is the maximum number of generations.

3   Proposed Approach

3.1   Antenna Structure

The antenna presented here consists of a number of linear dipole elements, as shown in Fig. 1 one driven element and five parasitic elements. Usually the driven element is based on the first resonance. The directors are typically somewhat

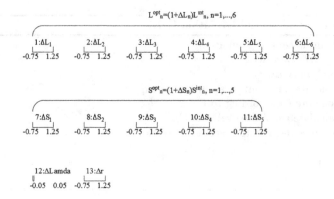

Fig. 2. The optimization parameters of antenna

smaller in length than the feed element, whereas the reflectors is longer. When the driven element is fed, the electromagnetic wave is excited to radiate, and then the current is induced in the parasitic elements. The total amplitude and phase of the induced currents is not determined solely by their lengths but also by their spacing.

Yagi-Uda are quite common in practice because they are simple to build, low-cost, and provide moderately desirable characteristics (including high gain, a unidirectional beam).

As shown in Fig. 2, 13 optimization parameters are selected, which are used as the particle of the population in the optimization.

$$\begin{cases} L_n(t+1) = (1 + \triangle L_n^{opt}(t+1))L_n(t), n = 1, \ldots, 6, \triangle L_n^{opt}(t+1) \in [-0.75, 1.25] \subset \Re \\ S_n(t+1) = (1 + \triangle S_n^{opt}(t+1))S_n(t), n = 1, \ldots, 5, \triangle S_n^{opt}(t+1) \in [-0.75, 1.25] \subset \Re \\ Lamda(t+1) = (1 + \triangle Lamda^{opt}(t+1))Lamda(t), \triangle Lamda^{opt}(t+1) \in [-0.05, 0.05] \subset \Re \\ r(t+1) = (1 + \triangle r^{opt}(t+1))r(t), \triangle r^{opt}(t+1) \in [-0.75, 1.25] \subset \Re \end{cases} \quad (4)$$

where $L_n(t)$ is the nth oscillator's length of t generation; $\triangle L_n^{opt}(t+1)$ is $L_n(t)$'s optimal quantity; $S_n(t)$ is the nth oscillator's spacing; $\triangle S_n^{opt}(t+1)$ is $S_n(t)$'s optimal quantity; $Lamda(t)$ is t generation wavelength; $\triangle Lamda^{opt}(t+1)$ is $Lamda(t)$'s optimal quantity; $r(t)$ is t generation oscillator's radius; and $\triangle r^{opt}(t+1)$ is $r(t)$'s optimal quantity.

3.2   Antenna Performance Parameters

In the optimization, the antenna's requirements are usually described by the scattering parameter and the radiation pattern. Gain and input impedance are the essentially parameters of antenna performance. Some other parameters such as side lobes, front-to-back ratio, cross-polarization will be involved in the complex designs, but the optimization procedure is similar.

$$E = Af(\theta, \phi) \quad (5)$$

$$F(\theta, \phi) = \frac{f(\theta, \phi)}{f_{max}} \tag{6}$$

where E is the electric field intensity of the far-field, A is the proportional coefficient, $f(\theta, \phi)$ is the directional function, $F(\theta, \phi)$ is the normalized direction function, and f_{max} is the maximum of $f(\theta, \phi)$. The relationship between radiation function and directivity is

$$D(\theta, \phi) = \frac{4\pi |F(\theta, \phi)|^2}{\int_0^{2\pi} \int_0^{\pi} |F(\theta, \phi)|^2 sin\theta d\theta d\phi} \tag{7}$$

and

$$G(\theta, \phi) = \frac{4\pi U(\theta, \phi)}{P_{in}} = \frac{4\pi U_{max}|F(\theta, \phi)|^2}{P_{in}} \tag{8}$$

where $U(\theta, \phi)$ is the radiation intensity in a certain direction, proportional to the square of the electric field intensity, U_{max} is the maximum of radiation intensity, P_{in} is the input power. Equation 8 shows that gain is determined by input power and radiation function. The radiation function depends on the far-field distribution of the antenna, and the input power depends on the impedance characteristics property and the current distribution of the antenna

$$Z_A = R_A + jX_A \tag{9}$$

$$P_{in} = \frac{1}{2} R_A |I_A|^2 \tag{10}$$

where Z_A is antenna impedance, I_A is current distribution. Further more, when the imaginary part of impedance is zero,

$$G(\theta, \phi) = \frac{4\pi U_{max}|F(\theta, \phi)|^2}{\frac{1}{2} \frac{1+S_{11}}{1-S_{11}} Z_0 |I_A|^2} = \alpha \frac{1 - S_{11}}{1 + S_{11}} (\frac{|F(\theta, \phi)|}{|I_A|})^2 \tag{11}$$

where $\alpha = \frac{8\pi U_{max}}{Z_0}$. It is clear from Eq. 11 that the relationship between the gain and the reflection coefficient is still complex.

4   Results and Discussion

4.1   Fitness Function

Important characteristics that define the antenna performance are gain, and input impedance. The quality of a design is expressed mathematically by a fitness function in DE optimization. think of the following Settings in the design of fitness function,

$$Fitness1(X) = |S_{11}(f_0)| \tag{12}$$

where $X = (L_1, \ldots, L_6, S_1, \ldots, S_5, Lamda, r)$, and f_0 is the center frequency. Hence, the impedance match and resonance can be obtained at f_0. Due to the symmetry of the resonant behaviour, $|S_{11}| < -10\,\text{dB}$ can be satisfied near resonance.

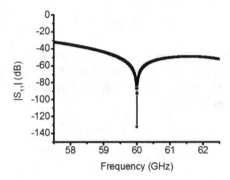

Fig. 3. Return loss, $Fitness1(X) = |S_{11}(f_0)|$, $X = (L_1, \ldots, L_6, S_1, \ldots, S_5, Lamda, r)$.

However, we have seen in the previous section the relationship between gain and S_{11} is complex. The impedance is depended on the current in the antenna and the distribution of the near field, whereas gain is depended on far field. The impedance can be tuned through the circuit technology. The gain is almost based on the theory of field. Compared with the impedance it is difficult to tune the gain. Therefore,

$$Fitness2(X) = \frac{1}{Gain(f_0, \theta, \phi)} \tag{13}$$

In this case, the fitness of an antenna is based solely on its maximum gain. Obviously, the other case is optimized both for high gain and low $|S_{11}|$ by means of weighted function.

$$Fitness3(X) = \alpha \frac{1}{Gain(f_0, \theta, \phi)} + \beta |S_{11}(f_0)| \tag{14}$$

Further more, the fitness function can be set as

$$Fitness4(X) = \sum_{i=1}^{n} \alpha_i \frac{1}{G(f_i, \theta, \phi)} + \sum_{i=1}^{n} \beta_i |S_{11}(f_i)| \tag{15}$$

where i represents the frequencies we are interested. Hence we can investigate the antenna performance more carefully. However, the fitness function will be more complex. It will cost much more optimization time because of the mutual effect between performance parameters.

In the case of $Fitness1(X) = |S_{11}(f_0)|$, Figs. 3, 4 and 5 present the simulation results of the optimized Yagi-Uda antenna. $|S_{11}(60\,\text{GHz})| < -100\,\text{dB}$ and the impedance match is obtained in the resonant frequency as shown in Fig. 3. Furthermore because of continuity and symmetry of the scatter parameter, over a certain bandwidth (58–62 GHz), the return loss is less than 10 dB. We can see that DE algorithm performs well. In the view of return loss, $Fitness1(X)$ is a good fitness function. However, the results of radiation pattern is terrible as shown in Figs. 4 and 5. In the direction of $\theta = 90°, \phi = 0°$, the gain is very small

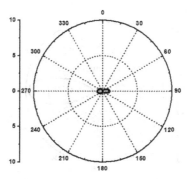

Fig. 4. 2D radiation pattern, $Fitness1(X) = |S_{11}(f_0)|$.

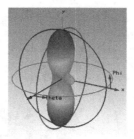

Fig. 5. 3D radiation patter, $Fitness1(X) = |S_{11}(f_0)|$.

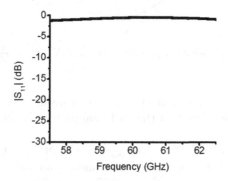

Fig. 6. Return loss, $Fitness2(X) = \frac{1}{Gain(f_0,\theta,\phi)}$.

and the side lobe level is too big. The antenna has smaller gain values but better impedance values. The gain is not good enough when the return loss meets the requirement. It means $Fitness1(X)$ cannot satisfy the goals of optimization.

In the case of $Fitness2(X) = \frac{1}{Gain(f_0,\theta,\phi)}$, the simulation results are shown in Figs. 6, 7 and 8. The gain is about 13 dB at 60 GHz as shown in Fig. 7. And Fig. 8 presents the stronger directivity is obtained. However the return loss is more than -10 dB in the 60 GHz as shown in Fig. 6. Furthermore, the return

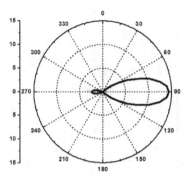

Fig. 7. 2D radiation pattern, $Fitness2(X) = \frac{1}{Gain(f_0,\theta,\phi)}$.

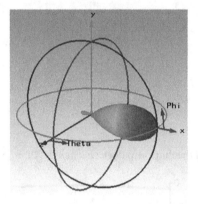

Fig. 8. 3D radiation pattern, $Fitness2(X) = \frac{1}{Gain(f_0,\theta,\phi)}$.

loss curve near 60 GHz is asymmetrical, which means this is likely to be an unstable solution. It is likely that the radiation pattern in the vicinity of 60 GHz is unstable.

In the case of $Fitness3(X) = \alpha\frac{1}{Gain(f_0,\theta,\phi)} + \beta|S_{11}(f_0)|$, the fitness of an antenna is calculated based on both its maximum gain and its input impedance. The weighting constants are set to $\alpha = 0.75, \beta = 0.25$ and so that gain receives more emphasis than the impedance terms of the fitness function.

Although the return loss is not as good as Fig. 3, the performance is significantly improved compared with Fig. 6 as shown in Fig. 9. From Figs. 10 and 11, it is observed more than 10 dB gain is obtained and the antenna focus on +x axis. The radiation pattern depends on the far field distribution, on the other hand the impedance more depends on current and near field. The choice of the α, β depends on the performance goals. Compared with high gain, the antenna's input impedance is easier to match. Hence, it is clear the ratio of gain should be more than that of return loss.

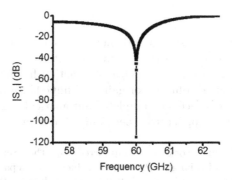

Fig. 9. Return loss, $Fitness3(X) = \alpha \frac{1}{Gain(f_0,\theta,\phi)} + \beta|S_{11}(f_0)|(\alpha = 0.75, \beta = 0.25)$.

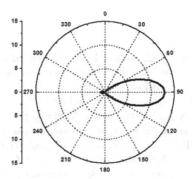

Fig. 10. 2D radiation pattern, $Fitness3(X) = \alpha \frac{1}{Gain(f_0,\theta,\phi)} + \beta|S_{11}(f_0)|(\alpha = 0.75, \beta = 0.25)$.

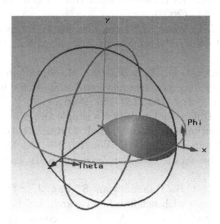

Fig. 11. 3D radiation pattern, $Fitness3(X) = \alpha \frac{1}{Gain(f_0,\theta,\phi)} + \beta|S_{11}(f_0)|(\alpha = 0.75, \beta = 0.25)$.

5 Conclusion

The Yagi-Uda antenna is designed and then it is optimized by using DE algorithm. All work is done by internal system with interactive simulation of C++ and CST. A detail analyse about the fitness function has been presented. It is observed that the optimal solution for gain and impedance has been improved. The whole optimization process is completed automatically. This method is simple, effective and can be applied to other kinds of antennas.

Acknowledgement. This work was supported by the Science and Technology Research Project of Jiangxi Provincial Education Department (Grant No. GJJ151115), the Distinguished Young Talents Plan of Jiangxi Province (Grant No. 20171BCB23075), and the Natural Science Foundation of Jiangxi Province (Grant No. 20171BAB202035).

References

1. Sotiroudis, S.P., Goudos, S.K., Gotsis, K.A., Siakavara, K., Sahalos, J.N.: Application of a composite differential evolution algorithm in optimal neural network design for propagation path-loss prediction in mobile communication systems. IEEE Antennas Wirel. Propag. Lett. **12**, 364–367 (2013)
2. Goudos, S.K., Gotsis, K.A., Siakavara, K., Vafiadis, E.E., Sahalos, J.N.: A multiobjective approach to subarrayed linear antenna arrays design based on memetic differential evolution. IEEE Trans. Antennas Propag. **61**(6), 3042–3052 (2013)
3. Pantoja, M.F., Bretones, A.R., Ruiz, F.G., Garcia, S.G., Martin, R.G.: Particle-swarm optimization in antenna design: optimization of log-periodic dipole arrays. IEEE Antennas Propag. Mag. **49**(4), 34–47 (2007)
4. Bozza, G., Pastorino, M., Raffetto, M., Randazzo, A.: Synthesis of metamaterial coatings for cylindrical structures by an ant-colony optimization algorithm. In: Proceedings of the 2006 IEEE International Workshop on Imagining Systems and Techniques, pp. 143–147 (2006)
5. Chen, P.Y., Chen, C.H., Wang, H., Tsai, J.H., Ni, W.X.: Synthesis design of artificial magnetic metamaterials using a genetic algorithm. Opt. Express **16**(17), 12806–12818 (2008)
6. Di Cesare, N., Chamoret, D., Domaszewski, M.: Optimum topological design of negative permeability dielectric metamaterial using a new binary particle swarm algorithm. Adv. Eng. Softw. **101**, 149–159 (2016)
7. Storn, R., Price, K.: Differential evolution-a simple and efficient heuristic for global optimization over continuous spaces. J. Glob. Optim. **11**(4), 341–359 (1997)
8. Zhao, J., Lv, L., Wang, H., Sun, H., Wu, R., Nie, J., Xie, Z.: Particle swarm optimization based on vector Gaussian learning. KSII Trans. Internet Inf. Syst. **11**(4), 2038–2057 (2017)
9. Wang, H., Sun, H., Li, C.H., Rahnamayan, S., Pan, J.S.: Diversity enhanced particle swarm optimization with neighborhood search. Inf. Sci. **223**, 119–135 (2013)
10. Sun, H., Wang, K., Zhao, J., Yu, X.: Artificial bee colony algorithm with improved special centre. Int. J. Comput. Sci. Math. **7**(6), 548–553 (2016)
11. Yu, G.: A new multi-population-based artificial bee colony for numerical optimization. Int. J. Comput. Sci. Math. **7**(6), 509–515 (2016)

12. Lv, L., Wu, L.Y., Zhao, J., Wang, H., Wu, R.X., Fan, T.H., Hu, M., Xie, Z.F.: Improved multi-strategy artificial bee colony algorithm. Int. J. Comput. Sci. Math. **7**(5), 467–475 (2016)

13. Wang, H., Wu, Z.J., Rahnamayan, S., Sun, H., Liu, Y., Pan, J.S.: Multi-strategy ensemble artificial bee colony algorithm. Inf. Sci. **279**, 587–603 (2014)

14. Zhou, X.Y., Wang, H., Wang, M.W., Wan, J.Y.: Enhancing the modified artificial bee colony algorithm with neighborhood search. Soft. Comput. **21**(10), 2733–2743 (2017)

15. Cui, Z., Sun, B., Wang, G., Xue, Y., Chen, J.: A novel oriented cuckoo search algorithm to improve DV-Hop performance for cyber-physical systems. J. Parallel Distrib. Comput. **103**, 42–52 (2017)

16. Zhang, M., Wang, H., Cui, Z., Chen, J.: Hybrid multi-objective cuckoo search with dynamical local search. Memet. Comput. (2017, in press). https://doi.org/10.1007/s12293-017-0237-2

17. Yu, G.: An improved firefly algorithm based on probabilistic attraction. Int. J. Comput. Sci. Math. **7**(6), 530–536 (2016)

18. Wang, H., Cui, Z., Sun, H., Rahnamayan, S., Yang, X.S.: Randomly attracted firefly algorithm with neighborhood search and dynamic parameter adjustment mechanism. Soft. Comput. **21**(18), 5325–5339 (2017)

19. Lv, L., Zhao, J.: The firefly algorithm with Gaussian disturbance and local search. J. Signal Process. Syst. (2017, in press). https://doi.org/10.1007/s11265-017-1278-y

20. Wang, H., Wang, W., Sun, H., Rahnamayan, S.: Firefly algorithm with random attraction. Int. J. Bio-Inspired Comput. **8**(1), 33–41 (2016)

21. Kaur, M., Sharma, P.K.: On solving partition driven standard cell placement problem using firefly-based metaheuristic approach. Int. J. Bio-Inspired Comput. **9**(2), 121–127 (2017)

22. Wang, H., Wang, W.J., Zhou, X.Y., Sun, H., Zhao, J., Yu, X., Cui, Z.: Firefly algorithm with neighborhood attraction. Inf. Sci. **382–383**, 374–387 (2017)

23. Wang, H., Zhou, X.Y., Sun, H., Yu, X., Zhao, J., Zhang, H., Cui, L.Z.: Firefly algorithm with adaptive control parameters. Soft. Comput. **21**(17), 5091–5102 (2017)

24. Cai, X., Gao, X.Z., Xue, Y.: Improved bat algorithm with optimal forage strategy and random disturbance strategy. Int. J. Bio-Inspired Comput. **8**(4), 205–214 (2016)

25. Bantin, C., Balmain, K.: Study of compressed log-periodic dipole antennas. IEEE Trans. Antennas Propag. **18**(2), 195–203 (1970)

26. Li, X., Zhang, X., Hei, Y.: Antenna Gain Imbalance detection method using Particle Swarm algorithm for MIMO systems. In: International Conference on Wireless Communications and Signal Processing (WCSP), pp. 1–6, October 2012

27. Pu, T.L., Huang, K.M., Wang, B., Yang, Y.: Application of micro-genetic algorithm to the design of matched high gain patch antenna with zero-refractive-index metamaterial lens. J. Electromagn. Waves Appl. **24**(8–9), 1207–1217 (2010)

28. Wang, H., Rahnamayan, S., Sun, H., Omran, M.G.H.: Gaussian bare-bones differential evolution. IEEE Trans. Cybern. **43**(2), 634–647 (2013)

29. Zhou, X.Y., Wu, Z.J., Wang, H., Rahnamayan, S.: Enhancing differential evolution with role assignment scheme. Soft. Comput. **18**(11), 2209–2225 (2014)
30. Wang, H., Rahnamayan, S., Wu, Z.J.: Parallel differential evolution with self-adapting control parameters and generalized opposition-based learning for solving high-dimensional optimization problems. J. Parallel Distrib. Comput. **73**(1), 62–73 (2013)
31. Wang, H., Wu, Z.J., Rahnamayan, S.: Enhanced opposition-based differential evolution for high-dimensional optimization problems. Soft. Comput. **15**(11), 2127–2140 (2011)

Research on Coordination Fresh Product Supply Chain Under New Retailing Model

Bo Yang[1]([✉]) and Dongbo Zhang[2]

[1] Huashang College, Guangdong University of Finance and Economics, Zengcheng,
Guangzhou City 511300, Guangdong Province, China
jxyb2008@126.com
[2] Department of Computer Science, Guangdong University of Science and Technology,
Dongguan City 523000, Guangdong Province, China
dongbozhang2013@qq.com

Abstract. Under the New Retailing mode, the sales model of fresh agricultural produce is significantly different from traditional model and O2O; therefore, it is necessary to study the coordination of fresh supply chain. This paper analyzed the fresh supply chain under the New Retailing, analyzed the supply chain coordination in detail based on the centralized decision-making mode, put forward some strategies that we should increase the proportion of online business, reduce the price of online sales, and build the information sharing platform under New Retailing mode.

Keywords: New Retailing · Fresh agricultural products
Supply chain coordination

1 Preface

In recent years, with the high-speed development of Chinese national economy, the living standard of the people is enhanced increasing. Market demand on the fresh agricultural products is increasing. In order to strong support the centralized trading activities of agricultural products online, and Realize the transformation and upgrading of traditional industries. 2013, The State Administration for Industry and Commerce issued the "opinions on accelerating the development of circulation industry ". Thus it can be seen, It's becoming a trend for improving the circulation efficiency of fresh agricultural products through network transaction and information construction. Under this background, Diversified sales models emerge at a historic moment. The Online to Offline (O2O) model draws wide attention of scholars and entrepreneurs because it has the advantages of efficient interactivity for electronic information, flexible transformation for the amount of network access and entity store usages, high-quality offline services, etc. 2015, The retail enterprises of Supermarket seized expanding their online markets, while traditional e-commerce enterprises expand energetically their offline markets. Although the Online to Offline (O2O) model has great potentials, it still has some problems.as consumer spending habits is not on all fours with us, the subsidies of cut prices is difficult to control. New Retailing is a new term in the development mode based on

© Springer Nature Singapore Pte Ltd. 2018
K. Li et al. (Eds.): ISICA 2017, CCIS 874, pp. 439–445, 2018.
https://doi.org/10.1007/978-981-13-1651-7_39

O2O, To a certain it inherited. The characteristics of O2O mode, and this also puts forward new requirements for it.

2 The Connotation of New Retailing

Jack Ma put forward the word "New Retailing" at the cloud conference of Alibaba in October 2016, he pointed out that there will be no e-commerce in the future, offline and online should be combined with logistics together, this kind of marketing way which combines logistics with online and offline business effectively is called New Retailing, it has some features of the current O2O. Different scholars give different views on the definition of New Retailing, some industry insiders point out that the New Retailing is to retail the data, some people believe that New Retailing is "online + off-line + logistics", its core is to take consumer as the center, to open up comprehensive data, including membership information, inventory, payment, service and so on [1]. Du and Jiang provide a clear definition of New Retailing, they point out that the enterprises rely on the Internet, use the technical means, such as large data, artificial intelligence, etc., upgrade the supply chain process of products to form a new industrial structure ecosystem, and combine online and offline services with logistics deeply, this kind of retail way is called New Retailing [2]. From the definition, we can see that the essence of New Retailing is to highlight the importance of logistics, this kind of dual channels supply chain improved the response speed of retailers in a certain extent, improved the overall profit of the supply chain, and can influence profits of whole supply chain, retailers, manufacturers, etc.

3 Problems Existing in Fresh Supply Chain Under New Retailing

In the past two years, the development of fresh e-commerce suppliers is rapid in China, especially in the O2O model, which has been developed rapidly, and even some people have suggested that "who get the marketing of fresh products must have the world". However, with the rapid development, there are still some problems in the fresh e-commerce. Such as the costs of developing and maintaining customer are high, logistics costs are high, and some areas are lack of cold chain logistics resources, customers' needs are diversified, product types cannot meet customer needs, the cost of supply chain is high, fresh consumption is higher in the process of logistics transportation, the supplier is unstable and so on.

Under the background of New Retailing, logistics become more important, but it still can't solve the problems existing in the fresh supply chain, such as price reduction subsidies behavior still existed in the enterprises under the New Retailing, the price war between e-commerce and enterprises is hard to avoid, its market share is high, the valuation is also high, but the enterprises who can make profits are very few. In the process of supply chain, there are also some problems, such as purchasing channels are unstable, short term capital shortage caused by unbalanced capital flow, high cost of logistics transportation, shortage of cold chain logistics resources and so on. Therefore, how to realize the effective coordination of fresh supply chain in order to increase profit has become an important research direction under the background of New Retailing.

4 Coordination Strategy of Fresh Supply Chain Under New Retailing

Due to the consumability of fresh agricultural products, the coordination of its supply chain is more important. Supply chain coordination refers to coordinate the logistics, information flow and cash flow between upstream and downstream in the supply chain, ensure the effective transfer in supply chain, reduce the uncertainty problems caused by asymmetric information, so as to reduce the conflicts of interest between each link of supply chain, eventually reduce supply chain cost, improve the overall efficiency of the supply chain.

4.1 Problem Description and Hypothesis

In order to simplify the study, this paper discussed the coordination of two pole fresh supply chain between a retailer and a distributor under the background of New Retailing in a centralized decision model.

Under the background of New Retailing, supply chain structure of fresh e-commerce is online shopping and offline distribution and consumption, order from consumers to retailers, orders from retailers to distributors, and directly distribute to consumers by distributors, at this point, the retailer's price range, sales price, order period and distributor's sales price are the decision variables, while the others are constant.

4.2 Supply Chain Coordination Under Centralized Decision-Making Mode

4.2.1 The Construction of Supply Chain Profit Model Under Centralized Decision-Making Model

From the form of New Retailing sales, fresh e-commerce sales (online sales) are closer to pre-sales, therefore, inventory deterioration losses are much smaller than traditional retail methods, cannot be considered. But because of the need for offline distribution, there will be an increase in distribution costs. Therefore, it is assumed that the market size of fresh products is α_1, retail price is p_1, product demand price elasticity is K, and distribution cost is c_r. Based on the demand of fresh agricultural products has the characteristic of high elasticity, this paper used the exponential demand model to construct the demand of fresh agricultural products q_1, then q_1 can be expressed as

$$q_1 = \alpha_1 p_1^{-k} \tag{1}$$

It is assumed that the ordering period of fresh product is T, the order quantity of fresh products is Q, an ordering period of fresh product aggregate is D, and the supplier cost is c_s then under the online sales, profit function of the supply chain can be expressed as

$$R_1 = \alpha_1 P_1^{-k} T (P_1 - C_s) - \alpha_1 C_t T \tag{2}$$

New Retailing is a way to combine online sales with offline sales, for the offline sales part, it is assumed that the market size of fresh products is α_2, retail price is p_2, this paper used the exponential demand model to construct the demand of fresh Agricultural products q_2, then q_2 can be expressed as

$$q_2 = \alpha_2 p_2^{-k} \tag{3}$$

The inventory stock $I(t) = 0$, according to the metamorphic inventory formula, can get $dI(t) = -q_2 dt - \lambda dt$ among them; λ is the logistics loss rate of fresh products. Through the differential equation to deform, then obtained

$$I(t) = \frac{q_2 e^{\lambda T} - q_2 e^{\lambda t}}{\lambda e^{\lambda t}} \tag{4}$$

Because of the order quantity of fresh products $Q = I(0)$, then

$$Q = \frac{q_2 e^{\lambda T} - q}{\lambda} \tag{5}$$

The ordering period of fresh product aggregate is $D = q_1 T$. On the basis of Formula 1 and Formula 5. Then the profit of fresh products under offline sales can be expressed as

$$R_2 = \left(p_2 T - C_s \frac{e^{\lambda T} - 1}{\lambda} \right) \alpha_2 p_2^{-k} \tag{6}$$

When the enterprises developed the New Retailing, in order to avoid the conflict between online and offline channels, enterprises often adopt a unified pricing strategy, which is $P_1 = P_2$, expressed as P. Then under the background of New Retailing, the profits of supply chain can be expressed as

$$R = \alpha_1 P^{-k} T (P - C_s) - \alpha_1 C_t T + \left(PT - C_s \frac{e^{\lambda T} - 1}{\lambda} \right) \alpha_2 P^{-k} \tag{7}$$

Let $\dfrac{dR}{dT} = 0$, then can get the best order period of enterprise is

$$T^* = \frac{1}{\lambda} \ln \left(\frac{\alpha_2 P + \alpha_1 P - \alpha_1 C_s - \alpha_1 C_t}{\alpha_2 C_s} \right) \tag{8}$$

In the same way, let $\dfrac{dR}{dP} = 0$, then the optimal price is

$$P^* = \frac{\alpha_1 TKC_s + \alpha_2 KC_s \frac{e^{\lambda T} - 1}{\lambda}}{(k-1)(\alpha_1 + \alpha_2)T} \tag{9}$$

4.2.2 The Optimal Strategy Under Centralized Decision Making Mode

(a) Increasing the proportion of online business is beneficial to reduce the cost of supply chain and increase profit

According to the above model, the deterioration inventory of fresh products in an order cycle can be expressed as ΔI

$$\Delta I = Q - D = \alpha_2 P_2^{-k} \frac{e^{\lambda T} - 1 - \lambda T}{\lambda} \tag{10}$$

Due to there are offline sales, there must be inventory, so $\frac{d\Delta I}{dT} \rangle 0$, it can be concluded that the optimal order cycle is needed to reduce the deteriorating stock of the whole supply chain. Therefore, it can change the market structure of the online and offline business to change the order cycle. Because the online business does not need inventory, increasing the proportion of online business can minimize inventory and reduce inventory loss, so that the best order cycle is shortened.

(b) Subsidies or lower prices on online businesses can increase supply chain profits

It is assumed that under the background of New Retailing, the online business and offline business use different prices, assuming that the optimal price for offline sales is P^*, and then the profit can be expressed as

$$R = \alpha_1 P_1^{-k} T (P_1 - C_s) - \alpha_1 C_t T + \left(P^* T - C_s \frac{e^{\lambda T} - 1}{\lambda} \right) \alpha_2 P^{*-k} \tag{11}$$

Thus, the optimal online sales price is introduced to $P_1^* = \frac{kC_s}{k-1}$, because $e^{\lambda} - 1 \rangle \lambda T$, it can be rolled out $P^* \rangle P_1$. It can be seen that the online business can improve the profits of the supply chain for the price reduction from P^* to P_1^*. At this point the best price reduction is $\Delta P = P^* - P_2$.

5 The Strategy of Fresh Supply Chain Coordination Mechanism Under New Retailing

The coordination mechanism of fresh supply chain is composed of four layers of information layer, business cooperation layer, cooperation layer and interest layer. The profit and business structure is only a small part of it; it also needs to take into account the cooperation of information, cooperation, and benefit sharing contract and other aspects, the specific strategies as follows.

(a) Use information technology to enhance information sharing

The supply chain coordination of New Retailing requires that all parties on the supply chain can effectively realize information sharing. The information must be authentic, reliable and timely, the whole process should be managed in all stages, and the tracking and effective control should be carried out. Therefore, the construction and use of ERP, POS, EOS, CRM and other information systems must be done. The replacement of e-commerce to New Retailing is not equal to the loss of information technology, but the more efficient application, it needs to integrate offline and online resources, and establish a public information platform.

(b) Build the trust mechanism of supply chain under New Retailing

The effective development of fresh retail will require close cooperation between the upstream and downstream of the supply chain, which requires a good relationship of trust between them. This includes trust between suppliers and retailers, and trust between retailers and customers. Therefore, building effective trust mechanism can form effective soft constraints between supply chain, strengthen effective collaboration between them, and ensure efficient operation of supply chain.

(c) Profit sharing contract of the members of the supply chain

The sharing contract of the members between the supply chain upstream and downstream can effectively solve the cooperation and competition relationships between them, it can guarantee the cooperation between upstream and downstream enterprises from economically and legal, to promote the overall reduction of logistics costs and reduce the bullwhip effect. It can also clarify the responsibility and location of the enterprises in the supply chain, ensure the cooperation is more efficient, and jointly improve the service level of the supply chain service.

6 Conclusion

Fresh agricultural products have a great market prospect under the background of New Retailing, and achieve the coordination of the supply chain to promote overall profit of supply chain; to improve the efficiency and profit of upstream and downstream members has the vital significance. For this, enterprises can increase the proportion of online business, subsidize online business, strengthen information technology construction to realize information sharing of supply chain, and build trust mechanism of supply chain.

Acknowledgment. This work was partially supported by Key platform construction leap plan and major project and achievement cultivation plan project, characteristic innovation project of Department of Education of Guangdong Province of China (Grant No. 2014GXJK174), Higher education innovation strong school project of Huashang College, Guangdong university of finance and economics (Grant No. HS2014CXQX10). We would like to thank the anonymous reviewers for their valuable comments that greatly helped us to improve the contents of this paper.

References

1. Liu, Y.: What is "New Retailing"? Alibaba is also hard to talk about currently. Kwangmyong Net. http://it.gmw.cn/2017-02/27/content_23830020.htm
2. Du, R., Jiang, K.: The new retailing: connotation, development motivation and key issues. Price Theory Pract. **2**, 139–141 (2017)
3. Zhang, J.: A new supply service model for retail delivery businesses. Science and engineering research center. In: Proceedings of 2016 International Conference on Advanced Manufacture Technology and Industrial Application (AMTIA 2016). Science and Engineering Research Center, June 2016
4. Yang, X.: Research on the coordination of fresh products O2O supply chain. Logist. Eng. Manag. **1**, 96–98 (2014)
5. Ma, W.M., Zhao, Z., Ke, H.: Dual-channel closed-loop supply chain with government consumption- subsidy. Eur. J. Oper. Res. **226**(2), 221–227 (2013)
6. Patel, P.C., Guides, M.J., Pearce, J.A.: The role of service operations management in new retail venture survival. J. Retail. **93**, 241–251 (2017)
7. Vandevijvere, S., Sushil, Z., Exeter, D.J., Swinburne, B.: Obesogenic retail food environments around New Zealand schools: a national study. Am. J. Prev. Med. **51**, e57–e66 (2016)
8. Grazioso, M.: Assessing the Retail food environment surrounding elementary schools across New York City (NYC) neighborhoods varying in their level of gentrification. J. Nutr. Educ. Behav. **48**(7) (2016)
9. Sun, D.: Research on game of fresh agriculture products distribution scheme in the supermarket-oriented mode, May 2017
10. Yan, B., Ye, B., Zhang, Y.-W.: Three-level supply chain coordination of fresh agricultural products under internet of things. Syst. Eng. **01** (2014). School of Economics and Trade, South China University of Technology
11. Wang, L., Dan, B.: Fresh-keeping and pricing strategy for fresh agricultural product based on customer choice. Chin. J. Manag. **03** (2014). Chongqing University
12. Sun, G.-H., Xu, L.: Option Contract of Two-echelon agricultural supply chain with random supply and demand. J. Ind. Eng. Eng. Manag. **02** (2014). School of Management Science and Engineering, Shandong University of Finance and Economic; Business School, Nankai University
13. Mei-na, H., Yu-ling, S., Kui-ran, S.: Ordering decision of fresh agricultural product supply chain with fairness concern. Ind. Eng. J. **02** (2014). School of Economics and Management, Nanjing University of Technology
14. Zhou, K.-Q., Hou, B., Zhao, H.-Z.: Coordination of fresh agricultural products supply chain with remaining subsidy contract under incomplete information. J. Chongqing Univ. Technol. (Soc. Sci.) **09** (2014). School of Mechanical Engineering, Chongqing University of Technology

Virtualization – Motion-Based Tracking

Real-Time RGBD Object Tracking via Collaborative Appearance and Motion Models

Danxian Chen[1], Zhanming Liu[1], Hefeng Wu[1(✉)], and Jin Zhan[2]

[1] School of Information Science and Technology,
Guangdong University of Foreign Studies, Guangzhou, China
`danxian.chen@foxmail.com, misswe@qq.com, wuhefeng@gmail.com`
[2] Institute of Computer Sciences, Guangdong Polytechnic Normal University,
Guangzhou, China
`jinerzhan@163.com`

Abstract. Visual object tracking remains an active and challenging topic in computer vision due to a great variety of intricate factors such as illumination variation, object deformation and background clutter. Recent research efforts have achieved impressive success in object tracking, but they commonly have to utilize complicated models requiring high computation cost, which renders these methods hardly suitable for many applications. Considering depth information of the scene can provide effective complement to color images, in this paper, we propose a novel and efficient method for tracking an object in RGBD videos by using collaborative appearance and motion models. Experimental results demonstrate that our method achieves superior tracking performance over several state-of-the-methods while running efficiently.

1 Introduction

Visual object tracking plays a fundamental and important role in many high-level computer vision applications, such as video surveillance, automatic driving and robotics. So far it remains a challenging problem due to a great variety of complex factors, e.g., large illumination variation, fast motion, object deformation, background clutter, etc. Recently, great progress has been made in object tracking since some sophisticated machine learning methods including deep learning-based models were introduced. Especially, with the great ability of representation learning and well-validated pre-trained models, deep learning-based object tracking methods has achieved surprising state-of-the-art performance [1,2]. However, these methods generally require high-performance computing resources, which limits their employment in many real-world applications lacking such computing circumstances.

Fortunately, progress in physical devices provides alternative ways for boosting object tracking performance in low computation requirements. Recently, camera devices like Kinect that can capture the depth information of the scene

© Springer Nature Singapore Pte Ltd. 2018
K. Li et al. (Eds.): ISICA 2017, CCIS 874, pp. 449–460, 2018.
https://doi.org/10.1007/978-981-13-1651-7_40

have made much progress and are easier to obtain than before. With the help of such devices, color images and depth images of the scene can be acquired simultaneously, and they are together commonly called RGBD images. In addition to RGB color information, the depth information can provide an effective auxiliary to help improve the performance on many computer vision tasks.

In this paper, we propose an efficient object tracking method for RGBD videos, by using collaborative appearance and motion models that take good advantage of color and depth information. Specifically, we build contour model, color model and motion model for the given target at initialization. Afterwards the three models will collaborate to achieve object tracking. The contour model that builds on the depth information plays an efficient role while tracking, and the color and motion models will help to overcome some challenging situations and achieve consistent tracking. Experiments conducted on challenging scenarios demonstrate that the proposed method can run in real-time and achieve favorable tracking performance when compared with several recent tracking methods.

The rest of this paper is organized as follows. Section 2 reviews the recent literature of object tracking, and Sect. 3 describes the proposed method in detail. The experimental results are presented in Sects. 4, and 5 concludes this paper.

2 Related Work

Visual object tracking is a critical component in many real world applications, and it has been an active research topic in computer vision and artificial intelligence for decades. Many works have been presented in the literature to address the challenging issues in this topic. We review below the recent advances in object tracking. Interested readers are recommended to the recent surveys [3–5] for a comprehensive study.

The visual appearance of the object to be tracked will change over time, and it poses many challenges to object tracking algorithms that large appearance variations happen due to complex factors such as severe object deformation, great illumination variations, abnormal motion and heavy occlusion. Therefore, building a robust object appearance representation model is an essential issue for object tracking methods, and the built object model should contain the intrinsic appearance feature of the object and also be able to tolerate the exterior variations over time. Image features like colors, textures and edges are commonly used for the object model representation, for example, building a histogram from them [6,7]. In [6], the histogram representation of the object was combined with a kernel method to formulate measuring histograms' similarity as a convex optimization problem and the mean shift algorithm was used as a nice tracking solution. However, the object appearances are often complex, so more delicate models are required. In [8], Gaussian mixture models were introduced to represent the object by considering multi-modal characteristics. Adam et al. [9] proposed to represent the object appearance in a spatially multi-modal manner using a fragments-based model, so that the model can better reflect the partial deformation of the object appearance. In [10], incremental principal component analysis was utilized to discover the primal elements in the object appearance. Sparse representation-based models [11–13] have also been studied for representing the object, generally through an over-complete dictionary. This kind of methods makes much progress in the tracking accuracy, but at the cost of high computation complexity due to the L_1 optimization rooted in these methods.

The trend of applying machine learning algorithms to build the object model has brought the advances of object tracking to a new high [14–18]. In [14], Grabner and Bischof used weak classifiers associated with Haar-like features to represent the object and employed the AdaBoost algorithm to select discriminative weak classifiers for locating the target. Later, they introduced semi-supervised learning into their online boosting tracking framework [19]. Babenko et al. [15] adapted the multiple instance learning algorithm to an online version for object tracking. In [20], the circulant matrix theory was combined with the ridge regression algorithm to learn the object model, where the learning process can be transformed into the frequency domain for fast computation. In these works, machine learning methods are applied to learn a discriminative object model based on hand-crafted image features, aiming to distinguish the object from the background. In recent years, inspired by the huge success of deep learning algorithms (e.g., convolutional neural networks) that can learn feature representation directly from raw images, many object tracking methods employed the pre-trained deep learning models to extract image features and combine them with previous algorithms to learn the object model [2,21,22]. Moreover, the feature extraction and model learning can be combined into the deep learning framework for joint optimization to achieve performance gain [1]. However, because the target is often given in the first frame, the problem of lacking enough examples of the target still lays serious obstacles for building a robust object model based on machine learning algorithms.

The recent success in object tracking is generally attributed to the introduction of sophisticated models that require high computational cost, which renders these methods hard to be applied in many real-world applications. Therefore, alternative efficient methods that take new capturing devices and input data into consideration are also necessary. In [23], the data obtained from a two-layered laser range sensor and a fisheye camera were combined to track people. Wu et al. [24] employed a multipoint infrared laser and a camera to detect and track people. Li et al. [25] used grayscale-thermal data to learn collaborative sparse representation for object tracking. In [26], RGBD data was used for superpixel-based object tracking by fusing the depth information, but this method requires sophisticated modeling and cannot run in real-time. In this paper, we study a new real-time object tracking method based on the RGBD data, by using the collaboration of several simple but effective models.

3 Proposed Method

The framework of our method is illustrated in Fig. 1. At initialization, three models, i.e., contour model, color model and motion model, are built for the given target. Then for each consecutive frame, these models are used in a collaborative way to perform object tracking.

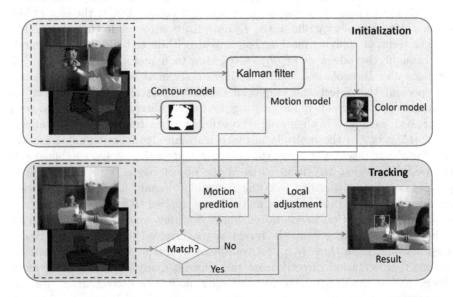

Fig. 1. Illustration of the proposed tracking framework.

Before delving into the detailed description, we first introduce some notations that will be used throughout the paper. For an RGBD video, each frame contains two images, i.e., an RGB image and a depth image, and we use I_i^c and I_i^d to denote the RGB image and depth image in the ith frame ($i \geq 1$). The width and height of a frame are denoted by W and H, respectively.

3.1 Model Initialization

In the first frame, the bounding box R is given, which contains the target to be tracked. Assume the upper-left corner, width and height of the given bounding box are (x_o, y_o), w and h, respectively. A depth image B of the background scene is obtained.

Contour Model: An extended bounding box R^e is obtained via extending the given bounding box R by 1/4 of the width and height, respectively. The width and height of R^e are denoted as w^e and h^e. A binary image F_1^b with the size of $W \times H$ is then created. With the pre-obtained background depth image B, each value of image F_1^b can be calculated by the following formula:

$$F_1^b(x,y) = \delta\left[(x,y) \in R^e \wedge |I_1^d(x,y) - B(x,y)| > T_d\right] \tag{1}$$

where $\delta[O]$ is an indicator function, i.e., $\delta[O]$ equals 1 if the statement O is true and 0 otherwise. T_d is a predefined threshold. Afterwards, an erosion and dilation operation is carried out to erase noise in the image F_1^b.

We then traverse the image F_1^b to find the largest eight-connected region. Let (x_1^c, y_1^c), w_1 and h_1 denote the center, width and height of the bounding box enclosing this region. We normalize this region into a contour template model M_c of size $s \times s$ (we set s to be 40 in experiments).

Color Model: A color model is built to represent the target appearance. Specifically, the R, G, B channels are divided into K bins separately, with a total of N bins ($N = K^3$), and a color histogram $\hat{Q}_1 = (\hat{q}_1, \hat{q}_2, ..., \hat{q}_N)$ is then built as the color model. The value of each bin can be calculated as:

$$\hat{q}_i = \sum_{(x,y)\in I_1^c} \delta\left[F_1^b(x,y) = 1 \wedge map(x,y) = i\right], \quad i = 1, ..., N \qquad (2)$$

where function $map(x,y) : R^2 \rightarrow \{1, ..., N\}$ maps the RGB value of pixel (x,y) to a bin number. In order to make the bin value to fall into the range $[0,1]$, we normalize the histogram value with the following formula:

$$\left\{\hat{q}_i = \frac{\hat{q}_i}{\hat{q}_{max}}\right\}_{i=1,...,N} \qquad (3)$$

where $\hat{q}_{max}$ is the maximum bin value in the unnormalized histogram, i.e., $\hat{q}_{max} = \max\{\hat{q}_i\}_{i=1}^N$.

Motion Model: A simple motion model using the Kalman filter is built for the given target. The Kalman filter [27] is an efficient auto-regressive filter, and it can be described by the state Eq. (4) and the observation Eq. (5):

$$X(i) = A(i)X(i-1) + W(i) \qquad (4)$$

$$Z(i) = H(i)X(i) + V(i) \qquad (5)$$

where $X(i)$ and $Z(i)$ are the state vector and the observation vector at the ith moment respectively, $A(i)$ is the transition matrix, and $H(i)$ is the measurement matrix. $W(i)$ and $V(i)$ are independently and identically distributed zero-mean Gaussian noise.

In this paper, we assume that the state vector $X(i) = (x_i, y_i, v_{xi}, v_{yi})^T$, where (x_i, y_i) denotes the target's location at the ith frame, v_{xi} and v_{yi} are the target's velocity in the x and y directions. The observation vector $Z(i) = (\tilde{x}_i, \tilde{y}_i)$. The matrices $A(i)$ and $H(i)$ are set as follows:

$$A(i) = \begin{bmatrix} 1 & 0 & dt & 0 \\ 0 & 1 & 0 & dt \\ 0 & 0 & 1 & 0 \\ 0 & 0 & 0 & 1 \end{bmatrix}, \quad H(i) = \begin{bmatrix} 1 & 0 & 0 & 0 \\ 0 & 1 & 0 & 0 \end{bmatrix} \qquad (6)$$

where dt is a constant parameter.

3.2   Collaborative Object Tracking

When a new frame i ($i \geq 2$) comes, the three models collaborate to track the target.

At the first stage, we obtain a binary image F_i^b:

$$F_i^b(x,y) = \delta \left[(x,y) \in R_i^e \wedge |I_i^d(x,y) - B(x,y)| > T_d \right] \tag{7}$$

where R_i^e is a region centered at (x_{i-1}^c, y_{i-1}^c) and with the size of $w^e \times h^e$. Then we traverse the image F_i^b to find the largest eight-connected region Ω', and normalize this region into a patch M_c' with the same size of the contour model M_c. The patch M_c' is matched with M_c using the formula below:

$$S_c = \sum_{(x,y) \in M_c} \delta \left[M_c'(x,y) = M_c(x,y) \right]. \tag{8}$$

If $S_c/s^2 > T_c$, where T_c is a constant threshold and $s \times s$ is the size of M_c, the target is considered to be located successfully. The target bounding box is the rectangle that encloses the region Ω', with the center (x_i^c, y_i^c), width w_i and height h_i. If this condition is not satisfied, the matching is considered to be failed, and the second stage will be performed.

At the second stage, the Kalman filter is first used to predict the target's location (x'_i^c, y'_i^c), and then the color model is applied to carry out local adjustment and get the final location (x_i^c, y_i^c).

Specifically, a probability map P is obtained:

$$P(x,y) = \begin{cases} \hat{q}_{map(x,y)}, & \text{if } (x,y) \in R_i^e \wedge F_i^b(x,y) = 1, \\ 0, & \text{otherwise.} \end{cases} \tag{9}$$

Then we find the centroid of the probability map as follows.
The 0th moment is:

$$M_{00} = \sum_x \sum_y P(x,y) \tag{10}$$

The 1st moments for x and y:

$$M_{10} = \sum_x \sum_y xP(x,y), \quad M_{01} = \sum_x \sum_y yP(x,y) \tag{11}$$

The centroid:

$$x_p = \frac{M_{10}}{M_{00}}, \quad y_p = \frac{M_{01}}{M_{00}} \tag{12}$$

We apply the following strategy to perform local adjustment of the target's location:

$$x_i^c = x'_i^c + sign(x_p - x'_i^c)\lambda_x, \quad y_i^c = y'_i^c + sign(y_p - y'_i^c)\lambda_y, \tag{13}$$

where $sign(z)$ equals 1 if $z > 0$, equals -1 if $z < 0$, and 0 otherwise. λ_x and λ_y are predefined constants.

3.3   Model Update

The target location (x_i^c, y_i^c) is used to update the Kalman filter model. Moreover, if the target is successfully tracked in the first stage, the contour model M_c will be replaced by the patch M_c', and the color model is also updated as:

$$\hat{Q}_i = (1 - \eta)\hat{Q}_{i-1} + \hat{Q}', \tag{14}$$

where $\hat{Q}'$ is the color histogram built in the current frame and the learning rate η is a constant parameter. If the target cannot be successfully tracked in the first stage, the contour and color models will not be updated.

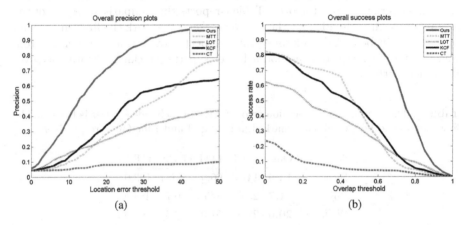

Fig. 2. Quantitative comparison on all test videos. (a) Precision in terms of center location error. (b) Successful rate in terms of overlap ratio.

4   Experiments

In this section, experiments are conducted to validate the performance of the proposed object tracking method. We test our tracker on four RGBD videos containing various challenging factors. Our method is compared with four recently presented tracking methods, including MTT [28], LOT [29], KCF [20] and CT [30]. All compared trackers are run with their default parameter settings. We use two measures from the OTB benchmark [31] to evaluate the tracking performance, i.e., the precision plot based on the center location error (CLE) and the success plot based on the overlap ratio (OLR). To be specific, denote the ground truth bounding box and the bounding box predicted by a tracker as r_g and r_p, respectively. CLE is the Euclidean distance between the centers of r_g and r_p, and the precision is calculated as the ratio between the number of frames with CLE less than a given threshold and the total number of frames. OLR is defined as $\frac{area(r_g \cap r_p)}{area(r_g \cup r_p)}$, and the success rate is calculated as the ratio between the number of frames with OLR greater than a given threshold and the total number of frames.

The proposed method is implemented in Matlab. In experiments, the important parameters are set as follows: $dt = 1$, $\lambda_x = 2.5$, $\lambda_y = 2.5$, and the learning rate $\eta = 0.005$. When running on a computer with Window 7 operating system, Intel i5 CPU and 4GB RAM, it can achieve about 30 frames per second averagely for images of size 640×480 without code optimization, which demonstrates that it can satisfy real-time processing requirements.

4.1   Quantitative Evaluation

The precision and success plots of the overall tracking performance on all the test videos are shown in Fig. 2, with respect to different thresholds for center location error and overlap ratio. Moreover, Table 1 reports the comparison results of the average CLE on each video for the tested trackers. We mark the first and second best results in bold and underline fonts, respectively. It can be observed from Fig. 2 and Table 1 that our method achieves superior tracking performance over the other methods.

Table 1. The average center location error (CLE) on each video (lower is better). The first and second best results are marked in bold and underline fonts, respectively.

Sequence	Ours	MTT	LOT	KCF	CT
Bear	**15.9**	141.3	<u>66.9</u>	216.1	133.0
Boy	**11.7**	29.0	445.7	<u>20.7</u>	483.9
Walking	**20.0**	28.5	51.0	<u>20.3</u>	161.3
Dancing	**14.3**	<u>34.2</u>	65.3	66.3	310.0

4.2   Qualitative Analysis

In this section, we will take qualitative evaluation to further analyze the performance of the compared tracking algorithms in some complex environments. Figure 3 shows example tracking results from two representative videos. The RGB color images with the tracking results of all compared methods are shown in the first row, while the second row exhibits the corresponding depth images.

In the Bear video shown in Fig. 3(a), the occlusion is the biggest disturbance factor. In the 44th frame, the bear will pass down the white box from top to bottom. We can see that, when the bear is occluded by the white box, the performance of our tracker and the CT tracker is better than that of others. The KCF tracker's predicted location is at the bottom of the bear, which considerably deviates from the actual location. In addition, the size of the predicted bounding box of the LOT tracker is significantly reduced, which is quite different from the actual size of the bear. Moreover, compared with the CT tracker, our tracker achieves better results. That is because our algorithm not only predicts the location of the bear through the Kalman filter to record the bear's motion

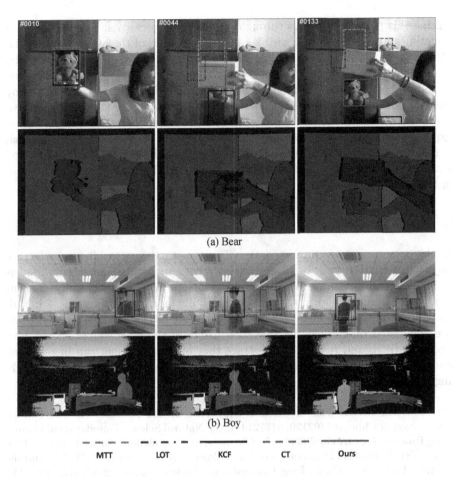

Fig. 3. Examples of the tracking results on representative RGBD videos: (a) Bear and (b) Boy. The predicted bounding boxes of the five trackers are presented with different colors. (Color figure online)

state, but also finely adjusts the predicted location based on the color appearance model, making our result closer to the ground truth even in the case of occlusion. In the 133rd frame, it is clear that only our algorithm can track the bear successfully. Resulted from the interference of the white box, the MTT, LOT and CT algorithms have been tracking the white box that occludes the bear, while the KCF algorithm has already lost the target. Our algorithm preserves the contour model when the target is occluded, and the contour model will continue to be updated after it matches at the predicted location, so our algorithm can persistently and accurately track the target after the target leaves from the occluding object.

In the Boy video, as shown in Fig. 3(b), a boy walks towards his seat in the laboratory, sits down and then stands up again to walk to another place.

During this period, he passes through some large objects that occlude him and sometimes his whole body is visible. At the beginning of this video (around the 15th frame), the LOT and CT trackers have lost the tracked target obviously. In the 52nd frame, the boy exposes the whole body, the sizes of the predicted bounding boxes of the MTT and KCF trackers are larger than that of the boy and are not able to properly include the boy's whole body. Our algorithm can continually update the contour model of the target, therefore, when the boy exposes the whole body, the algorithm will update the contour model, which makes the predicted result closer to the ground truth. Consequently, compared with the other methods, our method can capture and update the characteristics of the target better.

5   Conclusion

In this paper, we have presented an efficient method for tracking an arbitrary object in RGBD videos, by using collaborative appearance and motion models that take good advantage of color and depth information. Specifically, contour model, color model and motion model are built for the given target at initialization. These three models collaborate to achieve object tracking. Experiments are conducted on challenging scenarios to verify the effectiveness of the proposed method, and results demonstrate that our method can run in real-time and achieve favorable tracking performance compared with several recent tracking methods.

Acknowledgement. This research is supported by the National Natural Science Foundation of China (61402120, 61772144), the Natural Science Foundation of Guangdong Province (2014A030310348), the Characteristic Innovation (Natural Science) Program of the Education Department of Guangdong Province (2016KTSCX077), and the Startup Program in Guangdong University of Foreign Studies (299-X5122029). The corresponding author is Hefeng Wu.

References

1. Nam, H., Han, B.: Learning multi-domain convolutional neural networks for visual tracking. In: Proceedings of IEEE Conference on Computer Vision and Pattern Recognition (CVPR), pp. 4293–4302 (2016)
2. Li, H., Wu, H., Lin, S., Luo, X.: Coupling deep correlation filter and online discriminative learning for visual object tracking. J. Comput. Appl. Math. **329**, 191–201 (2018)
3. Yilmaz, A., Javed, O., Shah, M.: Object tracking: a survey. ACM Comput. Surv. **38**, 1–45 (2006)
4. Smeulders, A.W.M., Chu, D.M., Cucchiara, R., Calderara, S., Dehghan, A., Shah, M.: Visual tracking: an experimental survey. IEEE Trans. Pattern Anal. Mach. Intell. **36**, 1442–1468 (2014)
5. Li, X., Hu, W., Shen, C., Zhang, Z., Dick, A.R., van den Hengel, A.: A survey of appearance models in visual object tracking. ACM Trans. Intell. Syst. Technol. **4**, 58:1–58:48 (2013)

6. Comaniciu, D., Ramesh, V., Meer, P.: Kernel-based object tracking. IEEE Trans. Pattern Anal. Mach. Intell. **25**, 564–575 (2003)

7. Wu, H., Liu, N., Luo, X., Su, J., Chen, L.: Real-time background subtraction-based video surveillance of people by integrating local texture patterns. Signal Image Video Process. **8**, 665–676 (2014)

8. Jepson, A.D., Fleet, D.J., El-Maraghi, T.F.: Robust online appearance models for visual tracking. IEEE Trans. Pattern Anal. Mach. Intell. **25**, 1296–1311 (2003)

9. Adam, A., Rivlin, E., Shimshoni, I.: Robust fragments-based tracking using the integral histogram. In: Proceedings of IEEE Conference on Computer Vision and Pattern Recognition (CVPR), pp. 798–805 (2006)

10. Ross, D.A., Lim, J., Lin, R., Yang, M.: Incremental learning for robust visual tracking. Int. J. Comput. Vis. **77**, 125–141 (2008)

11. Mei, X., Ling, H.: Robust visual tracking and vehicle classification via sparse representation. IEEE Trans. Pattern Anal. Mach. Intell. **33**, 2259–2272 (2011)

12. Wang, D., Lu, H., Yang, M.: Online object tracking with sparse prototypes. IEEE Trans. Image Process. **22**, 314–325 (2013)

13. Zhan, J., Wu, H., Zhang, H., Luo, X.: Cascaded probabilistic tracking with supervised dictionary learning. Signal Process. Image Commun. **39**, 212–225 (2015)

14. Grabner, H., Bischof, H.: On-line boosting and vision. In: Proceedings of IEEE Conference on Computer Vision and Pattern Recognition (CVPR), pp. 260–267 (2006)

15. Babenko, B., Yang, M., Belongie, S.: Robust object tracking with online multiple instance learning. IEEE Trans. Pattern Anal. Mach. Intell. **33**, 1619–1632 (2011)

16. Hare, S., Saffari, A., Torr, P.H.S.: Struck: Structured output tracking with kernels. In: Proceedings of IEEE International Conference on Computer Vision (ICCV), pp. 263–270 (2011)

17. Liu, R., Zhong, G., Cao, J., Lin, Z., Shan, S., Luo, Z.: Learning to diffuse: a new perspective to design pdes for visual analysis. IEEE Trans. Pattern Anal. Mach. Intell. **38**, 2457–2471 (2016)

18. Li, H., Wu, H., Zhang, H., Lin, S., Luo, X., Wang, R.: Distortion-aware correlation tracking. IEEE Trans. Image Process. **26**, 5421–5434 (2017)

19. Grabner, H., Leistner, C., Bischof, H.: Semi-supervised on-line boosting for robust tracking. In: Proceedings of European Conference on Computer Vision (ECCV), pp. 234–247 (2008)

20. Henriques, J.F., Caseiro, R., Martins, P., Batista, J.: High-speed tracking with kernelized correlation filters. IEEE Trans. Pattern Anal. Mach. Intell. **37**, 583–596 (2015)

21. Wang, L., Ouyang, W., Wang, X., Lu, H.: Visual tracking with fully convolutional networks. In: Proceedings of IEEE International Conference on Computer Vision (ICCV), pp. 3119–3127 (2015)

22. Wang, N., Yeung, D.: Learning a deep compact image representation for visual tracking. In: Proceedings of the Conference on Neural Information Processing Systems (NIPS), pp. 809–817 (2013)

23. Hashimoto, M., Konda, T., Bai, Z., Takahashi, K.: Identification and tracking using laser and vision of people maneuvering in crowded environments. In: Proceedings of IEEE International Conference on Systems Man and Cybernetics, pp. 3145–3151 (2010)

24. Wu, H., Gao, C., Cui, Y., Wang, R.: Multipoint infrared laser-based detection and tracking for people counting. Neural Comput. Appl. **29**, 1405–1416 (2018)

25. Li, C., Cheng, H., Hu, S., Liu, X., Tang, J., Lin, L.: Learning collaborative sparse representation for grayscale-thermal tracking. IEEE Trans. Image Process. **25**, 5743–5756 (2016)
26. Yuan, Y., Fang, J., Wang, Q.: Robust superpixel tracking via depth fusion. IEEE Trans. Circuits Syst. Video Technol. **24**, 15–26 (2014)
27. Kalman, R.E.: A new approach to linear filtering and prediction problems. Trans. ASME-J. Basic Eng. **82**, 35–45 (1960)
28. Zhang, T., Ghanem, B., Liu, S., Ahuja, N.: Robust visual tracking via structured multi-task sparse learning. Int. J. Comput. Vis. **101**, 367–383 (2013)
29. Oron, S., Bar-Hillel, A., Levi, D., Avidan, S.: Locally orderless tracking. Int. J. Comput. Vis. **111**, 213–228 (2015)
30. Zhang, K., Zhang, L., Yang, M.: Fast compressive tracking. IEEE Trans. Pattern Anal. Mach. Intell. **36**, 2002–2015 (2014)
31. Wu, Y., Lim, J., Yang, M.H.: Online object tracking: a benchmark. In: Proceedings of IEEE Conference on Computer Vision and Pattern Recognition (CVPR), pp. 2411–2418 (2013)

Lip Password-Based Speaker Verification Without a Priori Knowledge of Speech Language

Yiu-ming Cheung[1,2(✉)] and Yichao Zhou[1,2]

[1] Department of Computer Science, Hong Kong Baptist University,
Hong Kong, China
{ymc,yczhou}@comp.hkbu.edu.hk
[2] HKBU Institute of Research and Continuing Education, Shenzhen, China

Abstract. Most recently, the lip password that embeds the password content into lip motion has been proposed for visual speaker verification (Liu and Cheung 2014). One merit of lip password is that it provides double security on the speaker verification, where only the target speaker saying the correct password can be accepted. Nevertheless, the previous work of lip password is based on identifying the distinguishing subunits of purely-digit password contents, thus limiting the application domain of lip password. To tackle this problem, we propose a novel visual speaker verification approach based on lip password without a priori knowledge of speech language, i.e. unknown language alphabet. We take advantage of the diagonal structure of sparse representation to preserve the temporal order of lip sequences by employ a diagonal-like mask in pooling stage and build a pyramid spatiotemporal features containing the structural characteristic under lip password. Experiments show the efficacy of the proposed approach comparing with the state-of-the-art ones.

Keywords: Speaker verification · Lip password · Language alphabet
Modality · Double security

1 Introduction

Visual speaker verification has recently attracted much attention, e.g. see [2, 4, 9, 19]. Lip movements contain both individual mouth appearance and dynamic behavioral information [4, 10], which can improve the performance of audio-based speaker verification systems especially in noisy environments [7].

Recently, Liu and Cheung [12] have proposed a new modality, named lip password that is composed of the lip movement of a target visual speaker saying a specific password without sound, for speaker verification. Subsequently, a lip password-based speaker verification system has been developed. As shown in Fig. 1, such a system can detect and reject not only the target user saying the wrong password, but also the impostor who knows the correct password. As shown in [12], lip password has at least four advantages: (1) The modality of

© Springer Nature Singapore Pte Ltd. 2018
K. Li et al. (Eds.): ISICA 2017, CCIS 874, pp. 461–472, 2018.
https://doi.org/10.1007/978-981-13-1651-7_41

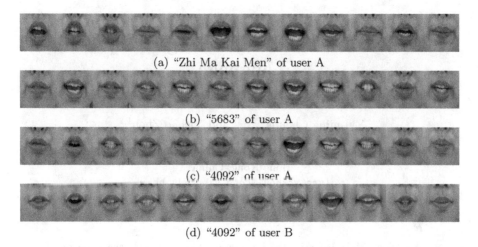

(a) "Zhi Ma Kai Men" of user A

(b) "5683" of user A

(c) "4092" of user A

(d) "4092" of user B

Fig. 1. Images of each row are sampled from the lip password sequences. Sub-figure (a~c) are different lip passwords spoken by the same user. If (c) is set as the correct lip password, then (a) and (b) are the case that the wrong password spoken by the target user, and (d) is the case that the impostor speaking the correct password. For a lip password-based speaker verification system, (a), (b) and (d) are all regarded as the impostor data and rejected.

lip motion is insusceptible to the background noise; (2) The acquisition of lip motion is insensitive to the distance to a certain degree; (3) Lip password can be performed silently; (4) It is applicable to speech impaired people. Nevertheless, the lip password studied in [12] is composed of digits from "0" to "9" only in English, which would limit the extensibility of lip password and its application domain. Actually, from the practical perspective, it is natural that different speakers saying the lip passwords in different languages. That is, any language can be used in lip password.

In this paper, we therefore propose a novel approach to lip password-based speaker verification with an arbitrary lip password, i.e. the password content to be a phrase of any languages. Since the language alphabet of the lip password is unknown, we design a representation of lip sequences which preserve the temporal order by employ a diagonal-like mask in pooling stage, and build a pyramid structure to form the spatiotemporal lip representation containing the structural characteristic under lip password.

Currently, most speaker verification methods only focus on verifying the identity of the speaker, ignoring the temporal information of password content. As shown in Fig. 1, since Fig. 1(a), (b) and (c) are spoken by the same person, the spatial-based features of these three sequences are similar. The proposed lip representation can keep this temporal order and improve the accuracy of verifying the password contents significantly. To further improve the accuracy of verifying the lip passwords, a pyramid structure of lips is proposed to contain more information in spatial domain.

In contrast to the previous related methods of detecting the mouth status and segmenting the whole lip sequences into several independent subsequences with the risk of cutting across potentially discriminating features [12,17], the proposed method directly utilizes the temporal order structure of sparse representation and does not rely on the detection of mouth status, whose computation is laborious and sensitive to the environment, e.g. illumination.

In order to evaluate the benefits of the proposed lip representation in lip password-based speaker verification system, we collect a database which contains both English and Chinese lip passwords. We empirically investigate the ability of state-of-the-art spatiotemporal lip features to verify the lip contents and speakers identity. Experimental results show that the proposed lip feature outperforms the state-of-the-art ones in all cases we have tried so far, especially when verifying the lip password contents.

2 Related Work

Several works have been done on visual speaker verification for its insusceptibility to noise and removability. For example, Saeed [15] extracted the static and dynamic lip features based on lip contour detection and utilized the Gaussian Mixture Model (GMM) to model the lip movements for person recognition. Wang and Liew [19] utilized the GMM to train the static features and Hidden Markov Model (HMM) for dynamic features. Also, Cetingul et al. [3] evaluated the explicit lip motion information comparing with lip intensity and geometry features through HMM for speaker identification. However, in most of those works, the users pronounce the same phrase for verification, which is the most constrained case as both duration and text are fixed. Moreover, the scenario of target user saying the phrase incorrectly is not considered.

Most recently, Liu and Cheung [12] have proposed a novel idea of lip password, which is composed of a password embedded in the lip movement and underlying characteristic of lip motion. They proposed a method based on appearance and contour features and multi-boosted HMMs learning approach for the lip password-based speaker verification system. Experiments in [12] have shown that their method performs favorably well to verify both of the private password information and the lip biometrics of speaker.

Furthermore, to improve the performance of speaker verification system, some works in the literature have performed the detection approaches to the mouth status of "closing" and segmented the whole lip sequences into several subsequences on those places [13,18]. For instance, Liu and Cheung [12] extracted the lip contour by lip localization and tracking first, e.g. see [5,11], and detect mouth status according to the area of lips. Karlsson and Bigun [8] used the optic flow to perform lip motion events detection and lip segmentation. Shaikh et al. [17] proposed a segmentation method based on the pair-wise pixel comparison of consecutive lip images. However, these methods partition the lip sequences into independent subsequences with the risk of cutting across potentially discriminating features, like the transformation between password units.

What is more, the computations are somewhat laborious and the results are sensitive to the environment such as illumination. In addition, some words with the mouth close deeply, e.g. "me", are departed.

In the literature, some works have utilized the spatiotemporal feature directly too. For instance, Chan et $al.$ [4] proposed the Local Ordinal Contrast Pattern with Three Orthogonal Planes (LOCP-TOP) for lip-based speaker verification. The LOCP is a texture descriptor that encodes the appearance of lip images. Meanwhile, by using the LOCP in TOP, it makes the final lip representation keep the dynamic information of lip movement. In [9], Lai et $al.$ used sparse coding under a hierarchical spatiotemporal structure to form the lip representation. The dictionary is learned from all users and max pooling in the hierarchical structure is performed in the sparse representation of lip subsequences. These lip representations densely extract lip features over time and space domains to avoid the segmentation and modeling of lip sequences, thus resulting in a better generative performance compared with the model-based approaches [9]. Nevertheless, this approach eliminates the location of sparse response, which is, however, critical for lip password. Also, the training of dictionary in [9] is performed on all users and time-consuming. By contrast, the proposed method uses the fixed dictionary chosen from the training data, and thus does not require dictionary learning on the dataset of all users.

3   The Proposed Method

For a lip password-based speaker verification, it is of great concern to extract lip features which are discriminative against different password contents and speaker's identity. In this paper, we propose a novel lip representation using the diagonal-like pooling based on sparse coding and a 3-layer pyramid structure designed for the lips. In the following, we first describe the details of the sparse learning scheme and the diagonal-like pooling method, and then introduce the 3-layer pyramid structure of lips.

3.1   Diagonal-Like Pooling

Given the correct lip password training data, the password-specified dictionary is formed as $D = \{d_1, \cdots, d_T\}$, where $d_i, i \in [1, T]$ is the vectorized ith frame in D. For a lip video with T frames, the input data is denoted as $X = \{x_1, \cdots, x_T\}$, where $x_i, i \in [1, T]$ is the vectorized ith frame in X. With the password-specified dictionary D, the sparse code of x_i can be obtained by Lasso algorithm [6]. We can get a matrix of coefficients $A = [\alpha_1, \ldots, \alpha_T]$ such that for every x_i, the corresponding column α_i is the solution of

$$\min_{\alpha_i} \|x_i - D\alpha_i\|_2^2, \text{ s.t.} \|\alpha_i\|_1 \leq \lambda, \tag{1}$$

where λ is the sparse regularization parameter.

The elements in this sparse matrix A can reflect the relation between input data X and dictionary D. The sparse values in α_i represent the weight of linear

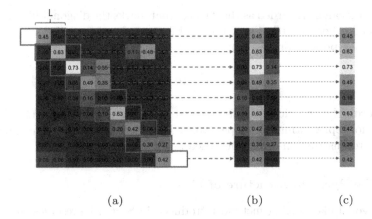

(a)   (b)   (c)

Fig. 2. The key idea of diagonal-like pooling. (a) $T \times T$ sparse representation of a video sequence with the matched lip password sequence as the dictionary, (b) Diagonal structure extraction using sliding window, and (c) Max pooling over time.

combination in D to construct x_i. The larger α_{ij} is, the more similar between d_i and x_j is. If the input data X is the correct password, the sparse values in its sparse matrix A should be near the diagonal of A. A special case is when $X = D$, then $A = I$ is the solution of Eq. (1), where I is the identity matrix. On the other hand, if the input data X is the wrong password spoken by the target user, it implies that X and D may have many large values in A, but not all near the diagonal of A. If we simply do max-pooling [16,20] over time, the location information will all be lost.

To address this problem, we introduce the diagonal-like pooling method as shown in Fig. 2, where we can observe that, if the input data is matched with the dictionary, the sparse representation gets most responses near the diagonal of the matrix A. The key idea is, by using the sliding windows through the diagonal elements, the value near the diagonal, which represents the relation between input data and dictionary within a very short period of time, is collected and other mismatched elements are removed. Then, doing max pooling over time can generate the features which keep the order information.

For more convenient calculation, let L denote the length of windows, we define a binary $T \times T$ matrix W, where each element of W is calculated as:

$$w_{ij} = \begin{cases} 1, \text{ if } |i - j| \le L, \\ 0, \text{ otherwise.} \end{cases} \qquad (2)$$

For example, if $T = 5$ and $L = 2$, the structure of W is:

$$\mathbf{W} = \begin{pmatrix} 1 & 1 & 0 & 0 & 0 \\ 1 & 1 & 1 & 0 & 0 \\ 0 & 1 & 1 & 1 & 0 \\ 0 & 0 & 1 & 1 & 1 \\ 0 & 0 & 0 & 1 & 1 \end{pmatrix}. \qquad (3)$$

This matrix can be regarded as the "mask" that marks the diagonal-like structure as "1" and others as "0". Then, the final feature $f = \{f_1, ..., f_T\}$ contained by diagonal-like pooling process can be written as:

$$f_i = \max_{j=1}^{T} (\alpha_{ij} w_{ij}). \tag{4}$$

Accordingly, the dictionary is selected from the training data that is most suitable for representing the lip password. That is, we select the sample with the smallest reconstruction error when used as the dictionary.

3.2   The Pyramid Structure of Lips

The diagonal-like pooling method introduced in Sect. 3.1 preserves the temporal order of features. To further enhance the representativeness of proposed lip feature, we build a pyramid spatial structures according to the characteristic of lip movement.

As shown in Fig. 3(a), the origin lip images are separated into three block group. The first layer is the whole mouth area. The second layer contains the upper lip, lower lip and the middle of the mouth containing teeth and tongue. It is observed that the lip movement has more vertical symmetry than horizontal symmetry. What is more, the shapes of upper lip and lower lip contain the information of the speaker's identity. The third layer is the center and four corners of the lip images. The center of the lip images, including teeth and tongue, can also provide some information of the password contents. These areas have overlap to each other.

All these blocks are regarded as the input sequential data and get the features by sparse coding and the proposed diagonal-like pooling shown in Sect. 3.1. Finally, the lip representation feature is a combination of those features of the blocks, as shown in Fig. 3. Denote D_k^l and f_k^l as the dictionary and corresponding feature of the kth block in the l layer using Eq. (4), the final feature can be written as:

$$F = \{f_1^1, f_1^2, ..., f_3^2, f_1^3, ..., f_5^3\}. \tag{5}$$

4   Experiments

4.1   Database

Most existing databases such as XM2VTSDB [14] and MVGL-AVD [3], despite their popularity in serving as benchmarks for traditional visual speaker verification tasks, are incompetent for our lip password-based system, because these databases contain the limited types of different password contents and languages. Under the circumstances, we constructed a database consisting of 8 kinds different lip passwords, including 2 kinds of English digits and 6 kinds of Chinese phrases, to evaluate the proposed algorithm. Some samples from the database are shown in Fig. 1. The database consists of 43 speakers with 19 males and

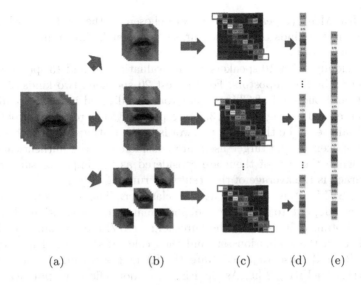

Fig. 3. The framework of the proposed approach. (a) Origin lip images are separated into (b) three groups of blocks; (c) Learning $T \times T$ sparse representations of each block; (d) Diagonal structures extraction using sliding window and max pooling over time; (e) The final lip feature.

24 females. Every speaker is asked to repeat all 8 kinds of passwords 75 times in total. Each utterance contains about 90 lip images of size 50×50 lasting for about 3 s. In our experiment, each lip sequence is sampled to 50 frames in order to reduce the computational complexity. The size of the second layer in pyramid structure is set as 20×50 and the size of the third layer is 30×30.

4.2 Experimental Protocol

Different from the speaker verification system that only detects and rejects the impostor with a fixed phrase, the lip password-based speaker verification system faces the challenge of three kinds of impostor scenarios, which are summarized as follows:

- **Target-Wrong**: The target speaker saying the incorrect password;
- **Impostor-Correct**: The impostor saying the correct password;
- **Impostor-Wrong**: The impostor saying the incorrect password.

Three impostor types above and the holistic impostor data (denoted as **All-Impostor**) are evaluated to show the performance of proposed algorithm.

Similar to the protocol in [14], we employ a protocol specific to lip password-based speaker verification.

Among 8 kinds of lip passwords spoken by 43 speakers, two kinds of English digits, "4092" and "5683" (repeat 20 times), and one kind of Chinese phrase,

"Zhi-Ma-Kai-Men" (repeat 10 times), are chosen as the lip password, respectively. The rest 5 kinds of Chinese phrases[1] (repeat 5 times) are only used as the wrong password.

We randomly select 10 speakers as the evaluation set and 13 speakers as the testing set to be the impostors. For the rest 20 speakers, two kinds of English passwords and one kind of Chinese password are divided into the training set, evaluation set and testing set with 3:3:4, respectively. The rest 5 kinds of Chinese passwords are set to be the wrong passwords in the testing set. In each trial, the password spoken by the target users are considered as the authenticated user samples, while the rest of them are considered as the impostor samples. The final accuracy is the average of the results by running 10 times.

Linear SVM [1] is adopted as the classifier. The half total error rate (HTER) is adopted to evaluate the performance of verification algorithms. HTER is obtained by setting the threshold of SVM to obtain Equal Error Rate (EER) in the evaluation set, and then calculated by the False Accepted Rate (FAR) and False Rejected Rate (FRR) in the testing set according to $HTER = (FAR + FRR)/2$ [9]. As the EER does not reflect the practical system performance when the testing data is unseen [14], the HTER is a more reasonable measurement of the performance for speaker verification system.

4.3   Experimental Results

To evaluate the influence of the choice of window size L and the utility of proposed pyramid structure, we extract the lip features under the different window sizes with both 3-layer structure and the single layer. The evaluation and testing results are shown in Fig. 4. We can observe that, when window length is set at 1, both the EER in the evaluation set and HTER in the testing set are very large compared to the cases with the other lengths. That is reasonable because even the same password spoken by the same person, the speed is hard be the same. When the window length is too small, it faces the problem of under-fitting. When $L = 2$, the error rate decreases rapidly and then slowly increases when increasing the value of L. This implies that the proposed method is not quite sensitive to the choice of window size as long as it is a reasonable value. Compared with the feature just using max pooling (denoted as "no mask" in Fig. 4), the proposed method also improves the performance, especially in the Target-Wrong scenario, which is very important for the lip password-based speaker verification system.

As observed in Fig. 4, the lip feature under the 3-layer pyramid structure outperforms the one under a single layer in almost all cases, especially in the Target-Wrong scenario. The results indicate that the proposed pyramid structure can preserve more discriminative features underlying the second and the third layer of lip sequences.

To sum up, the lip features under single layer and pyramid layers all obtain the best performance when $L = 2$ and better than the feature just using max

---

[1] "Kai-Men-Jian-Shan", "You-Qi-Wu-Li", "Gong-Xi-Fa-Cai", "Dong-Fang-Ming-Zhu", "Wu-Jing-Da-Cai".

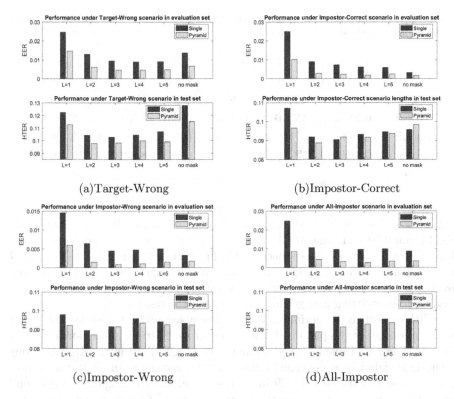

(a)Target-Wrong   (b)Impostor-Correct

(c)Impostor-Wrong   (d)All-Impostor

Fig. 4. Eval-set EER and Test-set HTER variation against the increasing window length L under four impostor scenario.

pooling. The diagonal-like pooling method significantly improves the performance, especially in the Target-Wrong scenario, which is of importance for the lip password-based speaker verification. What is more, the features formed from the 3-layer pyramid structure is more discriminative than one using a single layer only. The results of All-Impostor show the efficacy of the proposed approach.

4.4 Performance Comparison

To assess the effectiveness of the proposed method, two spatiotemporal features widely used in the verification system: the LOCP-TOP feature in [4] (LOCP-TOP in short) and the hierarchical pooling sparse lip representation in [9] (SC-HIER in short) are adopted for comparison. In [4], the LOCP is extracted in TOP to form the final lip representation. In [9], the lip representation is generated by using the hierarchical max pooling on the sparse representation, using a dictionary learned from all the users data.

For our approach, two kinds of features with the window length $L = 2$ under a single layer (our-single in short) and 3-layer pyramid (our-pyramid in short) structure, are investigated. Comparison results are shown in Table 1. It can be

Table 1. Feature performance comparison showing EER(Eval) and HTER(Test). For each scenario, the best result is shown in bold and the second one is in underline.

	Tar.-Wrong		Imp.-Correct		Imp.-Wrong		All-Imp.	
	Eval	Test	Eval	Test	Eval	Test	Eval	Test
LOCP-TOP	3.02%	17.39%	0.38%	10.10%	0.19%	10.10%	1.12%	10.68%
SC-HIER	**0.28%**	10.92%	**0.01%**	9.50%	**0.00%**	9.50%	**0.02%**	9.92%
our-single	1.30%	10.41%	0.89%	9.18%	0.64%	9.18%	1.05%	9.29%
our-pyramid	0.61%	**9.77%**	0.27%	**8.87%**	0.15%	**8.87%**	0.41%	**8.86%**

seen that the features based on sparse coding (our methods and SC-HIER) outperform the LOCP-TOP features. Further, although HIER has a very low EER (even 0.00% in Impostor-Wrong), the HTER in the testing dataset is even higher than our-single. This is because the lip representation generated by HIER is very sparse and has very high dimension, which makes it easy to be separated by the linear SVM and get very low verification error in the evaluation set, leading to over-fitting and get higher HTER in the testing set.

With respect to the performance in different scenarios, Target-Wrong, which is of importance for lip password-based speaker verification, looks more challenging. As shown in Table 1, the performance of LOCP-TOP in the Target-Wrong scenario is seriously worse than in the other two scenarios. In all four scenarios, our-pyramid gets the best HTER in the testing set and our-single gets the second best HTER, which outperform the other two kinds of lip representation. That is, the results have shown the promising discriminative power of the proposed lip representation.

5   Conclusion

This paper has proposed a novel speaker verification approach with unknown language alphabet. The proposed method works by generating lip feature of input data using diagonal-like max pooling on the sparse representation to preserve the temporal order of lip sequences. We also build a pyramid structure to form the spatiotemporal lip representation to catch the structural characteristic under lip password. It need not require the accurate alignment of feature sequences or detection on mouth status, whose computation is laborious. Experiments on different kinds of lip passwords have shown its promising result compared with the state-of-the-art ones.

Acknowledgment. This work was supported by the National Natural Science Foundation of China with the Grant Numbers: 61672444 and 61272366, by the Faculty Research Grant of Hong Kong Baptist University (HKBU) with the Project Code: FRG2/16-17/051, and by the SZSTI Grant: JCYJ20160531194006833.

References

1. Boser, B.E., Guyon, I.M., Vapnik, V.N.: A training algorithm for optimal margin classifiers. In: The Workshop on Computational Learning Theory, pp. 144–152 (1996)
2. Broun, C.C., Zhang, X., Mersereau, R.M., Clements, M.: Automatic speechreading with application to speaker verification. In: IEEE International Conference on Acoustics, Speech, and Signal Processing (ICASSP), vol. 1, pp. I-685. IEEE (2002)
3. Cetingul, H.E., Yemez, Y., Erzin, E., Tekalp, A.M.: Discriminative analysis of lip motion features for speaker identification and speech-reading. IEEE Trans. Image Process. **15**(10), 2879–2891 (2006)
4. Chan, C.H., Goswami, B., Kittler, J., Christmas, W.: Local ordinal contrast pattern histograms for spatiotemporal, lip-based speaker authentication. IEEE Trans. Inf. Forensics Secur. **7**(2), 602–612 (2012)
5. Cheung, Y.M., Liu, X., You, X.: A local region based approach to lip tracking. Pattern Recogn. **45**, 3336–3347 (2012)
6. Efron, B., Hastie, T., Johnstone, I., Tibshirani, R.: Least angle regression. Ann. Stat. **32**(2), 407–499 (2004)
7. Jourlin, P., Luettin, J., Genoud, D., Wassner, H.: Acoustic-labial speaker verification. Pattern Recognit. Lett. **18**(9), 853–858 (1997)
8. Karlsson, S.M., Bigun, J.: Lip-motion events analysis and lip segmentation using optical flow. In: IEEE Computer Society Conference on Computer Vision and Pattern Recognition Workshops (CVPRW), pp. 138–145 (2012)
9. Lai, J.Y., Wang, S.L., Liew, W.C., Shi, X.J.: Visual speaker identification and authentication by joint spatiotemporal sparse coding and hierarchical pooling. Inf. Sci. **373**, 219–232 (2016)
10. Li, M., Cheung, Y.M.: A novel motion based lip feature extraction for lip-reading. In: Proceedings of 2008 International Conference on Computational Intelligence and Security, pp. 361–365 (2008)
11. Li, M., Cheung, Y.M.: Automatic lip localization under face illumination with shadow considertion. Sig. Process. **89**(12), 2425–2434 (2009)
12. Liu, X., Cheung, Y.M.: Learning multi-boosted HMMs for lip-password based speaker verification. IEEE Trans. Inf. Forensics Secur. **9**(2), 233–246 (2014)
13. Liu, X., Cheung, Y.M., Tang, Y.Y.: Lip event detection using oriented histograms of regional optical flow and low rank affinity pursuit. Comput. Vis. Image Underst. **148**, 153–163 (2016)
14. Luettin, J., Maître, G.: Evaluation protocol for the extended M2VTS database (XM2VTSDB). IDIAP (1998)
15. Saeed, U.: Person identification using behavioral features from lip motion. In: IEEE International Conference on Automatic Face and Gesture Recognition and Workshops, pp. 131–136 (2011)
16. Serre, T., Wolf, L., Poggio, T.: Object recognition with features inspired by visual cortex. In: IEEE Computer Society Conference on Computer Vision and Pattern Recognition (CVPR), vol. 2, pp. 994–1000 (2005)
17. Shaikh, A.A., Kumar, D.K., Gubbi, J.: Automatic visual speech segmentation and recognition using directional motion history images and zernike moments. Vis. Comput. **29**(10), 969–982 (2013)
18. Shi, X.X., Wang, S.L., Lai, J.Y.: Visual speaker authentication by ensemble learning over static and dynamic lip details. In: IEEE International Conference on Image Processing, pp. 3942–3946 (2016)

19. Wang, S.L., Liew, A.W.C.: Physiological and behavioral lip biometrics: a comprehensive study of their discriminative power. Pattern Recogn. **45**(9), 3328–3335 (2012)
20. Yang, J., Yu, K., Gong, Y., Huang, T.: Linear spatial pyramid matching using sparse coding for image classification. In: IEEE Computer Society Conference on Computer Vision and Pattern Recognition (CVPR), pp. 1794–1801 (2009)

Human Motion Model Construction Based on Gene Expression Programming

Wei He[1], Shaoyang Hu[1(✉)], Shanni Li[2], Junlin Jin[1],
and Kangshun Li[1]

[1] College of Mathematics and Informatics, South China Agricultural University,
Guangzhou 510642, Guangdong, China
landerous@foxmail.com
[2] China Southern Capital Management Co., Ltd., Shenzhen, China

Abstract. In this paper, we propose a novel method based on *Gene Expression Programming (GEP)* to construct human motion model. Our approach better describes human motion features, which can be applied to improve the accuracy of human behavior recognition. On one hand, this method combines *Genetic Algorithm (GA)* and *Genetic Programming (GP)*, and overcomes the limitation of traditional high-dimension function approaching method, realizing the generalization of *Gene Expression Programming (GEP)* on Function Mining. On the other hand, it implements the human motion capture technique of Kinect sensor, interpolates data and increases the training data accuracy. In the experiments result, we use *GEP* to develop human trajectory dynamics model, which has characteristics like encoding and gene structure flexibility that can lead the trajectory simulation error much decline. Given that the result is better than traditional methods and able to maintain most of the human motion features, our human motion model can be applied to human behavior analysis area and other similar domains.

Keywords: Gene expression programming · Human motion model
Trajectory approaching

1 Introduction

Capturing human motion and building up human motion model has always been one of the hottest topics of Artificial Intelligent. Recently, with the rapid development on Machine Learning, Pattern Recognition, and Computer Vision, human behavior analysis and its applications have been grown further, which are mainly implemented in Smart City, Virtual Reality, Sports health, emergency events detection, etc. Human motion capture technique, broadly speaking, is to sample human motion data in time series. Functionally speaking, it can be divided into four parts, including initialization, tracking, pose estimation and recognition, or two parts, including two-dimensional capture and three-dimensional capture [1].

For human motion capture, we can describe it from outline or structure. From the outline perspective, Agarwal and Triggs [2] introduced a mapping between 2D human outline and 3D human pose, which was based on camera captured data. From the

K. Li et al. (Eds.): ISICA 2017, CCIS 874, pp. 473–485, 2018.
https://doi.org/10.1007/978-981-13-1651-7_42

structure perspective, O'Brien et al. [3] stated a human motion capture technique based on magnetic sensor, obtaining motion parameters in real-time. Ramakrishna et al. [4] put forward a technique based on the 2D image to represent 3D human joints structure. Obtaining movement information based on color image separation and video frames are time-consuming and space-consuming.

After the motion capture and data extraction, we are able to develop human motion model for specific human behavior. Early works in the domain focus on using the length and degree of freedom of each human joint to build up hierarchical model and then describe human behavior based on different segments of the body respectively [5, 8]. After the Principal Component Analysis (PCA), Yang and Tian [6] configured eigenvalues to model behaviors, and extracted static and dynamics feature from the coordinate distance on the same video frame and neighbor video frames correspondingly. Through the development of sensor and RGB-D camera [7], the state-of-the-art technique has allowed us to extract three dimension joint points data from the depth image. The modeling method that directly utilizes joint points data to depict human motion model has gradually attracted people's attention.

Park and Sheikh from CMU [9] constructed smooth human joint points trajectory in three-dimensional space based on image sequences, however, when the missing data percentage reached 5%, the average relative error became 13%. Dealing with the individual upper body joint points 3D data, Guo and Shen [10] put forward weight distributing kernel function based on SOM theory, so as to fit the motion trajectory. But the coefficient of determination is ranging from 0.904 to 0.942, and the mean square error is between 0.47 to 0.92. Recent approaches stated above perform badly when it comes to the more complex human motion model case.

Our purpose of constructing human behavior mainly focus on two aspects: on one hand, we implemented Kinect on 3D human motion data capture, which possesses advantage of low cost, convenient and high performance. It also overcomes the drawback of traditional human motion capture techniques, which mostly need subjects to wear sensing device and consider 2D model only. On the other hand, the simple function cannot represent the complicated joint points trajectory. To address the solution, we carry on *Gene Expression Programming*, which combines *Genetic Algorithm* and *Genetic Programming*.

Our approach makes full use of its advantages on complex function approaching and decreases the trajectory simulation error, whose result is better than recent works. *GEP* is able to solve the target equations effectively, performs well on function mining and develops numerical constants on equations [11–13]. The motion model configured by *GEP* can significantly decline the system complexity. Furthermore, the constructed trajectory model includes most motion sequence information, so that the features can contribute to human behavior recognition research.

2 Gene Expression Programming

GEP, a novel self-adaptive evolutionary algorithm, introduced by Ferreira whose application areas includes function parameter optimization, evolutionary modeling, neural network, classification and TSP problem. In *GA*, the individual is a linear string with fixed length and one with variant length and shape in *GP*, while it is a more complicated case in *GEP*. Individuals are first encoded into fixed length strings then represented by a string in different length and shape. *GEP* improves the expression limitation in *GA* for both the complexity of solution structure and genetic operations.

2.1 Gene Structure and Encoding in GEP

In *GEP*, Gene is made up of head and tail sections, where head consists of terminal symbols and function notations while tail has only terminal symbols. It is supposed that the head length is *h* (*depends on the problem*), the tail length is t. The relationship between the head with length *h* and the tail with length *t* can be represented as follows:

$$t = h \times (n - 1) + 1 \tag{1}$$

Where, n represents the number of parameters that needs most variables among the function notations set.

For instance, we consider a Gene constructed by {Q, +, −, *, /, a, b}, where Q represents rooting operation. Obviously, in the case where n equals 2, if we suppose that h equals 10 and t equal 11, the length of that Gene will be 10 + 11 = 21. When we take this Gene: Q * + * a * Qaababbaababaab (tail part has underlined) and transform it into corresponding Expression Tree (ET), only need to read characters in the Gene from left to right and follow the hierarchical order. From the rule above, this Gene can be visualized by Expression Tree and its expression $\sqrt{(a + a \times b) \times \sqrt{a} \times a}$.

2.2 GEP Adaptive Function

In evolutionary computation, the key to solve the problem is to design suitable adaptive function, which orients the evolution direction. For function modeling, the last programming solution will be an expression, and evaluation of the expression is to assess whether the data computed from the expression match the training data.

2.3 Operators in GEP

Selection Operator. *Selection operator* is the evolutionary operation that selects parents from the population, so as to generate new individual and increase the diversity of the population. This operator should follow the rule that the larger adaptive value is, the more chance individuals have to perform an evolutionary operation.

Mutation Operator. *Mutation* can happen in every position of a chromosome, whose structure should be also kept intact. In the head of the gene, any notation can mutate itself into function notation or terminal notation; while in the tail of the gene, any notation can only mutate into terminal notation. Following this rule of encoding in *GEP*, a chromosome structure can remain and we are able to foresee that the individuals constructed by mutation are correct in structure.

Three Insertion Sequence Operators. *Insertion Sequence (IS)* operator selects a sequence from a chromosome randomly and inserts it into the head of the corresponding gene. The insertion position is any position excluding the starting position. Furthermore, every element that exceeded the length of the head of that gene should be abandoned, which can be caused by the insertion operation.

Root Insertion Sequence (RIS) selects sequences from a chromosome and inserts them into the head of the corresponding gene. Also, exceeded elements should be abandoned.

Gene Insertion Sequence (GIS) selects complete gene, inserts it into the starting position, and the selected gene will be removed in the new gene.

Three Recombination Operators. *Single Point Recombination.* It switches elements after the position of a randomly selected point on two chromosomes, and constructs next two offspring genes.

Two Points Recombination. It switches elements between the position of two randomly selected points on two parents chromosomes and constructs next two offspring genes.

Gene Recombination. It switches the randomly selected gene of two chromosomes and constructs next two offspring genes.

3   Microsoft Kinect Sensor

In this paper, we extract data from Microsoft Kinect for Windows V2, as shown in Fig. 1(a), which consists of RGB camera (1080 p, 32 Hz), depth/infrared camera (512 × 484, 30 Hz) and microphone array. As a result, it is capable of exporting raw data like RGB image, depth image, infrared data, audio data, etc.

Kinect captures three-dimensional human joint points data from the depth image. The formation principle of depth image is mainly based on structured light measurement and encoding. Firstly, it projects structured light with a specific pattern (dots, line or plane) onto the subject surface. After receiving the reflected pattern, it calculates the subjects spatial information with the pattern position, its deformation, which is the theory of Triangulation.

After segmenting each part of the human body from the depth image, the sensor constructs joint points model on account of random forest recognition outcomes. Given that the border pixel will decrease the total accuracy of pose estimation, Shotton et al. [15] proposed a local estimation algorithm based on Mean shift method and weighted Gaussian kernel function, extracting multiple human joints 3D data in high confidence and in real-time. The experiment results show that this algorithm outputs coordinate data with high correlation to ground truth data and considerable accuracy. Compared to other types infrared cameras like those use Time of Flight (TOF), it is faster by 10× [16–18].

From the provided skeleton data, Livingston et al. [14] tested the device on monocular mode and obtained 100 groups joint points samples with the average relative error within 5.6 mm. Considering its low cost and convenient merits, it is proved that can be used for research in behavior recognition area.

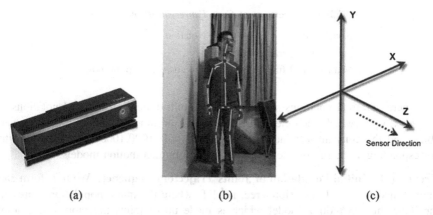

Fig. 1. Kinect for windows (a), capturing human motion data (b) and coordinate system (c)

In Kinect SDK v2, it provides 25 human joints real-time coordinate data. However, the extracted data has missing gaps. The reason varies from many cases, like the intrinsic shortage for indoors recognition, environment factors when we obtained, etc. As a result, we pre-processed the human joint points data with interpolation operation.

4 Experiments[1]

In this paper, we simulate four interpolation approaches and draw into conclusion with sufficient experiments that when it comes to interpolate in human joint points data, Piecewise Cubic Hermite Polynomial Interpolation outperforms Nearest-neighbour Interpolation, Piecewise Linear Interpolation, and Cubic Spline Interpolation. Considering its performance and stability, we adopt Piecewise Cubic Hermite Polynomial Interpolation for the missing data processing.

4.1 Computation on GEP Modeling

Procedure to construct human motion model includes a series operations of cross variation, Mutation, and Insertion. For each joint $J_i(i = 1, 2 ..., N)$, we denote the human trajectory model as $z_i = F_i(x, y, t)$, where $N = 25$ represents the number of joints. The *Gene Expression Programming* details are as follows:

Set Up Notation Set. As all human joints trajectories are the smooth curve on every motion, we configure the functions set F as F = {S, C, T, P, E, L}, where each character denoted a function:

S represents sine function, C represents cosine function, T is the tangent function, P is power function and E represents the logarithmic function with e, i.e. F = {sin, cos, tan, pow, exp, ln}. And the arithmetic operators are as Table 1 shown:

---

[1] The data extraction code is available for non-commercial research proposes.
Follow the links from https://github.com/erichhhhho/Kienct_Skeleton_Joints.

Table 1. Operators of *Gene Expression Programming*

Notation type	Notation
Arithmetic operation notation	$+, -, \times, \div$
Mathematical function	sin, cos, tan, exp, ln, pow

Where, terminal notations set includes the multi-dimension terminal notations set $T = \{t, x, y\}$ in which each element denotes each argument of multi-dimension variables, and the constant terminal notations set $C = \{-1000, 1000\}$. We substitute the corresponding value of arguments into the actual human motion models.

Creating the Initial Population of Joints Trajectory Sequence. We transform each joints trajectory into Expression Tree, i.e. ET, where the initial population represents the initial human motion model which is made up of joints trajectories, and joints trajectories is made up of function notations set and terminal notations set. The terminal notations set consists of time notation t in 3D joints time-series coordinate data, and 3D joints coordinate components.

The transformation of joints sequence expression into ET is as follows:

1. Randomly generate motion sequence, and set the first notation as the root node in ET.
2. Following the rule Breadth-First-Search (BFS), subsequently generate nodes in ET for function notation or terminal notation of each element, in which, the number of each child node is the number of parameters it should have. (terminal notation and non-argument function need 0 parameters, and the child node value is padding up following the notation order of the sequence)

In the following implementation, the motion sequence of each joint represents the Gene in the Evolutionary algorithm. For example, for the i_{th} joint J_i, the coordinate in specific behavior and instance t is $(J_{ix_t}, J_{iy_t}, J_{iz_t})$, which can be obtained from the 3D time-series data. Sequence $Q \times \times + J_{ix_t} Q + J_{iy_t} J_{iz_t} J_{ix_t} \underline{J_{ix_t} J_{iy_t} J_{iy_t} J_{iz_t} J_{ix_t} J_{iz_t} J_{iz_t} J_{iy_t}}$ is a legal motion sequence, in which Q represents the rooting operation, the underlined part is the tail of the Gene and the rest is the head section. Figure 2 is the corresponding ET with it and the Fig. 3 visualizes the procedure of the generation of ET.

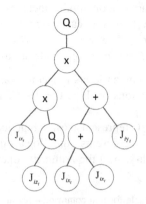

Fig. 2. Expression Tree (ET)

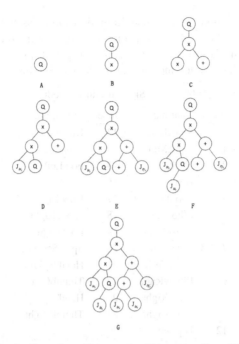

Fig. 3. Procedure of generating Expression Tree (ET)

Computing the Training Error and Testing Error. In this process, we first calculate the adaptive value in each trajectory models, which are the individuals in GEP. According to the result of the whole population, we can evaluate the models with the relative error or absolute error, which also represent the adaptive ability in GA. We select the specific number of individuals which accord with the evaluating condition as parents. If now the models satisfy the terminal condition, we output the result; Or we implement the evolutionary operation under the operators setting.

Evolutionary Operation. With the selected parents we implement cross variation, Mutation, and Insertion operation on individuals, so as to obtain the next generation.

Output the Result. If the model status satisfies the terminal condition or the threshold we set before, we stop the training and output the current optimal chromosome expression, i.e. the final trajectory model.

4.2 Dataset Details

Data Capture. We utilized Microsoft Kinect V2 SDK and OpenCV computer vision library to establish human motion capture system, which implemented monocular experiments based on four basic behaviors (waving left hand, waving right hand, kicking left leg and kicking right leg) and obtained 25 joint points data in real-time.

The data is based on a 3D Cartesian coordinate system, where the original point is the center of the infrared sensor, the positive direction of x, y, z axis are the device's left horizontal direction, vertical direction, front direction respectively (See Fig. 1(c)).

After the data collection, we obtained 50 motion sub-datasets for each behavior, in which each behavior consists of 25 human joint points data in 30 continuous instances and the coordinate unit is in meter. After connecting each joint point, we are able to draw "the skeleton plot" model and encode the joints (See Fig. 4 and Table 2).

Table 2. 25 Skeletal joints encoding

Index	Joint name	Index	Joint name
0	SpineBase	13	KneeLeft
1	SpineMid	14	AnkleLeft
2	Neck	15	FootLeft
3	Head	16	HipRight
4	ShoulderLeft	17	KneeRight
5	ElbowLeft	18	AnkleRight
6	WristLeft	19	FootRight
7	HandLeft	20	SpineShoulder
8	ShoulderRight	21	HandTipLeft
9	ElbowRight	22	ThumbLeft
10	WristRight	23	HandTipRight
11	HandRight	24	ThumbRight
12	HipLeft		

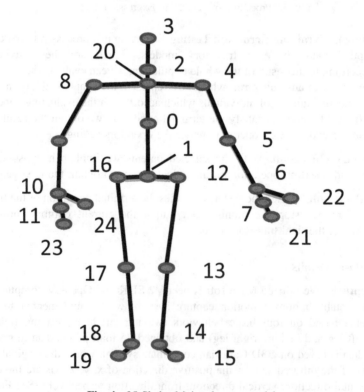

Fig. 4. 25 Skeletal joints human model

Data Pre-processing. With Piecewise Cubic Hermite Polynomial Interpolation approach, we pre-processed the captured motion time-series data samples. Stimulating the random loss of 30 s in specific joint point and behavior sequence, we then evaluate the four interpolation approaches with Mean Square Error (MSE):

$$MSE_j = \frac{1}{m} \sum_{i=1}^{m} [(x_{t_i} - \hat{x}_{t_i})^2 + (y_{t_i} - \hat{y}_{t_i})^2 + (z_{t_i} - \hat{z}_{t_i})^2]$$

$$\overline{MSE} = \frac{1}{25 \times n} \sum_{j=1}^{25 \times n} MSE_j$$

(2)

We conducted 8 times experiments on each joint point and chose m missing position in each experiment.

Where, MSE_j represents the MSE on j_{th} experiment.

To illustrate it better, in this paper, we take kicking right leg as an example, simulating data missing on the first to eighth instance one by one, conducting 25000 experiments.

Figures 5 and 6 are the performance comparison plot:

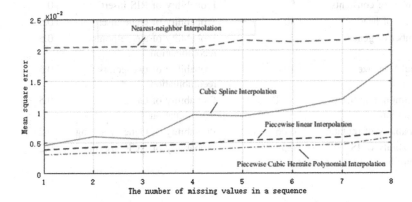

Fig. 5. Interpolation result on FootRight samples - mean square error

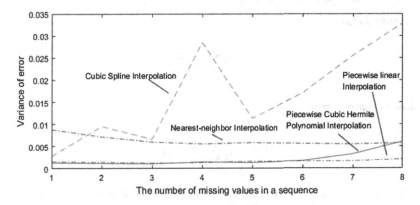

Fig. 6. Interpolation result on FootRight samples - error variance

5   GEP Implementation - Human Motion Model

In this section, on the basis of the time series interpolated sequence data we generated, we implement *GEP* to construct human motion model on different types of human behaviors. For instance, on kicking right leg behavior, we take 50 times motions of joint FootRight to develop joints trajectory model $z(x, y, t)$.

In each trajectory modeling process, we randomly divide the data into partitions, i.e. we consider k instances as the training set and N-k instances as the validation set.

After each division, we compute the statistical error parameters and divide it again for 10 times. Finally, we take the average error and MSE of each partition as the loss of *GEP* modeling. The parameters and operators setting of our *GEP* model are shown in Tables 2 and 3:

Table 3. Parameter setting of *Gene Expression Programming*

Parameters name	Parameters Value	Parameters name	Parameters Value
Number of constants	5	Probability of RIS insertion	0.5
Length of head	30	Probability of Gene insertion	0.5
Number of genes	5	Probability of single-point recombination	0.5
Population size	400	Probability of two-points recombination	0.5
Maximum number of generation	5000	Probability of Gene recombination	0.5
Probability of mutation	0.06	Probability of Gene switching	0.2
Probability of IS insertion	0.5	Probability of NCR insertion	0.2

The Evaluation Metric and statistical parameters we used to measure the *GEP* model performance are shown as below:

Relative Error

$$\delta = \sum_{i=1}^{N} \frac{|z_{t_i} - \hat{z}_{t_i}|}{z_{t_i}} \tag{3}$$

Mean Square Error

$$MSE = \frac{1}{N} \sum_{i=1}^{N} (z_{t_i} - \hat{z}_{t_i})^2 \tag{4}$$

After the experiments, we obtain 50 human motion models on joint FootRight for kicking right leg behavior (See Table 4).

From all 50 sample models construction stated in this Sect. 6.1, we can evaluate the performance of *GEP* now. It is shown that in the trajectory model of joint FootRight on kicking right leg, the average relative error is 0.0145844 and the mean square error is 0.0026030.

Using the human behavior sequence of joint KneeLeft on kicking left leg we captured, we develop models by *GEP* on the dataset, where the parameters setting is in Table 3. We divided the dataset and extracted the first 26 instance as training set when the last 4 instances as testing data. The trajectory motion model we obtained is as follows:

$$
\begin{aligned}
z(x, y, t) = {} & \cos((-275.74569580)) + \exp((-272.49028112)) \\
& + \cos(\cos((t/(94.61117469)))) + \sin((\tan((94.61117469)) \quad (5) \\
& + x)) + \cos(\cos(y)).
\end{aligned}
$$

In Table 4, we present the simulation result constructed by *GEP*:

Table 4. Simulation of one sample on Kicking left leg (Joint KneeLeft)

t	x/m	y/m	z/m	Simulation z'/m
0	0.146095	−0.326741	2.39079	2.36311
1	0.159648	−0.32887	2.39473	2.37893
2	0.166989	−0.33179	2.39949	2.38867
3	0.171385	−0.334395	2.40406	2.39554
4	0.174155	−0.342428	2.40144	2.40083
5	0.144635	−0.345083	2.35819	2.37466
6	0.078312	−0.282651	2.31867	2.31180
7	0.048378	−0.15173	2.2796	2.28461
8	7.50E−04	−0.0745968	2.23547	2.23965
9	−0.00630	−0.004632	2.2069	2.23523
10	−0.00344	9.97E−04	2.19686	2.24073
11	0.014933	−0.0867754	2.19748	2.26176
12	0.06601	−0.198467	2.24422	2.31540
13	0.12671	−0.321521	2.32764	2.37818
14	0.209081	−0.344676	2.40598	2.46089
15	0.245959	−0.343879	2.46045	2.49858
16	0.231109	−0.341763	2.45902	2.48723
17	0.219537	−0.343241	2.453	2.47892
18	0.206702	−0.344765	2.45555	2.46935
19	0.198436	−0.342189	2.46222	2.46412
20	0.192993	−0.340925	2.46271	2.46159
21	0.184122	−0.341288	2.4626	2.45575
22	0.176553	−0.339308	2.46401	2.45114

(continued)

Table 4. (*continued*)

t	x/m	y/m	z/m	Simulation z'/m
23	0.168045	−0.34117	2.46628	2.44561
24	0.16565	−0.341519	2.46619	2.44602
25	0.163631	−0.341832	2.46557	2.44680
26	0.162687	−0.341825	2.46546	2.44863
27	0.161672	−0.341829	2.46549	2.4504
28	0.161094	−0.342198	2.46496	2.45259
29	0.16092	−0.342652	2.46472	2.45519

6 Conclusion

In this paper, we presented a novel implementation of *Gene Expression Programming* in constructing human motion model. With the advantage of *GEP* on function mining and complex curve approaching, we are able to construct complicated joints trajectory model with complex function. Compared to the previous model construction process, in which factors like time and economic consumption are the bottleneck, we utilized Microsoft Kinect sensor, which is an RGBD camera and reliable data source, to extract human joint points data. Furthermore, we tried the GEP modeling on a dataset with four types of behavior. In each instance of the sequence, we simulate the value of z using the *GEP* model and consider 25 joints trajectory *GEP* models as a complete human behavior model. In the modeling result of each trajectory, the error magnitude is about 0.01.

As the shortcoming of the model, it would be the training time, which costs much time to optimize the *GEP* model.

For the future work, we will extend the modeling result on human behavior recognition area and try to approach the trajectory model in different function format, like $x(t)$, $y(t)$ and $z(t)$, to drop down the algorithm complexity. Moreover, we will investigate the possibility of building a brand new benchmark or implement *GEP* on other trajectory fitting cases.

Acknowledgement. We thank our supervisor, Professor Kangshun Li, for generously offering advice in both algorithm and implementation aspects, and supporting us laboratory for research. This work was supported by the National Natural Science Foundation of China (#61703170) and The Provincial Student's Training Program for Innovation and Entrepreneurship of Guangdong Education Department with the title "*The Study of Behavior Recognition based on Human Joint Points Model*" [19]. This work was also jointly supported by Natural Science Foundation of China (#61573157) as well as the Science and Technology Planning Project of the Guangdong Province, China (#2017A010101037).

References

1. Moeslund, T.B., Granum, E.: A survey of computer vision-based human motion capture. Comput. Vis. Image Underst. **81**(3), 231–268 (2001)
2. Agarwal, A., Triggs, B.: Monocular human motion capture with a mixture of regressors. In: IEEE Computer Society Conference on Computer Vision and Pattern Recognition-Workshops, 2005. CVPR Workshops, p. 72. IEEE (2005)
3. O'brien, J.F., et al.: Automatic joint parameter estimation from magnetic motion capture data. Georgia Institute of Technology (1999)
4. Ramakrishna, V., Kanade, T., Sheikh, Y.: Reconstructing 3D human pose from 2D image landmarks. In: Fitzgibbon, A., Lazebnik, S., Perona, P., Sato, Y., Schmid, C. (eds.) ECCV 2012. LNCS, vol. 7575, pp. 573–586. Springer, Heidelberg (2012). https://doi.org/10.1007/978-3-642-33765-9_41
5. Li, Z., Li, H.: 3D human motion model based on sports biomechanics. J. Syst. Simul. **10**, 2992–2994 (2006)
6. Yang, X., Tian, Y.L.: Effective 3D action recognition using EigenJoints. J. Vis. Commun. Image Represent. **25**(1), 2–11 (2014). ISSN 1047-3203
7. Zhang, Z.: Microsoft kinect sensor and its effect. IEEE Multimedia **19**(2), 4–10 (2012)
8. Shotton, J., et al.: Real-time human pose recognition in parts from a single depth image. In: CVPR. IEEE, June 2011
9. Park, H.S., Sheikh, Y.: 3D reconstruction of a smooth articulated trajectory from a monocular image sequence. In: 2011 International Conference on Computer Vision, Barcelona, pp. 201–208 (2011). https://doi.org/10.1109/iccv.2011.6126243
10. Guo, L., Shen, M.: Upper body spread pose and its motion trajectory prediction based on curve fitting. Chin. J. Ergon. (03), 75–81+85 (2013)
11. Ferreira, C., Gepsoft, U.: What is gene expression programming (2008)
12. Ferreira, C.: Gene expression programming in problem solving. In: Roy, R., Köppen, M., Ovaska, S., Furuhashi, T., Hoffmann, F. (eds.) Soft Computing and Industry, pp. 635–653. Springer, London (2002). https://doi.org/10.1007/978-1-4471-0123-9_54
13. Ferreira, C.: Function finding and the creation of numerical constants in gene expression programming. In: Benítez, J.M., Cordón, O., Hoffmann, F., Roy, R. (eds.) Advances in Soft Computing, pp. 257–265. Springer, London (2003). https://doi.org/10.1007/978-1-4471-3744-3_25
14. Livingston, M.A., et al.: Performance measurements for the Microsoft Kinect skeleton. In: Virtual Reality Short Papers and Posters (VRW), pp. 119–120. IEEE (2012)
15. Shotton, J., et al.: Real-time human pose recognition in parts from single depth images. Commun. ACM **56**(1), 116–124 (2013)
16. Ganapathi, V., Plagemann, C., Koller, D., Thrun, S.: Real time motion capture using a single time-of-flight camera. In: Proceedings of the CVPR, p. 1, 5, 7, 8 (2010)
17. Li, T.D., Wang, Y., He, Y., Zhu, G.Q.: Human single joint point repairment algorithm based on Kinect. (04), 96–98+120 (2016)
18. Wang, J., Liu, Z., Wu, Y., Yuan, J.: Mining actionlet ensemble for action recognition with depth cameras. In: IEEE Conference on Computer Vision and Pattern Recognition (CVPR 2012), Providence, Rhode Island, 16–21 June 2012, pp. 1290–1297 (2012)
19. Kangshun, L., Wei, H., et al.: Method of human behavior recognition based on GEP, CN 106909891A, CN 2017100595256, 30 June 2017

Research of Crowed Abnormal Behavior Detection Technology Based on Trajectory Gradient

Kangshun Li, Hongtao Huang$^{(\boxtimes)}$, Zebiao Zheng, and Yusheng Lu

College of Mathematics and Informatics, South China Agricultural University,
Guangzhou 510642, China
hongtao_fans@163.com

Abstract. Taking the characteristic value as the core, a population abnormality detection algorithm is used to process the crowd surveillance video. Using density detection, the density of the population is first obtained. Object-based feature extraction is used in low-density scenes, and pixel-based feature extraction in high-density scenes. So as to obtain the crowd of exercise intensity, trajectory gradient, entropy and local density and other characteristic value. Finally identify the abnormal behavior of the population based on characteristic value. The experimental results show that the characteristic value is obvious when the abnormality occurs. The algorithm's performance index is superior to the traditional crowd behavior recognition algorithm with high recognition rate.

Keywords: Characteristic value · Abnormal behavior detection
Trajectory gradient · Feature extraction

1 Population Abnormal Behavior Detection

With China's rapid economic development, population urbanization has become increasingly obvious. The urban population is increasing, led to growing number of public places (Including subway, airport, business district, stadium, etc.) in the city. Public safety issues related to social stability and people's life and property safety, it has become a global concern. In particular, during public holiday, the Crowded crowd is common, the Crowd abnormal behavior incidents occur constantly. Crowd as a Special management object, has been paid more and more attention by the community. So how to effectively monitor crowd behavior in real time, reducing the losses caused by Crowd abnormal behavior, is one of the most urgent problems nowadays.

Crowd abnormal behavior, often refers to the "abnormal" behavior that violates society civilized norms or group behavior and standards [1]. Common Crowd abnormal behavior include: Crowd fleeing, aggregation, disturbance, trampling, and processions. Crowd abnormal behavior, there are two main performance characteristic, abnormal density variation and movement patterns. Crowd motion patterns of abnormal characteristics, abnormal crowd behavior is mainly reflected in four aspects: the intensity of the crowd [2], path gradients, the crowd of crowd movements entropy, crowds of local density.

© Springer Nature Singapore Pte Ltd. 2018
K. Li et al. (Eds.): ISICA 2017, CCIS 874, pp. 486–500, 2018.
https://doi.org/10.1007/978-981-13-1651-7_43

Current study of crowd behavior has the following three methods: population analysis based on individual objectives, based on the analysis of local area adjacent to a group of people and global analysis method based on overall. Traditional behavioral detection implementation steps are the following:

In this paper, the crowd was used as the research object, and the crowd behavior analysis was carried out by using the abnormal behavior detection algorithm based on the trajectory gradient. Firstly, the moving object is obtained by codebook algorithm, and the population density is estimated by pixel statistics and texture analysis. Then, when the population density is low, the trajectory gradient algorithm is adopted. When the density is high, the optical flow method is used to obtain the four characteristic values such as exercise intensity, trajectory gradient, entropy and local density. Finally, by comparing the curve of each characteristic value of normal people's behavior and abnormal behavior, the threshold of each index of abnormal behavior is obtained, and the abnormal behavior of the population is identified according to these indexes.

2 Through the Analysis of Various Characteristics of the Collected Parameters, Ultimately, Detection and Classification of Normal and Abnormal People. Motion Detection

2.1 Background Modeling Method

This article uses the CodeBook background modeling method. The detailed background learning process is as follows.

(1) The first frame image in the preceding plurality of frame images is first converted into a gradation image and the initial codebook is set separately for each pixel of the frame gray scale image; and set the start learning threshold;

(2) For the image after the first frame image in the first few frame images taken out, the following operations are performed:
  The pixel value of the frame gray scale image is matched with the current codebook of the same position pixel of the previous frame gray scale image to detect whether or not the pixel gradation value is in the learning threshold range of one symbol of the codebook; If yes, the symbol member variable of the symbol is updated according to the pixel gray scale value of the frame gray scale image; If not, a new symbol is created based on the gray value of the pixel of the frame gray scale image;

(3) Whether or not the frame acquired in the detection 2 is the last frame image in the preceding frame image, and if not, the step 2 is continued when the next frame image is acquired; if so, the background learning is completed. And the background image information is acquired based on the respective codebooks of the respective pixel points obtained in step 2.

2.2   Crowds Density Detection

In this paper, population density detection is divided into two cases: sparse crowds density detection and dense crowds density detection.

(1) Population density statistic method based on foreground pixel.

The crowd density estimation method based on pixel statistic features can make a preliminary estimate of the target population image, the basic principle is that the population density is proportional to the proportion of edge pixel in picture. The classification idea of the population density grade is: statistics the sum of the edge pixels after the edge extraction, select the appropriate boundary value, divide the crowd into dense crowds and sparse crowds, then select the appropriate boundary value, divide the sparse crowds into lower density and low density. Thus, the crowd density estimation method based on pixel features roughly divides the dense crowds from the sparse crowds and further divides the sparse crowds into: low density and lower density (Table 1).

Table 1. Is a specific definition of the crowd density level.

Level	Number of persons	Density range $(1/m^2)$
Especially low	0–7	0–0.75
Low	8–16	0.83–1.58
Medium	17–29	1.67–2.42
High	30–39	2.5–3
Especially high	>40	>3

Data obtained by pixel statistics, after observed the distribution of points, these points are almost all in a straight line, if x is used to indicate the total number of pixels on the edge of the crowd, y indicates the number of people in the scene, their relationship can be expressed as: $y = ax + b$.

We can see that with the increase of crowd density, the edge pixel proportion of the motion foreground image regularly raised. Therefore, the foreground edge pixel proportion can show the crowd density basically, in the case of relatively small number of people, the more people in the area, the higher the foreground image edge.

(2) Texture feature analysis method based on gray level co-occurrence matrix.

To processes the foreground image of the current frame, the co-occurrence matrix is computed by using the method of gray co-occurrence Matrix, which is the most commonly used in statistical method. The texture features of the image are measured by the following five eigenvalues: ASM Energy, contrast, deficit moment, entropy, autocorrelation. The extracted texture features are inputted into the high-density population model, and the crowd density level is obtained through the output of the high-density population model. Concrete steps are as follows:

(a) For each frame image that needs to be detected for crowd density, first input to the low-density population model, using the method of pixel statistics, when the low-density population model gets the number of population results exceeds a certain value, then using high-density population estimation methods;

(b) Using gray level co-occurrence matrix to extract the texture features of the target foreground image of the frame image, constructing the gray-scale co-occurrence matrix of 0, 45, 90 and 135° in four directions, and the entropy, energy, deficit, contrast four characteristic parameters in four directions are calculated respectively, and the average value of four characteristic parameters is obtained. To form a vector as the eigenvector of the detection image;

(c) Input the texture characteristics of the computed image in step 2 into the previously trained neural network classifier. By classifying the classifier, analyzing the approximate population density of the current frame.

3   Research of Crowed Abnormal Behavior Detection Technology Based on Trajectory Gradient

3.1   The Detection Method of Abnormal Behavior of Population Based on Trajectory Gradient

The acquisition of crowd characteristics is the core of crowd behavior recognition, and feature extraction is an important basis for the identification and classification of crowd behavior.

Under the condition of low density population, the single optical flow method only extracts the velocity and direction of the characteristic corner point. It is judged by the two characteristics to judge the state of the population within the video surveillance, the accuracy is low and the crowd can not be judged well. In the high-density population, the hypothesis of the optical flow method will not be established due to shadows, boundaries and occlusion. Therefore, a single optical flow method is only suitable for anomalous detection in low-density populations. It is necessary to introduce a population disorder and density. The state of the crowd. To this end, this paper starts from the trajectory gradient to study the direction of the fastest growth of a certain point in the scalar field and proposes a corresponding improved algorithm - an anomaly behavior detection algorithm based on the trajectory gradient.

Specific steps are as follows:

The object-based feature extraction method is adopted at low density. First use the background difference to get the movement of the crowd, and then between the two blocks before the block between the block and the distance between the block to match the crowd image block, in order to achieve the tracking of people, record the individual trajectory and number of motion, by comparing the two Image information to obtain the human speed, direction and other eigenvalues, and then converted into the crowd within the scene of the exercise intensity, trajectory gradient and local density.

At high density based on the use of pixel-level feature extraction and texture feature extraction method. Lucas-Kanade's optical flow method is used to track the target points of the crowd, and the velocity and direction of the characteristic pixels are obtained by analyzing the vector field. The characteristics of the moving objects are not very effective. Furthermore, in the case of high-density population detection, an analysis method based on texture analysis and population behavior of light flow is introduced. Through the texture analysis, the density of the population and the entropy of the population are obtained. The velocity, direction, quantity and entropy of the four eigenvalue feature points are obtained by combining the two methods. The same is true for the motion intensity, trajectory gradient, local density and entropy of the population within the scene.

The extracted eigenvalues are defined as follows

(a) crowd exercise intensity: video surveillance in the crowd or the kinetic energy of the points and the sum.
(b) Trajectory gradient: In vector calculus, the scalar field gradient is a vector field. The gradient at a point in the scalar field points to the fastest direction of the scalar field, and the length of the gradient is the maximum rate of change. In the case of univariate real-valued functions, the gradient is only a derivative, or, for a linear function, that is, the slope of the line.
(c) Local density: local density of the population in video surveillance.
(d) Entropy: Entropy represents the degree of confusion in the system [3], the entropy in the image represents the size of the average information contained in the image, and the larger the entropy the greater the amount of information contained.

3.2   Abnormal Behavior Analysis of Low Density Crowd

Whether or not the frame acquired in the detection 2 is the last frame image in the preceding frame image, and if not, the step 2 is continued when the next frame image is acquired; if so, the background learning is completed. And the background image information is acquired based on the respective codebooks of the respective pixel points obtained in step 2. Figure 1 is the codebook algorithm to learn the separation of the foreground image.

At low density, with background removal and foreground extraction, we can get the current motion picture. Then the main processing steps are as follows:

Step one, individual target tracking. According to the Euclidean distance [4] of the pixel block between two frames, the motion pixel block of the previous frame matches the same pixel block of the current frame, so as to track the moving object. As shown in Fig. 1, moving individuals are identified.

$$Distance(X, Y) = (\sqrt{(x1 - x2)^2 + (y1 - y2)^2}) \tag{1}$$

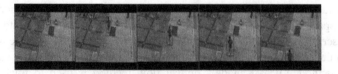

Fig. 1. Moving target tracking

Step two, the movement of individual target eigenvalue extraction. According to the difference between two adjacent frames of images, the relative pixel displacement of each moving object in the scene can be obtained. According to the conversion relationship between pixel displacement and actual displacement, the actual velocity of the moving target can be obtained as v; α is the conversion rate between the pixel displacement and the actual displacement:

$$v = distance(pix) * \alpha / \Delta t \tag{2}$$

The movement direction of the crowd is divided into eight directions by 45°. Through the relative displacement between two frames, the running direction of the moving object can be judged;

Step three, eigenvalue analysis and abnormal behavior determination. From Step 2, the eigenvalues of velocity, direction and density of the crowd are obtained, and the eigenvalues are transformed according to the following definition:

(a) Crowd exercise intensity definition:

$$E = \sum_{i=1}^{n} \frac{1}{2} * mv_i^2 \tag{3}$$

(b) Trajectory gradient definition:
The same moving target in a continuous moment of motion trajectory velocity into a two-dimensional matrix, f (x, y) that the moving object at coordinates (x, y) at the speed value.

$$gd = \text{gradf}(x, y) = \frac{df}{dx}i + \frac{df}{dy}j \tag{4}$$

And z = f (x, y) has a first order partial derivative in the plane and i and j are unit vectors.

(c) Local density:

$$D = \text{number of scenes } y / \text{scene area } S \tag{5}$$

Under normal circumstances, the movement of each individual in the crowd is disorderly. When anomaly occurs, the crowd's speed of movement generally accelerates, and the crowd's exercise intensity varies greatly. The intersection of the extension line of the movement direction of each individual in the crowd is the position where the abnormality occurs.

Formula 2 is to sum up the motion energy of all the tracking targets in the scene to obtain the crowd exercise intensity E of the current scene. E0 is the average exercise intensity of the crowd within the first 50 frames, and D0 is the average of changes in the local density within the scene under normal conditions.

When the change in local density $\Delta D > D0$, that is, the local density changes suddenly, the crowd exists in the scene of aggregation or diffusion phenomenon. At this point the crowd abnormal state. This is an abnormal population density change.

If the local density change $\Delta D \leq D0$, this time need to be combined with other eigenvalues for analysis. As shown in Fig. 2, the method of histogram is used to classify the direction of the trajectory gradient. The population direction is divided into eight directions by 45°. The number of each direction is calculated, and the percentage of the trajectory is calculated and the maximum dg_max is taken. When dg_max > 0.7 and $\Delta E /E0 > 3$, there is a sudden crowd running in one direction in the scene, and it is determined to be abnormal.

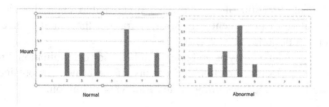

Fig. 2. Trace gradient histogram

3.3   Abnormal Behavior Analysis of High Density Crowd

In the case of the high density of the population in the pixel statistics, l−k optical flow method and texture feature analysis are used to obtain the movement eigenvalues of the population. In order to improve the operation efficiency of the program, the paper takes every 15 frames. In high density, the main steps are as follows (Fig. 3):

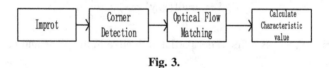

Fig. 3.

The Angle point detection method in this paper is the Harris point detection [5], and the corner point represents the change of the pixel gradient of the image. Assuming that the image window [6] is shifted by (u, v), the resulting grayscale change is E(u, v), then the sum of the derivatives of the adjacent domain direction can be obtained, as shown below:

$$E(u, v) = \sum_{x,y} w(x, y)[I(x+u, y+u) - I(x, y)]^2 \tag{6}$$

Where x and y are the image coordinates, w(x, y) is the window function, and I(x + u, y + v) is the gray value of the translation, and I(x, y) is the gray value before translation. To find a window with a corner point, search for a window with a larger pixel gray level, so expect to maximize the following:

$$I(x+u,y+v) = I(x,y) + I_x * u + I_y * v + O(u^2 + v^2) \tag{7}$$

Ix and Iy are partial derivatives respectively.

The results of the detection of the Harris point are shown in Fig. 4, which is the original image, and Fig. 5 is the corner test result:

Fig. 4. Original image   **Fig. 5.** Corner detection image

(2) Optical flow matching. A certain number of image feature points were obtained from the previous corner detection.

(A) assume that the motion picture P(x, y, t) is a continuous function of x, y, and t. Assume that the object has a pixel at time t, position (x, y), and its strength value is P(x, y, t). U(x, y) and v(x, y) represent the horizontal and vertical velocity components of the object at that point.

(B) at the t + dt moment, the image point moves from (x, y) to the position (x + dx, y + dy), and the intensity value becomes P(x + dx, y + dy, t + dt), where dx = udt and dy = VDT respectively indicate the horizontal and vertical displacement. According to the conservation theory of image intensity proposed by Horn-Schunk [7], the same target point at t + dt moment image point (x + dx, y + dy) intensity P(x + dx, y + dy, t + dt) is equal to the t moment image point (x, y) strength P(x, y, t).

$$P(x+dx, y+dy, t+dt) = P(x,y,t) \tag{8}$$

(C) according to the function P (x, y, t) in the variable x, y, and the continuity of t, in (x, y, t) using Taylor series to the right of the formula (6), high-order item and at the same time divided by the dt side of the equation, can get the following formula:

$$Pxdxdt + Pydydt + Pt = 0 \tag{9}$$

It can also be expressed in the following formula:

$$Px(x,y,t)u(x,y) + Py(x,y,t)v(x,y) + Pt(x,y,t) = 0 \tag{10}$$

The formula (14) is the basic optical flow constraint equation, which indicates that the change rate of gray to time is equal to the dot product of the space gradient and the velocity of the light stream. Because there are two unknowns, u and v in the fundamental equation, only the basic equation can't solve the flow field.

(3) Calculate the characteristic value of optical flow field. Make Px(x, y, z) = Ix, u(x, y) = u, Py(x, y, z) = Iy, v(x, y) = v, Pt(x, y, t) = It, can obtain:

$$Ix * u + Iy * v = -It \tag{11}$$

A small window size m * m, the pixel movement within it is consistent, and the following equation is obtained:

$$\begin{aligned}
Ix1 * u + Iy1 * v &= -It1 \\
Ix2 * u + Iy2 * v &= -It2 \\
Ix3 * u + Iy3 * v &= -It3 \\
&\cdots\cdots \\
Ixn * u + Iyn * v &= -Itn
\end{aligned} \tag{12}$$

The above overdetermined equations can be expressed as:

$$\begin{bmatrix} I_{x1} & I_{y1} \\ I_{x2} & I_{y2} \\ \cdot & \cdot \\ \cdot & \cdot \\ \cdot & \cdot \end{bmatrix} \begin{bmatrix} u \\ v \end{bmatrix} = - \begin{bmatrix} I_{t1} \\ I_{t2} \\ \cdot \\ \cdot \\ \cdot \end{bmatrix} \tag{13}$$

$$A\vec{u} = \vec{b} \tag{14}$$

The above equation is a system of equations of m * m, and only u and v are unknown. Obviously, the system needs to find the optimal solution. We use least square method to solve it, and we get as follow.

$$\vec{u} = \left(A^T A\right)^{-1} A^T \vec{b} \tag{15}$$

The above formula can be resolved.

$$\begin{bmatrix} u \\ v \end{bmatrix} = \begin{bmatrix} \sum I_x^2 & \sum I_x I_y \\ \sum I_x I_y & \sum I_y^2 \end{bmatrix}^{-1} \begin{bmatrix} -\sum I_x I_t \\ -\sum I_y I_t \end{bmatrix} \tag{16}$$

The sum of these is done in the m * m rectangle frame defined previously. Ix, Iy is the first partial derivative of the image x and y direction. In practical application, the above calculation is carried out in the surrounding area of the target pixel when calculating the light flow of the target pixels. People's research has found that the closer to the center's pixel point contribution is less than the further point, and this is where the weight is added to the upper equation:

$$\begin{bmatrix} u \\ v \end{bmatrix} = \begin{bmatrix} \sum wI_x^2 & \sum wI_xI_y \\ \sum wI_xI_y & \sum wI_y^2 \end{bmatrix}^{-1} \begin{bmatrix} -\sum wI_xI_t \\ -\sum wI_yI_t \end{bmatrix} \tag{17}$$

The above weight w generally adopts the gaussian function of a nuclear center at the center of the rectangle. So you get u and v.

Texture analysis gets people's entropy. By calculating the grayscale image, the symbiosis matrix of the image is obtained, and then the partial eigenvalue of the matrix is obtained through the calculation of the symbiosis matrix [8], which represents some texture characteristics of the image. These eigenvalues are random measures of the information contained in the image. When all the values in the symbiosis matrix are equal or the pixel values show the greatest randomness, entropy is the largest. Therefore, the entropy value indicates the complexity of the image grayscale distribution, and the larger the entropy value, the more complex the image. I, j is the coordinate of the point, p of I, j is the pixel value of the point, and Ent is the entropy.

$$\text{Ent} = -\sum_i \sum_j p(i,j) log p(i,j) \tag{18}$$

Step 2: analysis of eigenvalue and abnormal behavior decision:

By analysis and calculation steps a crowd movement speed, direction, such as density, entropy eigenvalue, in the same way we can get the crowd exercise intensity, E dg track gradient, local density D and entropy Ent. E0 is average crowd exercise intensity in the top 50 frames, when the scene within the local density change value $\Delta D > D0$, mutations, namely local density phenomenon crowd gathered or spread in the scene. In this case, the crowd state is abnormal, which belongs to the abnormal density of the population. If the local density change $\Delta D \leq D0$, this time need to be combined with other eigenvalue analysis. By using same way to get the biggest percentage dg_max track gradient, an exception occurs can determine the crowd when dg_max > 0.6, and $\Delta > 3$ E/E and ΔEnt > 0.1.

4   Experiments and Results

The occurrence of abnormal behavior of the crowd is mainly reflected in the movement intensity of the crowd, the trajectory gradient of the crowd, the local density of the population and the entropy of the crowd. The abnormal behavior of the crowd is generally believed to be a crowd gathering, fleeing, running. Here are two scenarios for the experiment. Scene 4.1, scene 4.2.

scene 4.1

scene 4.2

(a) In the case of low density, the characteristic value fluctuates in the abnormal scenario.

In the scenario of low density population, the abnormal behavior of the population can be judged by a single eigenvalue, which can lead to greater errors. Therefore, in scene 4.1, in order to determine whether an abnormality occurs in the scene, it is necessary to combine the distribution of the intensity, density and trajectory gradient of the crowd. Figure 6 is the characteristic curve. Scene 4.1 sends out the diffusion of the crowd. At the time of 3000 frames, there is an anomaly in the scene, the movement of the crowd has become larger, the population density has decreased, and the trajectory gradient tends to be 4. This can be judged as an exception.

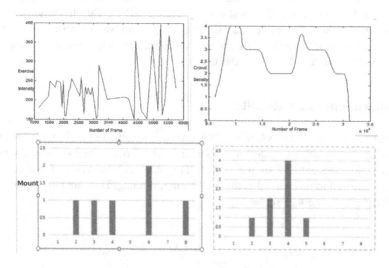

Fig. 6. Change of eigenvalues

(b) in the high density situation, the characteristic value fluctuates in the abnormal scenario

In the high-density crowd scenario, the population trajectory gradient is divided into eight directions by 45°, and the number of each direction is calculated and divided into the upper class. According to the distribution of crowd movement in eight directions, it was divided into three levels to describe the degree of chaos in the direction of movement.

	Converge	Normal	Disorder
Directional distribution	<2 directions	2~6 directions	>6 directions
Ent	Ent ≤ 3.02500	3.02500 < Ent ≤ 3.12000	Ent > 3.12000

The tendency of crowd disorder to be consistent and chaotic is the necessary condition for the abnormal behavior of the crowd. Table 2 is the test result of the crowd movement entropy of scene 4.2.

Table 2. Test results of crowd movement entropy of scene 4.2

Frame number	50	100	150	200	250
Ent	3.01928	3.05269	3.06398	3.05974	3.0567
Frame number	300	320	340	360	380
Ent	3.05588	3.07382	3.08713	3.07777	3.07398
Frame number	400	420	440	460	480
Ent	3.07462	3.11765	3.13619	3.1275	3.08058

表错误!文档中没有指定样式的文字。.2

Frame number	100	200	300	400	480	490
Density	12	12	13	12	14	9
Frame number	500	510	520	530	540	
Density	6	3	1	1	0	

Similarly, the movement intensity of the crowd in the scene can be combined with the results shown in Fig. 7.

As shown in Fig. 8, it is the combination of the three characteristics of the crowd, the intensity of movement, crowd entropy and crowd density. It can be concluded that when the crowd is abnormal, the characteristics of the population fluctuate greatly. Scene 4.2 is the sudden flight of the crowd. At the time of 500 frames, the crowd began to flee, the crowd of exercise intensity greatly, $\Delta E/E > 3$, population density in the changes from high to low density $\Delta D > D0$, entropy change $\Delta Ent > 0.1$. At this time, the abnormal transmission can be judged according to the threshold, and it is in accordance with the abnormal density of the population and abnormal change of population.

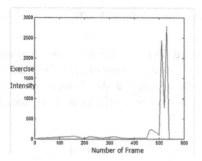

Fig. 7. Shows the experimental data sets

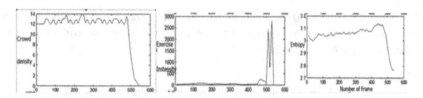

Fig. 8. The fluctuation curve of three eigenvalues

(c) Threshold determination method

The first step is to obtain the range of characteristic values of the characteristic values of the motion state of the crowd in the normal crowd motion of each frame. In the second step, statistical analysis was used to obtain the optimal threshold. This threshold is used as the basis for the detection of abnormal behavior of the population, as well as the critical mass of normal movement and abnormal movement of the population.

Thus, in the low population density of the scene, when the scene within the local density change value $\Delta D > D0$, namely local density mutation, phenomenon crowd gathered or spread in the scene. This is a variation of population density. If the local density change $\Delta D \leq D0$, when dg_max > 0.7 and $\Delta E/E > 3$, the scene within the crowd suddenly to a running direction, an exception occurs in the scene.

In the high population density of the scene of the moment, when the scene within the local density change value $\Delta D > D0$, namely the local density of mutations, occurs in the scene crowd gathered or diffusion phenomenon. In this case, the crowd state is abnormal, which belongs to the abnormal density of the population. If the local density change $\Delta D \leq D0$, when dg_max > 0.6, and $\Delta > 3$ E/E and $\Delta Ent > 0.1$, an exception occurs within the scene.

In this experiment, video data of the sudden diffusion of different directions was adopted, and the total frame number of the sample was 311. This algorithm is compared with the abnormal behavior detection [9] and the group abnormal behavior recognition based on KOD energy characteristics [10]. The results are shown in the following table (Table 3).

Table 3. Comparison of this algorithm with other algorithms

		Normal	Abnormal
The algorithm of the article	. Frame number	224	87
	Recognition rate	90%	96%
Abnormal behavior detection of high - density population	Frame number	214	78
	Recognition rate	84%	89%
The group abnormal behavior recognition based on KOD energy characteristics	Frame number	219	79
	Recognition rate	82%	90%

The experimental results show that the proposed algorithm is superior to algorithm one and algorithm two. There is a great improvement in the recognition accuracy of abnormal behavior.

5   Conclusion

In this paper, an anomaly behavior detection algorithm based on trajectory gradient is proposed. The algorithm uses the crowd as the research object to analyze the population behavior of the collected crowd video. By dividing the population density, the population of different grade density adopts different methods. The eigenvalues of the population were extracted, and the curves of the individual eigenvalues were compared with the normal population behavior and the abnormal behavior. The thresholds of the indexes of the abnormal behavior were obtained, and the emergency behaviors of the population were identified according to these indexes. Experiments show that the performance of the abnormal behavior detection algorithm based on the trajectory gradient is superior to each other, and the recognition rate of abnormal behavior is high.

Acknowledgments. We thank our advisor, Professor Li Kangshun, for generously offers help in both software and hardware facilities, and supports us laboratory for research on our project. This work is supported by Ministry of Education of the People's Republic of China for the National Student's Training Program for Innovation and Entrepreneurship, "The Detection of Crowd Behavior Based on Deep Learning". This work was jointly supported by Natural Science Foundation of Guangdong Province of China (#2017A010101037), and Natural Science Foundation of China (#61573157 and #61703170).

References

1. Wei, Y., Zhuang, X., Fu, Q.: Research progress on the crowd abnormal recognition technology. Comput. Syst. Appl. **25**(9), 10–16 (2016)
2. Zhang, P.: Crowd status analysis and abnormal behavior detection. Civil Aviation University of China (2016)
3. Zhang, J.: Anomaly detection of crowd based on motion entropy. Modem Comput. (07), 40–43 (2013)

4. He, C.-Y., Wang, P., Zhang, X.-H., et al.: Abnormal behavior detection of small and medium crowd based on intelligent video surveillance. J. Comput. Appl. **36**(6), 1724–1729 (2016)
5. Chenguang, G., Xianglong, L., Linfeng, Z., et al.: A fast and accurate corner detector based on Harris algorithm. In: International Symposium on Intelligent Information Technology Application, pp. 49–52. IEEE (2009)
6. Shen, M., Song, H.: Optic flow target tracking method base on croner detection. Electron. Devices **30**(4), 1397–1399 (2007)
7. Deng, X., Tong, Q., Wen, Z., et al.: The comparison and analysis of moving target detection based on optical flow of Horn-Schunk and Kalman filtering technology. Int. J. Adv. Comput. Technol. (2013)
8. Xu, S.-F., Wu, S.-L., Li, H.: An analysis on human skin texture based on gray-level co-occurrence matrix. Acta Lebihreei Sinica **20**(3), 324–328 (2011)
9. Hu, B.: Detection of abnormal crowd event in high density video. Anhui University (2013)
10. Duan, J.-J., Gao, L., Fan, Y., et al.: Abnormal crowd behavior recognition based on KOD energy model. Appl. Res. Comput. **30**(12), 3836–3839 (2013)

A Novel Monitor Image De-hazing for Heavy Haze on the Freeway

Chunyu Xu[1], Yufeng Wang[2,3(✉)], and Wenyong Dong[2]

[1] School of Computer and Information Engineering,
Nanyang Institute of Technology, Nanyang 473000, Henan, China
[2] School of Computer Science, Wuhan University, Wuhan 430072, Hubei, China
wangyufeng@whu.edu.cn
[3] School of Software, Nanyang Institute of Technology,
Nanyang 473000, Henan, China

Abstract. On the freeway, the serious fog and haze weather frequently appears on some road sections due to the geographical factors. The haze seriously damages the image quality of the road monitoring system. In this paper, we proposed a novel monitor image de-hazing algorithm (IDHA) for the heavy haze on the freeway. IDHA can accurately segment the haze monitor image into two regions (road region and non-road region), according to the prior knowledge of edges learned by the other fine weather monitor image of the same camera. An improved guided filtering method with dark channel prior and an improved adaptive histogram equalization algorithm is used on these two regions, respectively. Experiments show that the proposed algorithm IDHA can significantly outperform the dark channel prior algorithm and histogram algorithm on the running time and the de-haze effect on the heavy haze monitor image.

Keywords: Haze removal · Dark channel prior · Monitor haze image
Region segmentation

1 Introduction

On the freeway, the traffic monitor image is captured by cameras installing on the monitor point, and it is often affected by the changeable outdoor weather. The heavy haze weather conditions may cause traffic monitor image degradation, such as, image contrast degradation and image fuzzification. As a result, it can degenerate the performance for vehicle identification and ultimately affect the effective operation of the whole traffic video processing system. Image de-hazing technology can significantly improve the quality degradation of images caused by haze and fog, and can effectively assist traffic monitoring system. Therefore, the de-hazing method plays a significant role in the many fields of the video monitoring system.

In the past decades, many researchers have devoted their attention to solving the problem of restoring haze images [1]. All the research methods fall into two

© Springer Nature Singapore Pte Ltd. 2018
K. Li et al. (Eds.): ISICA 2017, CCIS 874, pp. 501–511, 2018.
https://doi.org/10.1007/978-981-13-1651-7_44

categories: the ones based on image enhancement techniques, the others based on image restoring techniques [2]. For the image enhancement techniques, the histogram equalization has frequently employed the method. Kim [3] proposed the partially overlapped sub-block histogram (POSHE) algorithm, in which, a low-pass filter-type mask is used to get a non-overlapped sub-block histogram-equalization function to produce the high contrast associated with local histogram equalization but with the simplicity of global histogram equalization. Wang [4] proposed a self-adaptive algorithm for haze image that improves the contrast of haze images adaptively in the fuzzy region by use of the mean and standard deviation, and the luminance and contrast by use of contrast-limited adaptive histogram equalization. Wu [5] proposed the gray level mapping optimization mathematical model for classical histogram equalization, in order to solve the shortcoming of low operation efficiency of the optimal contrast image enhancement method by linear programming.

Others focus on the image restoration technique based on the physical model. Oakley and Satherley [6] introduces a method for reducing the degradation in situations that the scene geometry is known. Tan [7] assumes that the local area of the ambient light is constant, and achieve the haze removal by maximizing the local contrast. Since the methods are designed to enhance image contrast, they can not solve the real scene albedo from the physical model. These methods introduce some shortcomings, such as super-saturability for background color and Halo effect for the depth of field discontinuity. To solve them, Fattal [8] estimate the scene albedo by using independent component analysis and assume that the transmission and surface shading are locally uncorrelated. While there are varying degrees of success, the majority of existing methods suffer from shortcomings [9]. Many researchers are concerned about the dark channel prior algorithm, which is proposed by Dr. He [10]. The dark channel prior is a kind of statistics of the haze-free outdoor images [10]. Image restoration is clearer and the loss of image information is lesser for haze image that has enough dark pixels using dark channel prior theory. However, this approach can not well handle monitoring heavy haze images on the freeway and may fail in the cases that the assumption is broken.

In this paper, we proposed a novel monitor image de-hazing algorithm (IDHA) for the heavy haze accumulation on the freeway. It firstly uses the prior knowledge of the other fine weather monitor image on the same camera to separate the haze monitor image into the road regions and non-road regions. And then, for the non-road region, it employs an improved adaptive histogram equalization algorithm based on the bilinear interpolation and median filtering to avoid the color distortion. For the road region, IDHA employs an improved guided filtering method with dark channel prior to amend the estimation of the atmosphere transmittance. Finally, two image restoration regions are merged and generated a new haze removal image. Experimental results show that the proposed IDHA can effectively work, and its performance is super than or equal to some state-of-art algorithms.

The remainder of this paper is organized as follows. Section 2 presents an overview of the dark channel prior algorithm. In Sect. 3, we present the proposed approach IDHA. Section 4 presents the experimental studies. Finally, Sect. 5 gives the conclusion and future directions.

2   The Theory of Image De-haze

2.1   Atmospheric Scattering Model

The atmospheric physical model [10] is widely used in computer vision to describe the information of haze images. It consists of two parts: attenuation model and atmospheric light model. This model is shown as follows:

$$I(x) = J(x) t(x) + A(1 - t(x)) \tag{1}$$

where x is a pixel's position in the image, $I(x)$ is a 3-D vector of the color at x in the captured monitor image with haze, $J(x)$ is a 3-D vector of the light reflected color by the scene point at x, t is a transmission rate of the haze, A is a 3-D vector of the global atmospheric light intensity value. We can obtain the restored images without haze by estimated parameters t and A.

2.2   Dark Channel Prior

The dark channel prior theory is a statistical rule obtained from the observation of haze images. For any input image J, its dark channel can be expressed as:

$$J_{dark}(x) = \min_{y \in \Omega(x)} \left(\min_{c \in \{r,g,b\}} J_c(y) \right) \tag{2}$$

where J_c represents each channel of color image, and $c \in \{r, g, b\}$, $\Omega(x)$ represents a window centered on pixel x. If J is an outdoor clear image without haze, according to the dark channel prior theory [10]. The strength of J_{dark} is always very low except for the sky region, and it is close to 0. J_{dark} is called the dark primary color of J.

3   Proposed Approach

For the monitor image on the freeway, the road region plays a key role in the image haze removal. The road regions contain a great deal of valuable information for the traffic monitoring system. Considering the characteristics of monitor image, we proposed a novel image de-hazing method IDHA, the basic flow diagram of IDHA was shown in the Fig. 1.

In IDHA, the haze image is first segmented into two regions (road region R and non-road region R') by using the prior knowledge contained in the clear image of the same rang. For each kind of regions, a different method is employed to remove the haze. For the case of road region, an improved guided filtering method with a dark channel prior algorithm is used to estimate the atmosphere transmittance aim to enhance the vehicle image on the road. For the case of the non-road region, an improved adaptive histogram equalization algorithm based on the bilinear interpolation and median filtering is used to eliminate the haze. After haze removal, these two regions merged and generated the final restored image J. The key details of IDHA can be found in Sects. 3.1–3.3.

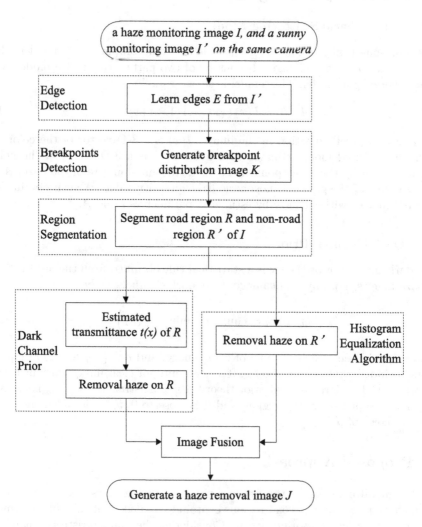

Fig. 1. The basic frame diagram of IDHA

3.1   Region Segmentation and Extraction

Region segmentation and extraction are the cornerstones of IDHA. In order to avoid the inaccurate image segmentation of the haze image I, IDHA learns the breakpoint distribution K by the other fine weather monitor image I' of the same camera. Assuming that the given monitor haze image is I, and the other fine weather monitor image is I', we segment the haze image I by the following steps:

1. Edge Detection. The EDPF [11] algorithm is employed to rapidly identify the edges E of the road regions in I', which is a set of the edge lines as $E = \{\varphi_1, \varphi_2, \cdots, \varphi_n\}, n \in \mathbb{R}$, and $\varphi_i = \{p_j(x_j, y_j) | j = 1...m\}$. Where p_j is the endpoint of the edge lines.

2. Breakpoints Detection. If the total pixel number in the eight direction neighborhood of $p_j(x_j, y_j)$ is equal to 2, $p_j(x_j, y_j)$ will be defined as a breakpoint. Repeat this method for every edge φ_i, IDHA will generate the breakpoint distribution K.

3. Edge Joining. Given a random breakpoint P. If there have any other breakpoints in the $M \times M$ neighborhood of P, such as $P_1, P_2, \cdots, P_n$. The gradient G between P and any other breakpoint $P_i, (i = 1, 2, \cdots, m)$ is given as:

$$G = \sum_{i=1}^{n} \left[(Gr_{i+1,j} - Gr_{i,j})^2 + (Gr_{i,j+1} - Gr_{i,j})^2 \right]^{\frac{1}{2}} \tag{3}$$

where $Gr_{i,j}$ is the gray scale value of the (x_i, y_i) in the connection path of the two breakpoints on the monitor image. The maximum value of the gradient in the connection path between P and other break points P_i will be selected as the final edge connection path.

4. Edge Growth. If there have no other breakpoints in the $M \times M$ neighborhood of P, the edge line will be mapped to a symmetric position centered at breakpoint P. Meanwhile, the location of breakpoint P is updated to the new endpoint. This new endpoint is set as a new starting point for the new edge growth until it meets other edge points in the process of edge growth, and then stop.

After Edge Joining and Growth method, in order to reduce fuzziness of image edge detection and adjust the size of the morphological structural elements, morphological expansion and corrosion operation has been used. Because of the similarity between I and I', we can easily transfer the region boundary information from I' to the haze image I, and segment the haze monitor image into two regions (road region and non-road region). Since the characteristics of these two regions are different, different haze removal algorithm is applied to these two regions, respectively.

3.2   De-haze Method for Road Region

On the road region R, IDHA employs an improved guided filtering method based on dark channel prior to remove the haze. In order to get the accurate transmittance $t(x)$, the dark channel prior theory has been used. Firstly, we normalize the atmospheric Scattering Model by A^c, the Eq. (1) is transformed into:

$$\frac{I^c(x)}{A^c} = t(x)\frac{J^c(x)}{A^c} + 1 - t(x) \tag{4}$$

where the superscript c represents the meaning of $R/G/B$ 3 channels. Assume that the transmittance $t(x)$ is constant in the local region, we minimized Eq. (4), and get the Eq. (5)

$$\min_{y\in\Omega(x)}\left(\min_{c\in\{r,g,b\}}\frac{I^c(y)}{A^c}\right) = t(x)\min_{y\in\Omega(x)}\left(\min_{c\in\{r,g,b\}}\frac{J^c(y)}{A^c}\right) + 1 - t(x) \tag{5}$$

According to the dark channel prior theory, the dark channel value in the road region:

$$\min_{y\in\Omega(x)}\left(\min_{c\in\{r,g,b\}}J^c(y)\right) \to 0 \tag{6}$$

The Eq. (6) is brought into the Eq. (5), we can get:

$$t(x) = 1 - \min_{y\in\Omega(x)}\left(\min_{c\in\{r,g,b\}}\frac{I^c(y)}{A^c}\right) \tag{7}$$

The global atmospheric light intensity A of the atmospheric Scattering Model is set the top 0.1% Highest brightness pixel in the dark channel value. So, the restored image is:

$$J(x) = \frac{I(x) - A}{t(x)} + A \tag{8}$$

Equation (8) is based on the constant of $t(x)$ in the local region. However, the transmittance at the junction edge of the foreground and background is different. Hence, the restored image produce block effects and light blooming. In order to eliminate this problem, an improved guided filtering method is used to further optimize transmittance $t(x)$. The guided filter is as follow:

$$q_i = \frac{1}{|w|}\sum_{i\in W_k}(a_k I_i + b_k) \tag{9}$$

where q_i is a filtered output image, I_i is an Oriented filtering image, W_k is a block area with a radius of r, (a_k, b_k) is a constant parameter on the W_k, w is the pixel number. So we have:

$$\tilde{t}_x = \frac{1}{|w|}\sum_{x\in w_k}(a_k t(x) + b_k) \tag{10}$$

where $\tilde{t}_x$ is the final estimated transmittance.

3.3 De-haze Method for Non-road Region

The non-road region R' usually contains large sky areas. The de-hazing algorithm based on the dark channel produce many color distortion on the non-road region of the restored image. In order to avoid the color distortion, IDHA employs an improved adaptive histogram equalization algorithm on the non-road region. Histogram equalization method can make the histogram of the haze image evenly distributed, aim to achieve the purpose of contrast enhancement. The steps as follow: Firstly, IDHA select a proper size of the sliding window and processed the haze image using histogram equalization algorithm. The calculated value is given to the center point of the sliding window. Then, move the sliding window to the next pixel, and repeat this operation until the end of every pixel traversal. Because the method is enhanced in the local area, and each pixel is taken into account, so it has a good adaptability. The processing results can reduce block effect and color distortion.

4   Experiments Studies

4.1 Compare with Other Haze Removal Algorithms

Four different haze monitor images were selected for the de-hazing experiment, which is evaluated from two aspects of qualitative and quantitative analysis. In order to verify the performance of IDHA, We compare three algorithms on the visual contrast and runtime. These compared algorithms are He [10], He [12] and Kim [3]. In our experiments, the simulation is done on an Intel(R) Core(TM) i5 3.30 GHz computer with 8 GB RAM, our codes are implemented in Matlab.

Figure 2 shows the comparison results of the algorithm of He [10], He [12], Kim [3] and our algorithm IDHA. From Fig. 2(b), we can see that the algorithm of He [10] can have a better visual effect on the haze removal, but it doesn't work well in the non-road region. The restored image of the non-road region contains a lot of noise, and the computational cost of this algorithm is large and time-consuming. He [12] adopts the guided filter to replace the optimized transmission rate of the soft cut out operation, and the operation speed is significantly improved, but it reduces the quality of the restored image. The algorithm of Kim [3] can effectively avoid blocking effect in the non-road region, but the road region of the restored image is distorted, the whole visual effects of restored images are not very good. Compared with these algorithms, the restored image of our algorithm IDHA is more natural and clear in both road regions and non-road region. IDHA can eliminate the texture of road area and block phenomenon. Meanwhile, the restored image has a real color and high resolution, which can provide a satisfactory visual effect.

4.2 Time Comparison

Table 1 has shown the running time of these algorithms on the different size of haze monitor image with different resolution. Table 2 has shown the running time of these algorithms on the different size of haze monitor image with the same resolution. From Tables 1 and 2, we can see that our algorithm IDHA is faster than the algorithm of He [10] and Kim [3]. The running time of IDHA is similar to the algorithm He [12], but the performance of IDHA is better than He [12] in restoring the real color of image scene and higher resolution. Therefore, the IDHA not only greatly improves the processing speed of the de-hazing algorithm, but also effectively avoids the block effects and light blooming in the haze removal image, and the visual effect of the restored image more natural and clear.

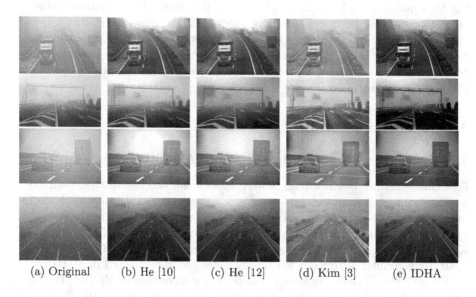

(a) Original   (b) He [10]   (c) He [12]   (d) Kim [3]   (e) IDHA

Fig. 2. Comparison of de-haze effect of different algorithms

Table 1. Comparison of the running time with different resolution and different image

No.	Size	Running time (s)			
		He [10]	He [12]	Kim [3]	IDHA
1	399 × 300	45.6235	4.9356	5.5326	4.9786
2	500 × 375	65.2546	7.0235	8.1265	6.8742
3	600 × 400	79.2879	8.5632	9.0125	8.9853
4	1024 × 591	157.4562	16.7425	19.3468	16.4352

4.3   Image Quality Evaluation

The variance is one of the important indicators to measure the image information, it reflects the degree of gray deviation from the mean gray level. Large variance means that the level of image gray dispersion is higher. When all the

Table 2. Comparison of the running time with same resolution and different image

No.	Size	Running time (s)			
		He [10]	He [12]	Kim [3]	IDHA
1	600 × 400	76.6235	8.4356	8.8326	7.9786
2	600 × 400	78.2532	8.6535	9.1265	8.2742
3	600 × 400	79.2825	8.5621	9.2125	8.3853
4	600 × 400	75.4535	8.4425	8.7453	8.1352

Table 3. Comparison of image evaluation indexes

Index	Algorithm	No. 1	No. 2	No. 3	No. 4
Variance	He [10]	0.0918	0.0723	0.0506	0.0654
	He [12]	0.0803	0.0618	0.0496	0.0546
	Kim [3]	0.0964	0.0757	0.0589	0.0718
	IDHA	0.0756	0.0603	0.0455	0.0512
Average gradient	He [10]	0.0305	0.0256	0.0203	0.0235
	He [12]	0.0298	0.0243	0.0197	0.0228
	Kim [3]	0.0335	0.0287	0.0224	0.0254
	IDHA	0.0257	0.0225	0.0185	0.0205
Image fuzzy entropy	He [10]	0.8536	0.6723	0.5718	0.6718
	He [12]	0.8721	0.6523	0.5624	0.6524
	Kim [3]	0.8865	0.6759	0.5789	0.6429
	IDHA	0.8452	0.6853	0.5823	0.6548
Information entropy	He [10]	13.0716	11.4785	12.8728	10.5718
	He [12]	13.1356	11.5873	12.9736	10.7756
	Kim [3]	12.0718	10.9732	12.0718	10.0743
	IDHA	13.6542	11.9632	13.0368	10.9716

gray levels are approximately equal in the image, the amount of information contained in the image tends to be maximum. The image information entropy is an important index to measure the information abundance of the image and it shows the ability to detail between images.

Table 3 has shown the comparison of image evaluation indexes about the restore image by the algorithm of He [10], the algorithm of He [12], the algorithm of Kim [3], and the algorithm of IDHA from Fig. 2(a). From Table 3, we can see that the variance of He [10], He [12] and Kim [3] is slightly larger than the algorithm of IDHA in this paper, which shows that the probability of all gray levels in the processed image is relatively large. The average gradient and the image fuzzy entropy of this paper are similar to the other three algorithms, which shows that the average gradient and image fuzzy entropy of the algorithm of He [10], the algorithm of He [12], the algorithm of Kim [3] and our IDHA are similar. The information entropy of the algorithm of Kim [3] is the least, and the algorithm of He [10] is similar to the algorithm of He [12], the algorithm of IDHA is the most. It shows that the image of the algorithm of IDHA carries most information than the other three algorithms.

5   Conclusion and Future Work

In this paper, a simple and effective monitor image de-hazing algorithm (IDHA) is proposed. It segments the haze image into two regions (road region and non-road region) according to the prior knowledge of the edge lines learned by the other fine weather monitor image on the same camera. For the road and non-road region, Two efficient de-hazing algorithms are used to remove the haze. The experimental results show that our IDHA algorithm significantly outperforms other three state-of-the-art de-hazing algorithms in both performance and efficiency.

In the future, some open issues are worthy of further study. First, we should seek new haze model to improve the performance of de-haze algorithms. Second, we should develop a new real time de-haze algorithm to narrow the gap between the practical application and the academic experiment. Last but not least, we should study the evaluation mechanism of the performance of the de-hazing algorithm.

References

1. Wu, D., Zhu, Q.S.: The latest research progress of image dehazing. Acta Autom. Sinica **41**(2), 221–239 (2015)
2. Gonzalez, R.C., Woods, R.E.: Digital image processing. Prentice Hall Int. **28**(4), 484–486 (2010)
3. Kim, J.Y., Kim, L.S., Hwang, S.H.: An advanced contrast enhancement using partially overlapped sub-block histogram equalization. IEEE Trans. Circuits Syst. Video Technol. **11**(4), 475–484 (2001)
4. Wang, H., He, X.-H., Yang, X.M.: An adaptive foggy image enhancement algorithm based on fuzzy theory and clahe. Microelectron. Comput. **29**(1), 32–40 (2012)
5. Wu, C.-M.: Studies on mathematical model of histogram equalization. Acta Electron. Sinica **41**(3), 598–602 (2013)
6. Oakley, J.P., Satherley, B.L.: Improving image quality in poor visibility conditions using a physical model for contrast degradation. IEEE Trans. Image Process. **7**(2), 167–179 (1998)
7. Tan, R.T.: Visibility in bad weather from a single image. In: IEEE Conference on Computer Vision and Pattern Recognition (CVPR), pp. 1–8 (2008)
8. Fattal, R.: Single image dehazing. ACM Trans. Graph. **27**(3), 1–9 (2008)
9. Yang, W., Tan, R.T., Feng, J., Liu, J., Guo, Z., Yan, S.: Deep joint rain detection and removal from a single image. In: The IEEE Conference on Computer Vision and Pattern Recognition (CVPR), pp. 1357–1366 (2017)
10. He, K., Sun, J., Tang, X.: Single image haze removal using dark channel prior. IEEE Trans. Pattern Anal. Mach. Intell. **33**(12), 2341–2353 (2010)

11. Akinlar, C., Topal, C.: EDPF: a real-time parameter-free edge segment detector with a false detection control. Int. J. Pattern Recogn. Artif. Intell. **26**(01), 898–915 (2012)
12. He, K., Sun, J., Tang, X.: Guided image filtering. IEEE Trans. Pattern Anal. Mach. Intell. **35**(6), 1397–1409 (2013)

Real-Time Tracking with Multi-center Kernel Correlation Filter

Taoe Wu[1], Zhiqiang Zhao[2], Zongmin Cui[2(✉)], Anyuan Deng[2], and Xiao Yang[2]

[1] Basic Teaching Department, Gongqing Institute of Science and Technology, Gongqingcheng, China
[2] School of Information Science and Technology, Jiujiang University, Jiujiang, China
cuizm01@gmail.com

Abstract. Recently, visual object tracking based on kernel correlation filtering has achieved great success. Application of robust feature, such as the Histogram of Oriented Gradients, is an important reason for the success of the kernel correlation filtering. However, the extraction of the HOG feature may bias the estimation of the target. To overcoming such kind of deviation, this paper proposes a real-time tracker with a multi-center strategy based on the kernel correlation filtering. Finally, abundant experimental results show that the multi-center kernel correlation filtering tracker of this paper has been made great progress relative the kernel correlation filtering tracker.

Keywords: Object tracking · Kernel · Correlation filtering
Multi-center

1 Introduction

Visual object tracking [14] is widely used in compute vision, such as robot, manless driving, etc. With the development of the discriminative technology in computer vision, the vision object tracking based on discriminative technology attracts the attention of researchers. The main idea of the tracking method based on discriminative technology is that it learns the target by distinguishing the target from the background. Abdechiri et al. [1] propose a new multiple instance learning method with a chaotic appearance model. Felzenszwalb et al. [7] propose a hard negative mining to do discriminative training on partially labeled data. The key point of these tracking methods based on discriminative is that they need substantial negative samples for training. However, the large number of negative sample determines the speed of learning directly. Thus, early discriminative training is either to learn a small number of negative samples or to train with off-line for the purpose of the requirement on speed.

With the development of the correlation filter [4] in computer vision, Bolme et al. [2] apply the correlation filter to visual object tracking successfully. The main idea of correlation filter is that it trains and detects the target with signal processing technology. The signal, which is most similar to the target, gets the

K. Li et al. (Eds.): ISICA 2017, CCIS 874, pp. 512–521, 2018.
https://doi.org/10.1007/978-981-13-1651-7_45

peak of the responses, and the rest of the signals get lower responses. Henriques et al. [9] introduce the tricks of kernel and circulant matrices to the correlation filtering, which can learn a large number of negative samples one-time. Such learning method solves the complexity of obtaining negative samples and greatly reduces the computational cost.

Another success of kernel correlation filter is that it uses a robust feature, namely the histogram of oriented gradients (HOG) [7]. The kernel correlation filter with HOG features is widely used in visual object tracking. Chen et al. [3] propose a kernel correlation filter based on patch strategy and introduce the HOG feature to their multiple features also. Li and Zhu [10] propose a adaptive kernel correlation filter tracker based on the HOG feature to capture the scale variation of the target. Danelljan et al. [5,6] propose a discriminative scale space correlation filter based on the HOG feature to capture the scale variation of the target also. Zhang and Suganthan [17] propose a co-trained kernelized correlation filters with the HOG feature.

A good feature can describes and models the target adequately. Compared with other features based on pixel, such as gray scale, histogram and so on, the HOG feature can describes the gradient and the texture variation of the target better. Generally, this paper extracts the HOG features from the target by the way of a dense grid of rectangular. These rectangular are called as cells, which are set as $4 * 4$ pixels in the trackers based on correlation filter. The drawback of this cell structure is that it may cause an offset of one or two pixels from the center of the target. With the continuous of the tracking process, such offset will produce a cumulative error to the center of the target. Therefore, this paper proposes an improve kernel correlation filtering with multi-centers. This paper uses multi detections on the target from different centers to overcome the problem of the offset (Fig. 1).

Next, the trick of KCF is briefly described, and then the proposed method in this paper is discussed detailed. Finally, the experiment results and conclusions are given.

2   Kernel Correlation Filter

The main idea of the KCF [9] tracker is that a large number of negative samples are accessed rapidly and are used for training by a tracking-by-detection scheme. The reason for the success of the KCF tracker is that it uses several tricks, which include kernel, circulant matrices, ridge regression and so on.

2.1   Kernel

To linear regression, the goal of training is to find a linear function $f(z) = w^T z$ that minimizes the squared error over samples x_i and their regression targets y_i,

$$\min_{w} \sum_{i} (f(x_i) - y_i)^2 + \lambda \|w\|^2 \tag{1}$$

where λ is a regularization parameter.

KCF use a non-linear regression, ridge regression, to replace the traditional linear regression. First, it create a mapping between linear regression and non-linear regress by kernel trick

$$w = \sum_i \alpha_i \varphi(x_i)$$ (2)

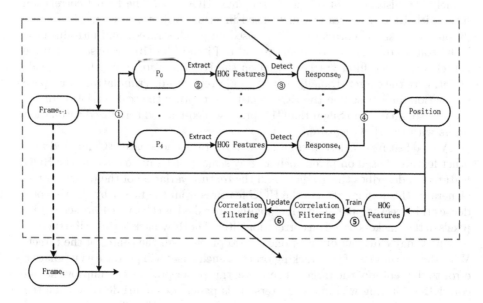

Fig. 1. The framework of the method in this paper.

After this conversion, the constant for regression is transferred from w to α, which is said to be in the dual space [15]. This paper use kernel function κ to definite the dot-products, $\kappa(x, x') = \varphi^T(x)\varphi(x')$. Generally, the kernel function has many forms in KCF, and the Gaussian function is used in this paper.

$$\kappa(x, x') = exp(-\frac{1}{\sigma^2}\|x - x'\|^2)$$ (3)

Under the condition of kernel function, the regression function can be deduced as

$$f(z) = w^T z = \sum_{i=1}^{n} \alpha_i \kappa(z, x_i)$$ (4)

2.2   Ridge Regression

The solution to the kernelized version of ridge regression is given by Rifkin et al. [12]

$$\alpha = (K + \lambda I)^{-1} y$$ (5)

where K is the kernel matrix and α is the vector of coefficents α_i. Due to the data of K are formed from cyclic shift of the base sample, the K is circulant. Thus, a fast solution can be obtained by diagonalize Eq. (7) as in the linear case

$$\hat{\alpha} = \frac{\hat{y}}{\hat{k}^{xx} + \lambda} \tag{6}$$

where k^{xx} is a kernel auto-correlation and is the first row of the kernel matrix K. Under the premise of Gaussian kernel, $k^{(xx')}$ can be represented as

$$k^{(xx')} = exp\left(-\frac{1}{\sigma^2}\left(\|x\|^2 + \|x'\|^2 - 2\mathcal{F}^{-1}(\hat{x}^* \odot \hat{x}')\right)\right) \tag{7}$$

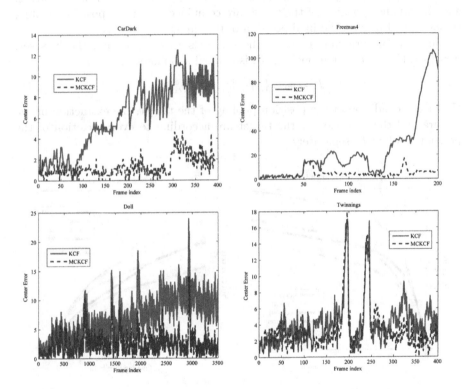

Fig. 2. The center errors between the KCF and the MCKCF.

3   Multi-center Kernel Correlation Filter

This section elaborate the multi-center kernel correlation filter detailed. In the HOG feature, the size of cell is set as 4 * 4. The KCF tracker predicts the exact position of the target based on the location of the cell. The exact position of the target, located at any position of the cell, will yield the same position of the target. Thus, there is a certain offset between the actual position of the target and the predicted position of the target. This paper proposes a multi-center strategy to deal with the offset problem above. The main idea of the multi-center strategy is that a position pooling of the target is used to capture the accurate position of the target. The detail process of the MKCF tracker is shown in Fig. 2. In summary, the tracker includes six parts: (1) The position pooling of the target; (2) Feature extraction; (3) Detection; (4) Position prediction; (5) Training; (6) Updating.

(1) The position pooling of the target

To solve the offset problem above, four differently positions of the target, which are the diagonally adjacent positions from the original position of the target, and the original position of the target are combined to form a position pooling to detect and predict the final position of the target.

Suppose the original center of the target is $p_0 = [x, y]$ and the bias set, $S = \{-1, 1\}$, the position pooling, P, is defined as follows

$$P = p_0 \cup \{p_0 + [b_1, b_2]|_{b_1, b_2 \in S}\} \tag{8}$$

Under the condition of the position pooling of the target, the extraction of the feature and the detection of the target are according to every position of the position pooling P, separately.

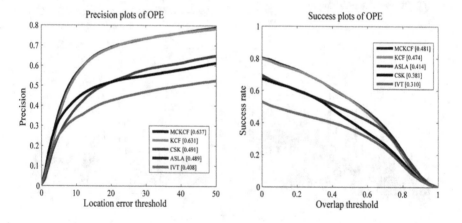

Fig. 3. Plots of OPE. The performance score for each tracker is shown in the legend.

(2) Feature extraction

The HOG features are commonly used in computer vision. The HOG feature, which is based on the cell, extracts the direction information of the cell and generates a gradient histogram. Such feature is very useful for collect the texture information and edge information of the target. This paper uses the HOG feature [1], which has a good calculation performance. The size of cell is set as $4 * 4$.

(3) Detection

After the HOG feature is obtained, a kernel matrix between all training samples and all candidate patches is established, $K^{z_i} = C(k^{xz_i})$, and k^{xz_i} is the correlation filtering between x and z_i. Thus, the regression function in frequency domain to all candidate patches is computed as

$$\hat{f}_i(z_i) = \hat{k}(xz_i) \odot \hat{\alpha} \qquad (9)$$

every $\hat{f}_i(z_i)$ is a full detection response, which is a combination of k^{xz_i} after they are weighted by α. On this foundation, the detection responses in real domain can be obtained

$$R_i = \mathcal{F}^{-1}(\hat{f}_i(z_i)) \qquad (10)$$

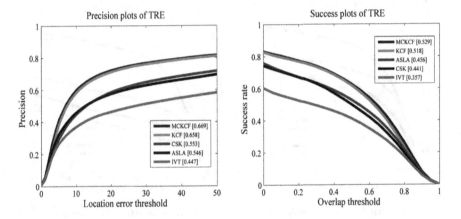

Fig. 4. Plots of TRE. The performance score for each tracker is shown in the legend.

(4) Position prediction

After get all responses, R, in real domain, the final position of the target can be obtained the max value of R.

(5) Training

After obtain the accurate position of the target, the HOG features are extracted from the target according to step (2). Then, the kernel auto-correlation, k^{xx}, with Gaussian kernel in frequency domain according to Eq. (8) is trained. At last, α can be obtained by the kernel auto-correlation, k^{xx}, in frequency domain with Eq. (8).

(6) Updating

The dual space coefficients, α, can be updated by the following equation

$$\hat{\alpha}_t = (1 - \zeta)\hat{\alpha_{t-1}} + \zeta\hat{\alpha}_t \tag{11}$$

where ζ is the learning rate for updating.

The other coefficient in Eq. (8), $\hat{k}_t^{xx}$, is updated by the following equation.

$$\hat{k}_t^{xx} = (1 - \zeta)\hat{k}_{t-1}^{xx} + \zeta\hat{k}_t^{xx} \tag{12}$$

where x is the new sample extracted from the newest position of the target from step (4).

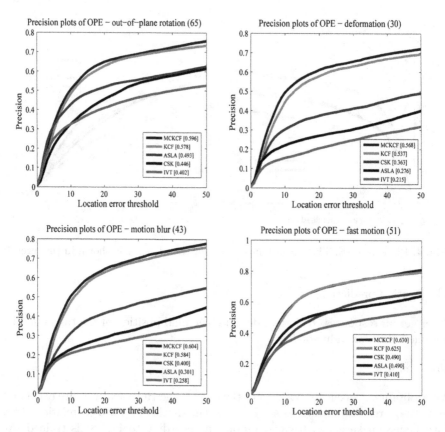

Fig. 5. Precision plots of OPE for OPR, DEF, MB and FM.

4   Experiments and Results

In this section, abundant experiments are used to verify the superiority of the MCKCF tracker in this paper. All experimental results are derived from the native Matlab on an Inter i7 CPU (2.8 GHz) PC with 16 GB memory. The size of cell is set as 4. The tracker in this paper with multi-center strategy is abbreviated as MCKCF.

Four sequences are used to verify the performance of these two trackers in Fig. 2. The criteria for evaluation is the center errors between the tracking results and the ground truths of the target. As shown in Fig. 2, we can that the MCKCF tracker can track the target more accurate than the KCF tracker under the strategy of multi-center pooling.

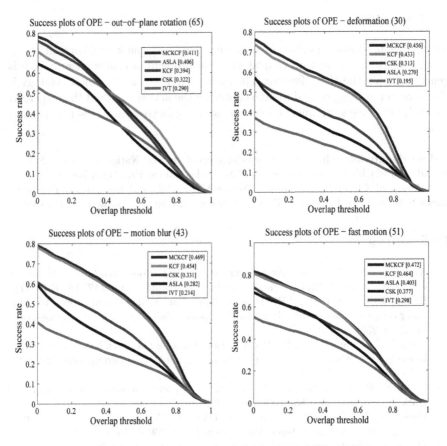

Fig. 6. Success plots of OPE for OPR, DEF, MB and FM.

In order to evaluate the performance of these two trackers comprehensively, this paper do some tests on the full dataset of object tracking benchmark (OTB)

[16]. The tracking algorithms compared with our algorithm include the ASLA [11], IVT [13], CSK [8], and the KCF trackers. Figures 3 and 4 display the precision and the success plots of the one-pass evaluation (OPE) and the temporal robustness evaluation (TRE), respectively. The performance score for all trackers is shown in the legend. It is obviously that the precision plots and the success plots of the MCKCF tracker are better than other trackers in both the OPE plot and the TRE plot. In addition, this paper lists 4 challenging factors in object benchmark [16], which include out-of-plane rotation (OPR), deformation (DEF), motion blur (MB) and fast motion (FM), for visual object tracking. Figures 5 and 6 display the precision the precision and the success plots of OPE, respectively. All these figures show that the multi-center correlation filtering tracker is better than other trackers.

5   Conclusion

This paper proposes a multi-center strategy based on the kernel correlation filtering and the HOG feature to deal with the problem of center offset, which is generated from the HOG feature. The main idea of the multi-center strategy is that a center pooling is used to help capturing the actual center of the target. At the end, extensive experiments demonstrate the effectiveness of the multi-center strategy.

Acknowledgment. This research was supported by the National Natural Science Foundation of China [grant number 61762055]; the Jiangxi Provincial Natural Science Foundation of China [grant number 20161BAB202036]; and the Jiangxi Provincial Social Science "13th Five-Year" Planning Project of China [grant number 16JY19].

References

1. Abdechiri, M., Faez, K., Amindavar, H.: Visual object tracking with online weighted chaotic multiple instance learning. Neurocomputing **247**, 16–30 (2017)
2. Bolme, D.S., Beveridge, J.R., Draper, B.A., Lui, Y.M.: Visual object tracking using adaptive correlation filters. In: Computer Vision and Pattern Recognition, pp. 2544–2550 (2010)
3. Chen, W., Zhang, K., Liu, Q.: Robust visual tracking via patch based kernel correlation filters with adaptive multiple feature ensemble. Neurocomputing **214**, 607–617 (2016)
4. Chen, Z., Hong, Z., Tao, D.: An experimental survey on correlation filter-based tracking. Comput. Sci. **53**(6025), 68–83 (2015)
5. Danelljan, M., Hager, G., Khan, F.S., Felsberg, M.: Discriminative scale space tracking. IEEE Trans. Pattern Anal. Mach. Intell. **39**(8), 1561 (2016)
6. Danelljan, M., Häger, G., Khan, F.S., Felsberg, M.: Accurate scale estimation for robust visual tracking. In: British Machine Vision Conference, pp. 65.1–65.11 (2014)
7. Felzenszwalb, P.F., Girshick, R.B., Mcallester, D., Ramanan, D.: Object detection with discriminatively trained part-based models. IEEE Trans. Pattern Anal. Mach. Intell. **32**(9), 1627 (2010)

8. Henriques, J.F., Caseiro, R., Martins, P., Batista, J.: Exploiting the circulant structure of tracking-by-detection with kernels. In: Fitzgibbon, A., Lazebnik, S., Perona, P., Sato, Y., Schmid, C. (eds.) ECCV 2012. LNCS, vol. 7575, pp. 702–715. Springer, Heidelberg (2012). https://doi.org/10.1007/978-3-642-33765-9_50
9. Henriques, J.F., Rui, C., Martins, P., Batista, J.: High-speed tracking with kernelized correlation filters. IEEE Trans. Pattern Anal. Mach. Intell. **37**(3), 583 (2015)
10. Li, Y., Zhu, J.: A scale adaptive kernel correlation filter tracker with feature integration. In: European Conference on Computer Vision, pp. 254–265 (2014)
11. Lu, H., Jia, X., Yang, M.H.: Visual tracking via adaptive structural local sparse appearance model. In: IEEE Conference on Computer Vision and Pattern Recognition, pp. 1822–1829 (2012)
12. Rifkin, R., Yeo, G., Poggio, T.: Regularized least-squares classification. Acta Electronica Sin. **190**(1), 93–104 (2003)
13. Ross, D.A., Lim, J., Lin, R.S., Yang, M.H.: Incremental learning for robust visual tracking. Int. J. Comput. Vis. **77**(1–3), 125–141 (2008)
14. Smeulders, A.W.M., Chu, D.M., Cucchiara, R., Calderara, S., Dehghan, A., Shah, M.: Visual tracking: an experimental survey. IEEE Trans. Pattern Anal. Mach. Intell. **36**(7), 1442–1468 (2013)
15. Smola, A.J.: Learning with kernels-support vector machines. Lect. Notes Comput. Sci. **42**(4), 1–28 (2008)
16. Wu, Y., Lim, J., Yang, M.H.: Object tracking benchmark. IEEE Trans. Pattern Anal. Mach. Intell. **37**(9), 1834 (2015)
17. Zhang, L., Suganthan, P.N.: Robust visual tracking via co-trained kernelized correlation filters. Pattern Recognit. **69**, 82–93 (2017)

Virtualization – Image Recognition

The Reorganization of Handwritten Figures Based on Convolutional Neural Network

Xingzhen Tao[✉], Wenxiang Wang, and Lei Lu

College of Information Engineering, Jiangxi College of Applied Technology,
Ganzhou 341000, Jiangxi, China
348627805@qq.com

Abstract. Due to the coming of the era of big data, and the computer processing power has been greatly improved. This will provide favorable conditions for the development of the convolutional neural network and then it will become an important object in the field of computer vision. Firstly, it summarized the development of the convolutional neural network and enumerated some successful models of convolutional neural network. Secondly, it introduced the working principle of convolutional neural network in detail, and also analyzed the operation mode of convolutional layer and sampling layer. Finally, it realize the recognition of handwritten figures Based on Convolutional Neural Network, and the experimental result shows, 196 were correct and 4 were wrong when the samples are 200, the recognition rate was 98%.

Keywords: CNN · Convolutional layer · Pooling layer
Recognition of handwritten figures

1 The Introduction

Since 2006, the concept of deep learning by Hinton was put forward in the first time [1], many of the artificial neural network model based on deep learning has also been put forward. The convolutional neural network is one of the most successful algorithms in deep learning algorithms. After decades of development, the convolutional neural network has achieved a series of breakthrough research results in image classification and target detection [3], image semantic segmentation. Its powerful feature learning and classification ability have attracted widely attention and have important analysis and research values. Because the network does not require complex preprocessing of images, it can enter the original image directly, avoiding complicated feature extraction and data reconstruction, so it is widely used in the field of image pattern recognition. The convolutional neural network can recognize displacement, zoom and Torsion invariance by means of local receptive field and weight sharing, pooling. In addition, the convolutional neural network reduces the number of parameters that needed to be trained by the sensory region and weight sharing, which greatly reduces the computation. There are also dramatic improvements in the performance of graphics processors, which take less time to train and test than traditional algorithms. The convolutional neural network has received more and more attention in the application of artificial neural network [2, 3].

© Springer Nature Singapore Pte Ltd. 2018
K. Li et al. (Eds.): ISICA 2017, CCIS 874, pp. 525–531, 2018.
https://doi.org/10.1007/978-981-13-1651-7_46

Handwritten digit recognition is a hot topic in image recognition, but the accuracy of digital recognition is often higher than that of ordinary images. There are some difficulties in handwritten digit recognition, such as small amount of information, little difference between different numbers, and lack of context in digital combination [4]. The algorithm of handwritten digit recognition based on convolution neural network can obtain high accuracy.

2   The Development of Convolutional Neural Network

Training deep neural networks is very difficult before unsupervised pre-training. Until the 1960s, the biological studies of Hubel showed that visual information was stimulated through multiple layers of respective fields from the retina to the brain [5]. In 1984, the Japanese scholar k.f. ukushima and others were based on the concept of respective field and they had the first theoretical model of the convolutional neural network, Neocognitron. It is a multi-layer neural network model of self-organization, and the response of each layer is stimulated by the local respective field in the previous layer. The identification of patterns is not affected by location, smaller shape change, and size.

In 1998, Lecun and others, they based on the reverse propagation algorithm of gradient to design lenet-5, and they successfully conducted supervised training for the model. This model has excellent performance in some pattern recognition tasks, especially in handwriting character recognition tasks. The successful application of lenet-5 in the field of handwriting character recognition has attracted the attention of academia In the convolutional neural network. In the same period, the convolutional neural network has been gradually developed in the aspects of speech recognition, object detection and face recognition.

With the development of big data and the improvement of computer processing power, it provides favorable conditions for the comprehensive development of the convolutional neural network, and then, various excellent models of convolutional neural network are proposed. In 2012, Alex.net [6] was put forward by Krizhevsky and then it won the champion by virtue of the accuracy beyond 11% of the second in large Image database, ImageNet, this makes the convolutional neural network became the focus of the academia. After AlexNet, new models of the convolutional neural network were proposed, such as VGG (Visual GeometryGroup) of Oxford University, Google's GoogLeNet, Microsoft's ResNet [7]. These networks refresh the record set by AlexNet on the ImageNet [8, 9].

3   The CNN Algorithm Realized

3.1   The Description of CNN Algorithm

The implementation of CNN algorithm mainly includes 8 layers: data input layer, the first convolutional layer, pooling layer, the second convolutional layer, pooling layer, the third convolutional layer, fully connected layer and output layer. The algorithm flow chart is shown in Fig. 1.

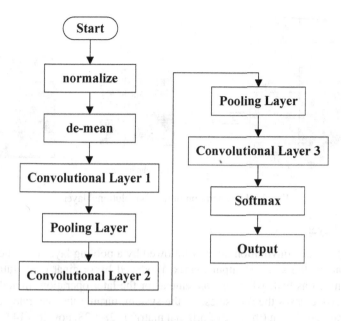

Fig. 1. The CNN algorithm

1. Data Input Layer

During the process of data input, In order to improve the operation rate and robustness, it is necessary to handle data with removing mean value normalization value method.

(1) Normalization: normalizes the 32 * 32 bitmap to 0–255;
(2) Mean Removal: If the sample has a non-zero mean, being inconsistent with the non-zero mean of the tested part, it may lead to a decrease in the recognition rate. The process of mean removal can increase the robustness of the system.

2. Convolution Layer C1

The difference about the convolutional neural network compared to other neural networks is that it joins the convolution layer in front of the input layer [6]. The work of the convolutional layer is through a learnable convolution kernel (feature matrix) and the previous layer of image matrix. According to certain sequence and its pixel of corresponding position multiplied. When the convolution is completed, a new image matrix will be generated, and the convolutional layer will complete the work of the previous feature extraction. The operation of the convolutional layer is shown in the Fig. 2.

Convolution layer is the core of convolutional neural network. Convolution kernel is equivalent to a filter. Different filters have different features. The features of the picture can be got through diverse convolution kernel. For example, as to handwritten figure recognition, one convolutional kernel extracts "—" and another one extracts "1", so the number is more likely to be identified as "7".

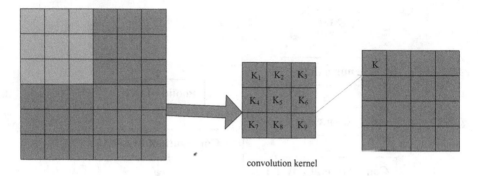

convolution kernel

Fig. 2. The operation of the convolutional layer

3. Pooling Layer

Generally, each convolution layer is followed by a pooling layer to reduce dimension. In general, the size of output matrix, produced by original convolution layer, reduces as much as half, which is convenient for the later operation. In addition, the pooling layer increases the robustness of the system, turning the accurate description into a general description (the size of original matrix is 28 * 28, now it is 14 * 14. Some information must be lost, preventing the overfitting in some degree).

4. Convolution Layer C2

Similar to the previous one, we extract the features from the previous features and express the original samples at a deeper level. Note: this is not a full outer join. This isn't a full outer join. This isn't a full outer join. X represents a connection, and a blank represents a non-connection. The connection of C2 is shown in Fig. 3.

	0	1	2	3	4	5	6	7	8	9	10	11	12	13	14	15
0	X				X	X	X			X	X	X	X		X	X
1	X	X			X	X	X			X	X	X	X			X
2	X	X	X			X	X	X			X			X	X	X
3		X	X	X		X	X	X	X			X			X	X
4			X	X	X		X	X	X	X		X	X			X
5			X	X	X			X	X	X	X			X	X	X

Fig. 3. The connection of C2

5. Pooling Layer

The pooling layer usually follows the convolution layer and then samples the feature graph according to the certain sampling rules. The function of the lower sampling layer has two main functions [7]:

(1) the dimension of feature graph is reduced;

(2) maintain the invariant characteristics of the features to a certain extent. There are many methods of pooling, the two most common methods are mean pooling and maximum pooling. The mean pooling is the sum of all values in the pool domain, and take its mean value as the eigenvalue of the downsampling feature graph.

6. Convolution Layer (fully join)

There are 120 convolution kernels. This is full outer join. The matrix is convoluted into a number, which is convenient for the network to judge.

7. Fully Connected Layer

Like the hidden layer in MLP, the description of high-dimensional spatial data is obtained.

8. Output Layer

The RBF network is generally used here, and the center of each RBF is the symbol of each category. The larger the network output is, the more dissimilar the network is. And the minimum output value is the result of discriminant analysis for the network.

3.2 Handwritten Figures Recognition Realized by CNN Algorithm

Handwriting recognition is a classic recognition problem, since handwritten digits always to be small, but the calculation is large, so handwritten recognition has always been a hot subject [10]. In order to realize the recognition of handwritten figures, we must train the CNN recognition algorithm and then begin to recognize figures after meeting the standard of training frequency. The algorithm flow chart is shown in Fig. 4.

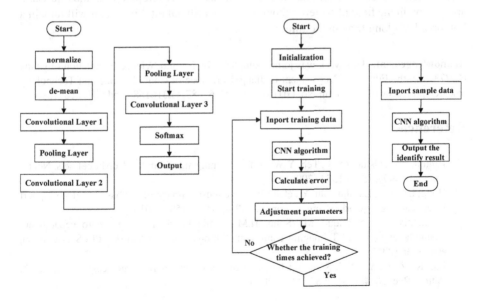

Fig. 4. The flow chart of handwritten figure recognition

4   Experiment and Result

In this experiment, the training sample is 800, and the tested sample is 200. By inspecting the recognition rate of tested sample, we check the accuracy of the algorithm. For the sake of simplicity, we choose one convolution layer. Experimental Result is shown in Fig. 5.

Name ▲	Value	Min	Max
bias_f1	<1x100 double>	-0.25...	0.2263
classify	10	10	10
count	196	196	196

Fig. 5. The result of experiment

In the 200 samples, 196 were correct and 4 were wrong, so the recognition rate was 98%.

5   Conclusions

In the background of big data development, Google, Microsoft, IBM, baidu and other high-tech companies with big data have poured resources into deep learning technology. At present, the convolutional neural network has achieved great progress in object detection, speech recognition, attitude estimation and so on. It also makes achievements in the field of image classification. There are still many problems to be solved in the convolutional neural network, but this does not affect its development. In the future and application in the field of pattern recognition and artificial intelligence, it will remain a hot topic for a long time in the future.

Acknowledgement. This work was jointly supported by Natural Science Foundation of China (61773296), the Education Department of Jiangxi Province of China Science and Technology research projects with the Grant No. GJJ151433, GJJ161687, GJJ161688 and GJJ161691.

References

1. Hinton, G.E., Osindero, S., Teh, Y.W.: A fast learning algorithm for deep belief nets. Neural Comput. **18**(7), 1527–1554 (2006)
2. Hong, L., Fang, L., Shuyuan, Y., et al.: Re-mote sensing image fusion based on deep support value learning networks. Chin. J. Comput. **39**(8), 1583–1596 (2016)
3. Ypsilantis, P.P., Siddique, M., Sohn, H.M., et al.: Predicting response to neoadjuvant chemotherapy with PET imaging using convolutional neural networks. PLoS One **10**(9), e0137036 (2015)
4. Liu, R., Zhao, Y., Wei, S., et al.: Indexing of CNN features for large scale image search. Pattern Recogn. **48**(10), 2983–2992 (2016)

5. Yang, W., Chen, Y., Liu, Y., et al.: Cascade of multi-scale convolutional neural networks for bone suppression of chest radiographs in gradient domain. Med. Image Anal. **35**, 421–433 (2017)
6. Wang, X., Wang, M., et al.: Based on the improved convolution neural network LeNet 5 model identification method. Appl. Res. Comput. **7**(35) (2017)
7. Liang, X., Xu, C., Shen, X., et al.: Human parsing with contextualized convolutional neural network. IEEE Trans. Anal. Mach. Intell. **39**(1), 115–127 (2017)
8. Zhao, Z., Kumar, A.: Accurate periocular recognition under less constrained environment using semantics-assisted convolutional neural network. IEEE Trans. Inf. Forensics Secur. **12**(5), 1017–1030 (2017)
9. Lu, H., Zhang, Q.: The summarize for the application of depth convolution neural networks in computer vision. J. Data Acquis. Proc. 1–17 (2016)
10. Ashiquzzaman, A., Tushar, A.K.: Handwritten Arabic numeral recognition using deep learning neural networks. In: IEEE International Conference on Imaging, Vision & Pattern Recognition, pp. 1–4. IEEE (2017)

Comparison of Machine Learning Algorithms for Handwritten Digit Recognition

Shixiao Wu[1,2(✉)], Wanyun Wei[3], and Libing Zhang[2]

[1] School of Electronic Information, Wuhan University, Wuhan 430072, China
343564602@qq.com
[2] Department of Information Engineering, Wuhan Business University,
Wuhan 430056, China
[3] Department of Electromechanical Engineering,
Lanzhou Resource & Environment Voc-Tech College, Lanzhou 730000, China

Abstract. This paper adopts 10 machine learning algorithms to present the classification results of handwritten digit recognition on Minist dataset. These algorithms include k-nearest neighbors, support vector machine (SVM), decision trees (DT), random forest (RF), naive bayes, multilayer perception (MLP), logistic regression with neural network, artificial neural network (ANN), back-propagation (BP), convolutional neural network (CNN) and so on. We execute the experiments through matlab2015b and anaconda (python 3.6), and the result (accuracy and run-time) shows that SVM and RF achieve better performance. They has the accuracy of 98.08% and 97% separately, less running-time is taken compared with other methods. All the experiment are executed in CPU environment, without GPU. We also execute CNN algorithm for handwritten digit recognition in GPU (Nvidia GeForce GTX 1060), finally find that this algorithm achieves the best performance and the best classification result, the accuracy is up to 99%.

Keywords: Machine learning · Handwritten digit recognition
Comparison

1 Introduction

Handwritten Digit recognition is well known in OCR and pattern recognition [1]. They can deal with problems like Zipcode recognition, bank check processing, form data entry, etc. For the Zipcode recognition, Wang and Srihari believe that acquisition, binarization, location, and preliminary segmentation should be performed [2]. Metric that judges a recognition system always include recognition accuracy and elapsed time, memory and so on. Feature extraction and classifier selection largely effect the performance of the recognition [3, 4].

Compared with on-line handwritten digit recognition [5], off-line recognition still plays the leader [6]. In this paper, we only focus on the classifier performance and off-line handwritten digit recognition. There are many existing techniques for handwritten digit recognition. LeCun et al., apply large BP networks to solve real image-recognition problems [7], Matan et al., adopt space displacement neural network (SDNN) to recognize

© Springer Nature Singapore Pte Ltd. 2018
K. Li et al. (Eds.): ISICA 2017, CCIS 874, pp. 532–542, 2018.
https://doi.org/10.1007/978-981-13-1651-7_47

handwritten multi-digit string [8], Hinton et al., use linear auto-encoders to recognize handwritten digit from grey-level images. UNIPEN database is also a famous testbed in isolated handwritten character recognition [9]. Statistical methods such as fisher discriminant analysis and PCA [10, 11], machine learning methods like MLP, RF, ANN, CNN, BP, NB, SVM, etc., are well-known solutions [3]. Lotfi and Benyettou apply probabilistic neural networks for handwritten digits [12]. Le Cun et al., make a comparison about various classifiers, such as Baseline Linear Classifier, Baseline Nearest Neighbor or Classifier, Pairwise Linear Classifier, Principal Component Analysis and Polynomial Classifier, Radial Basis Function Network, Multilayer Neural Network, LeNet network, Tangent Distance Classifier (TDC), Optimal Margin Classifier (OMC) and so on. Cheng-Lin Liu et al., has estimated the performance of different classifier, such as MLP, RBF classifier, PC, and LVQ classifier, DLQDF, SVM, etc. In this paper, we use the accuracy and elapsed time as metric to compare the performance of 10 machine learning solutions, also we will mention main parameter settings.

The next arrangement are as follows: we will make a short description for MNIST database in Sect. 2, we introduce 10 mentioned machine learning solutions in Sect. 3, we give the experiment figures and tables result in Sect. 4, we make a conclusion in Sects. 5 and 6 we give the project supports directions.

2 MNIST

Generally, the MINIST Database are composed of 60,000 training images and 10,000 testing images. For a larger set available from NIST, MNIST is a subset [13]. NIST's Special Database 3 (SD3) and Special Database 1 (SD1) help to construct the MINST. Among the complete set of samples, they believe that the result should be independent of the choice of training set and test, data from different sources (SD3 and SD1) are collected to mix the NIST database. For the 60,000 training images and 10,000 testing images, SD3 and SD1 take the half. This paper adopt all the training images and testing images to test the performance of the classifier. As the experiments we have made, different number of images, different configuration of the computer, even the running status of the CPU (like run a lot of procedures) will largely effect the performance.

3 Machine Learning Methods

In this section, we will talk about 10 machine learning methods, which include CNN, RF, SVM (Poly kernel, rbf kernel, linear kernel), KNN (5NN, 9NN), ANN, BP, MLP, NB, Logistic Regression (LR) with NN, DT and so on. We will introduce them shortly as below:

3.1 CNN

Traditionally, CNN consists of input layer, convolutional layer, pooling/downsampling layer and fully-connected layer [14]. The convolutional layer detects local conjunctions of features from the previous layer, the pooling layer merge semantically similar features into one. The main difference from other deep architectures is that CNN is designed to use minimal amounts of pre-processing [15]. The massive amount of convolution operations and large memory requirement are two bottlenecks common in CNN-based inferencing.

Among all the methods in CPU environments, CNN provides the highest accuracy, up to 99%, however, the elapsed time of CNN is about 7389 s, they take too much time to train. This test is executed in matlab deep learning toolbox, 60000 pictures are used to train and 10000 pictures are used to test, each of them has the size 28 * 28. For CNN layers, input layer, 2 convolution layers, 2 sub sampling layers are included, alpha is 1, batch-size is 50, numepochs is 1.

We also execute CNN algorithms in GPU environment, therein we install nvidia driver 384.90, cuda 9.0, cudnn v5, caffe. With the help of Nvidia GeForce GTX 1060 3G, we complete the test in 1 min and finally get the same classification result. Deep learning method CNN performs better.

3.2 RF

A random forest is composed of a collection of tree-structured classifiers, therein each of them is independent identically distributed random vector, and they casts a unit vote separately for the most popular class [16]. For RandomForestClassifier, parameters is set by n_estimators = 150, criterion = "gini", max_depth = 32, max_features = "auto" (Table 1 and Fig. 1).

Table 1. SVM for MNIST (kernel = 'poly', degree = 2)

	Precision	Recall	F1-score	Support
0	0.9730	0.9918	0.9823	980
1	0.9920	0.9885	0.9903	1135
2	0.9561	0.9700	0.9630	1032
3	0.9598	0.9693	0.9645	1010
4	0.9774	0.9705	0.9739	982
5	0.9773	0.9641	0.9707	892
6	0.9740	0.9781	0.9760	958
7	0.9735	0.9660	0.9697	1028
8	0.9627	0.9528	0.9577	974
9	0.9601	0.9534	0.9567	1009
Avg/total	0.9707	0.9707	0.9707	10000

3.3 SVM

By projecting data into feature space and then searching the optimal separate hyperplane, SVM can transform the non-linear problems into linear problems [3]. Basically, SVM solve binary classifier problems, but LIBSVM is developed to cover the multi-class problems [17]. Its kernel function includes poly kernel, rbf kernel, linear

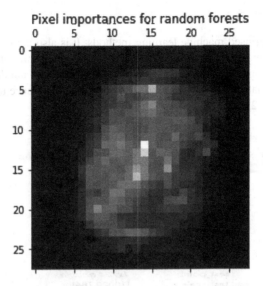

Fig. 1. Pixel importances for RF (dot: graph is too large for cairo-renderer bitmaps. Scaling by 0.0741169 to fit)

kernel and sigmoid kernel. We compared the first three kernel function, finally find that poly kernel has the best performance and less time. For parameter setting, kernel = 'poly', degree = 2. Also we tried 3 kernel, two other kernel include linear and rbf, training images is 60000, and training labels is 60000. Running time: is 510.06473140107244 s. This code is provided by efe (Table 2).

Table 2. SVM for MNIST (kernel = 'poly', degree = 2)

	Precision	Recall	F1-score	Support
0	0.9789	0.9929	0.9858	980
1	0.9886	0.9938	0.9912	1135
2	0.9777	0.9767	0.9772	1032
3	0.9782	0.9772	0.9777	1010
4	0.9827	0.9837	0.9832	982
5	0.9798	0.9776	0.9787	892
6	0.9874	0.9812	0.9843	958
7	0.9795	0.9747	0.9771	1028
8	0.9774	0.9764	0.9769	974
9	0.9751	0.9703	0.9727	1009
Avg/total	0.9806	0.9806	0.9806	10000

3.4   KNN

KNN is an unsupervised machine learning methods. It is also the most simple method for machine learning, they determine the final classification through the affiliation of the great majority among the nearest neighbor. They have the fatal disadvantage, large quantity of computation takes more time. The value of K influence the performance of the classifier (Fig. 2).

Name ▲	Value
accuracy	0.9691
c1	6
dist	60000x1 double
dist_tmp	60000x1 double
i	10000
idx1	22425
j	60000
num_correct	9691
t1	3.9831e+03
test_classify_label	10000x1 double
test_label	10000x1 double
test_point	1x784 double
test_scale	[10000,784]
test_set	10000x784 double
tmp	1x784 double
train_label	60000x1 double
train_point	1x784 double
train_scale	[60000,784]
train_set	60000x784 double

Fig. 2.   KNN (k = 1), result and parameter setting (DT)

3.5   ANN

Bajpai et al., believes that ANN is one type of network which treat the node as "artificial neurons" [18]. Inspired in the natural neurons, this artificial neuron is a computational model, which highly abstract the complexity of real neurons. The neuron is activated when natural neurons receive strong signals, and then inputs and outputs of the natural neurons are computed through some mathematical function. This kind of network always include input layer, hidden layer, and output layer. For handwritten digit recognition, the input and output layer has 784 and 10 nodes separately, the number of hidden layer units is 300. For parameter settings, alpha = 0.1; (learning rate), beta = 0.01 (scaling factor for sigmoid function).

3.6   BP

Based on Deepest-Descent technique, Buscema et al., believe that BP is one kind of ANN [19]. If the hidden units has an appropriate number, they can simulate complex computation and minimize the error of nonlinear functions under this situation. Although they have flexible structure, the learning speed is slow and local minimum is easy to come out (Fig. 3).

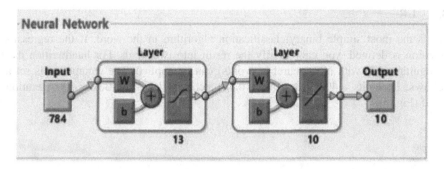

Fig. 3. BP network settings (DT)

3.7   MLP

For performing a wide variety of estimation tasks, MLP is a non-parametric technique [20]. The most widely used algorithm for training MLP is Error back propagation (EBP). MLP is one kind of ANN. For parameter setting, learning_rate is 0.5, weight_decay is 0, momentum is 0, minibatch sample size is 1, the number of iterations between displaying info is 100, the maximum number of iterations is 100000, the number of iterations between testing is 10. The final result is when testing iterations reaches 100000, mean loss is 0.06542, mean accuracy is 96.22%, the running time is 8680.518088.

3.8   NB

Given the value of the class variable, all attributes are independent in NB network [21]. The conditional independence assumption in the real word is feasible, so it employs competitive performance. Train set Accuracy: 83.545%, test set Accuracy: 84.26% (Fig. 4).

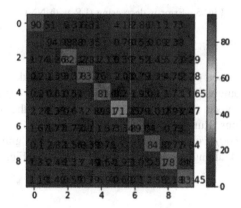

Fig. 4. Naive_Bayes_feature_confusion

3.9   LR

LR is the most simple binary-classification algorithm in the word. If the regression function is defined, you can classify the result into two kinds. For handwritten digit recognition, LR with NN (neural networks) can be applied. The parameters is set as follows: HidUnits = 400, learnRate = 0.1, batchSz = 100, miniBSz = 100, iterations is 20 (Fig. 5).

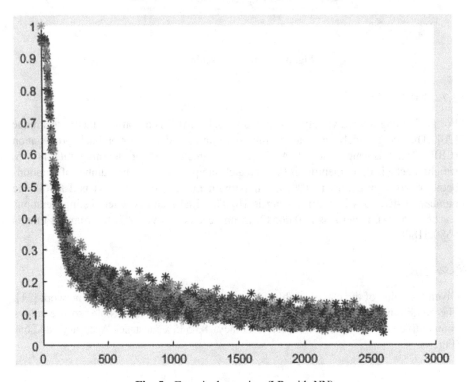

Fig. 5. Error is decreasing (LR with NN)

3.10   DT

Decision trees is composed of intermediate nodes and leaf nodes [22]. Conditions are adopted to label the outgoing edges from intermediate nodes, decisions or actions are used to label the leaf node. A leaf is reached by starting at the root then navigating down on true conditions. Running time in python (spider) is 249.32678459503586 s, Accuracy is 0.87 (±0.00). For parameter settings, criterion = "gini", max_depth = 32, max_features = 784. Training images is 60000 (Table 3 and Fig. 6).

Table 3. The running-time and accuracy for ten machine learning.

	Precision	Recall	F1-score	Support
0	0.9106	0.9357	0.9230	980
1	0.9546	0.9630	0.9588	1135
2	0.8708	0.8488	0.8957	1032
3	0.8259	0.8525	0.8408	1010
4	0.8749	0.8829	0.8789	982
5	0.8401	0.8307	0.8354	892
6	0.8912	0.8894	0.8903	958
7	0.9098	0.9027	0.9062	1028
8	0.8286	0.8039	0.8161	974
9	0.8564	0.8573	0.8569	1009
Avg/total	0.8781	0.8783	0.8781	10000

4 Experiment and Result

We introduce the algorithms in Sect. 3 shortly, this part we will give the final experiment result. We execute the code in python 3.5 and matlab 2015b. We execute CNN,
the deep learning toolbox is needed. ANN, MLP, KNN, BP, CNN, LR with NN, SVM
are executed in matlab 2015b, DT, RF, NB is executed in python 3.5. In Table 4, we
will show the result for ten machine learning algorithms for handwritten digit
recognition.

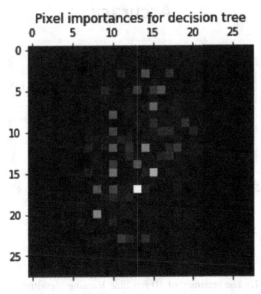

Fig. 6. Pixel importances for DT (dot: graph is too large for cairo-renderer bitmaps. Scaling by
0.0741169 to fit)

Table 4. The running-time and accuracy for ten machine learning.

Algorithm	Accuracy	Running time (seconds)
5NN	96.88%	3639.7
9NN	96.59%	3649.2
ANN	97.87%	7997.149
SVM (poly kernel)	98.08%	324.371
SVM (rbf kernel)	94.46%	839.468
SVM (linear kernel)	93.98%	482.513
MLP	96.44%	8680.518
CNN	98.85%	7389.859
LR with NN	96.81%	1187.660
NB	84.26%	1620
BP	92.04%	312.425
DT	87%	240.642
RF	97%	332.275

We find that the final running-time is highly related with the configuration and manipulation of the computer. If the same algorithm is executed in different software (python/matlab), the running time seems different. But through the accuracy and the running time, you can get the result that RF and SVM has the less time and higher accuracy (Figs. 7 and 8).

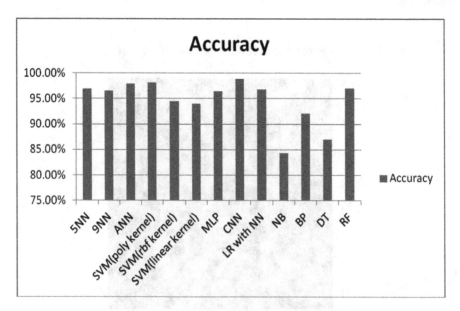

Fig. 7. The accuracy of 10 machine learning algorithms (CPU)

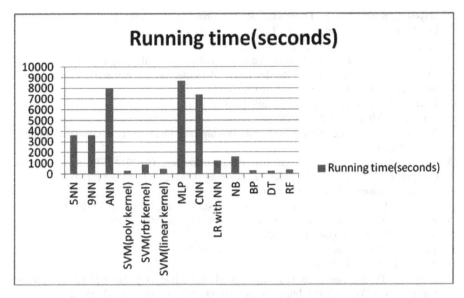

Fig. 8. Running time of 10 machine learning algorithms (CPU)

5   Conclusions

In this paper, we compare 10 machine learning algorithms for handwritten digit recognition in CPU environment. After the experiment, we find that RF and SVM take the less time and higher accuracy. The running time of the algorithms is largely effected by the status and the configuration of the computer. We also execute the experiment in GPU mode, we have Nvidia 384.90 driver, cuda 9.0 and cudnn v5, caffe, then get the same result in CPU but better speed, no more than 1 min.

Acknowledgments. This work was financially supported by Wuhan Teaching and Learning Research Programme (2017113).

References

1. Liu, C.-L., Nakashima, K., Sako, H., Fujisawa, H.: Pattern Recogn. **36**, 2271 (2003)
2. Wang, C.-H., Srihari, S.N.: Int. J. Comput. Vision **2**, 125 (1988)
3. Niu, X.-X., Suen, C.Y.: Pattern Recogn. **45**, 1318 (2012)
4. Lauer, F., Suen, C.Y., Bloch, G.: Pattern Recogn. **40**, 1816 (2007)
5. Kherallah, M., Haddad, L., Alimi, A.M., Mitiche, A.: Pattern Recogn. Lett. **29**, 580 (2008)
6. Arica, N., Yarman-Vural, F.T.: IEEE Trans. Syst. Man Cybern. Part C (Appl. Rev.) **31**, 216 (2001)
7. LeCun, Y., et al.: Advances in Neural Information Processing Systems, pp. 396–404 (1990)
8. Matan, O., Burges, C.J., LeCun, Y., Denker, J.S.: Advances in Neural Information Processing Systems, pp. 488–495 (1992)

9. Guyon, I., Schomaker, L., Plamondon, R., Liberman, M., Janet, S.: Proceedings of the 12th IAPR International Conference on Pattern recognition. Conference B: Computer Vision and Image Processing, vol. 2, pp. 29–33. IEEE (1994)

10. Duda, R.O., Hart, P.E., Stork, D.G.: Pattern Classification. Wiley, New York (2012)

11. Yu, N., Jiao, P.: 2012 IEEE Fifth International Conference on Advanced Computational Intelligence (ICACI), pp. 689–693. IEEE (2012)

12. Lotfi, A., Benyettou, A.: J. Artif. Intell. **4**, 288 (2011)

13. LeCun, Y.: MNIST OCR data (2013)

14. Bag, S.: Deep learning localization for self-driving cars, Ph.D. thesis. Rochester Institute of Technology (2017)

15. Qiu, X., Zhang, L., Ren, Y., Suganthan, P.N., Amaratunga, G.: 2014 IEEE Symposium on Computational Intelligence in Ensemble Learning (CIEL), pp. 1–6. IEEE (2014)

16. Breiman, L.: Mach. Learn. **45**, 5 (2001)

17. Chang, C.-C., Lin, C.-J.: ACM Trans. Intell. Syst. Technol. (TIST) **2**, 27 (2011)

18. Bajpai, S., Jain, K., Jain, N.: Int. J. Soft Comput. Eng. (I-JSCE) **1** (2011)

19. Buscema, M.: Subst. Use Misuse **33**, 233 (1998)

20. Verma, B.: IEEE Trans. Neural Netw. **8**, 1314 (1997)

21. Zhang, H.: AA **1**, 3 (2004)

22. Dobra, A.: Decision trees. In: Liu, L., Özsu, M. (eds.) Encyclopedia of Database Systems. Springer, New York (2016). https://doi.org/10.1007/978-1-4899-7993-3_553-2

A User Identification Algorithm for High-Speed Rail Network Based on Switching Link

Wenxiang Wang and Xingzhen Tao[✉]

College of Information Engineering, Jiangxi College of Applied Technology,
Ganzhou 341000, Jiangxi, China
406873165@qq.com

Abstract. With the rapid growth of LTE user, the overload problem of high-speed rail network is more and more serious, and identifying the public network users who access to the rail network is the key and difficult point to solve the overload problem of high-speed rail network, especially for the low speed scenario where the existing speed detection algorithm effect is poorer. To this end, this paper proposes a high-speed rail network user identification algorithm based on the switching link, which realized the user's identification by extracting cell level switch link, and simulation results show that the algorithm can effectively intercept 96% of public network users who access to the high-speed rail network, and there was no significant difference in both high speed and low speed scene of the performance of the algorithm.

Keywords: High-speed rail network · User identification · Switch link · LTE

1 Introduction

In the rapid development of high-speed railway, with the speed increasing and raising the impermeability of the carriages, the communication on high-speed rail faces many challenges. The influence of Doppler-effect on wireless communication is more obvious in high-speed railway [1], and the switching is more frequent in the high speed travel which will increases the scheduling load of base station. The highly impermeable high-speed railway compartments have great effect on the attenuation of cell phone signal [2], for example, the wear loss value of different CRH models of the LTE signal is between −20 dbm and −30 dbm. At the same time, the high density of user distribution on high-speed rail poses a challenge to the load of base station.

To this end, telecom operators will generally build special network for high-speed rail. In order to reduce the frequency of the cell switching, the high-speed rail network has adopted the combined cell plan, and about 10 RRU are merged into one combined cell [3]. However, after the adoption of the combined cell, the load problem in the cell is faced with challenge, and the telecom operator adopts a double-layer network strategy for the LTE network [4], which adopts the F+D double layer network scheme. However, with the growing number of LTE users, Internet users in dense urban areas will also be able to access to the high-speed rail network, and the network will still be overwhelmed when a train passes by. On the other hand, due to deletion of neighbor cell between the

© Springer Nature Singapore Pte Ltd. 2018
K. Li et al. (Eds.): ISICA 2017, CCIS 874, pp. 543–550, 2018.
https://doi.org/10.1007/978-981-13-1651-7_48

rail network and public network, the public users can't switch to the public network once they have accessed to the rail network [5]. To improve the user experience of high speed rail network and public network, it must be timely to discover and eliminate the public users in the rail network, and distinguishing the public users from all users who access to the rail network is particularly critical.

In the user identification of high-speed rail network, Huawei has proposed a special network user recognition algorithm based on speed detection, and was widely deployed in the national high-speed rail line in 2016 [6]. The algorithm has achieved an ideal effect in high-speed scene, effectively reducing the number of high-load special network in line. However, in the low-speed scenario near the downtown station or in the special bend, the algorithm is still difficult to identify the network and the public network users, and these areas are often the dense area of public network users. What's more, there are a few studies reported on the identification of high-speed rail users. Therefore, it is of great practical significance to study the user identification algorithm of high-speed railway network, and it is also valuable on theoretical research.

2   High Speed Railway Network User Identification Algorithm Based on Switching Link

In order to solve the poor recognition problem of user recognition algorithm based on speed detection in low speed scene, this paper proposes a special rail network user identification algorithm based on switching link. First, we introduce the LTE network architecture. LTE network can be divided into evolutionary core network (EPC) and evolutionary access network (E-UTRAN), as is shown in Fig. 1, E-UTRAN consists of eNodeB which is responsible for the wireless network access function, while EPC

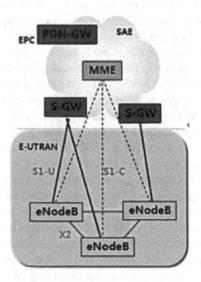

Fig. 1.   Network architecture of LTE

consists of core network equipment such as S-GW, MME and so on, including S-GW bearing business, MME responsible for mobility management.

In the high-speed rail network, users access the LTE network through eNodeB, and MME is responsible for the mobile management of the terminal. A switching lists consisting of service cells and historical service cells can be established on the MME, which can be represented by matrix CL:

$$CL = \begin{bmatrix} C_1 & HC_{11} & HC_{12} & T_1 \\ C_2 & HC_{21} & HC_{22} & T_2 \\ \cdots & & & \\ C_N & HC_{N1} & HC_{N2} & T_N \end{bmatrix} \tag{1}$$

Among them, each row of CL is on behalf of a user's switch information, the first column C represents the current service cells, the second column HC_1 represents the last history switching cells, the third column HC_2 represents the history of the first two switching service cells, T represents user categories, 1 for high speed rail network users, -1 for public users, 0 for unknown users. Switch list can reflect the user's basic movement, and through the analysis of trajectory can basically determine whether the user is a high-speed rail network user or the public one, which is the basic principle of the user identification algorithm for high-speed rail network based on switching link.

Because of the complexity of users' behavior, the previous switching list can't be directly used to determine user categories. There are three accessed ways for users in the eNodeB: switch [7], re-election [8] and redirect [9]. The last serving cell of switching access user can be read for historical cell, which can form a list of switching cells, while the history severing cells of users who access to the network by re-election and redirection are unable to be read. Therefore, the switching list that MME gets in the actual process is usually incomplete, because the history service cells of re-election and redirect users will not be able to be read and set to zero.

To this end, this paper presents the concept of switching link to solve the problem that switching list is invalid in identify users of high speed rail network. In each service cell, although it can't ensure that each user's switch link is complete, there are some user distribution with complete or relatively complete switch link in the many users according to the statistics law. The cell level switch link can be obtained by weighting statistics of the full or relative complete user switching link. The matrix CH is used to represent the cell switching link:

$$CH = \begin{bmatrix} CC_1 & CH_{11} & CH_{12} \\ CC_2 & CH_{21} & CH_{22} \\ \cdots & & \\ CC_k & CH_{k1} & CH_{k2} \end{bmatrix} \tag{2}$$

In the above, each row of CH represents a switch link of a rail network cell, and there are a total of k rail network cells, the first column CC represents the current service cell, the second column CH_1 is the weighted result of history link 1, the third column CH_2 is the weighted result of history link 2, then we have:

$$\begin{cases} CH_{k1} = \sum_{i=1}^{N} HC_{i1}|_{Ci=CC_k} \\ CH_{k2} = \sum_{i=1}^{N} HC_{i2}|_{Ci=CC_k} \end{cases} \tag{3}$$

In high speed rail network, the rail network does not add adjacent cells of public network cells, so the inside public users of the rail network could not access ENodeB by switching mode, and can only access by re-electing or redirecting, so these public users switch link is missing. When there is no railway user in a high-speed rail network, it is not possible to extract the whole cell switch link because of the lack of the switch link of the public network user. When there are some railway users in this network, the completed switching link can be extracted according to the probability statistics. Therefore, it is possible to determine whether there are some railway users in the high-speed rail network cell through judging the integrity of the cell level switching link. If the community level switch link is completed, namely $CH_{k1} \neq 0$ or $CH_{k2} \neq 0$, it can be judged that there are some railway users; If the cell switch link is not completed, namely $CH_{k1} = 0$ and $CH_{k2} = 0$, it can be judged that there are no railway users. Then the MME can give instructions to remove the public network users from the cell. The flow chart of the user recognition algorithm based on switching link is shown in Fig. 2.

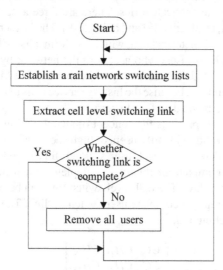

Fig. 2. Flow chart of user identification algorithm for high-speed rail network based on switching link

Because railway users only exist in high-speed rail network cell in the short time when a train pass, and in most time, there is no railway user in idle state, so the algorithm can clear the public users that access to the rail network in most time; Although in the process of a train passing through, public network users can use this period of time to access to the rail network. But it is too short, and the access priority is lower than switching users, so it is limited for the public users to access to the rail network in this period of time. Therefore, it can be seen that the user identification algorithm for high-speed rail network based on the switching link can effectively limit the public user to access to the railway network by theoretical analysis.

3   Simulation Experiment

In order to verify the effectiveness of the user identification algorithm based on switching link, a simulation experiment is set up according to the high-speed railway operation scenario, as is shown in Fig. 3. There was a high-speed line of 80 km, covered with 20 rail network cells (or base station), whose both ends of line were the railway stations, near the station it was urban district with intensive public users, and there was a village with thick neighborhood in the middle of the line. High-speed railway runs at a low speed near the station, while in the other parts of the line it runs at a high speed, with a low speed scene of 40 km/h and 250 km/h at high speed for the simplified model. User Settings: there are 1000 users in the train, and 100 persons are in each area of the city and village.

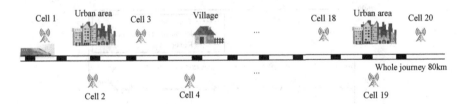

Fig. 3. Simulation model of high-speed rail network

The simulation process is carried out according to the flowchart in Fig. 4, and the simulation results are analyzed from both the net user and the cell level user. First of all, in the process of a train passing, using MME statistics, the number of the entire network users using the link before and after recognition algorithm, is shown in Fig. 5. In order to compare Fig. 5, the accessed public network users shall be displayed during the process of simulation, which can be seen from Fig. 5 where public users were basically knocked out since the user identify algorithm is used, and form experimental data statistics we got that the public user interception rate reached 96% during the entire simulation process. Secondary, we carried out the cell level user analysis, Fig. 6 lists the real-time user numbers of the 20 railway network cells before and after employing the user identification algorithm based on switching link at 400 s and 1000 s. Therefore, when there was no train through idle time, public users were all be blocked, and when the train passed, the public users were mostly to be intercepted.

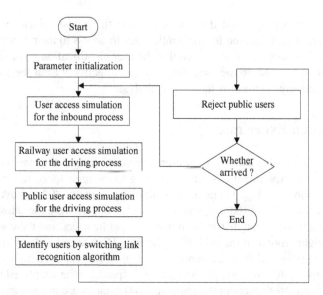

Fig. 4. Flow chart of simulation process

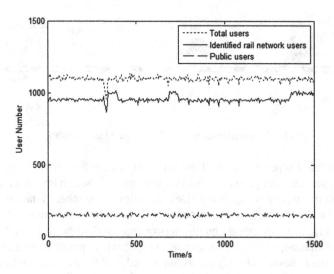

Fig. 5. Statistical graph of user number in the whole network

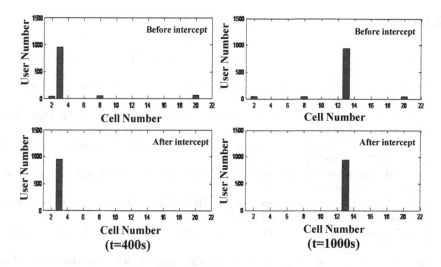

Fig. 6. Statistical graph of user number in cells

The simulation results show that the high-speed rail network user identification algorithm based on link recognition can effectively intercept public user access both in high speed and low speed scene, busy or idle time, and the whole interception rate achieved 96%, which verified the effectiveness of user identification algorithm for high speed rail network based on switching link.

4 Conclusion

For public users and railway users in the high-speed rail network, it is difficult to identify them, especially for low speed scene, this paper puts forward a kind of high-speed rail network user identification algorithm based on switching link. From the perspective of theory and engineering practice, the algorithm can completely intercept the public users when the cell is idle, and intercept the majority of public users when the cell is busy. Then a set of simulation experiment is set up, and the results show that the interception rate for the public network user is 96%. Theoretical analysis and simulation experiments show that the user identification algorithm based on switching link identification has high interception rate as well as high application value.

Acknowledgement. This work was jointly supported by Natural Science Foundation of China (61773296), the Education Department of Jiangxi Province of China Science and Technology research projects with the Grant No. GJJ151433, GJJ161687, GJJ161688 and GJJ161691.

550 W. Wang and X. Tao

References

1. Li, J., Shi, C.: The research and analysis of the doppler frequency shift of high-speed railway. Inf. Technol., 100–102 (2008)
2. Zhang, M., Li, Y., et al.: Analysis of penetration loss of high-speed railway carriage. Mob. Commun., 21–25 (2011)
3. Li, H.: Optimization and practice of high-speed rail network planning. Beijing Univ. Posts Telecommun. (2014)
4. Lei, L., Tian, C.: The dual frequency network of high-speed rail network increases user perception. China New Telecommun., 14 (2017)
5. Wang, M·Study on optimization method of TD-TLE high-speed rail network. Mob. Commun., 67–71 (2014)
6. Li, B., Huang, Q., et al.: The application of HCS in WCDMA high-speed rail network optimization. Telecommun. Technol., 53–57 (2016)
7. Chen, J., Peng, M., et al.: Research on switching technology of TD-TLE system. In: ZTE Commun., 54–58 (2011)
8. Zheng, L., Shen, Z., et al.: The optimization algorithm analysis of LTE community based on access probability. Electron. Technol. Appl., 103–106 (2012)
9. Luo, C., Song, H.: The study of the scheme for the redirection of TD-LTE multimode terminal. Inf. Commun., 166–167 (2014)

A New Language Evolution Model
for Chinese Spatial Preposition

Qi Rao[1] and Youjie Zheng[2(✉)]

[1] MOE Key Laboratory of Computational Linguistics, Peking University,
Beijing, China
qirao@pku.edu.cn
[2] School of Foreign Studies, Huanggang Normal University, Huanggang, China
youjiezheng@163.com

Abstract. Chinese has a history of literature more than three thousand years, it is very valuable to observe the evolution process of language. However, until now there is no such a computation model to describe the evolution process of Chinese, particular for Chinese functional words. In this paper, we propose a new model (NSkip-gram) to describe representations of Chinese functional words. By training Chinese data from early Chinese to Mandarine Chinese, we get the vector space representation of words. The experimental results of the statistical analysis on the preposition "yu(于)" reveal that, the preposition "yu(于)" has lost the ability of co-occurrence with the location argument gradually. And this implies the proposed NSkip-gram model can describe the situations of "yu(于)" in Chinese evolution process well, and it can also be applied to other Chinese functional words.

1 Introduction

In the historical process of language study in different periods, researchers have paid considerable attention to the problem of language evolution. In recent years, this problem has attracted many researchers from other related research fields, such as biological genetics adjacent to the discipline of linguistics, frequent involvement of brain and cognitive sciences, animal behavior and computational science. The research issues related to the evolution of language in fact has become one of the important scientific problems in many disciplines researchers pursue. The root of change in language and music is the 11th of the most challenging 125 scientific problems that need to be solved in the following 1/4 century [5].

Language evolves across time. This change usually occurs at all level, such as phonetics, semantics and grammar, e.g. some new words are created, and some words 'die out' (many irregular verbs are being replaced by their regularized counterparts), existing words adopt additional senses [14]. How to capture the process and details of language evolution are very important. But for a long time, the existing linguistic researches mainly focus on reporting the results of

© Springer Nature Singapore Pte Ltd. 2018
K. Li et al. (Eds.): ISICA 2017, CCIS 874, pp. 551–560, 2018.
https://doi.org/10.1007/978-981-13-1651-7_49

the language evolution but seldom notice the evolution process. Nevertheless, the evolution of language is a complex process, and it is very necessary to do some quantitative analysis on how it evolves.

It is gratifying that in recent years some studies have focused on the quantitative research of language evolution [3,10,12,13,18]. But these quantitative researches mainly focus on shifts in words meaning. Chinese has been lasting for more than 3000 years, which has been proved to be the best sample for observing superficial linguistic features of language changes, demonstrating procedures of language changes and exploring causes and mechanisms of language changes. In those elements of Chinese, prepositions, as a closed set in numbers and the important component in syntax as well as high frequency in usage, is one of the most important window to observe language changes.

Different from the previous researches on language evolution, in this paper, we propose a new language evolution model named as NSkip-gram which is a hybrid model of traditional Skip-gram model and neural network model, and do a quantitative study on the linguistic change by tracking individual shifts in syntactic functional words. We investigate the evolution process of the spatial prepositions in Chinese and choose "yu(于)" as the case study object. We want to capture the movement periods of the words and identify periods of rapid change, and we simultaneously identify words that have changed and also the specific periods during which they changed.

The rest of this paper is organized as follows. Section 2 introduces some related works on the language evolution study. Section 3 goes into details of describing the proposed NSkip-gram model for Chinese spatial prepositions. Experimental results and their analyses are given in Sect. 4. Section 5 concludes the paper.

2   Related Work

There are three complex processes which interact with each other in language evolution, the ability for perceiving and learning language, the dynamics of a centain language in a population, and the biological evolution of the ability to use language [11]. Language evolution therefore involves both biological evolution and cultural evolution, and the co-evolution between biology and culture. Computer models have been proved to be an indispensable tool to simulate the evolving process of the language.

The first attempt on language evolution study which involves computer model emerged in the seventies and eighties of 20th century. And since then, the study on evolution of language has attracted more and more researchers' attention [1,7,8,17,19]. Bailey has proposed the "wave" models to abstract the depiction of language evolution, while it has two principles: the first one is that language evolution follows an "S-shaped" curve in time, with new forms replacing established ones slowly in the beginning of language change, then in the middle stages of a change accelerating their replacements, and finally as the old forms become rare, slowing their advances once again; and the second one is that the differences of the usage rate for a new form in different contexts reflect the importance of the new form which appears in different contexts and different time [3]. But this is only a graphical simulation, and it is not possible to import the process of language evolution into a situation that can be calculated. Altmann and Kroch [12] have proposed a specific mathematical function underlying the "S-shaped" curve of linguistic change. While some other researchers believe that it is appropriate to make a statistical analysis on the changing percentages of alternating forms over time for language evolution [20], and the most representative one is the cumulative function of the normal distribution [2]. The equation of the logistic curve is as given as follows:

$$P = \frac{e^{k+st}}{1 + e^{k+st}} \tag{1}$$

where P is the fraction of the advancing form, t is the time variable, and s and k are constants.

In recent years, the launch of a number of large-scale historical corpus (e.g. Google Books N gram corpus) has provided the possibility for this old problem to breed new ideas and methods, and a number of new work has emerged on language evolution. Sagi et al. used a variation of Latent Semantic Analysis to identify semantic change of specific words from early to modern English [18]. Wijaya and Yeniterzi utilized a Topics-over Time model and K-means clustering to identify periods during which selected words move from one topic/cluster to another [21]. Gulordava and Baroni used co-occurrence counts of words from 1960 s and 1990 s to detect semantic change [7]. Mihalcea and Nastase proposed a supervised learning approach to predict the time period which a word belongs to according to its surrounding context, and they found that the words identified by the model are consistent with evaluations from human raters [16]. Juola compared language from different time periods and quantified the change [9]. Lijffijt et al. and Saily et al. did the stability study variation in noun/pronoun frequencies, and lexical in a historical corpus [15].

These existing studies greatly promote the development of further research on language evolution, but also have some limitations. Firstly, these studies mainly focus on the English and German, but rarely care about Chinese, which has the longest history of literature and rich corpus. Secondly, all these studies focus on the field of words, and lack of the attention on the usage frequencies of function words in the language. As a result of this, in this paper, we propose a new model to do the study on Chinese language evolution in a quantitative way, and we choose the spatial preposition used most frequently and believe it will give a valuable interpretation for how the language changes.

3 Spatial Preposition Evolution Model

3.1 Events and Event Representation

ACE 2005 is a dataset which has a detailed definition of the category of events, separating eight categories of events (type) and 33 seed categories (subtype). As is shown in Table 1:

Table 1. Event type

Type	Subtype
Movement	Transport
Life	Born, Marry, Divorce, Injure, Die
Conflict	Attack, Demonstrate
Contact	Meet, Phone-Write
...	...

Events are triggered by event words and event structure elements, usually verbs work as a trigger. Figure 1 in combination with the role of Chinese labeling specifications for events to ACE shows that the birth is the event trigger word. The trigger events type is $<Life>$, subtype is $<Be-Born>$, three constituent elements of the event are "Deng Xiaoping, August 22, 1904, Sichang Guanan", respectively corresponding to the class (Life/Be - Born) event template tags of the three elements: "Person, Time, Place".

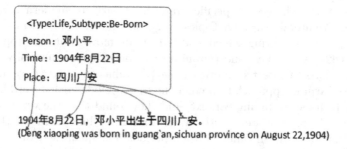

Fig. 1. The event element constitutes an example.

Grammarians believe that prepositions can clearly mark the role of the argument between verbs and nouns. In the Chinese example given in Fig. 1, the preposition "yu(于)" plays the role of identifying the premises, which is the bridge that the space is able to map in the syntactic. In the past few years, however, there has been a lot of disagreement about the limits of the number of argument types in the language. But the Location argument is necessary for each language, it answers the "where" question in the event representation.

3.2 Versatility of "yu(于)" in Early Chinese

The Chinese language has a history of more than 3,000 years, which is often divided into four periods: early Chinese, middle Chinese, the modern Chinese and Mandarin Chinese. In the early Chinese, the theory of introduction is diverse, as shown in Table 2.

Table 2. The probability distribution of "yu" and different arguments in the early Chinese

Token	Argument	Percentage of Distribution
	Location argument	0.87
Yu (于)	Time argument	0.08
	Target argument	0.05

We concern how to quantify the linguistic shift in a given word co-occurrences in the context change and across time. To model word evolution, we construct a time series $S(w)$ for each word $w \in D$. Each point $S_t(w)$ corresponds to statistical information extracted from corpus snapshot c_t that reflects the usage of w at time t.

3.3 Argument Evolution Model

How to count the probability of occurrence of the real text sequence in a given language is a basic question in NLP. Traditional statistical language models transfer the joint probability of the real text sequence into the product of a series of conditional probabilities, which predicts the conditional probability $p(w_t | w_1, w_2, \ldots, w_{t-1})$ in given previous words by the following formula:

$$P(S) = P(w_1, w_2, \ldots w_r) = \prod_{t=1}^{T} p(w_t | w_1, w_2, \ldots, w_{t-1}) \tag{2}$$

But there are maximal parameter spaces for this method, which makes the usefulness of the model very weak.

As a result of this, researchers usually use the simplified version of the model—Ngram model. However, there is an inherent problem for the Ngram model. The problem is "lexical gaps" in the word representation, which means any word is isolated from another, and there is no way to present correlative semantic relationships among words at all, even the same is true of the synonymous words such as "microphone" and "mike". The word in nature is regarded as an isolated atomic unit in the Ngram model, which corresponds to the mathematical form of discrete "one-hot" vector. For example, as for a dictionary of size 5 ["shuang ye (antumn maple leaves)", "hong (red)", "yu (over)", "eryue (February)", "hua (flowers)" 霜叶红于二月花], the "one-hot" vector that **"yu(于)"** corresponds to is $[0, 0, 1, 0, 0]$. In practical applications, the "one-hot" vector usually shows the curse of dimensionality.

Some other researchers attempt to use continuous dense vectors to describe the features of words, the continuous dense vector is called distributed representations of words [22]. The concept is defined as Vector Space Model (VSM), which is based on statistical semantic hypothesis. At the same time, linguistics researches show that words that occur in similar contexts tend to have similar meanings. We can construct a word-context matrix in terms of Distributional

Hypothesis. In the co-occurrence matrix, there are still problems in both the curse of dimensionality and data sparsity, therefore, how to reduce the dimensions in the matrix is what researches concern, such as Latent Semantic Analysis (LSA) [6], Latent Dirichlet Allocation (LDA) [4], and even within neural language models there exist various architectures for learning word vectors.

Mikolov et al. [17] proposed the word2vec word for learning model, by the model, people can find the word vector space exists the phenomenon of translation invariant, such as C(king) − C(queen) ≈ C(man) − C(woman), word2vec learned word vector can effectively capture the word king and queen of implied semantic relationship between man and woman.

Neural language model can predict a set of future word given some history of previous words. In neural language model, words are projected from a sparse, $1 - of - V$ encoding onto a lower dimensional vector space via a hidden layer, V is the size of the vocabulary. This allows for better representation of semantic properties of words compared to traditional language models. Thus, words that are semantically close to one another would have word vectors that are likewise "close" in the vector space. That is, the concept is learned by the vector representations.

Since the neural network model has powerful learning ability and can train the model very fast, in this paper, we follow the popular Skip-gram model, then combine it with neural network model and propose a new model named as NSkip-gram model (see Fig. 2) as the training model for Chinese spatial prepositions, which can be effectively efficient estimation of word vectors from large-scale corpus.

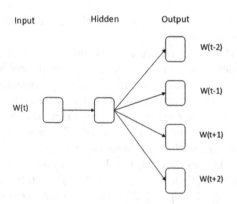

Fig. 2. Architecture of a NSkip-gram model.

From Fig. 2, we can see that, in a NSkip-gram model, each word in the corpus is used to predict a window of surrounding words. The word representations are found in the hidden layer which can be gotten by the word vector $w(t)$. And the weights of the connections in the NSkip-gram model can be trained by a traditional neural network model.

4   Experimental Study

4.1   Corpus

The CCL corpus contains 10 million characters, from early Chinese to Mandarin Chinese. From the corpus, we randomly selected 100000 characters from each period to build a continuous and homogeneous corpus of Chinese historical corpus.

4.2   Model Training and Results

We lower-case all words after sampling and restrict the vocabulary to words that occurred at least 5 times in the corpus. For the model, we use a window size of 3 and dimensionality of 100 for the word vectors. Within each period, we iterate over epochs until convergence, where the measure of convergence is defined as the average angular change in word vectors between epochs. That is, if $S(p)$ is the vocabulary set for period p, and $x_w(p, e)$ is the word vector for word w in a given period p and epoch number e, we continue iterating over epochs.

$$Stop = \begin{cases} 1, if \frac{1}{|S(p)|} \sum_{w \in S(p)} across \frac{x_w(p,e) \cdot x_w(p,e-1)}{\|x_w(p,e)\| \|x_w(p,e-1)\|} \\ 0, otherwise \end{cases} \tag{3}$$

The learning rate in our NSkip-gram model is set to 0.01 at the start of each epoch and linearly decreased to 0.0001.

Once the word vectors for period p have converged, we initialize the word vectors for period $p + 1$ with the previous period's word vectors and train on the $p + 1$ data until convergence. We use an open source implementation in the gensim package, training took approximately 1 days on a 3.5 GHz machine.

4.2.1   Similarity Comparison on the Preposition Vectors

Since cosine similarity is to measure the similarity by finding the cosine of the angle between the vectors which has widely been used in the comparison work in text mining, here, in order to measure the similarity of the preposition vectors, we use this metric to find the change of the preposition "yu(于)".

$$sim(W_A, W_B) = \cos(\theta) = \frac{W_A \cdot W_B}{\|W_A\| \|W_B\|} \tag{4}$$

where $sim(W_A, W_B)$ is the similarity of the preposition vectors, W_A and W_B are two word vectors which can be gotten from our previous model.

For the sake of comparing the evolution results at different time periods, we further choose the other 3 popular spatial prepositions "zai(在)", "dao(到)", and "cong(从)", and compare their results with "yu(于)". The value of similarity is bigger, the percentage of change during the language evolution is smaller, and the priority of element selection is higher. The comparison on the similarity results of word vectors with different prepositions is shown in Table 3.

From Table 3, we can see that, the similarity result of preposition "yu(于)" has the smallest value compared with the other 3 prepositions, while the similarity results of the other 3 prepositions are very close and all of them are more than 0.9. It means that, with time going by, the percentage of change for the preposition "yu(于)" during the language evolution is relatively smaller than the others, and it is more stable in the usage for the representation of displacement events.

Table 3. The results of similarity distribution of "yu" and other 3 popular prepositions from the early Chinese to Mandarin Chinese

Spatial preposition	Similarity
Yu (于)	0.371
Zai (在)	0.9331
Dao (到)	0.9665
Cong (从)	0.9662

An interesting direction of research could involve analysis and characterization of the different types of change. We further inferring the syntactic type of change, our proposed time series construction methods require linguistic knowledge and resources. And our datasets span different time scales.

4.2.2   Similarity Comparison on the Arguments of Prepositions

We firstly count the preposition "yu(于)" that appear with role r as subject in the part-of-speech tagged corpus. The resulting collection of verbs is then ranked by computing their pointwise mutual information with the subject role r.

In a given time period the distribution of argument similarity $sim(a, a_0)$ can be defined as,

$$sim(a, a_0) = a \cdot sim_{DIST}(a, a_0) \tag{5}$$

where a is the known argument, a_0 is an unknown similar type argument, $0 \leq a \leq 1$, and sim_{DIST} represents the way of distribution of corpus and normalizes the similarity.

In this paper, we regard the distribution orientation of the preposition "yu(于)" on on the selection of arguments as the weight of the similarity degree of all the group distribution in a certain time period, and the formula is expressed as follows:

$$SR_{sim}(pr, r, a_0) \sum_{a \in Seen(pr,r)} weight(pr, r, a) \cdot sim(a, a_0) \tag{6}$$

Here, pr represents the preposition, and SR can be presented by a quad $<p, r, a, SR>$, in which R stands for the syntactic relationship or the relationship among semantic roles, R is a parameter, and usually it is a noun or noun phrase. SR represents the probability distribution of the space prepositions "yu(于)" and different types of elements in the corpus in a given time period $p \in [0, 1]$.

We make a statistical analysis on the corpus and get the probability distribution of different types of argument with the preposition "yu(于)" during different time periods.

The experimental results as shown in the Fig. 3 as follows:

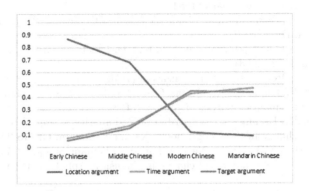

Fig. 3. The evolution tendency of "yu" and three types of arguments' co-occurrence.

From Fig. 3, we can find that, the distribution probabilities of time argument and target argument are always very close and increasing steadily across time. It is obviously in Fig. 3 that the distribution probability of the location argument is decreasing across time. And especially, it decreased very fast during the time period from Middle Chinese to Modern Chinese, which means with time going, the preposition "yu(于)" lost the ability of co-occurrence with the location argument gradually. And this implies our proposed NSkip-gram model can describe the situations of "yu(于)" in Chinese evolution process well, which can also be applied to other Chinese functional words.

5   Conclusions

In this paper we provide a NSkip-gram model to analyze the evolution process in Chinese written language across time through word vectors from different periods' Chinese corpus. The experimental results show the preposition "yu(于)" has lost the ability of co-occurrence with the location argument gradually as the time goes by. An interesting direction of research could involve analyzing different types of change.

In the future work, we attempt to automated reasoning the type of change, such as semantic or syntactic, broadening or narrowing. It may be a case that words that undergo a broadening in sense exhibit regularities in how they move in the vector space.

Acknowledgment. This work is supported by China Postdoctoral Science Foundation (No. 2017M610004), and the National High Technology Research and Development Program of China (863 Program, No. 2015AA015402).

References

1. Adams, R.P., Mackay, D.J.C.: Bayesian online changepoint detection. Statistics (2007)
2. Aldrich, J.H., Nelson, F.D.: Linear Probability, Logit, and Probit Models. Sage Publications, Thousand Oaks (1984)
3. Bailey, C.J.N.: Variation and Linguistic Theory. Center for Applied Linguistics, Washington, D.C. (1973)
4. Blei, D.M., Ng, A.Y., Jordan, M.I.: Latent Dirichlet allocation. J. Mach. Learn. Res. **3**, 993–1022 (2003)
5. Chomsky, N.: Remarks on Nominalization (1970)
6. Deerwester, S., Dumais, S.T., Furnas, G.W., Landauer, T.K., Harshman, R.: Indexing by latent semantic analysis. J. Assoc. Inf. Sci. Technol. **41**(6), 391–407 (1990)
7. Gulordava, K., Baroni, M.: A distributional similarity approach to the detection of semantic change in the Google books Ngram corpus. In: GEMS 2011 Workshop on Geometrical MODELS of Natural Language Semantics, pp. 67–71 (2011)
8. Houston, A.C.: Continuity and change in English morphology: the variable (ING) (1985)
9. Juola, P.: The time course of language change. Comput. Humanit. **37**(1), 77–96 (2003)
10. Kim, Y., Chiu, Y., Hanaki, K., Hegde, D., Petrov, S.: Temporal analysis of language through neural language models. Comput. Sci. **6**(3), 153–178 (2014)
11. Kirby, S., Hurford, J.R.: The emergence of linguistic structure: an overview of the iterated learning model. In: Cangelosi, A., Parisi, D. (eds.) Simulating the Evolution of Language. Springer, London (2002). https://doi.org/10.1007/978-1-4471-0663-0_6
12. Kroch, A.S.: Reflexes of grammar in patterns of language change. Lang. Var. Change **1**(3), 199–244 (1989)
13. Kulkarni, V., Al-Rfou, R., Perozzi, B., Skiena, S.: Statistically significant detection of linguistic change. In: International Conference on World Wide Web, pp. 625–635 (2015)
14. Lieberman, E., Michel, J.B., Jackson, J., Tang, T., Nowak, M.A.: Quantifying the evolutionary dynamics of language. Nature **449**(7163), 713 (2007)
15. Lijffijt, J., Säily, T., Nevalainen, T.: CEEcing the baseline: lexical stability and significant change in a historical corpus. In: Helsinki Corpus Festival (2012)
16. Mihalcea, R., Nastase, V.: Word epoch disambiguation: finding how words change over time. In: Meeting of the Association for Computational Linguistics: Short Papers, pp. 259–263 (2013)
17. Mikolov, T., Yih, W.T., Zweig, G.: Linguistic regularities in continuous space word representations. In: HLT-NAACL (2013)
18. Sagi, E., Kaufmann, S., Clark, B.: Tracing semantic change with latent semantic analysis (2012)
19. Tabor, W.: Syntactic innovation: a connectionist model. Ph.D. thesis, Stanford University (1994)
20. Tuckey, J.W.: Exploratory Data Analysis. Addison-Wesley Pub. Co., Boston (1977)
21. Wijaya, D.T., Yeniterzi, R.: Understanding semantic change of words over centuries. In: International Workshop on Detecting and Exploiting Cultural Diversity on the Social Web, pp. 35–40 (2011)
22. Bengio, Y., Ducharme, R., Vincent, P., Jauvin, C.: Neural probabilistic language model. J. Mach. Learn. Res. **3**, 1137–1155 (2003)

Research on Location Technology Based on Mobile Reference Nodes

Xuefeng Yang[✉], Lin Li, and Yue Liu

Jiangxi College of Applied Technology, Ganzhou City 341000, Jiangxi Province, China
303568969@qq.com

Abstract. The localization process of sensor networks using mobile reference nodes is discussed in detail, and errors in each phase are analyzed. Whether the selection of reference node path is appropriate, whether the unknown nodes can finish localization in closely related monitoring area, the reference node moving distance, and the time and energy consumed, concerning the above problem this paper proposed a covering algorithm to set the location of the radio beacon in advance, and the path traversal of these points planning. Simulation results show that the proposed mechanism can make the mobile reference node traverse the whole monitoring area as much as possible and avoid the obstacles effectively.

Keywords: Sensor network localization · Mobile path · Virtual force
Covering algorithm

1 Introduction

One of the key problems in sensor network localization is how to make each unknown node know the location information of enough reference nodes at the minimum cost, and its location relation with these reference nodes. The reference node density and the amount of communication seems to always be a pair of irreconcilable contradictions, how to solve the core problem of this chapter starts from the need for compromise between the two. In addition, once the job is done, then the reference node will have the same status as the normal node. Unless the network topology changes massively, the GPS device it carries is no longer so important. Therefore, the more initial reference nodes are allocated, the greater the waste. For these reasons, it is economically the most cost-effective if a single reference node can be used to complete the location.

This paper discusses the moving reference node to traverse the monitoring region for the unknown node positioning method, and path of the reference node selection problem and traversal efficiency were analyzed, the simulation results are given.

2 Mobile Routing

In the localization mechanism based on mobile reference nodes, an interesting question is how the reference node determines its optimal mobile path, and the best timing (or location) of the broadcast beacon data. But the difficulty is that we do not know where

© Springer Nature Singapore Pte Ltd. 2018
K. Li et al. (Eds.): ISICA 2017, CCIS 874, pp. 561–570, 2018.
https://doi.org/10.1007/978-981-13-1651-7_50

the unknown nodes are in advance, and thus can not determine the moving path of the reference nodes according to the distribution. Nobody has put forward the solutions are given in this paper [1], mobile path selection heuristic rules: first, the unknown nodes closer to the mobile node's trajectory, the position estimation is more accurate, mobile route reference nodes to try to choose in the nearest place from the unknown node, that is said that the grid generated by the path dividing control area to be as small as possible. Second, each unknown node must receive more than three non collinear beacons.

This paper proposes a universal and operable solution. On the basis of virtual force method and traveling salesman problem, a method of beacon location selection and path planning is proposed. The main idea is; first determine the reference node beacon relative topological relations between the position, is when the reference node according to the topology of radio beacon, the coverage area of any point can receive more than three beacons, so as to ensure any unknown nodes and reference nodes within the scope of coverage, can determine their own virtual coordinates; method with the help of sensor network layout, quickly find the corresponding topological relations in the monitoring region is the launch point; then we based on the classical traveling salesman algorithm to find the optimal path traversal of the launch point.

2.1   Radio Beacon Location Determination

It is because without knowing the coordinates of the unknown nodes, to receive beacons at least three times, only to ensure that any point in the monitoring area can be the reference node communication range covering at least three times. The use of covering algorithm to find the reference node broadcasts a beacon position, because this is the reference node for all unknown nodes in monitoring area (Communication) coverage, the equivalent of the coverage problem in sensor nodes of the monitoring area coverage (detection). And, if a monitoring area coverage to cover, in other words, to make any point in monitoring area received a mobile beacon node reference, reference node is required to broadcast beacon, then, to make the unknown node in the control area at least three times the reference node positioning mobile beacon the reference node must broadcast at least once.

The best location reference node broadcasts a beacon to advance in its first mobile calculated. It is known that the length of the monitoring area is x, y, the communication radius R of the node, The area covered by each regular hexagon is $\dfrac{3\sqrt{3}}{3}R^2$, in the best regular hexagon coverage mode [2]. Then, the reference nodes need to broadcast a secondary beacon $xy \bigg/ \left(\dfrac{3\sqrt{3}}{2}R^2 \right) = \dfrac{2\sqrt{3}xy}{9R^2}$, as shown in Fig. 1.

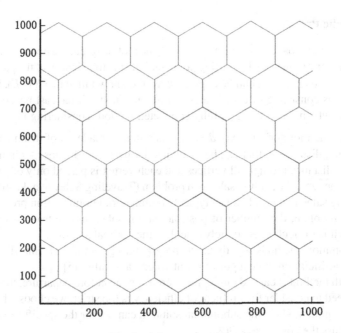

Fig. 1. A heavy covering

To ensure that any unknown node in the monitoring area can receive at least three positioning beacons, the reference node should at least $\dfrac{2\sqrt{3}xy}{3R^2}$ pan beacon. These beacons are regarded as virtual nodes, and the initial coordinates are set at random. Then, the virtual force [3] algorithm is used to find the best location of these nodes.

That is, in the use of virtual force for beacon broadcast position, the first virtual force calculation of the virtual node does not exist between the upper limit of communication problems, so the threshold C_{th} can get bigger, here take $3R$; communication coverage range depends on the communication radius of nodes, so the attraction and repulsion transform threshold D_{th} take $\sqrt{3}R$. Instead, the purpose is to make a certain overlapping coverage area. If the monitoring area is larger, it is necessary to appropriately increase the parameters W_{Rss} describing the repulsive force so that these beacons can be separated to ensure the coverage of the monitoring area.

Assuming that the reference node has a communication radius of 100, we need to set up 150 beacon broadcast locations in the 10001000 monitoring area. The following figure is the location of the broadcast beacon obtained by the virtual force algorithm. As can be seen from the diagram, the black point is calculated as a beacon broadcast position, and almost any location in the monitoring area can obtain at least three positioning beacons.

2.2 Ergodic Process

When the beacon broadcast position is located, the mobile reference node traverses these points in turn and returns to the starting point, making the total length of the pass shortest. To solve the traversal path can be expressed as: the known map $G = (V, E)$, $V(x, y)$ is a set of vertices consisting of coordinates of reference nodes broadcast beacons, E is an edge set (in which the graph G is fully connected without considering the presence of obstacles in the monitoring area), $d = (d_{i,j})$ is a matrix made up of distances between vertices i and distances j that can be computed by their coordinates. A circuit that has the shortest distance through all vertices and each vertex is passed only once [4].

This is the typical traveling salesman problem (Traveling Salesman, Problem, TSP), the traveling salesman problem is a typical combinatorial optimization problem, and is a NP hard problem, the number of possible path number and vertex n is exponential growth, so it is difficult to accurately calculate the optimal solution.

The common methods are dynamic programming, simulated annealing, genetic algorithm, artificial immune algorithm, ant colony algorithm [5].

The path method is carried out in the reference node before departure, the algorithm does not need to consider the problem of efficiency is too strict, we choose the dynamic programming method for the suboptimal solution can be, for the specific algorithm we do not discuss the complexity $O(2^n)$.

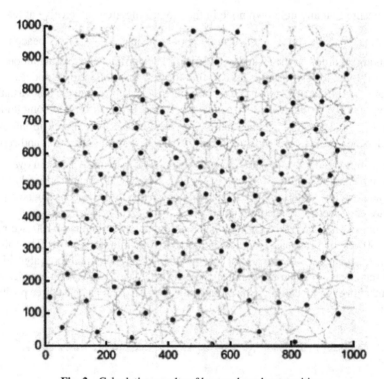

Fig. 2. Calculation results of beacon broadcast position

The path calculated by the beacon broadcast position in Fig. 2 is shown in Fig. 3 below. Its final path length is 13792.

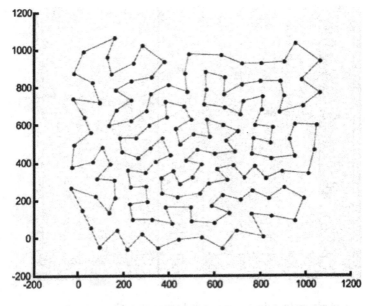

Fig. 3. Path calculation results

3   Simulation Experiment

We simulate the above algorithm in Matlab environment to verify its effectiveness. The main parameters are listed below:

Monitor area size: 500500
Node communication radius: 100
Unknown node number: 80

Reference node broadcast beacon number: the formula $\dfrac{2\sqrt{3}xy}{3R^2}$, ideally, requires 30 beacon broadcasts. Path attenuation disturbance: 3% means that about 20% of the ranging error is in the range where RSSI is used for ranging.

3.1   The Location of Beacon Broadcast Points Is Random

If the location of beacon broadcast is randomly chosen, the 30 beacon broadcasting can not meet the requirement of location at all, and many unknown nodes can not be located. Gradually increasing the number of beacon broadcasts to 80 times, this situation has been improved, it can be seen that the location of broadcast points is not planned in advance, resulting in low positioning efficiency. Figures 4 and 5 are the positioning pattern and relative positioning error of the 100 broadcast.

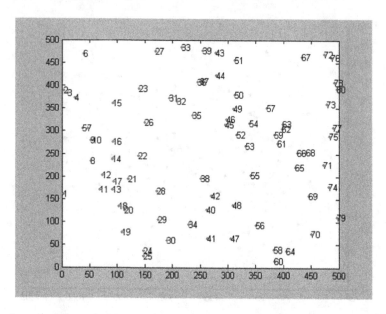

Fig. 4. Location pattern (positioning pattern and error without beacon broadcast point planning)

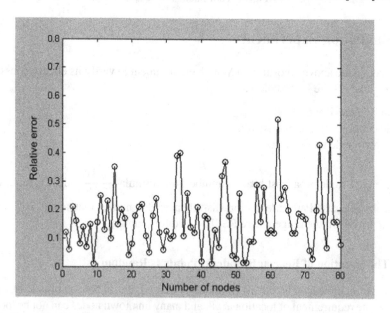

Fig. 5. Error distribution (positioning pattern and error without beacon broadcast point planning)

3.2 Error Comparison Before and After Path Planning

When the number of beacon broadcasts is 50 times unchanged, we compare the location error before and after path selection. As shown in Fig. 6, since there are only 50 broadcasts of beacons, without planning the path, some nodes cannot be located, and we do not have statistics on the location errors of these nodes. It can be seen from the figure, not for path planning, only more than 50 node localization; path planning, there are very few (3–5) nodes cannot locate the node location, and the accuracy is also maintained at the level of 18%.

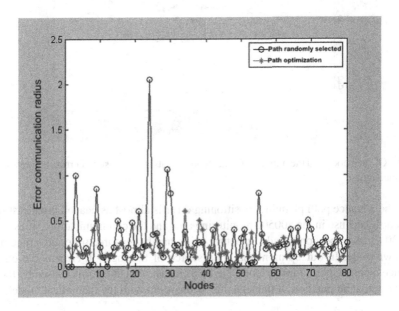

Fig. 6. The positioning error (comparison of errors before and after path selection)

The number of beacon broadcasts is increased from 50 to 120, and the influence of beacon broadcasting times on the location error before and after path selection is discussed. As shown in Fig. 7, the ordinate is the average location error of 80 unknown nodes, and there is no statistical non positioning node. Visible in path selection, the number of effects of beacon broadcast positioning error is quite large, and in less than a certain value, the positioning error was not stable; when the beacon radio frequency is bigger than a certain value (here is about 90), the positioning error basically increases with the number of broadcast reduced.

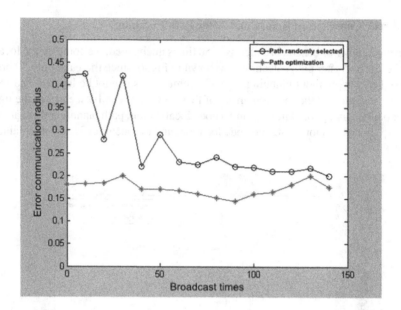

Fig. 7. Of each node and the number of beacon broadcasts (comparison of errors before and after path selection)

If the advance path planning, positioning error is almost as beacon broadcast times of change, because in the 500500 monitoring area and the communication radius is 100, only 30 times to beacon broadcast optimization had better three times coverage of the unknown node. In our opinion, the main source of the location error is the error of distance measurement. Furthermore, the RSSI measurement error is caused by the fact that the actual attenuation of the signal is inconsistent with the theoretical model.

3.3   Comparing the Total Distance Between Reference Nodes Before and After Path Planning

Since the reference node path length determines the positioning time required, and the energy consumption of mobile devices equipped with the reference node, so the following verification using the path planning scheme of TSP can be reduced as the path length of the reference node need to traverse.

The same as the experimental conditions above, the number of beacon broadcasts is increased from 50 to 120, and the total path length of the reference nodes before and after path planning is investigated. As can be seen from Fig. 8, the total length of the randomly chosen path increases with the number of beacon broadcasts, while the total path length of the optimized path increases very little.

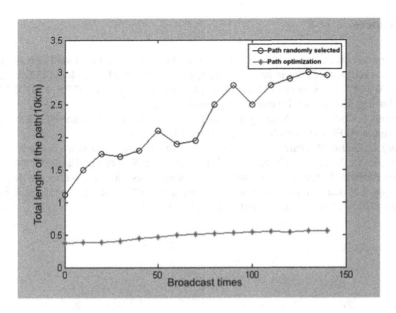

Fig. 8. Comparison of total path lengths of reference nodes

4   Conclusion

The location mechanism of mobile node is based on the reference time for the typical cost, on the other hand, it is the reference node and positioning mechanism which is commonly used in the discussion of the placement problem is in fact consistent, from this perspective, it is the problem of space usually multiple reference nodes are placed into a single mobile reference node in the iteration process time.

In the mechanism, each unknown node does not need to propagate location information, and all traffic is concentrated on the reference node. Since the reference nodes are deployed in mobile robots, or by unmanned vehicles carrying small, movable with respect to the energy required to send data, energy consumption is almost negligible, and we have reason to believe that the reference node can be obtained from the energy supplement loading platform where it. To save a lot of expensive GPS systems, the cost is only a small mobile platform and the time it takes to traverse the monitoring area.

In addition, this chapter mentions the positioning mechanism, the mobile node is only a reference, in fact, still can consider location accuracy, energy consumption and time requirement, using several mobile reference nodes, each reference node is responsible for positioning of independent regions, also can not the same path traversal the same area, and to accelerate the speed of location, or improve the positioning accuracy.

Acknowledgement. This work was jointly supported by Natural Science Foundation of China (61773296), the Education Department of Jiangxi Province of China Science and Technology research projects with the Grant No. GJJ151433, GJJ161687, GJJ161688 and GJJ161691.

References

1. Chakrabarty, K., Iyengar, S.S., Qi, H., Cho, E.: Grid coverage for surveillance and target location in distributed sensor networks. IEEE Trans. Comput. **51**(12), 1448–1453 (2002)
2. Lee, W.C.Y.: Lee's model cellular radio path loss prediction. In: IEEE 42nd Vehicular Technology Conference, Denver, CO, pp. 36–51 (1992)
3. Guvenc, I., Chong, C.C.: A survey on TOA based wireless localization and NLOS mitigation techniques. IEEE Commun. Surv. Tutor. **11**(3), 107–124 (2009)
4. Jianqi, L., Qinruo, W., Jiafu, W., et al.: Towards real-time indoor localization in wireless sensor networks. In: Xingang, L., Yang, L.T., Min, C., et al., pp. 877–884 (2012)
5. Gour, P., Sarje, A.: Localization in wireless sensor networks with ranging error. In: Buyya, R., Thampi, S.M. (eds.) Intelligent Distributed Computing. AISC, vol. 321, pp. 55–69. Springer, Cham (2015). https://doi.org/10.1007/978-3-319-11227-5_6

Author Index